D0752965

Biomechanics

Mechanical Properties of Living Tissues

Second Edition

Springer
New York
Berlin
Heidelberg
Barcelona
Hong Kong
London
Milan
Paris
Singapore
Tokyo

Other titles by the same author:

Biomechanics: Motion, Flow, Stress and Growth (1990)

Biomechanics: Circulation, second edition (1996)

Y.C. Fung

Biomechanics

Mechanical Properties of Living Tissues

Second Edition

With 282 Illustrations

 Springer

Y.C. Fung
Department of Bioengineering
University of California, San Diego
9500 Gilman Drive
La Jolla, CA 92093-0412
USA

Cover illustration: Branch point of a capillary blood vessel (see Fig. 5.7:2).

Library of Congress Cataloging-in-Publication Data
Fung, Y.C. (Yuan-cheng), 1919–
    Biomechanics: mechanical properties of living tissues / Yuan-
Cheng Fung. — 2nd ed.
        p.   cm.
    Includes bibliographical references and index.
    ISBN 0-387-97947-6 (New York). — ISBN 3-540-97947-6 (Berlin)
    1. Tissues.   2. Biomechanics.   3. Rheology (Biology).   I. Title.
QP88.F87   1993
612´.014—dc20   92-33749

Printed on acid-free paper.

© 1993 Springer-Verlag New York, Inc.
All rights reserved. This work may not be translated or copied in whole or in part without the
written permission of the publisher (Springer-Verlag New York, Inc., 175 Fifth Avenue, New York,
NY 10010, USA), except for brief excerpts in connection with reviews or scholarly analysis.
Use in connection with any form of information storage and retrieval, electronic adaptation,
computer software, or by similar or dissimilar methodology now known or hereafter developed
is forbidden.
The use of general descriptive names, trade names, trademarks, etc., in this publication, even if
the former are not especially identified, is not to be taken as a sign that such names, as understood
by the Trade Marks and Merchandise Marks Act, may accordingly be used freely by anyone.

Production managed by Bill Imbornoni; manufacturing supervised by Jacqui Ashri.
Typeset by Asco Trade Typesetting Ltd., Hong Kong.
Printed and bound by Edwards Brothers, Inc., Ann Arbor, MI.
Printed in the United States of America.

9 8

ISBN 0-387-97947-6   SPIN 10922736
ISBN 3-540-97947-6

Springer-Verlag New York Berlin Heidelberg
A member of BertelsmannSpringer Science+Business Media GmbH

Dedicated to
Chia-Shun Yih,

and
Luna, Conrad, and
Brenda Fung

Preface to the Second Edition

The objective of this book remains the same as that stated in the first edition: to present a comprehensive perspective of biomechanics from the stand point of bioengineering, physiology, and medical science, and to develop mechanics through a sequence of problems and examples. My three-volume set of *Biomechanics* has been completed. They are entitled: *Biomechanics: Mechanical Properties of Living Tissues*; *Biodynamics: Circulation*; *and Biomechanics: Motion, Flow, Stress, and Growth*; and this is the first volume. The mechanics prerequisite for all three volumes remains at the level of my book *A First Course in Continuum Mechanics* (3rd edition, Prentice-Hall, Inc., 1993).

In the decade of the 1980s the field of Biomechanics expanded tremendously. New advances have been made in all fronts. Those that affect the basic understanding of the mechanical properties of living tissues are described in detail in this revision. The references are brought up to date. Among the new topics added are the following: the coagulation of blood, thrombus formation and dissolution, cellular mechanics, deformability of passive leukocytes, mechanics of the endothelial cells in a continuum, news about new types of collagen, new methods of testing mechanical properties of soft tissues, the relationship between continuum mechanics and the structure and ultrastructure of tissues, the cross-bridge theory of muscle contraction, experimental evidences for sliding elements in muscle cells, the constitutive equation of myocardium, the residual stresses in organs, the constitutive equations of soft tissues based on their zcro-stress states, the constitutive equation of the individual layers of a multilayered tissue such as the blood vessel wall, the influence of stress and strain on the remodeling of living tissues, the tensorial Wolff's law, the triphasic theory of cartilage, tissue engineering, and a perspective of biomechanics of the future. And, in keeping with our tradition of emphasizing problem-formulation and problem-solving as a means to learn a subject, many new problems are added.

For this edition, I wish to record my thanks to Drs. Mohan D. Deshpande, J.P. du Plessis, L.J.M.G. Dortmans, A.A.F. van de Ven, A.A.H.J. Sauren, and R. Ponnalagar Samy for sending me errata and discussions. To Professors Aydin Tözeren, Richard Skalak, Shu Chien, Geert Schmid-Schönbein, John Pinto, Andrew McCulloch, Robert Nerem, Van Mow, Michael Lai, and Savio Woo, I am grateful for frequent discussions and advices. I have enjoyed working with my former students, Drs. Paul Zupkas, Shu Qian Liu, Ghassan Kassab, Jianbo Zhou, Jack Debes, and Hai Chao Han, and Visiting Professors Qi Lian Yu, Jia Ping Xie, Rui Fang Yang, ShanXi Deng, and Jun Tomioka whose results are referred to here. To them, and to Perne Whaley who has collaborated with me over twenty years on manuscript preparation, I am very grateful.

La Jolla, California   Yuan-Cheng Fung

Preface to the First Edition

The motivation for writing a series of books on biomechanics is to bring this rapidly developing subject to students of bioengineering, physiology, and mechanics. In the last decade biomechanics has become a recognized discipline offered in virtually all universities. Yet there is no adequate textbook for instruction; neither is there a treatise with sufficiently broad coverage. A few books bearing the title of biomechanics are too elementary, others are too specialized. I have long felt a need for a set of books that will inform students of the physiological and medical applications of biomechanics, and at the same time develop their training in mechanics. We cannot assume that all students come to biomechanics already fully trained in fluid and solid mechanics; their knowledge in these subjects has to be developed as the course proceeds. The scheme adopted in the present series is as follows. First, some basic training in mechanics, to a level about equivalent to the first seven chapters of the author's *A First Course in Continuum Mechanics* (Prentice-Hall, Inc. 1977), is assumed. We then present some essential parts of biomechanics from the point of view of bioengineering, physiology, and medical applications. In the meantime, mechanics is developed through a sequence of problems and examples. The main text reads like physiology, while the exercises are planned like a mechanics textbook. The instructor may fill a dual role: teaching an essential branch of life science, and gradually developing the student's knowledge in mechanics.

To strike a balance between biological and physical topics in a single course is not easy. Biology contains a great deal of descriptive material, whereas mechanics aims at quantitative analysis. The need to unify these topics sometimes renders the text nonuniform in style, stressing a mathematical detail here and describing an anatomy there. This nonuniformity is more pronounced at the beginning, when the necessary background material has to be introduced.

A special word needs to be said about the exercises. Students of mechanics thrive on exercises. We must constantly try to formulate and solve problems. Only through such practice can we make biomechanics a living subject. I do not wish to present this book as a collection of solved problems. I wish to present it as a way of thinking about problems. I wish to illustrate the use of mechanics as a simple, quantitative tool. For this reason many problems for solution are proposed in the text; some are used as a vehicle to inform the readers of some published results, others are intended to lead the reader to new paths of investigation. I followed this philosophy even at the very beginning by presenting some problems and solutions in the Introductory Chapter 1. I think colleagues who use this as a textbook would appreciate this, because then they can assign some problems to the students after the first lecture.

With our limited objective, this book does not claim to be a compendium or handbook of current information on the selected topics, nor a review of literature. For those purposes a much larger volume will be needed. In this volume we develop only a few topics that seem related and important. A comprehensive bibliography is not provided; the list of references is limited to items quoted in the text. Though the author can be accused of quoting papers and people familiar to him, he apologizes for this personal limitation and hopes that he can be forgiven because it is only natural that an author should talk more about his own views than the views of others. I have tried, however, never to forget mentioning the existence of other points of view.

Biomechanics is a young subject. Our understanding of the subject is yet imperfect. Many needed pieces of information have not yet been obtained; many potentially important applications have not yet been made. There are many weaknesses in our present position. For example, the soft tissue mechanics developed in Chapter 7, based on the concept of quasilinear viscoelasticity and pseudo-elasticity, may someday be replaced by constitutive equations that are fully nonlinear but not too complex. The blood vessel mechanics developed in Chapter 8 is based on a two-dimensional average. Our discussion of the muscle mechanics in Chapter 9–11 points out the deficiency in our present knowledge on this subject.

I wish to express my thanks to many authors and publishers who permitted me to quote their publications and reproduce their figures and data in this book. I wish to mention especially Professors Sidney Sobin, Evan Evans, Harry Goldsmith, Jen-shih Lee, Wally Frasher, Richard Skalak, Andrew Somlyo, Salvatore Sutera, Andrus Viidik, Joel Price, Savio Woo, and Benjamin Zweifach who supplied original photographs for reproduction.

This book grew out of my lecture notes used at the University of California, San Diego over the past ten years. To the students of these classes I am grateful for discussions. Much of the results presented here are the work of my colleagues, friends, and former students. Professors Sidney Sobin,

Benjamin Zweifach, Marcos Intaglietta, Arnost and Kitty Fronek, Wally Frasher, Paul Johnson, and Savio Woo provided the initial and continued collaboration with me on this subject. Drs. Jen-shi Lee, Pin Tong, Frank Yin, John Pinto, Evan Evans, Yoram Lanir, Hyland Chen, Michael Yen, Donald Vawter, Geert Schmid-Schoenbein, Peter Chen, Larry Malcom, Joel Price, Nadine Sidrick, Paul Sobin, Winston Tsang, and Paul Zupkas contributed much of the material presented here. The contribution of Paul Patitucci to the numerical handling of data must be especially acknowledged. Dr. Yuji Matsuzaki contributed much to my understanding of flow separation and stability. Professor Zhuong Feng-Yuan read the proofs and made many useful suggestions. Eugene Mead kept the laboratory going. Rose Cataldi and Virginia Stephens typed the manuscript. To all of them I am thankful.

Finally, I wish to thank the editorial and production staffs of Springer-Verlag for their care and cooperation in producing this book.

La Jolla, California   Yuan-Cheng Fung

Contents

Chapter 3
The Flow Properties of Blood   66

Chapter 4
Mechanics of Erythrocytes, Leukocytes, and Other Cells   109

Chapter 8
Mechanical Properties and Active Remodeling of Blood Vessels   321

Chapter 9
Skeletal Muscle

Chapter 10
Heart Muscle

Chapter 11
Smooth Muscles

CHAPTER 1

Introduction: A Sketch of the History and Scope of the Field

1.1 What Is Biomechanics?

Biomechanics is mechanics applied to biology. The word "mechanics" was used by Galileo as a subtitle to his book *Two New Sciences** (1638) to describe force, motion, and strength of materials. Through the years its meaning has been extended to cover the study of the motions of all kinds of particles and continua, including quanta, atoms, molecules, gases, liquids, solids, structures, stars, and galaxies. In a generalized sense it is applied to the analysis of any dynamic system. Thus thermodynamics, heat and mass transfer, cybernetics, computing methods, etc., are considered proper provinces of mechanics. The biological world is a part of the physical world around us and naturally is an object of inquiry in mechanics.

Biomechanics seeks to understand the mechanics of living systems. It is a modern subject with ancient roots and covers a very wide territory. In this book we concentrate on physiological and medical applications, which constitute the majority of recent work in this field. The motivation for research in this area comes from the realization that biology can no more be understood without biomechanics than an airplane can without aerodynamics. For an airplane, mechanics enables us to design its structure and predict its performance. For an organism, biomechanics helps us to understand its normal function, predict changes due to alterations, and propose methods of artificial intervention. Thus diagnosis, surgery, and prosthesis are closely associated with biomechanics.

---

* The full title is: *Discorsi e Dimostrazioni Matematiche*, intorno à due nuoue Scienze Attenenti alla Mecanica & i Movimenti Locali, del Signor Galileo Galilei Linceo.

1.2 Historical Background

To explain what our field is like, it is useful to consider its historical background.

The earliest books containing the concepts of biomechanics were probably the Greek classic *On the Parts of Animals* by Aristotle (384–322 B.C.), and the Chinese book, *Nei Jing* (內經, or *Internal Classic*) written by anonymous authors in the Warring Period (472–221 B.C.). Aristotle presents a comprehensive description of the anatomy and function of internal organs. His analysis of the peristaltic motion of the ureter in carrying urine from the kidney to the bladder is remarkably accurate. But he mistook the heart as a respiratory organ, probably because in his dissection of corpses of war a day or two after the battles, he never saw blood in the heart. Nei Jing discusses the concept of circulation in man and in the universe. It states that "the blood vessels are where blood is retained" (in the chapter 脉要精微論) and that "all blood in the vessels originates from the heart" (in the chapter 五臟生成篇). In 營衛生會第十八, it says that "the blood and chi circulate without stopping. In 50 steps, they return to the starting point. Yin succeeds Yang, and vice versa, like a circle without an end." It recognizes harmony or disharmony between man and his environment as the causes of health and diseases, and on that basis discusses the details of accupuncture and pulse wave diagnosis.

Modern development of mechanics, however, received its impetus from engineering. Biomechanics was initiated along with the development of mechanics. The following are early contributors to biomechanics:

> Galileo Galilei (1564–1642)
> William Harvey (1578–1658)
> René Descartes (1596–1650)
> Giovanni Alfonso Borelli (1608–1679)
> Robert Boyle (1627–1691)
> Robert Hooke (1635–1703)
> Isaac Newton (1642–1727)
> Leonhard Euler (1707–1783)
> Thomas Young (1773–1829)
> Jean Poiseuille (1797–1869)
> Herrmann von Helmholtz (1821–1894)
> Adolf Fick (1829–1901)
> Diederik Johannes Korteweg (1848–1941)
> Horace Lamb (1849–1934)
> Otto Frank (1865–1944)
> Balthasar van der Pol (1889–1959)

A brief account of the contributions of these people may be of interest. William Harvey, of course, is credited with the discovery of blood circulation. He made this discovery in 1615. Without a microscope, he never saw the capillary blood vessels. This should make us appreciate his conviction in logical reasoning even more deeply today because, without the capability of seeing the passage from the arteries to the veins, the discovery of circulation must be regarded as "theoretical." The actual discovery of capillaries was made by Marcello Malpighi (1628–1694) in 1661, 45 years after Harvey made the capillaries a logical necessity.

Galileo was 14 years older than William Harvey, and was a student of medicine before he became famous as a physicist. He discovered the constancy of the period of a pendulum, and used the pendulum to measure the pulse rate of people, expressing the results quantitatively in terms of the length of a pendulum synchronous with the beat. He invented the thermoscope, and was also the first one to design a microscope in the modern sense in 1609, although rudimentary microscopes were first made by J. Janssen and his son Zacharias in 1590.

Young Galileo's fame was so great and his lectures at Padua so popular that his influence on biomechanics went far beyond his personal contributions mentioned above. According to Singer (*History*, p. 237), William Harvey should be regarded as a disciple of Galileo. Harvey studied at Padua. (1598– 1601) while Galileo was active there. By 1615 Harvey had formed the concept of circulation of blood. He published his demonstration in 1628. The essential part of his demonstration is the result not of mere observation but of the application of Galileo's principle of measurement. He showed first that the blood can only leave the ventricle of the heart in one direction. Then he measured the capacity of the heart, and found it to be about two ounces.* The heart beats 72 times a minute, so that in one hour it throws into the system $2 \times 72 \times 60$ ounces = 8640 ounces = 540 pounds = 234 kg! Where can all this blood come from? Where can it all go? He concludes that the existence of circulation is a necessary condition for the function of the heart.

Another colleague of Galileo, Santorio Santorio (1561–1636), a professor of medicine at Padua, used Galileo's method of measurement and philosophy to compare the weight of the human body at different times and in different circumstances. He found that the body loses weight by mere exposure, a process which he assigned to "insensible perspiration." His experiments laid the foundation of the modern study of "metabolism." (See Singer, *History*, p. 236.)

The physical discoveries of Galileo and the demonstrations of Santorio and of Harvey gave a great impetus to the attempt to explain vital processes in terms of mechanics. Galileo showed that mathematics was the essential key to science, without which nature could not be properly understood. This outlook inspired Descartes, a great mathematician, to work on physiology. In a work published posthumously (1662 and 1664), he proposed a physiological theory upon mechanical grounds. According to Singer, this work is the first important modern book devoted to the subject of physiology.

---

* We know today that the resting cardiac output is very close to one total blood volume per minute in nearly all mammals.

Descartes did not have any extensive practical knowledge of physiology. On theoretical grounds he set forth a very complicated model of animal structure, including the function of nerves. Subsequent investigations failed to confirm many of his findings. These errors of fact caused the loss of confidence in Descartes' approach—a lesson that should be kept in mind by all theoreticians.

Other attempts a little less ambitious than Descartes's were more successful. Giovanni Alfonso Borelli (1608–1679) was an eminent Italian mathematician and astronomer, and a friend of Galileo and Malpighi. His *On Motion of Animals* (*De Motu Animalium*) (1680) is the classic of what is variously called the "iatrophysical" or "iatromathematical" school, ιατρός, Gr., physician. He was successful in clarifying muscular movement and body dynamics. He treated the flight of birds and the swimming of fish, as well as the movements of the heart and of the intestines.

Robert Boyle studied the lung and discussed the function of air in water with respect to fish respiration. Robert Hooke gave us Hooke's law in mechanics and the word "cell" in biology to designate the elementary entities of life. His famous book *Micrographia* (1664) has been reprinted by Dover Publications, New York (1960).

Newton was seven years younger than Hooke. Newton did not write about biomechanics, but his calculus, his laws of motion, and his constitutive equation for a viscous fluid are the foundation of biomechanics. Leonhard Euler generalized Newton's laws of motion to a partial differential equation for a continuum. Euler wrote the first definitive paper on the propagation of pulse waves in arteries in 1775. He was unable to solve these equations, however, and the solution would have to wait for the arrival of George Friedrich Bernhard Riemann (1826–1866).

Thomas Young studied the formation of human voice, identified it as vibrations, connected it with the elasticity of materials, gave us the legacy of Young's modulus, then developed the wave theory of light, and a theory of color vision, as well as the solution to a practical problem of astigmatism of lenses. Poiseuille improved the mercury manometer to measure the blood pressure in the aorta of the dog. Then he set out to determine the pressure–flow relationship of pipe flow. He understood the significance of turbulences on this relationship, and decided to use micropipettes in his experiments in order to be assured that the flow was laminar. His results, published in 1843, were so precise that they played a decisive role in establishing the *no-slip condition* as the proper boundary condition between a viscous fluid and a solid wall. His empirical relationship, now known as Poiseuille's law, is used extensively in cardiology.

To von Helmholtz (Fig. 1.2:1) might go the title "Father of Bioenginering." He was professor of physiology and pathology at Königsberg, professor of anatomy and physiology at Bonn, professor of physiology at Heidelberg, and finally professor of physics in Berlin (1871). He wrote his paper on the "law of conservation of energy" in barracks while he was in military service fresh out of medical school. His contributions ranged over optics, acoustics,

Figure 1.2:1 Portrait of Hermann von Helmholtz. From the frontispiece to *Wissen-schaftliche Abhandlungen von Helmholtz*. Leipzig, Johann Ambrosius Barth, 1895. Photo by Giacomo Brogi in 1891.

thermodynamics, electrodynamics, physiology, and medicine. He discovered the focusing mechanism of the eye and, following Young, formulated the trichromatic theory of color vision. He invented the phakoscope to study the changes in the lens, the ophthalmoscope to view the retina, the ophthalmometer for measurement of eye dimensions, and the stereoscope with interpupillary distance adjustments for stereo vision. He studied the mechanism of hearing and invented the Helmholtz resonator. His theory of the permanence of vorticity lies at the very foundation of modern fluid mechanics. His book *Sensations of Tone* is popular even today. He was the first to determine the velocity of the nerve pulse, giving the rate 30 m/s, and to show that the heat released by muscular contraction is an important source of animal heat.

The other names on the list are equally familiar to engineers. The physiologist Fick was the author of Fick's law of mass transfer. The hydrodynamicists Korteweg (1878) and Lamb (1898) wrote beautiful papers on wave propagation in blood vessels. Frank worked out a hydrodynamic theory of circulation. Van der Pol (1929) wrote about the modeling of the heart with nonlinear oscillators, and was able to simulate the heart with four Van der Pol oscillators to produce a realistic looking electrocardiograph.

This list perhaps suffices to show that there were, and of course are, people who would be equally happy to work on living subjects as well as inanimate objects. Indeed, what a scientist picks up and works on may depend a great deal on chance, and the biological world is so rich a field that one should not permit the opportunities there to slip by unnoticed. The following example about Thomas Young may be of interest to those who like to ponder about the threads of development in scientific thought. When he tried to understand the human voice, Thomas Young turned to the mechanics of vibrations. Let us quote Young himself (from his "Reply to the *Edinburgh Reviewers*" (1804), see *Works*, ed. Peacock, Vol. i, pp 192–215):

> When I took a degree in physic at Göttingen, it was necessary, besides publishing a medical dissertation, to deliver a lecture upon some subject connected with medical studies, and I choose for this Formation of the Human Voice, ... When I began the outline of an essay on the human voice, I found myself at a loss for a perfect conception of what sound was, and during the three years that I passed at Emmanuel College, Cambridge, I collected all the information relating to it that I could procure from books, and I made a variety of original experiments on sounds of all kinds, and on the motions of fluids in general. In the course of these inquiries I learned to my surprise how much further our neighbours on the Continent were advanced in the investigation of the motions of sounding bodies and of elastic fluids than any of our countrymen. And in making some experiments on the production of sounds, I was so forcibly impressed with the resemblance of the phenomena that I saw to those of the colours of thin plates, with which I was already acquainted, that I began to suspect the existence of a closer analogy between them than I could before have easily believed."

This led to his 'Principle of Interferences" (1801) which earned him lasting fame in the theory of light. How refreshing are these remarks! How often do we encounter the situation "at a loss for a perfect conception of what ... was." How often do we leave some vague notions untouched!

1.3 What's in a Name?

Biomechanics is mechanics applied to biology. We have explained in Sec. 1.1 that the word "mechanics" has been identified with the analysis of any dynamic system. To people who call themselves workers in applied mechanics, the field includes the following topics:

Stress and strain distribution in materials
Constitutive equations which describe the mechanical properties of materials
Strength of materials, yielding, creep, plastic flow, crack propagation, fracture, fatigue failure of materials; stress corrosion
Dislocation theory, theory of metals, ceramics
Composite materials

Flow of fluids: gas, water, blood, and other tissue fluids
Heat transfer, temperature distribution, thermal stress
Mass transfer, diffusion, transport through membranes
Motion of charged particles, plasma, ions in solution
Mechanisms, structures
Stability of mechanical systems
Control of mechanical systems
Dynamics, vibrations, wave propagation
Shock waves, and waves of finite amplitude

It is difficult to find anything living that does not involve some of these problems.

On the other hand, how did the word biology come about? The term *biology* was first used in 1801 by Lamarck in his *Hydrogéologie* (see Merz, *History*, p. 217). Huxley, in his *Lecture on the Study of Biology* [South Kensington (Dec. 1876), reprinted in *American Addresses*, 1886, p. 129] gave the following account of the early history of the word:

> About the same time it occurred to Gottfried Reinhold Treviranus (1776–1837) of Bremen, that all those sciences which deal with living matter are essentially and fundamentally one, and ought to be treated as a whole; and in the year 1802 he published the first volume of what he also called *Biologie*. Treviranus's great merit lies in this, that he worked out his idea, and wrote the very remarkable book to which I refer. It consists of six volumes, and occupied its author for twenty years—from 1802 to 1822. That is the origin of the term "biology"; and that is how it has come about that all clear thinkers and lovers of consistent nomenclature have substituted for the old confusing name of "natural history," which has conveyed so many meanings, the term "biology," which denotes the whole of the sciences which deal with living things, whether they be animals or whether they be plants.

In the present volume, we address our attention particularly to continuum mechanics in physiology. By physiology we mean the science dealing with the normal functions of living things or their organs. Originally the term had a much more broad meaning. William Gilbert (1546–1603), personal physician to Queen Elizabeth, wrote a book (1600) called *On the Magnet and on Magnetic Bodies and Concerning the Great Magnet, the Earth, a New Physiology*, which was the first major original contribution to science published in England (Singer, *History*, p. 188). It earned the admiration of Francis Bacon and of Galileo. Note the last word in the title, "physiology." The word *physiologia* was, in fact, originally applied to the material working of the world as a whole, and not to the individual organism.

1.4 Mechanics in Physiology

In Sec. 1.2 we listed a slate of giants in applied mechanics. We can equally well list a slate of giants in physiology who clarified biomechanics. For example, following William Harvey (1578–1658), we have

Marcello Malpighi (1628–1694)
Stephen Hales (1677–1761)
Otto Frank (1865–1944)
Ernest Henry Starling (1866–1926)
August Krogh (1874–1949)
Archibald Vivian Hill (1886–1977)

To outline their contributions, we should remember that there are great ideas which, when perfectly understood, seem perfectly natural. The idea of blood circulation is one of them. Today we no longer remember what William Harvey had to fight against in 1615 when he conceived the principle of blood circulation. There was the great Galen, whose medical teachings had been accepted and unquestioned for 15 centuries. There was the fact that no connection between arteries and veins had ever been seen. There was the fact that the arterial and venous bloods do look different. To fight against these and other difficulties, Harvey took the quantitative method promulgated by Galileo. He concluded that the existence of circulation is a necessary condition for the function of the heart. All other difficulties were temporarily put aside. Thus Harvey won, illustrating at once the principle that a crucial argument can be settled by a detailed quantitative analysis.

There was much more to be discovered after Harvey. For example, it was not clear how blood completes the circulation. Of this, James Young (1930, p. 1) wrote:

> It has often been observed, by men ill-versed in the history of scientific developments, that great new ideas, when developed, might easily have been inferred from others accepted long before. For example, when Erasistratos has told us that the heart's valves ensure a one-way course of the blood, and that all the blood of the body can be driven out by the opening of one artery by aid of the *horror vacui*, and Celsus has informed us that the heart is a muscular viscus, one might have inferred the circulation of the blood without waiting for Harvey. Even so one might imagine that Harvey, to complete his system, might have inferred the presence of definite vessels of communication between the arteries and the veins instead of an indefinite soakage through the "porosities of the tissues." But he did not do so, nor did any one else for thirty years of keen discussion. The discovery of the capillaries was reserved for the work of Malpighi, who was trying to clear his views about the structure of the lungs.

Marcello Malpighi (1628–1694) communicated his discovery in two letters addressed to Alfonso Borelli in 1661. Let us quote him from his second letter [see Young's translation (1930)]:

> In the frog lung owing to the simplicity of the structure, and the almost complete transparency of the vessels which admits the eye into the interior, things are more clearly shown so that they will bring the light to other more obscure matters . . . .

Observation by means of a microscope reveals more wonderful things than those viewed in regard to mere structure and connection: for while the heart is still beating the contrary (i.e., in opposite directions in the different vessels) movement of the blood is observed in the vessels—though with difficulty—so that the circulation of the blood is clearly exposed . . . .

Thus by this impulse the blood is driven in very small arteries like a flood into the several cells, one or other branch clearly passing through or ending there. Thus the blood, much divided, puts off its red colour, and, carried round in a winding way, is poured out on all sides till at length it may reach the walls, the angles, and the absorbing branches of the veins . . . .

What better introduction to pulmonary circulation is there! With the anatomy known, biomechanical analysis can begin.

Of the other people in the list, Stephen Hales (Fig. 1.4:1) was the man who measured the arterial blood pressure in the horse, and correlated it to hemorrhage. He made wax casts of the ventricles at the normal distending pressure in diastole and then measured the volume of the cast to obtain an estimate of the cardiac output. With his measurements he was able to estimate the forces in ventricular muscle. He measured the distensibility of the aorta and used it to explain how the intermittent pumping of the heart can be converted to a smooth flow in the blood vessels. The explanation was that the distension of the aorta functions like the air chamber in a fire engine, which converts intermittent pumping into a steady jet. (Air chamber was rendered as *Windkessel* in the first German translation of his book;

Figure 1.4:1 Stephen Hales.

hence the famous theory by that name.) He introduced the concept of peripheral resistance in blood flow, and showed that the main site of this resistance was in the minute blood vessels in the tissue. He even went further to show that hot water and brandy have vasodilatational effects.

Otto Frank worked out a hydrodynamic theory of circulation. Starling proposed the law for mass transfer across biological membrane and clarified the concept of water balance in the body. Krogh won his Nobel prize on the mechanics of microcirculation. Hill won his Nobel prize on the mechanics of the muscle. Their contributions form the foundation on which biomechanics rests.

1.5 What Contributions Has Biomechanics Made to Health Science?

Biomechanics has participated in virtually every modern advance of medical science and technology. Molecular biology may appear far removed from biomechanics, but in its deeper reaches one has to understand the mechanics of the formation, design, function, and production of the molecules. Surgery seems to be an activity unrelated to mechanics, yet healing and rehabilitation are intimately related to the stress and strain in the tissues.

Biomechanics has helped solving clinical problems in the cardiovascular system with the invention and analysis of prosthetic heart valves, heart assist devices, extracorporeal circulation, the heart–lung machines, and the hemodialysis machines. It played a major role in advancing the art of heart transplantation and artificial heart replacement. It has helped solving problems of postoperative trauma, pulmonary edema, pulmonary atelectasis, arterial pulse-wave analysis, phonoangiography, and the analysis of turbulent noise as indications of atherosclerosis or stenosis in arteries.

Atherosclerosis has been studied intensely as a hemodynamic disorder because the locations of atherosclerotic plaques seem to correlate with certain features of blood flow. Recent investigation has been focused on the stress acting in the endothelial cells and the response of the endothelial cells to the stress.

A most vigorous development of biomechanics is associated with orthopedics, because the most frequent users of the surgery rooms in the world are patients with musculoskeletal problems. In orthopedics, biomechanics has become an everyday clinical tool. Fundamental research has included not only surgery, prosthesis, implantable materials, and artificial limbs, but also cellular and molecular aspects of healing in relation to stress and strain, and tissue engineering of cartilage, tendon, and bone.

The biomechanics of trauma, injury, and rehabilitation is becoming more important to modern society. Because people injured by automobile accidents and other violences are younger, the economic impact on the society is bigger.

In the long run, the most important contribution of modern biomechanics to medicine probably lies in its promotion of a better understanding of physiology. The methodology and standards of mechanics, developed in the age of industrialization, can be adopted to deal with the complex problems of health science and technology. Thus, the system analysis, rheology of biological tissues, mass transfer through membranes, interfacial phenomena, and microcirculation, which are traditional strongholds of biomechanics, are becoming common sense in medicine.

1.6 Our Method of Approach

In the tradition of physics and engineering, our approach to the study of problems in biomechanics consists of the following steps:

(1) Study the morphology of the organism, anatomy of the organ, histology of the tissue, and the structure and ultrastructure of the materials in order to know the geometric configuration of the object we are dealing with.
(2) Determine the mechanical properties of the materials or tissues that are involved in a problem. In biomechanics this step is often very difficult, either because we cannot isolate the tissue for testing, or because the size of available tissue specimens is too small, or because it is difficult to keep the tissue in the normal living condition. Furthermore, biological tissues are often subjected to large deformations, and the stress–strain relationships are usually nonlinear and hisotry dependent. The nonlinearity of the constitutive equation makes its determination a challenging task. Usually, however, one can determine the mathematical form of the constitutive equation of the material quite readily, with certain numerical parameters left to be determined by physiological experiments named in steps 6 and 7 below.
(3) On the basis of fundamental laws of physics (conservation of mass, conservation of momentum, conservation of energy, Maxwell's equations, etc.) and the constitutive equations of the materials, derive the governing differential or integral equations.
(4) Understand the environment in which an organ works in order to obtain meaningful boundary conditions.
(5) Solve the boundary-value problems (differential equations with appropriate initial and boundary conditions) analytically or numerically, or by experiments.
(6) Perform physiological experiments that will test the solutions of the boundary-value problems named above. If necessary, reformulate and resolve the mathematical problem to make sure that the theory and experiment do correspond to each other, i.e., that they are testing the same hypotheses.
(7) Compare the experimental results with the corresponding theoretical ones. By means of this comparison, determine whether the hypotheses made in the theory are justified, and, if they are, find the numerical values of the undetermined coefficients in the constitutive equations.

(8) A theory so validated can be used to predict the outcome of other boundary-value problems associated with the same basic equations. Then one can use the method to explore practical applications of the theory and experiments.

The most serious frustration to a biomechanics worker is usually the lack of information about the constitutive equations of living tissues. Without the constitutive laws, no analysis can be done. On the other hand, without the solution of boundary-value problems the constitutive laws cannot be determined. Thus, we are in a situation in which serious analyses (usually quite difficult because of nonlinearity) have to be done for hypothetical materials, in the hope that experiments will yield the desired agreement. If no agreement is obtained, new analyses based on a different starting point would become necessary.

It is for this reason that we emphasize the constitutive equations in this book. All biological solids and fluids are considered. Among the biofluids blood is treated in greater detail. Among biosolids the blood vessels, muscles, bone, and cartilage are given special attention. I give smooth muscles a separate chapter because of their extreme importance, although we really know very little about their mechanics.

1.7 Tools of Investigation

To carry out the steps listed in the preceding section, proper tools are needed. The field can advance only when tools are available.

Table 1.7:1 lists some topics and some tools. The table is short, but the idea could be elaborated. For example, consider *Geometry*. We need to know the shape and dimensions of the organs, the structure of the tissues, and the composition of the materials. If we are studying a soft tissue, we need to know the quantity, distribution, and curvature of its collagen and elastin fibers. If it contains smooth muscle cells, we want to know the cell dimensions and orientations, and the dense-body spacing in them. For cells, we may also need geometric data on their cell membranes, their architecture, nucleus, nucleolus, vacuoles, Golgi complex, endoplasmic reticula, and chromesomes. To obtain these geometric parameters, one must use the tools of anatomy and histology, such as the optical, scanning, and transmission electron microscopes, phase contrast and interference microscopes, confocal microscopes, lasers, scanning tunneling microscopes, and the mathematics of stereology, the theory of morphometry, and fractals. Sometimes the mathematical theory of topology could be brought to bear (e.g., in the determination of the distribution of arterioles and venules in the lung, the islands-in-the-ocean argument was used, see *Biodynamics: Circulation*, p. 348, Fung 1984). Somtimes the group theory will help (e.g., in identifying the polyhedral model of the pulmonary alveoli, see *Biomechanics: Motion, Flow, Stress, and Growth*, p. 405, Fung, 1990). Sometimes a physical argument must be used to avoid mistakes (e.g., in

TABLE 1.7:1 Topics and Tools of Living Tissue Research

Topics	Tools
Geometry	Morphometry, histology, EM, computer automation
Materials	Biochemistry, histochemistry, molecular mechanics
Biology	Cell biology, extracellular matrix, pharmacology, immunology, gene expressions, growth factors
Mechanical properties	Constitutive equations, strength, failure modes
Basic principles	Physics, chemistry, biology
Medicine, surgery, trauma, rehabilitation, individual cases	Boundary-value problems
Tissue engineering	Remodeling, pathology, healing, artificial tissue substitutes
Design, invention	Artificial organs, prosthesis, clinical and commercial devices

rejecting the spheres-on-parallel-tubes model of the lung, see Fung, 1990, ibid., p. 404.) Sometimes a method from a seemingly unrelated field may help (e.g., the use of the *Strahler system* of describing the rivers and rivulets in geography to classify the patterns of blood vessels or nerve networks). In most cases it is essential to use computers to collect and organize the morphometric data. Automation of the data gathering process may turn a tedious job into pleasure. Thus, to handle *geometry* a wide variety of tools are at hand; but more are needed. Many tasks that cannot be done today are waiting for better techniques to appear.

Next, consider *Materials*. To meet the needs to understand the structure of materials in biology, the tools of chemistry, biochemistry, histochemistry, polymer chemistry, and molecular biology may have to be brought in. Chromatography and immunochemical techniques are used. X-ray, nuclear magnetic resonance, and positron emission methods are also used to clarify the structure of biological materials.

Biological factors influence biological properties of cells and tissues. Enzymes, growth factors, and toxins are of concern. The properties of cells and extracellular matrix must be understood. Pharmacology and immunology play an important role.

For the determination of the *Mechanical Properties of Tissues*, special equipment are needed. Many testing machines used in the determination of the constitutive equations of living tissues are described in this book.

The *basic principles* of biomechanics includes the axioms of physics and mathematics. Any *ad hoc* hypotheses or principles must be given great attention and handled with great care. No unmentioned hypothesis should be allowed to sneak in.

With the basic information and principles assembled, one can then formulate *boundary-value problems* to be solved. Speical problems of medicine, surgery, trauma, and rehabilitation are individual boundary-value problems. The ability to solve the boundary-value problems must be cultivated. The greater the ability, the more power biomechanics will have for applications. Here the analytical theory, closed-form solutions, numerical methods, computational approach, finite-element, boundary element, and boundary integral methods, modal or eigenfunction expansions, or Green's function methods can each hold sway. Experimental solutions by testing real organs or their physical, biological, or analog models are of great value.

Tissue engineering is unique to biology. Under stress, a living tissue can change its shape, grow or shrink in size, and modify its chemical, cellular, and extracellular structures. The material composition and the mechanical properties may change with time. The zero stress state of the tissue will be a function of time. This is an expected function of life. To make use of this feature for the advantage of an individual, and to stave off any detrimental effects, is engineering. To master tissue engineering, we need tools to identify the dynamic changes as functions of time.

Finally, with the basic information in and method of solution at hand, one can then let the imagination run, and invent, or develop, or manufacture, or sell according to one's wishes. Thus we see that the tools of biomechanics is extremely varied.

1.8 What Contributions Has Biomechanics Made to Mechanics?

Comparing the contents of biomechanics outlined in the preceding sections with the classical theories of hydrodynamics and elasticity, we see that many aspects of biomechanics are beyond the scope of classical theories. The bodies studied in hydrodynamics and elasticity usually have very simple shapes, or are at least known. The classical media (e.g., air, water, steel) have well-known constitutive equations. The basic equations and admissible boundary conditions of the classical theory have been identified. The classical disciplines are concerned mainly with the methods of solution of boundary-value problems. In contrast, the main energy of biomechanical research has been spent on he steps leading to the formulation of the boundary-value problems, especially the establishment of constitutive equations. Thus biomechanics is still very young. I am sure that the greatest contribution of biomechanics to mechanics lies in its youthfulness. It brings mechanics back to a new formative stage, and asks it to be young again.

1.9 On the Law of Laplace

Biomechanics thrives on problem solving. Some problems are presented at the end of the chapter. Study Figs. P1.4 and P1.5 and solve Probs. 1.4 and 1.5. Equations (1) and (2) of Prob. 1.4 are known as the law of Laplace. Engineers may be surprised at the prominent role played by this law in physiology.

Originally, this law states a relationship between pressure, surface tension, and the curvature of a liquid surface. Consider the surface of a liquid column in a capillary tube standing in a bath. On assuming that the surface tension is constant in every direction, the law of Laplace reads:

$$p = T\left(\frac{1}{r_1} + \frac{1}{r_2}\right), \tag{1}$$

in which p denotes the transmural pressure acting on the surface, T is the surface tension, and r_1, r_2 are the principal radii of curvature of the surface. This equation can be applied to a thin membrane.

If the membrane tension in the wall is not the same in every direction, or if the wall is not very thin, then Eq. (1) is not applicable. A more general result is the following: Consider a curved membrane with principal radii of curvature r_1 and r_2. Let the principal axes of the membrane stress resultants be coincident with the principal radii of curvature, so that the tension in the directions of the principal curvatures $1/r_1$ and $1/r_2$ be T_1 and T_2, respectively. Then there is no shear stress resultant in these directions. Hence, when the membrane is subjected to an internal pressure p_i and an external pressure p_o, the equation of equilibrium is

$$p_i - p_o = \frac{T_1}{r_1} + \frac{T_2}{r_2}. \tag{2}$$

This equation is valid as long as the membrane is so thin that bending rigidity can be neglected. It is valid even if the membrane is made of an inelastic material. If $T_1 = T_2$, then Eq. (2) reduces to Eq. (1). If $T_1 = T_2$ and $r_1 = r_2$, then Eq. (2) reduces to Eq. (1) of Prob. 1.4. When $T_1 = T$ and $1/r_2 = 0$, then Eq. (2) reduces to Eq. (2) of Prob. 1.4 for an arbitrary value of membrane tension T_2 in the axial direction of the cylinder. For a thin membrane only the pressure difference counts, hence $p_i - p_o$ can be replaced by p.

Equations (1)—(2) do not apply to thick-walled shells such as the heart and the arteries. (The ratio of the wall thickness to the inner radius of the heart, the veins, large arteries, and arterioles are, respectively, about 25%, 3%, 20%, and 100%.) In thick-walled shells the stress is not necessarily uniformly distributed in the wall. A more detailed analysis is necessary. In the case of a thick-walled circular cylinder, the solution was obtained by Lamé and Clapeyron in 1833 (see Todhunter and Pearson, 1886, 1960, Vol. 1, Sec. 1022). It is known that if there were no residual stress, then the circumferential stress in the wall tends to be concentrated toward the inner wall. However, when the circumferential stress is integrated throughout the thickness of the wall, an equation of equilibrium can be obtained that resembles Eqs. (1) and (2), namely, for a cylinder,

$$T_1 = h\langle\sigma_\theta\rangle = p_i r_i - p_o r_o. \tag{3}$$

Here $h = r_o - r_i$ denote the wall thickness, $\langle\sigma_\theta\rangle$ denotes the average of the circumferential stress σ_θ, T_1 is the circumferential stress resultant, and r_i, r_o

are the inner and outer radii of the cylinder, respectively. Since this is an equation of equilibrium, it can be derived directly from statics as follows. Draw a free-body diagram for half of a circular cylinder with unit length in the axial direction. The total force acting radially outward due to internal pressure is $p_i \cdot 2r_i \cdot 1$; that due to external pressure is $-p_o \cdot 2r_o \cdot 1$. The total force due to the circumferential stress σ_θ is $2T_1 = 2\langle\sigma_\theta\rangle \cdot h \cdot 1$. Since in a cylinder the axial forces have no influence on the balance of forces in the radial direction, we obtain the equation of equilibrium in the radial direction, Eq. (3), immediately.

Equation (3) reduces to Eq. (2) of Prob. 1.4 in the case of a thin-walled cylinder, and is valid even if the cylinder is made of nonhomogeneous, non-linear, inelastic material. It is applicable to finite deformation as long as r_i, r_o represent the final radii.

Similarly, a consideration of the longitudinal equilibrium of a circular cylinder with closed ends leads directly to the equation

$$T_2 = h\langle\sigma_x\rangle = \frac{p_i r_i^2 - p_o r_o^2}{r_o + r_i},$$ (4)

where T_2 is the longitudinal membrane stress. To derive Eq. (4), we note that on a free-body diagram which includes an end of the cylinder, the axial force due to internal pressure is $\pi r_i^2 p_i$, that due to the external pressure is $-\pi r_o^2 p_o$, whereas the resultant axial force in the wall is equal to the mean stress $\langle\sigma_x\rangle$ multiplied by the area $\pi r_o^2 - \pi r_i^2$, or the product of the membrane stress T_2 and the circumference $\pi(r_i + r_o)$. The vanishing of the sum of all axial forces leads to Eq. (4).

An entirely analogous consideration of a spherical shell yields the average circumferential stress

$$T = h\langle\sigma_\theta\rangle = \frac{p_i r_i^2 - p_o r_o^2}{r_o + r_i},$$ (5)

which again reduces to Eq. (1) of Prob. 1.4 in the case of a very thin shell. Equations (3)–(5) are valid for arbitrary materials.

It may be interesting to comment on the history of the law of Laplace itself. It is generally recognized that Thomas Young obtained this law earlier than Laplace. According to Merz (1965), Young presented a derivation of this formula, based entirely on the existence of surface tension, to the Royal Society on the 20th of December, 1804, in a *Memoir on the cohesion of the fluids*. In December, 1805, Laplace read before the Institute of France his theory of capillary attraction, in which the formula was derived on the basis of a special hypothesis of molecular attraction. In a supplement to his memoir, which appeared anonymously in the first number of the *Quarterly Review* (1809, no. 1, p. 109) Young, evidently annoyed that some of his results had been reproduced without acknowledgment, reviewed the treatise of Laplace and said: "The point on which M. Laplace seems to rest the most material part of his claim to originality is the deduction of all the phenomena of capillary action from the simple consideration of molecular attraction. To us, it does not appear that the fundamental principle from which he sets out is at all a necessary consequence of the established properties of matter; and we conceive that this mode of stating the question is but partially justified by the coincidence of the results derived from it with experiment, since he has not demonstrated that a similar coincidence might not be obtained by proceeding on totally different grounds" (Merz, 1965, p. 20).

This historical incident is interesting in that it shows how precariously fame rests. It also exhibits an age-old conflict in the scientific method of approach. Young's derivation was phenomenological, quite general, and had few assumptions, but was regarded by many as "abstract." Laplace's derivation was complicated, and was based on a special assumption about molecular attraction, yet was accepted by many. Even today, such a divergence of approach is encountered again and again. To some, a formula is "understood" when it is derived from a molecular model, however imperfect. To others, the abstract (mathematical and phenomenological) approach is preferred.

Problems

1.1 To practice the use of free-body diagram, consider the problem of analyzing the stress in man's back muscles when he does hard work. Figure P1.1 shows a man shoveling snow. If the snow and shovel weigh 10 kg and have a center of gravity located at a distance of 1 m from the lumbar vertebra of his backbone, what is the moment about that vertebra?

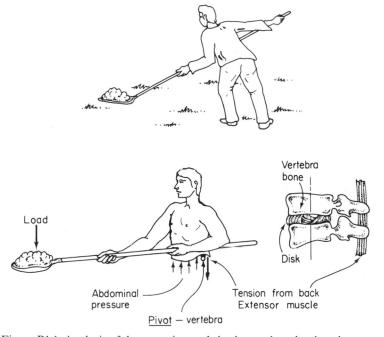

Figure P1.1 Analysis of the stress in man's back muscle as he shovels snow.

The construction of a man's backbone is sketched in the figure. The disks between the vertebrae serve as the pivots of rotation. It may be assumed that the intervertebral disks cannot resist rotation. Hence the weight of the snow and shovel has to be resisted by the vertebral column and the muscle. Estimate the loads in his back muscle, vertebrae, and disks.

Lower back pain is such a common affliction that the loads acting on the disks of patients were measured with strain gauges in some cases. It was found that no agreement can be obtained if we do not take into account the fact that when one lifts a heavy weight, one tenses up the abdominal muscles so that the pressure in the abdomen is increased. A free-body diagram of the upper body of a man is shown in Fig. P1.1. Show that it helps to have a large abdomen and strong abdominal muscles.

See Ortengren et al. (1978) for in vivo measurements of disk and intra-abdominal pressures. See Schultz et al (1991) for an in-depth discussion.

1.2 Compare the bending moment acting on the spinal column at the level of a lumbar vertebra for the following cases:

(a) A secretary bends down to pick up a book on the floor (i) with the knees straight and (ii) with the knees bent.

(b) A water skier skiis (i) with the arms straight and (ii) with elbows hugging the sides.

Discuss these cases quantitatively with proper free-body diagrams.

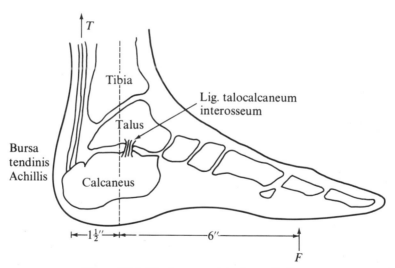

Figure P1.3 The bones and tendons of a foot.

1.3 Consider the question of the strength of human tissues and their margins of safety when we exercise. Take the example of the tension in the Achilles tendon in our foot when we walk and when we jump. The bone structure of the foot is shown in Fig. P1.3. To calculate the tension in the Achilles tendon we may consider the equilibrium of forces that act on the foot. The joint between the tibia and talus bones may be considered as a pivot. The strength of tendons, cartilages, and bones can be found in Yamada's book (1970).

Solution. Assume that the person weighs 68 kg and that the dimensions of the foot are as shown in the figure. Let T be the tension in the Achilles tendon. Refer to Fig. P1.3. We see that the balance of moments requires that

$$T \times 1.5 = 34 \times 6,$$

which gives $T = 136$ kg. If the cross-sectional area of the Achilles tendon is 1.6 cm^2, then the tensile stress is

$$136/1.6 = 85 \text{ kg/cm}^2.$$

This is the stress acting in the Achilles tendon when one is poised to jump but remaining stationary. To actually jump, however, one must generate a force on the feet larger than the body weight. Equivalently, one must generate a kinetic energy initially. At the highest point of his jump the initial kinetic energy is converted to the potential energy. Thus

$$mgh = \tfrac{1}{2}mv^2,$$

where m is the mass of the body, g is the gravitational acceleration, h is the height of jump measured at the center of gravity of the body, and v is the initial velocity of jump measured at the center of gravity of the body.

If the person weighs 68 kg, and the height of jump h is 61 cm, then

$$v = 346 \text{ cm/sec.}$$

To do this jump, one must bend the knees, lowering the center of gravity, then suddenly jump up. During this period, the only external forces acting are gravity and the forces acting on the soles. The impulse imposed on the feet must equal to the momentum gained during this period. Let F be the additional force on the feet, and t be the time interval between the instant of initation of the jump and the instant leaving the ground, then the impulse is $F \cdot t$ and the momentum is mv, and we have

$$Ft = mv.$$

How long t is depends on the gracefulness of the jump. Some experimental results by Miller and East (1976) show that t may be about 0.3 sec. Let us assume that $t - 0.3$ sec. Then, since $mg = 68$ kg, $F = 80$ kg.

Thus, if the person uses both feet in this jump the total force acting on each foot is $(68 + 80)/2 = 74$ kg. Correspondingly, the tension in the Achilles tendon is $74 \times (6/1.5) = 296$ kg. The stress is, again assuming a cross-sectional area of 1.6 cm^2, equal to 185 kg/cm^2.

How strong are our tendons? From Yamada's book, Table 70, p. 100, we have the value 5.6 ± 0.09 kg/mm^2 for the ultimate tensile stress for the calcaneal (Achilles) tendon in the age group 10 to 29 years. Comparing the computed stress of 1.83 kg/mm^2 with the ultimate stress, we see that the factor of safety is quite small. Such a small factor of safety is compatible with life only because the living organism has an active self-repairing and self-renewal capability.

1.4 Consider a thin-walled spherical balloon made of uniform material and having a uniform wall thickness (Fig. P1.4). When it is inflated with an internal pressure

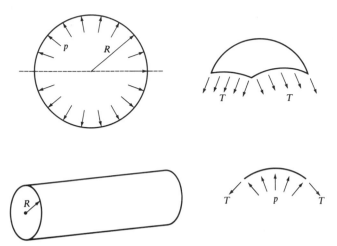

Figure P1.4 A thin-walled spherical balloon and a circular cylindrical shell subjected to internal pressure.

p, the radius of the sphere is R and the tension in the wall is T per unit length. Derive the condition of equilibrium:

$$p = \frac{2T}{R}.$$ (1)

Consider next a circular cylindrical tube inflated with an internal pressure p to a radius R. Show that the circumferential tension per unit length, T (the so-called hoop tension), is related to p and R by

$$p = \frac{T}{R}.$$ (2)

Resistance to bending of the wall is assumed to be negligible.

1.5 Palpation is used commonly to estimate the internal pressure in an elastic vessel such as a balloon, artery, eyeball, or aneurysm. But when we palpate, what are we measuring? The pressure? Or the resultant force? Show that in general the force or

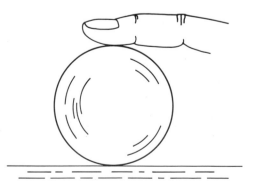

Figure P1.5 Palpation of a blood vessel.

pressure acting on the finger will be affected by the tension in the vessel wall. Show also, however, that if you push just so much that the membrane (vessel wall) is flat (has infinitely large radius of curvature), then you feel exactly the internal pressure in the vessel. The tacit assumption made in this analysis is that the resistance to bending of the wall is negligible. See Fig. P1.5.

1.6 The simplest theory of the cardiovascular system is the *Windkessel* theory proposed by Otto Frank in 1899. The idea was stated earlier by Hales (1733) and Weber (1850). In this theory, the aorta is represented by an elastic chamber, and the peripheral blood vessels are replaced by a rigid tube of constant resistance. See Fig. P1.6(a). Let $\dot{Q}$ be the inflow (cm^3/sec) into this system from the heart. Part of this inflow is sent to the peripheral vessels, and part of it is used to distend the elastic chamber. If p is the blood pressure (pressure in the aorta or elastic chamber), the flow in the peripheral vessel is equal to p/R, where R is called peripheral resistance. For the elastic chamber, its change of volume is assumed to be proportional to the pressure. The rate of change of the volume of the elastic chamber with respect to time t is therefore proportional to dp/dt. Let the constant of proportionality be written as K. Show that on equating the inflow to the sum of the rate of change of volume of the elastic chamber and the outflow p/R, the differential equation governing the pressure p is

$$\dot{Q} = K\frac{dp}{dt} + \frac{p}{R}. \tag{1}$$

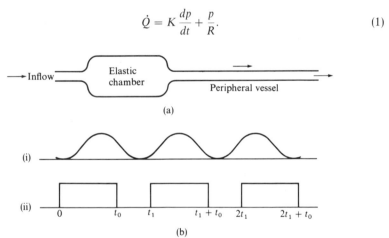

(a)

(i)

(ii)

(b)

Figure P1.6 (a) A simple representation of the aorta. (b) Some pulse waves.

Show that the solution of this differential equation is

$$p(t) = \frac{1}{K}e^{-t/(RK)}\int_0^t \dot{Q}(\tau)e^{\tau/(RK)}\,d\tau + p_0e^{-t/(RK)}, \tag{2}$$

where p_0 is the value of p at time $t = 0$. Using this solution, discuss the pressure pulse in the aorta as a function of the left ventricle contraction history $\dot{Q}(t)$ (i.e., the pumping history of the heart). Discuss in particular the condition of periodicity and the steady-state response of the aorta to a periodically beating heart. This equation works remarkably well in correlating experimental data on the total blood flow $\dot{Q}$ with the blood pressure p, particularly during diastole. Hence in spite of the simplicity of the underlying assumptions, it is quite useful.

Note: The derivation of Eq. (1) is already outlined in the statement of the problem. For the last part, it would be interesting to discuss the solutions when $\dot{Q}(t)$ is one of the idealized periodic functions shown in Fig P1.6(b):

$$\dot{Q}(t) = a \sin^2 \omega t, \tag{3}$$

$$\dot{Q}(t) = a\{[1(t) - 1(t - t_0)] + [1(t - t_1) - 1(t - t_1 - t_0)]$$
$$+ [1(t - 2t_1) - 1(t - 2t_1 - t_0)] + \cdots\}$$

$$= \sum_{n=0}^{\infty} a[1(t - nt_1) - 1(t - nt_1 - t_0)]. \tag{4}$$

1.7 Show that the action of the air chamber in a hand-pumped fire engine is governed by the same equations as those of Problem 1.6. Hence the name *Windkessel*.

References

Fung, Y. C. (1984) *Biodynamics: Circulation*. Springer-Verlag, New York.

Fung, Y. C. (1990) *Biomechanics: Motion, Flow, Stress, and Growth*. Springer-Verlag, New York.

Galileo Galilei (1638) *Discorsi e Dimostrazioni matematiche, intorno á due nuove Scienze, Attenenti alla Mecanica & i Movimenti Locali*. Elzevir, Leida. Translated into English by H. Crew and A. de Salvio under the title *Dialogues Concerning Two New Sciences*. MacMillan, London, 1914. Reissued by Dover Publications, New York, 1960.

Merz, J. T. (1965) *A History of European Thought in the Nineteenth Century*. Dover Publications, New York (reproduction of first edition, W. Blackwood and Sons, 1904).

Miller, D. I. and East, D. J. (1976) Kinematic and kinetic correlates of vertical jumping in woman. In *Biomechanics V-B*, P. V. Komi (ed.) University Park Press, Baltimore, pp. 65–72.

Ortengren, R., Andersson, G., and Nachemson, A. (1978) Lumbar back loads in fixed working postures during flexion and rotation. In *Biomechanics VI-B*, E. Asmussen and K. Jorgensen (eds.) University Park Press, Baltimore, pp. 159–166.

Schultz, A. B. and Ashton-Miller, J. A. (1991) Biomechanics of the human spine. In *Basic Orthopaedic Biomechanics*, (ed. by V. C. Mow and W. C. Hayes) Raven Press, New York Chap. 8, pp. 337–374.

Singer, C. J. (1959) *A Short History of Scientific Ideas to 1900*. Oxford University Press, New York.

Todhunter, I. and Pearson, K. (1960) *A History of the Theory of Elasticity, and of the Strength of Materials from Galilei to Lord Kelvin*. Cambridge University Press (1886, 1893); Dover Publications, New York.

Wolff, H. S. (1973) Bioengineering—A many splendored thing—but for whom? In *Perspectives in Biomedical Engineering*, Proceedings of a symposium, (ed. by R. M. Kenedi). University Park Press, Baltimore, pp. 305–311.

Yamada, H. (1970) *Strength of Biological Materials*. Williams and Wilkins, Baltimore (translated by F. G. Evans).

Young, J. (1930) Malpighi's "de Pulmonibus." *Proc. Roy. Soc. Med.* **23**, Part 1, 1–14.

CHAPTER 2

The Meaning of the Constitutive Equation

2.1 Introduction

In the biological world, atoms and molecules are organized into cells, tissues, organs, and individual organisms. We are interested in the movement of matter inside and around the organisms. At the atomic and molecular level the movement of matter must be analyzed with quantum, relativistic, and statistical mechanics. At the cellular, tissue, organ, and organism level it is usually sufficient to take Newton's laws of motion as an axiom. The object of study of this book is at the animal, organ, tissue, and cell level. The smallest volume we shall consider contains a very large number of atoms and molecules. In these systems it is convenient to consider the material as a continuum.

In mathematics, the real number system is a continuum. Between any two real numbers there is another real number. The classical definition of a material continuum is an isomorphism of the real number system in a three-dimensional Euclidean space: between any two material particles there is another material particle. Each material particle has a mass. The mass density of a continuum at a point P is defined by considering a sequence of volumes ΔV enclosing P. If the mass of particles in ΔV is denoted by ΔM, and if the ratio $\Delta M/\Delta V$ tends to a limit $\rho(P)$ when ΔV tends to zero, then $\rho(P)$ is the mass density of the continuum at P.

Current physics does not conceive the spatial distribution of elementary particles, atoms, and molecules in a living organism as an isomorphism of the real number system. Hence a material continuum based on the classical definition is not compatible with the concept of particle physics. It is necessary to modify the classical definition of a continuum before it can be applied to biology.

Our modification is based on the fact that observations of living organisms can be made at various levels of size: e.g., at the level of the naked eye, or within the limit of optical microscopy, or within the limit of the electron microscopes, or at the limit of the scanning tunneling microscopes. Each of these instruments defines a limit of size below which little information can be obtained. And the images of a biological entity look very different at different levels of magnification. This suggests that we can define a continuum of the real world with a specific bound on the lower scale of size. Thus we may define the mass density of a material at a point P in a three-dimensional space as the limit of the ratio $\Delta M/\Delta V$ when ΔV tends to a finite lower bound which is not zero. ΔM is the mass of the particles in a volume ΔV which encloses the point P. Since ΔV is not allowed to tend to zero, the limiting value $\Delta M/\Delta V$ may be equal to $\rho(P) \pm$ a tolerable error whose bounds must be specified and accepted.

In Sec. 2.2 we discuss the concepts of stress as the force per unit area acting on a surface separating two sets of material particles on the two sides of the surface. Stress is defined by the limit of the ratio $\Delta T/\Delta S$, where ΔT is the force due to particles on the positive side of the surface of area ΔS acting on the particles on the negative side of the surface. In taking the limit we again restrict ΔS to be small but finite, with a specified lower bound of size. The limiting value of $\Delta T/\Delta S$ would have a prescribed acceptable bound of error.

If a material system has a mass density and a stress vector definable in this manner, then we say that the system is a material continuum. Thus, our definition of material continuum is based on a specific lower bound of size and a set of specific accepted bounds of errors.

Since the lower bound of size is a part of the definition, a different choice of the lower bound of size may result in a different continuum. For example, the whole blood may be considered as a continuum at the scale of the heart, large arteries, and large veins, but must be regarded as a two phase fluid, with plasma and blood cells as two separate phases in capillary blood vessels and arterioles and venules. At a smaller scale one may identify the red cell membrane as a continuum, the red cell content as another. Take the example of the lung, one may identify the lung parenchyma as a continuum when the lower bound of size is of the order of 1 cm. If the lower bound is 1 μm we may identify the alveolar wall as a continuum, and the small blood vessel wall as another continuum. At an even smaller scale the collagen and elastin fibers, the ground substances, and the cells in the blood vessel wall may be considered as separate continuous media.

Once a material continuum is identified in this way, we can make a classical copy of the continuum as follows. For the classical copy, the ratios $\Delta M/\Delta V$, $\Delta T/\Delta S$ agree with those of the real system within specific bounds when ΔV, ΔS are limited to sizes specified in the definition of the real system, but limiting values of $\Delta M/\Delta V, \Delta T/\Delta S$ are assumed to exist as $\Delta V \to 0, \Delta S \to 0$. The classical copy is isomorphic with the real number system. We can then analyze the stress, strain, motion, and mechanical properties of the classical copy. The specified acceptable errors can then be used to evaluate the errors of the solutions.

Treating systems of very large numbers of atoms and molecules by continuum approach greatly simplifies the analysis. Sometimes certain parts of the continuum can be considered as a rigid body, or as lumped masses. If so, then the analysis is simplified further. Usually the partial differential equations of motion and continuity can be approximated by finite elements to make practical calculations. But the greatest advantage of the continuum approach is the ability to express the mechanical properties of the system by constitutive equations. This is the topic to be discussed in this chapter.

While the reader is referred to standard works on continuum mechanics for a more detailed presentation, we shall outline very briefly the concepts, definitions, and notations we shall use in this work.

2.2 Stress

If we want to determine the strength of a tendon, we can test a large or small specimen. A large specimen can sustain a large force, a smaller specimen can take only a small force. Obviously, the size of the test specimen is incidental; only the force relative to the size is important. Thus, we are led to the concept that it is the stress (force per unit cross-sectional area) that is related to the strength of the material.

Let the cross-sectional area of a tendon be A, and let the force that acts in the tendon be F. The ratio

$$\sigma = F/A \tag{1}$$

is the *stress* in the tendon. In the International System of Units (SI units), the basic unit of force is the *newton* (N) and that of length is the *meter* (m). Thus the basic unit of stress is *newton per square meter* (N/m^2) or *pascal* (Pa, in honor of Pascal). 1 MPa = 1 N/mm^2. A force of 1 N can accelerate a body of mass 1 kg to 1 m/sec^2. A force of 1 dyne (dyn) can accelerate a body of mass 1 gram to 1 cm/sec^2. Hence, 1 dyn = 10^{-5} N. Some conversion factors are listed below:

$$1 \text{ pound force} \doteq 4.448 \text{ N}$$
$$1 \text{ pound per square inch (p si)} \doteq 6894 \text{ N/m}^2 = 6.894 \text{ kPa}$$
$$1 \text{ dyn/cm}^2 \doteq 0.1 \text{ N/m}^2 = 0.1 \text{ Pa}$$
$$1 \text{ atmosphere} \doteq 1.013 \times 10^5 \text{ N/m}^2 = 1.013 \text{ bar}$$
$$1 \text{ mm Hg at } 0°C \doteq 133.32 \text{ N/m}^2 = 1 \text{ torr} \sim \tfrac{1}{7.501} \text{ kPa}$$
$$1 \text{ cm H}_2\text{O at } 4°C \doteq 98 \text{ N/m}^2$$
$$1 \text{ poise (viscosity)} = 0.1 \text{ N sec/m}^2.$$

More generally, the concept of stress expresses the interaction of the material in one part of the body on another. Consider a material continuum

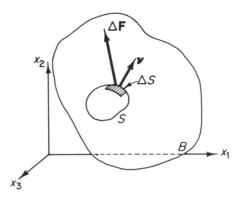

Figure 2.2:1 Stress principle.

B occupying a spatial region V (Fig. 2.2:1). Imagine a small surface element of area ΔS. Let us draw, from a point on ΔS, a unit vector* $\mathbf{v}$ normal to ΔS. Then we can distinguish the two sides of ΔS according to the direction of $\mathbf{v}$. Let the side to which this normal vector points be called the positive side. Consider the part of material lying on the positive side. This part exerts a force $\Delta \mathbf{F}$ on the other part, which is situated on the negative side of the normal. The force $\Delta \mathbf{F}$ depends on the location and size of the area and the orientation of the normal. We introduce the assumption that as ΔS tends to zero, the ratio $\Delta \mathbf{F}/\Delta S$ tends to a definite limit, $d\mathbf{F}/dS$, and that the moment of the force acting on the surface ΔS about any point within the area vanishes in the limit. The limiting vector will be written as

$$\overset{v}{T} = d\mathbf{F}/dS, \tag{2}$$

where a superscript v is introduced to denote the direction of the normal v of the surface ΔS. The limiting vector $\overset{v}{\mathbf{T}}$ is called the *traction*, or the *stress vector*, and represents the force per unit area acting on the surface.

The general notation for stress components is as follows. Consider a little cube in the body (replacing S in Fig. 2.2:1 by a cube, as shown in Fig. 2.2:2). It has 6 surfaces. Erect a set of rectangular cartesian coordinates $\mathbf{x}_1, \mathbf{x}_2, \mathbf{x}_3$. Let the surface of the cube normal to $\mathbf{x}_1$ be denoted by ΔS_1. Let the stress vector that acts on the surface ΔS_1 be $\underset{1}{\mathbf{T}}$. Resolve the vector $\underset{1}{\mathbf{T}}$ into three components in the direction of the coordinate axes and denote them by $\tau_{11}, \tau_{12}, \tau_{13}$. Then $\tau_{11}, \tau_{12}, \tau_{13}$ are the stress components acting on the small cube. Similarly, we may consider surfaces $\Delta S_2, \Delta S_3$ perpendicular to $\mathbf{x}_2$ and $\mathbf{x}_3$, the stress vectors acting on them, and their components in $\mathbf{x}_1, \mathbf{x}_2, \mathbf{x}_3$ directions. We can arrange the components in a square matrix:

---

* All vectors are printed in bold face in this book.

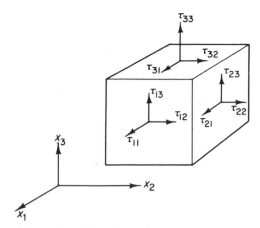

Figure 2.2:2 Notations of stress components.

| | Components of Stresses | | |
	1	2	3
Surface normal to x_1	τ_{11}	τ_{12}	τ_{13}
Surface normal to x_2	τ_{21}	τ_{22}	τ_{23}
Surface normal to x_3	τ_{31}	τ_{32}	τ_{33}

This is illustrated in Fig. 2.2:2. The components $\tau_{11}, \tau_{22}, \tau_{33}$ are called *normal stresses*, and the remaining components τ_{12}, τ_{13}, etc., are called *shearing stresses.*

It is important to emphasize that a stress will always be understood to be the force (per unit area) that the part lying on the positive side of a surface element (the side on the positive side of the outer normal) exerts on the part lying on the negative side. Thus, if the outer normal of a surface element points in the positive direction of the x_2 axis and τ_{22} is positive, the vector representing the component of normal stress acting on the surface element will point in the positive x_2 direction. But if τ_{22} is positive while the outer normal points in the negative x_2 axis dirction, then the stress vector acting on the element also points to the negative x_2 axis direction (see Fig. 2.2:3).

Similarly, positive values of τ_{21}, τ_{23} will imply shearing stress vectors pointing to the positive x_1, x_3 axes if the outer normal agrees in sense with the x_2 axis, whereas the stress vectors point to the negative x_1, x_3 directions if the outer normal disagrees in sense with the x_2 axis, as illustrated in Fig. 2.2:3.

We now give without proof four important formulas concerning stresses. (See Y. C. Fung, 1993, *A First Course in Continuum Mechanics*.) These formulas will be given in indicial notation, in which all free index of a variable such as x_i stands for $i = 1$, or 2, or 3. x_i stands for x_1, or x_2, or x_3. τ_{ij} stands for $\tau_{11}, \tau_{12}, \tau_{13}, \tau_{21}, \tau_{22}, \tau_{23}, \tau_{31}, \tau_{32}, \tau_{33}$ since $i = 1$, 2, or 3; $j = 1$, 2, or 3.

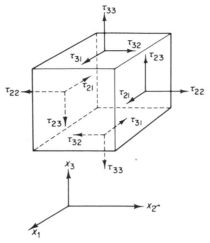

Figure 2.2:3 Senses of positive stress components.

Repetition of an index in a single term means a summation over the whole range of that index. Thus $x_i x_i$ means the sum $x_1 x_1 + x_2 x_2 + x_3 x_3$. First, it can be shown that knowing the components of a stress tensor τ_{ij} with respect to a rectangular Cartesian frame of references, we can write down at once the stress vector acting on any surface with unit over outer normal vector v whose components are v_1, v_2, v_3. This stress vector is denoted by $\overset{v}{\mathbf{T}}$, with components $\overset{v}{T_1}, \overset{v}{T_2}, \overset{v}{T_3}$ given by Cauchy's formula

$$\overset{v}{T_i} = v_1 \tau_{1i} + v_2 \tau_{2i} + v_3 \tau_{3i}. \tag{3}$$

Next, for a body in equilibrium, the stress components must satisfy the differential equations

$$\frac{\partial \tau_{11}}{\partial x_1} + \frac{\partial \tau_{12}}{\partial x_2} + \frac{\partial \tau_{13}}{\partial x_3} + X_1 = 0,$$

$$\frac{\partial \tau_{21}}{\partial x_1} + \frac{\partial \tau_{22}}{\partial x_2} + \frac{\partial \tau_{23}}{\partial x_3} + X_2 = 0, \tag{4}$$

$$\frac{\partial \tau_{31}}{\partial x_1} + \frac{\partial \tau_{32}}{\partial x_2} + \frac{\partial \tau_{33}}{\partial x_3} + X_3 = 0,$$

where X_1, X_2, X_3 are the components of the body force (per unit volume) acting on the body. Third, the stress tensor is always symmetric:

$$\tau_{12} = \tau_{21}, \qquad \tau_{23} = \tau_{32}, \qquad \tau_{31} = \tau_{13}. \tag{5}$$

Finally, let a system of rectangular cartesian coordinates x_1, x_2, x_3 be transformed into another such system x'_1, x'_2, x'_3 by the transformation

$$x'_k = \beta_{ki} x_i \equiv \beta_{k1} x_1 + \beta_{k2} x_2 + \beta_{k3} x_3 \qquad (k = 1, 2, 3), \tag{6}$$

where β_{ki} denotes the direction cosine of the $\mathbf{x}'_k$ axis with respect to the $\mathbf{x}_i$ axis. Then the stress components are transformed according to the *tensor transformation law*:

$$\tau'_{km} = \tau_{ji}\beta_{kj}\beta_{mi}. \tag{7}$$

In these formulas, the summation convention of the index is used: any repetition of an index in a single term means summation over that index. For example, Eq. (7) stands for

$$\tau'_{km} = \beta_{k1}\beta_{m1}\tau_{11} + \beta_{k1}\beta_{m2}\tau_{12} + \beta_{k1}\beta_{m3}\tau_{13} + \beta_{k2}\beta_{m1}\tau_{21} + \beta_{k2}\beta_{m2}\tau_{22}$$
$$+ \beta_{k2}\beta_{m3}\tau_{23} + \beta_{k3}\beta_{m1}\tau_{31} + \beta_{k3}\beta_{m2}\tau_{32} + \beta_{k3}\beta_{m3}\tau_{33}$$

for any combination of $k = 1, 2,$ or $3,$ and $m = 1, 2,$ or $3.$

2.3 Strain

Deformation of a solid that can be related to stresses is described by strain. Take a string of an initial length L_0. If it is stretched to a length L as shown in Fig. 2.3:1(a), it is natural to describe the change by dimensionless ratios such as $L/L_0, (L - L_0)/L_0, (L - L_0)/L.$ Use of dimensionless ratios eliminates the absolute length from consideration. The ratio L/L_0 is called the *stretch ratio* and is denoted by the symbol λ. The ratios

$$\varepsilon = \frac{L - L_0}{L_0}, \qquad \varepsilon' = \frac{L - L_0}{L} \tag{1}$$

are *strain measures*. Either of them can be used. Numerically, they are different. For example, if $L = 2, L_0 = 1,$ we have $\varepsilon = 1, \varepsilon' = \frac{1}{2}.$ We shall have reasons (to be named later) also to introduce the measures

$$e = \frac{L^2 - L_0^2}{2L^2}, \qquad \mathscr{E} = \frac{L^2 - L_0^2}{2L_0^2}. \tag{2}$$

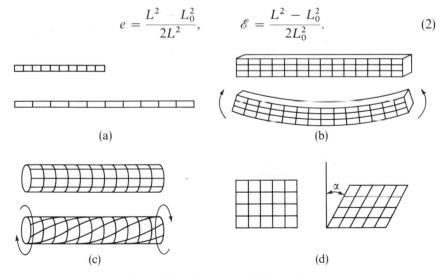

(a)   (b)

(c)   (d)

Figure 2.3:1 Patterns of deformation.

If $L = 2$, $L_0 = 1$, we have $e = \frac{3}{8}$, $\mathscr{E} = \frac{3}{2}$. But if $L = 1.01$, $L_0 = 1.00$, then $e \doteq 0.01$, $\mathscr{E} \doteq 0.01$, $\varepsilon \doteq 0.01$, and $\varepsilon' \doteq 0.01$. Hence in infinitesimal elongations all the strain measures named above are equal. In finite elongations, however, they are different.

To illustrate shear, consider a circular cylindrical shaft as shown in Fig. 2.3:1(c). When the shaft is twisted, the elements in the shaft are distorted in a manner shown in Fig. 2.3:1(d). In this case, the angle α may be taken as a strain measure. It is more customary, however, to take $\tan \alpha$, or $\frac{1}{2} \tan \alpha$, as the *shear strain*; the reasons for this will be elucidated later.

The selection of proper strain measures is dictated basically by the stress–strain relationship (i.e., the constitutive equation of the material). For example, if we pull on a string, it elongates. The experimental results can be presented as a curve of the tensile stress σ plotted against the stretch ratio λ, or strain e. An empirical formula relating σ to e can be determined. The case of infinitesimal strain is simple because the different strain measures named above all coincide. It was found that for most engineering materials subjected to an infinitesimal strain in uniaxial stretching, a relation like

$$\sigma = Ee \tag{3}$$

is valid within a certain range of stresses, where E is a constant called *Young's modulus*. Equation (3) is called *Hooke's law*. A material obeying Eq. (3) is said to be a *Hookean material*. Steel is a Hookean material if σ lies within certain bounds that are called *yield stresses*. Corresponding to Eq. (3), the relationship for a Hookean material subjected to an infinitesimal shear strain is

$$\tau = G \tan \alpha, \tag{4}$$

where G is another constant called the *modulus of rigidity*. The range of validity of Eq. (4) is again bounded by yield stresses. The yield stresses in tension, in compression, and in shear are generally different.

Deformations of most things in nature are much more complex than those discussed above. We therefore need a general method of treatment. Let a body occupy a space S. Referred to a rectangular cartesian frame of reference, every particle in the body has a set of coordinates. When the body is deformed, every particle takes up a new position, which is described by a new set of coordinates. For example, a particle P located originally at a place with coordinates (a_1, a_2, a_3) is moved to the place Q with coordinates (x_1, x_2, x_3) when the body moves and deforms. Then the vector **PQ** or, **u**, is called the *displacement vector* of the particle (see Fig. 2.3:2). The components of the displacement vector are, clearly,

$$u_1 = x_1 - a_1, \qquad u_2 = x_2 - a_2, \qquad u_3 = x_3 - a_3. \tag{5}$$

If the displacement is known for every particle in the body, we can construct the deformed body from the original. Hence, a deformation can be described by the displacement field. Let the variables (a_1, a_2, a_3) identify a particle in the original configuration of the body, and let (x_1, x_2, x_3) be the

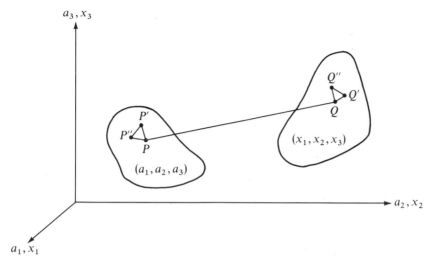

Figure 2.3:2 Deformation of a body.

coordinates of that particle when the body is deformed. Then the deformation of the body is known if x_1, x_2, x_3 are known functions of a_1, a_2, a_3:

$$x_i = x_i(a_1, a_2, a_3) \qquad (i = 1, 2, 3). \qquad (6)$$

This is a transformation (mapping) from a_1, a_2, a_3 to x_1, x_2, x_3. We assume that the transformation is one-to-one; i.e., the functions in Eq. (6) are single-valued, continuous, and have a unique inverse,

$$a_1 = a_i(x_1, x_2, x_3), \qquad (7)$$

for every point in the body.

A rigid-body motion induces no stress. Thus, the displacements themselves are not directly related to the stress. To relate deformation with stress we must consider the stretching and distortion of the body. For this purpose, it is sufficient if we know the change of distance between any arbitrary pair of points.

Consider an infinitesimal line element connecting the point $P(a_1, a_2, a_3)$ to a neighboring point $P'(a_1 + da_1, a_2 + da_2, a_3 + da_3)$. The square of the lengths ds_0 of PP' in the original configuration is given by the Pythagoras rule because the space is assumed to be Euclidean:

$$ds_0^2 = da_1^2 + da_2^2 + da_3^2. \qquad (8)$$

When P and P' are deformed to the points $Q(x_1, x_2, x_3)$ and $Q'(x_1 + dx_1, x_2 + dx_2, x_3 + dx_3)$, respectively, the square of the length ds of the new element QQ' is

$$ds^2 = dx_1^2 + dx_2^2 + dx_3^2. \qquad (9)$$

By Eqs. (6) and (7), we have

$$dx_i = \frac{\partial x_i}{\partial a_j} da_j, \qquad da_i = \frac{\partial a_i}{\partial x_j} dx_j. \tag{10}$$

Hence, on introducing the *Kronecker delta*, δ_{ij}, which has the value 1 if $i = j$, and zero if $i \neq j$, we may write

$$ds_0^2 = \delta_{ij} da_i da_j = \delta_{ij} \frac{\partial a_i}{\partial x_l} \frac{\partial a_j}{\partial x_m} dx_l dx_m, \tag{11}$$

$$ds^2 = \delta_{ij} dx_i dx_j = \delta_{ij} \frac{\partial x_i}{\partial a_l} \frac{\partial x_j}{\partial a_m} da_l da_m. \tag{12}$$

The difference between the squares of the length elements may be written, after several changes in the symbols for dummy indices, either as

$$ds^2 - ds_0^2 = \left(\delta_{\alpha\beta} \frac{\partial x_\alpha}{\partial a_i} \frac{\partial x_\beta}{\partial a_j} - \delta_{ij} \right) da_i da_j, \tag{13}$$

or as

$$ds^2 - ds_0^2 = \left(\delta_{ij} - \delta_{\alpha\beta} \frac{\partial a_\alpha}{\partial x_i} \frac{\partial a_\beta}{\partial x_j} \right) dx_i dx_j. \tag{14}$$

We define the strain tensors

$$E_{ij} = \frac{1}{2} \left(\delta_{\alpha\beta} \frac{\partial x_\alpha}{\partial a_i} \frac{\partial x_\beta}{\partial a_j} - \delta_{ij} \right), \tag{15}$$

$$e_{ij} = \frac{1}{2} \left(\delta_{ij} - \delta_{\alpha\beta} \frac{\partial a_\alpha}{\partial x_i} \frac{\partial a_\beta}{\partial x_j} \right), \tag{16}$$

so that

$$ds^2 - ds_0^2 = 2E_{ij} da_i da_j, \tag{17}$$

$$ds^2 - ds_0^2 = 2e_{ij} dx_i dx_j. \tag{18}$$

The strain tensor E_{ij} was introduced by Green and St.-Venant and is called *Green's strain tensor.* The strain tensor e_{ij} was introduced by Cauchy for infinitesimal strains and by Almansi and Hamel for finite strains and is known as *Almansi's strain tensor.* In analogy with terminology in hydrodynamics, E_{ij} is often referred to as *Lagrangian* and e_{ij} as *Eulerian.*

E_{ij} and e_{ij} thus defined are tensors. They are symmetric; i.e.,

$$E_{ij} = E_{ji}, \qquad e_{ij} = e_{ji}. \tag{19}$$

An immediate consequence of Eqs. (17) and (18) is that $ds^2 - ds_0^2 = 0$ implies $E_{ij} = e_{ij} = 0$ and vice versa. But a deformation in which the length of every line element remains unchanged is a rigid-body motion. Hence, the necessary and sufficient condition that a deformation of a body be a rigid-body motion is that all components of the strain tensor E_{ij} or e_{ij} be zero throughout the body.

If the components of displacement u_i are such that their first derivatives are so small that the squares and products of the partial derivatives of u_i are negligible compared with the first order terms, then e_{ij} reduces to *Cauchy's infinitesimal strain tensor*,

$$\varepsilon_{ij} = \frac{1}{2}\left(\frac{\partial u_j}{\partial x_i} + \frac{\partial u_i}{\partial x_j}\right). \tag{20}$$

In unabridged notation, writing u, v, w for u_1, u_2, u_3 and x, y, z for x_1, x_2, x_3, we have

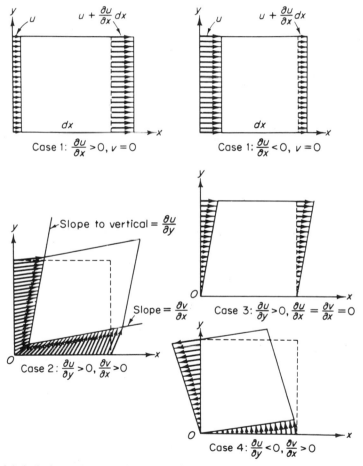

Figure 2.3:3 Deformation gradients and interpretation of infinitesimal strain components.

$$\varepsilon_{xx} = \frac{\partial u}{\partial x}, \qquad \varepsilon_{xy} = \frac{1}{2}\left(\frac{\partial u}{\partial y} + \frac{\partial v}{\partial x}\right) = \varepsilon_{yx},$$

$$\varepsilon_{yy} = \frac{\partial v}{\partial y}, \qquad \varepsilon_{xz} = \frac{1}{2}\left(\frac{\partial u}{\partial z} + \frac{\partial w}{\partial x}\right) = \varepsilon_{zx}, \qquad (21)$$

$$\varepsilon_{zz} = \frac{\partial w}{\partial z}, \qquad \varepsilon_{yz} = \frac{1}{2}\left(\frac{\partial v}{\partial z} + \frac{\partial w}{\partial y}\right) = \varepsilon_{zy}.$$

In the infinitesimal displacement case, the distinction between the Lagrangian and Eulerian strain tensor disappears, since then it is immaterial whether the derivatives of the displacements are calculated at the position of a point before or after deformation.

It is important to visualize the geometric meaning of the individual strain components. The illustrations shown in Fig. 2.3:3 will be helpful. For greater details the reader is referred to the *First Course*, Chapter 5.

Note that our definition of the shear strains ε_{xy}, ε_{yz}, ε_{zx} given in Eq. (21) makes ε_{ij} as defined in Eq. (20) a *tensor*. Only when ε_{ij} is a tensor can the tensor transformation law such as Eq. (7) of Sec. 2.2, the tensor equations (17), (18), and the constitutive equations presented in Secs. 2.6–2.8 be valid. In older engineering mechanics books the shear strains are defined as twice as large as those defined in Eq. (21), and the tensor quality of the strain is ruined. This differences should be borne in mind in reading engineering literature.

2.4 Strain Rate

For fluids in motion we must consider the velocity field and the rate of strain. If we refer the location of each fluid particle to a frame of reference O-xyz; then the field of flow is described by the velocity vector field $\mathbf{v}(x, y, z)$, which defines the velocity at every point (x, y, z). In terms of components, the velocity field is expressed by the functions

$$u(x, y, z), \qquad v(x, y, z), \qquad w(x, y, z),$$

or, if index notations are used, by $v_i(x_i, x_2 x_3)$.

For continuous flow, we consider the continuous and differentiable functions $v_i(x_1, x_2, x_3)$. To study the relationship of velocities at neighboring points, let two particles P and P' be located instantaneously at x_i and $x_i + dx_i$, respectively. The difference in velocities at these two points is

$$dv_i = \frac{\partial v_i}{\partial x_j} dx_j, \qquad (1)$$

where the partial derivatives $\partial v_i / \partial x_j$ are evaluated at the particle P. Now

$$\frac{\partial v_i}{\partial x_j} = \frac{1}{2}\left(\frac{\partial v_i}{\partial x_j} + \frac{\partial v_j}{\partial x_i}\right) - \frac{1}{2}\left(\frac{\partial v_j}{\partial x_i} - \frac{\partial v_i}{\partial x_j}\right). \qquad (2)$$

If we define a *strain rate* tensor V_{ij} and a *vorticity tensor* Ω_{ij} as

$$V_{ij} \equiv \frac{1}{2}\left(\frac{\partial v_l}{\partial x_j} + \frac{\partial v_j}{\partial x_i}\right), \tag{3}$$

$$\Omega_{ij} \equiv \frac{1}{2}\left(\frac{\partial v_j}{\partial x_i} - \frac{\partial v_i}{\partial x_j}\right), \tag{4}$$

then Eq. (1) becomes

$$dv_i = V_{ij}\,dx_j - \Omega_{ij}\,dx_j. \tag{5}$$

It is evident that V_{ij} is symmetric and Ω_{ij} is antisymmetric; i.e.,

$$V_{ij} = V_{ji}, \qquad \Omega_{ij} = -\Omega_{ji}. \tag{6}$$

Equation (3) is similar to Eq. (2.3:20). Its geometric interpretation is also similar. Therefore, the analysis of the velocity field is similar to the analysis of an infinitesimal deformation field. Indeed, if we multiply v_i by an infinitesimal interval of time dt, the result is an infinitesimal displacement $u_i = v_i\,dt$. Hence, whatever we learned about the infinitesimal strain field can immediately be extended to the *rate of change* of strain, with the word *velocity* replacing the word *displacement*.

2.5 Constitutive Equations

The properties of materials are specified by constitutive equations. A wide variety of materials exist. Thus, we are not surprised that there are a great many constitutive equations describing an almost infinite variety of materials. What should be surprising, therefore, is the fact that three simple, idealized stress–strain relationships, namely, the nonviscous fluid, the Newtonian viscous fluid, and the Hookean elastic solid, give a good description of the mechanical properties of many materials around us. Within certain limits of strain and strain rate, air, water, and many engineering structural materials can be described by these idealized equations. Most biological materials, however, cannot be described so simply.

A constitutive equation describes a physical property of a material. Hence it must be independent of any particular set of coordinates of reference with respect to which the components of various physical quantities are resolved. Hence a constitutive equation must be a tensor equation: Every term in the equation must be a tensor of the same rank.

2.6 The Nonviscous Fluid

A *nonviscous fluid* is one for which the stress tensor is of the form

$$\sigma_{ij} = -p\delta_{ij}, \tag{1}$$

where δ_{ij} is the Kronecker delta, which has the value 1 if $i = j$, and zero if $i \neq j$, and p is a scalar called *pressure*. The pressure p in an *ideal gas* is related to the density ρ and temperature T by the *equation of state*

$$\frac{p}{\rho} = RT, \tag{2}$$

where R is the gas constant. For a real gas or a liquid, it is often possible to obtain an equation of state:

$$f(p, \rho, T) = 0. \tag{3}$$

An anomaly exists in the case of an *incompressible fluid*, for which the equation of state is merely

$$\rho = \text{const.} \tag{4}$$

Thus, the pressure p is left as an arbitrary variable for an incompressible fluid.

2.7 The Newtonian Viscous Fluid

A *Newtonian viscous fluid* is a fluid for which the shear stress is linearly proportional to the strain rate. For a Newtonian fluid the stress–strain relationship is specified by the equation

$$\sigma_{ij} = -p\delta_{ij} + \mathcal{D}_{ijkl}V_{kl}, \tag{1}$$

where σ_{ij} is the stress tensor, V_{kl} is the strain rate tensor, $\mathcal{D}_{ijkl}$ is a tensor of viscosity coefficients of the fluid, and p is the *static pressure*. The term $-p\delta_{ij}$ represents the state of stress possible in a fluid at rest (when $V_{kl} = 0$). The static pressure p is assumed to depend on the density and temperature of the fluid according to an equation of state. For a Newtonian fluid we assume that the elements of the tensor $\mathcal{D}_{ijkl}$ may depend on the temperature but not on the stress or the rate of deformation. The tensor $\mathcal{D}_{ijkl}$, of rank 4, has $3^4 = 81$ elements. Not all these constants are independent. A study of the theoretically possible number of independent elements can be made by examining the symmetry properties of the tensors σ_{ij}, V_{kl} and the symmetry that may exist in the atomic constitution of the fluid.

If a tensor has the same array of components when the frame of reference is rotated or reflected, then it is said to be an *isotropic tensor*. It can be shown that, in a three-dimensional Euclidean space, there are two independent isotropic tensors of rank 4, namely, $\delta_{ij}\delta_{kl}$ and $\delta_{ik}\delta_{jl} + \delta_{il}\delta_{jk}$. If the tensor D_{ijkl} is isotropic, then it can be expressed in terms of two independent constants λ and μ,

$$\mathcal{D}_{ijkl} = \lambda\delta_{ij}\delta_{kl} + \mu(\delta_{ik}\delta_{jl} + \delta_{il}\delta_{jk}). \tag{2}$$

On substituting Eq. (2) into Eq. (1), we obtain an isotropic constitutive equation:

$$\sigma_{ij} = -p\delta_{ij} + \lambda V_{kk}\delta_{ij} + 2\mu V_{ij}. \tag{3}$$

A material whose constitutive equation is isotropic is said to be an *isotropic material*. For an isotropic Newtonian fluid, a contraction of Eq. (3) gives

$$\sigma_{kk} = -3p + (3\lambda + 2\mu)V_{kk}. \tag{4}$$

If it is assumed that the mean normal stress $\frac{1}{3}\sigma_{kk}$ is independent of the rate of dilation V_{kk}, then we must set

$$3\lambda + 2\mu = 0; \tag{5}$$

and the constitutive equation becomes

$$\sigma_{ij} = -p\delta_{ij} + 2\mu V_{ij} - \tfrac{2}{3}\mu V_{kk}\delta_{ij}. \tag{6}$$

This formulation is due to George G. Stokes, and a fluid that obeys Eq. (6) is called a *Stokes fluid*, for which one material constant μ, the *coefficient of viscosity*, suffices to define its property.

If a fluid is *incompressible*, then $V_{kk} = 0$, and we have the constitutive equation for an incompressible viscous fluid:

$$\sigma_{ij} = -p\delta_{ij} + 2\mu V_{ij}. \tag{7}$$

If $\mu = 0$, we obtain the constitutive equation of the nonviscous fluid:

$$\sigma_{ij} = -p\delta_{ij}. \tag{8}$$

Newton's concept of viscosity may be explained in the simplest case of a shear flow with a uniform velocity gradient as sketched in Fig. 2.7:1. Newton proposed the relationship

$$\tau = \mu \frac{du}{dy} \tag{9}$$

for the shear stress τ, where μ is the coefficient of viscosity. In the centimeter–gram–second system of units, in which the unit of force is the dyne, the unit of μ is called a *poise*, in honor of Poiseuille. In the SI system, the unit of viscosity is newton–second per square meter (Ns/m^2). 1 poise (P) is 0.1 N s/m^2.

The viscosities of air and water are small, being 1.8×10^{-4} poise for air and 0.01 poise for water at atmospheric pressure and $20°C$. At the same temperature the viscosity of glycerine is about 8.7 poise. The viscosity of liquids decreases as temperature increases. The viscosity of gases increases with increasing temperature.

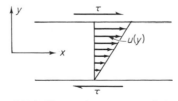

Figure 2.7:1 Newtonian concept of viscosity.

2.8 The Hookean Elastic Solid

A *Hookean elastic solid* is a solid that obeys Hooke's law, which states that
the stress tensor is linearly proportional to the strain tensor; i.e.,

$$\sigma_{ij} = C_{ijkl} e_{kl}, \tag{1}$$

where σ_{ij} is the stress tensor, e_{kl} is the strain tensor, and C_{ijkl} is a tensor of
elastic constants, or moduli, which are independent of stress or strain.

A great reduction in the number of elastic constants is obtained when the
material is *isotropic*, i.e., when the constitutive equation is isotropic, and the
array of elastic constants C_{ijkl} remains unchanged with respect to rotation and
reflection of coordinates. By an argument similar to the discussion of D_{ijkl} in
Sec. 2.7, we see that an isotropic material has exactly two independent elastic
constants, for which the Hooke's law reads

$$\sigma_{ij} = \lambda e_{\alpha\alpha} \delta_{ij} + 2\mu e_{ij}. \tag{2}$$

The constants λ and μ are called the *Lamé constants.* In the engineering literature, the second Lamé constant μ is practically always written as G and
identified as the *shear modulus*.

Writing out Eq. (2) *in extenso*, and with x, y, z as rectangular cartesian
coordinates, we have Hooke's law for an isotropic elastic solid:

$$\sigma_{xx} = \lambda(e_{xx} + e_{yy} + e_{zz}) + 2G e_{xx},$$

$$\sigma_{yy} = \lambda(e_{xx} + e_{yy} + e_{zz}) + 2G e_{yy},$$

$$\sigma_{zz} = \lambda(e_{xx} + e_{yy} + e_{zz}) + 2G e_{zz}, \tag{3}$$

$$\sigma_{xy} = 2G e_{xy}, \quad . \quad \sigma_{yz} = 2G e_{yz}, \quad \sigma_{zx} = 2G e_{zx}.$$

These equations can be solved for e_{ij}. But customarily, the inverted form is
written as

$$e_{ij} = \frac{1+v}{E} \sigma_{ij} - \frac{v}{E} \sigma_{kk} \delta_{ij} \tag{4a}$$

or

$$e_{xx} = \frac{1}{E}[\sigma_{xx} - v(\sigma_{yy} + \sigma_{zz})], \qquad e_{xy} = \frac{1+v}{E} \sigma_{xy} = \frac{1}{2G} \sigma_{xy},$$

$$e_{yy} = \frac{1}{E}[\sigma_{yy} - v(\sigma_{zz} + \sigma_{xx})], \qquad e_{yz} = \frac{1+v}{E} \sigma_{yz} = \frac{1}{2G} \sigma_{yz}, \tag{4b}$$

$$e_{zz} = \frac{1}{E}[\sigma_{zz} - v(\sigma_{xx} + \sigma_{yy})], \qquad e_{zx} = \frac{1+v}{E} \sigma_{zx} = \frac{1}{2G} \sigma_{zx}.$$

The constants E, v, and G are related to the *Lamé constants* λ and G (or μ).
E is called *Young's modulus*, v is called *Poisson's ratio*, and G is called the
modulus of elasticity in shear, or *shear modulus*. They can be written out as
follows:

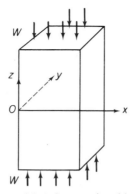

Figure 2.8:1 Stresses in a block.

$$\lambda = \frac{2Gv}{1 - 2v} = \frac{G(E - 2G)}{3G - E} = \frac{Ev}{(1 + v)(1 - 2v)},$$

$$G = \frac{\lambda(1 - 2v)}{2v} = \frac{E}{2(1 + v)},$$

$$v = \frac{\lambda}{2(\lambda + G)} = \frac{\lambda}{(3K - \lambda)} = \frac{E}{2G} - 1,$$

$$E = \frac{G(3\lambda + 2G)}{\lambda + G} = \frac{\lambda(1 + v)(1 - 2v)}{v} = 2G(1 + v).$$

(5)

It is very easy to remember Eq. (4). Apply it to the simple block as illustrated in Fig. 2.8:1. When the block is compressed in the z direction, it shortens by a strain:

$$e_{zz} = \frac{1}{E} \sigma_{zz}.$$

(6)

At the same time, the lateral sides of the block will bulge out somewhat. For a linear material the bulging strain is proportional to σ_{zz} and is in a sense opposite to the stress: A compression induces lateral bulging; a tension induces lateral shrinking. Hence we write

$$e_{xx} = -\frac{v}{E} \sigma_{zz}, \qquad e_{yy} = -\frac{v}{E} \sigma_{zz}.$$

(7)

This is the case in which σ_{zz} is the only nonvanishing stress. If the block is also subjected to σ_{xx}, σ_{yy}, as is illustrated in Fig. 2.2:3, and if the material is isotropic and linear (so that causes and effects are linearly superposable), then the influence of σ_{xx} on e_{yy}, e_{zz} and σ_{yy} on e_{xx}, e_{zz} must be the same as the influence of σ_{zz} on e_{xx}, e_{yy}. Hence we have

$$e_{zz} = \frac{1}{E} \sigma_{zz} - \frac{v}{E} \sigma_{xx} - \frac{v}{E} \sigma_{yy},$$

(8)

which is one of the equations of (4). Other equations in (4) can be explained in a similar manner. For shear, the stress σ_{ij} and the strain e_{ij} ($i \neq j$) are directly proportional.

2.9 The Effect of Temperature

In the preceding sections, the constitutive equations are stated at a given temperature. The viscosity of a fluid, however, varies with temperature (think of the motor oil in your car) as does the elastic modulus of a solid. In other words, $\mathscr{D}_{ijkl}$ in Eq. (2.7:1) and C_{ijkl} in Eq. (2.8:1) are functions of temperature and are coefficients determined under an isothermal experiment (with temperature kept uniform and constant).

If the temperature field is variable, Hooke's law must be modified into the *Duhamel–Neumann form*. Let the elastic constants C_{ijkl} be measured at a uniform constant temperature T_0. Then if the temperature changes to T, we put

$$\sigma_{ij} = C_{ijkl}e_{kl} - \beta_{ij}(T - T_0), \tag{1}$$

in which β_{ij} is a symmetric tensor, measured at zero strains. For an isotropic material, the second order tensor β_{ij} must also be isotropic. It follows that β_{ij} must be of the form $\beta\delta_{ij}$, hence

$$\sigma_{ij} = \lambda e_{kk}\delta_{ij} + 2Ge_{ij} - \beta(T - T_0)\delta_{ij}. \tag{2}$$

Here λ and G are Lamé constants measured at constant temperature. The inverse can be written as

$$e_{ij} = \frac{1 + v}{E}\sigma_{ij} - \frac{v}{E}\sigma_{kk}\delta_{ij} + \alpha(T - T_0)\delta_{ij}. \tag{3}$$

The constant α is the *coefficient of linear expansion*.

2.10 Materials with More Complex Mechanical Behavior

Nonviscous fluids, Newtonian fluids, and Hookean elastic solids are abstractions. No real material is known to behave exactly as one of these, although in limited ranges of temperature, stress, and strain, some materials may follow one of these laws accurately. They are the simplest laws we can devise to relate the stress and strain, or strain rate. They are not expected to be exhaustive.

Almost any real material has a more complex behavior than these simple laws describe. Among fluids, blood is non-Newtonian. Household paints and varnish, wet clay and mud, and most colloidal solutions are non-Newtonian. For solids, most structural materials are, fortunately, Hookean in the useful range of stresses and strains; but beyond certain limits, Hooke's law no longer applies. For example, virtually every known solid material can be broken (fractured) one way or another, under sufficiently large stresses or strains; but to break is to disobey Hooke's law. Few biological tissues obey Hooke's law. Their properties will be discussed in the following chapters.

2.11 Viscoelasticity

When a body is suddenly strained and then the strain is maintained constant afterward, the corresponding stresses induced in the body decrease with time. This phenomena is called *stress relaxation*, or *relaxation* for short. If the body is suddenly stressed and then the stress is maintained constant afterward, the body continues to deform, and the phenomenon is called *creep*. If the body is subjected to a cyclic loading, the stress–strain relationship in the loading process is usually somewhat different from that in the unloading process, and the phenomenon is called *hysteresis*.

The features of hysteresis, relaxation, and creep are found in many materials. Collectively, they are called features of *viscoelasticity*.

Mechanical models are often used to discuss the viscoelastic behavior of materials. In Fig. 2.11:1 are shown three mechanical models of material behavior, namely, the *Maxwell model*, the *Voigt model*, and the *Kelvin model* (also called the *standard linear solid*), all of which are composed of combina-

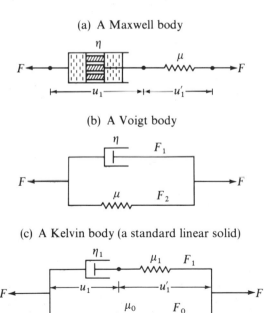

Figure 2.11:1 Three mechanical models of viscoelastic material. (a) A Maxwell body, (b) a Voigt body, and (c) a Kelvin body (a standard linear solid).

tions of linear springs with spring constant μ and dashpots with coefficient of viscosity η. A *linear spring* is supposed to produce instantaneously a deformation proportional to the load. A *dashpot* is supposed to produce a velocity proportional to the load at any instant. Thus if F is the force acting in a spring and u is its extension, then $F = \mu u$. If the force F acts on a dashpot, it will produce a velocity of deflection $\dot{u}$, and $F = \eta \dot{u}$. The shock absorber on an airplane's landing gear is an example of a dashpot. Now, in a *Maxwell model*, shown in Fig. 2.11:1(a), the same force is transmitted from the spring to the dashpot. This force produces a displacement F/μ in the spring and a velocity F/η in the dashpot. The velocity of the spring extension is $\dot{F}/\mu$ if we denote a differentiation with respect to time by a dot. The total velocity is the sum of these two:

$$\dot{u} = \frac{\dot{F}}{\mu} + \frac{F}{\eta} \qquad \text{(Maxwell model)}. \qquad (1)$$

Furthermore, if the force is suddenly applied at the instant of time $t = 0$, the spring will be suddenly deformed to $u(0) = F(0)/\mu$, but the initial dashpot deflection would be zero, because there is no time to deform. Thus the initial condition for the differential equation (1) is

$$u(0) = \frac{F(0)}{\mu}. \qquad (2)$$

For the *Voigt model*, the spring and the dashpot have the same displacement. If the displacement is u, the velocity is $\dot{u}$, and the spring and dashpot will produce forces μu and $\eta \dot{u}$, respectively. The total force F is therefore

$$F = \mu u + \eta \dot{u} \qquad \text{(Voigt model)}. \qquad (3)$$

If F is suddenly applied, the appropriate initial condition is

$$u(0) = 0. \qquad (4)$$

For the Kelvin model (or standard linear model), let us break down the displacement u into u_1 of the dashpot and u_1' for the spring, whereas the total force F is the sum of the force F_0 from the spring and F_1 from the Maxwell element. Thus

$$\text{(a) } u = u_1 + u_1', \qquad \text{(b) } F = F_0 + F_1, \qquad (5)$$
$$\text{(c) } F_0 = \mu_0 u, \qquad \text{(d) } F_1 = \eta_1 \dot{u}_1 = \mu_1 u_1'.$$

From this we can verify by substitution that

$$F = \mu_0 u + \mu_1 u_1' = (\mu_0 + \mu_1)\, u - \mu_1 u_1.$$

Hence

$$F + \frac{\eta_1}{\mu_1}\, \dot{F} = (\mu_0 + \mu_1)u - \mu_1 u_1 + \frac{\eta_1}{\mu_1}\, (\mu_0 + \mu_1)\dot{u} - \eta_1 \dot{u}_1.$$

Replacing the last term by $\mu_1 u_1'$ and using Eq. (5a), we obtain

$$F + \frac{\eta_1}{\mu_1} \dot{F} = \mu_0 u + \eta_1 \left(1 + \frac{\mu_0}{\mu_1} \right) \dot{u}. \tag{6}$$

This equation can be written in the form

$$F + \tau_\varepsilon \dot{F} = E_R(u + \tau_\sigma \dot{u}) \qquad \text{(Kelvin model)}, \tag{7}$$

where

$$\tau_\varepsilon = \frac{\eta_1}{\mu_1}, \qquad \tau_\sigma = \frac{\eta_1}{\mu_0} \left(1 + \frac{\mu_0}{\mu_1} \right), \qquad E_R = \mu_0. \tag{8}$$

For a suddenly applied force $F(0)$ and displacement $u(0)$, the initial condition is

$$\tau_\varepsilon F(0) = E_R \tau_\sigma u(0). \tag{9}$$

For reasons that will become clear below, the constant τ_ε is called the *relaxation time for constant strain*, whereas τ_σ is called *the relaxation time for constant stress*.

If we solve Eqs. (1), (3), and (7) for $u(t)$ when $F(t)$ is a *unit-step function* $\mathbf{1}(t)$, the results are called *creep functions*, which represent the elongation produced by a sudden application at $t = 0$ of a constant force of magnitude unity. They are:
Maxwell solid:

$$c(t) = \left(\frac{1}{\mu} + \frac{1}{\eta} t \right) \mathbf{1}(t), \tag{10}$$

Voigt solid:

$$c(t) = \frac{1}{\mu} (1 - e^{-(\mu/\eta)t}) \mathbf{1}(t), \tag{11}$$

standard linear solid:

$$c(t) = \frac{1}{E_R} \left[1 - \left(1 - \frac{\tau_\varepsilon}{\tau_\sigma} \right) e^{-t/\tau_\sigma} \right] \mathbf{1}(t), \tag{12}$$

where the unit-step function $\mathbf{1}(t)$ is defined as [see Fig. 2.11:2(a)]

$$\mathbf{1}(t) = \begin{cases} 1 & \text{when } t > 0, \\ \frac{1}{2} & \text{when } t = 0, \\ 0 & \text{when } t < 0. \end{cases} \tag{13}$$

A body that obeys a load-deflection relation like that given by Maxwell's model is said to be a *Maxwell solid*. Since a dashpot behaves as a piston moving in a viscous fluid, the above-named models are called models of viscoelasticity.

Interchanging the roles of F and u, we obtain the *relaxation function* as a response $F(t) = k(t)$ corresponding to an elongation $u(t) = \mathbf{1}(t)$. The relaxation function $k(t)$ is the force that must be applied in order to produce

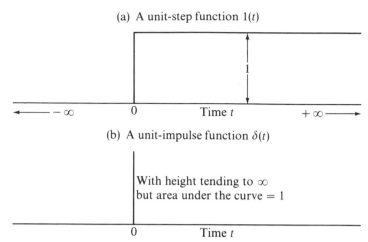

(a) A unit-step function $1(t)$

(b) A unit-impulse function $\delta(t)$

With height tending to ∞
but area under the curve $= 1$

Figure 2.11:2 (a) A unit-step function $1(t)$. (b) A unit-impulse function $\delta(t)$. The central spike has a height tending to ∞ but the area under the curve remains to be unity.

an elongation that changes at $t = 0$ from zero to unity and remains unity thereafter. They are
Maxwell solid:

$$k(t) = \mu e^{-(\mu/\eta)t}1(t), \tag{14}$$

Voigt solid:

$$k(t) = \eta\delta(t) + \mu1(t), \tag{15}$$

standard linear solid:

$$k(t) = E_R\left[1 - \left(1 - \frac{\tau_\sigma}{\tau_\varepsilon}\right)e^{-t/\tau_\varepsilon}\right]1(t). \tag{16}$$

Here we have used the symbol $\delta(t)$ to indicate the *unit-impulse function or Dirac-delta function*, which is defined as a function with a singularity at the origin (see Fig. 2.11:2(b)):

$$\delta(t) = 0 \qquad \text{(for } t < 0, \text{ and } t > 0),$$
$$\int_{-\varepsilon}^{\varepsilon} f(t)\delta(t)\,dt = f(0) \qquad (\varepsilon > 0), \tag{17}$$

where $f(t)$ is an arbitrary function, continuous at $t = 0$. These functions, $c(t)$ and $k(t)$, are illustrated in Figs. 2.11:3 and 2.11:4, respectively, for which we add the following comments.

For the Maxwell solid, the sudden application of a load induces an immediate deflection by the elastic spring, which is followed by "creep" of the dashpot. On the other hand, a sudden deformation produces an immediate reaction by the spring, which is followed by stress relaxation according to an exponential law [see Eq. (14)]. The factor η/μ, with the dimension of time, may be called the *relaxation time*: it characterizes the rate of decay of the force.

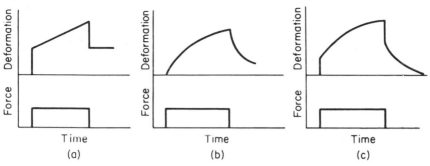

Figure 2.11:3 Creep functions of (a) a Maxwell, (b) a Voigt, and (c) a standard linear solid. A negative phase is superposed at the time of unloading.

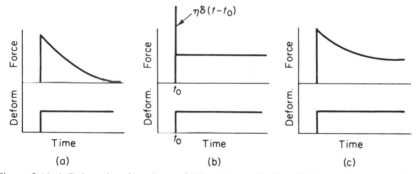

Figure 2.11:4 Relaxation functions of (a) a Maxwell, (b) a Voigt, and (c) a standard linear solid.

For the Voigt solid, a sudden application of force will produce no immediate deflection, because the dashpot, arranged in parallel with the spring, will not move instantaneously. Instead, as shown by Eq. (11) and Fig. 2.11:3, a deformation will be gradually built up, while the spring takes a greater and greater share of the load. The dashpot displacement relaxes exponentially. Here the ratio η/μ is again a relaxation time: it characterizes the rate of relaxation of the dashpot.

For the standard linear solid, a similar interpretation is applicable. The constant τ_ε is the time of relaxation of load under the condition of constant deflection [see Eq. (16)], whereas the constant τ_σ is the time of relaxation of deflection under the condition of constant load [see Eq. (12)]. As $t \to \infty$, the dashpot is completely relaxed, and the load-deflection relation becomes that of the springs, as is characterized by the constant E_R in Eqs. (12) and (16). Therefore, E_R is called the *relaxed elastic modulus*.

Maxwell introduced the model represented by Eq. (3), with the idea that all fluids are elastic to some extent. Lord Kelvin showed the inadequacy of the Maxwell and Voigt models in accounting for the rate of dissipation of energy in various materials subjected to cyclic loading. Kelvin's model is commonly called the standard linear model because it is the most general relationship that includes the load, the deflection, and their first (commonly called "linear") derivatives.

More general models may be built by adding more and more elements to the Kelvin model. Equivalently, we may add more and more exponential terms to the creep function or to the relaxation function.

The most general formulation under the assumption of linearity between cause and effect is due to Boltzmann (1844–1906). In the one-dimensional case, we may consider a simple bar subjected to a force $F(t)$ and elongation $u(t)$. The elongation $u(t)$ is caused by the total history of the loading up to the time t. If the function $F(t)$ is continuous and differentiable, then in a small time interval $d\tau$ at time τ the increment of loading is $(dF/d\tau)\,d\tau$. This increment remains acting on the bar and contributes an element $du(t)$ to the elongation at time t, with a proportionality constant c depending on the time interval $t - \tau$. Hence, we may write

$$du(t) = c(t - \tau)\,\frac{dF(\tau)}{d\tau}\,d\tau. \tag{18}$$

Let the origin of time be taken at the beginning of motion and loading. Then, on summing over the entire history, which is permitted under Boltzmann's hypothesis, we obtain

$$u(t) = \int_0^t c(t - \tau)\,\frac{dF(\tau)}{d\tau}\,d\tau. \tag{19}$$

A similar argument, with the role of F and u interchanged, gives

$$F(t) = \int_0^t k(t - \tau)\,\frac{du(\tau)}{d\tau}\,d\tau. \tag{20}$$

These laws are linear, since doubling the load doubles the elongation, and vice versa. The functions $c(t - \tau)$ and $k(t - \tau)$ are the *creep* and *relaxation functions*, respectively.

The Maxwell, Voigt, and Kelvin models are special examples of the Boltzmann formulation. More generally, we can write the relaxation function in the form

$$k(t) = \sum_{n=0}^{N} \alpha_n e^{-t\nu_n}, \tag{21}$$

which is a generalization of Eq. (16). If we plot the amplitude α_n associated with each characteristic frequency ν_n on a frequency axis, we obtain a series of lines that resembles an optical spectrum. Hence $\alpha_n(\nu_n)$ is called a *spectrum of the relaxation function*. The example shown in Fig. 2.11:5 is a *discrete spectrum*. A generalization to a *continuous spectrum* may sometimes be desired. In the case of a living tissue such as mesentery, experimental results on relaxation, creep, and hysteresis cannot be reconciled unless a continuous spectrum is assumed.

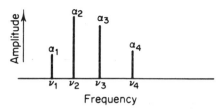

Figure 2.11:5 A discrete spectrum of the relaxation function.

Let us conclude this section by generalizing Eqs. (19) and (20) into tensor equations relating stress and strain tensors of a viscoelastic solid. All we need to say here is that if F and u are considered to be stress and strain tensors, then Eqs. (19) and (20) can be considered as constitutive equations of a viscoelastic solid if c and k are tensors defining the creep and relaxation characteristics of the material. We do have to be careful about the time derivatives $\dot{F}$ and $\dot{u}$ if the deformation is large, because then the time derivatives of the components of stress and strain of any material particle (such as the small cubic element shown in Fig. 2.2:3) may depend not only on how fast the element stretches, but also on how fast it rotates. To account for the stress rate (i.e., the rate of change of the stress tensor) in finite deformation turns out to be quite complex.* Hence, for the moment let us limit our consideration to a small deformation, with infinitesimal displacements, strains, and velocities. Under this restriction we can express our idea more explicitly. Let us denote the stress and strain tensors by σ_{ij} and e_{ij} (which are functions of space $\mathbf{x}$ (i.e., x_1, x_2, x_3) and time t). Then a material is said to be *linear viscoelastic* if $\sigma_{ij}(\mathbf{x}, t)$ is related to $e_{ij}(\mathbf{x}, t)$ by the following convolution integral:

$$\sigma_{ij}(\mathbf{x}, t) = \int_{-\infty}^{t} G_{ijkl}(\mathbf{x}, t - \tau) \frac{\partial e_{kl}}{\partial \tau}(\mathbf{x}, \tau) \, d\tau, \tag{22}$$

or its inverse,

$$e_{ij}(\mathbf{x}, t) = \int_{-\infty}^{t} J_{ijkl}(\mathbf{x}, t - \tau) \frac{\partial \sigma_{kl}}{\partial \tau}(\mathbf{x}, \tau) \, d\tau. \tag{23}$$

G_{ijkl} is called the *tensorial relaxation function*. J_{ijkl} is called the *tensorial creep function*. Note that the lower limit of integretion is taken as $-\infty$, which is to mean that the integration is to be taken before the very beginning of motion. If the motion starts at time $t = 0$, and $\sigma_{ij} = e_{ij} = 0$ for $t < 0$, then Eq. (22) reduces to

$$\sigma_{ij}(\mathbf{x}, t) = e_{kl}(\mathbf{x}, 0+) G_{ijkl}(\mathbf{x}, t) + \int_{0}^{t} G_{ijkl}(\mathbf{x}, t - \tau) \frac{\partial e_{kl}}{\partial \tau}(\mathbf{x}, \tau) \, d\tau, \tag{24}$$

where $e_{ij}(x, 0+)$ is the limiting value of $e_{ij}(x, t)$ when $t \to 0$ from the positive

* See Fung (1965), *Solid Mechanics*, p. 448 for details.

side. The first term in Eq. (24) gives the effect of initial disturbance: it arises from the jump of $e_{ij}(x, t)$ at $t = 0$. If the strain history contains other jumps at other instants of time, then each jump calls for an additional term similar to the first term in Eq. (24).

2.12 Response of a Viscoelastic Body to Harmonic Excitation

Since biological tissues are all viscoelastic, and since one of the simplest ways to experimentally determine the viscoelastic properties is to subject the material to periodic oscillations, we shall discuss this case in greater detail. Consider a quantity x, which varies periodically with frequency ω (radians per second) according to the rule

$$x = A \cos(\omega t + \varphi). \tag{1}$$

This is a *simple harmonic motion*; A is the *amplitude* and φ is the *phase angle*. We may consider x to be the projection of a rotating vector on the real axis (see Fig. 2.12:1). Since a vector is specified by two components, it can be represented by a complex number. For example, the vector in Fig. 2.12:1 can be specified by the components $x = A \cos(\omega t + \varphi)$ and $y = A \sin(\omega t + \varphi)$, and hence by the complex number $x + iy$. But

$$e^{i(\omega t + \varphi)} = \cos(\omega t + \varphi) + i \sin(\omega t + \varphi), \tag{2}$$

so the rotating vector can be represented by the complex number

$$x + iy = A e^{i(\omega t + \varphi)} = B e^{i\omega t}, \tag{3}$$

where

$$B = A e^{i\varphi}. \tag{3a}$$

Equation (1) is the real part of Eq. (3), and the latter is said to be the *complex representation* of Eq. (1). B is a complex number whose absolute value is the

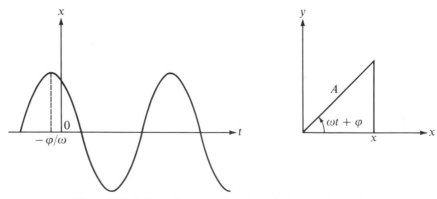

Figure 2.12:1 Complex representation of a harmonic motion.

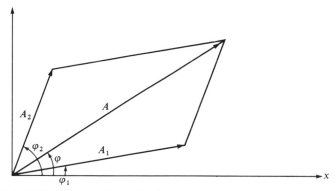

Figure 2.12:2 Vector sum of two simple harmonic motions of the same frequency.

amplitude, and whose polar angle $\varphi = $ arc $\tan(\text{Im } B/\text{Rl } B)$ is the *phase angle* of the motion.

The vector representation is very convenient for "composing" several simple harmonic oscillations of the same frequency. For example, if

$$x = A_1 \cos(\omega t + \varphi_1) + A_2 \cos(\omega t + \varphi_2) = A \cos(\omega t + \varphi), \tag{4}$$

then x is the real part of the resultant of two vectors as shown in Fig. 2.12:2.

Now, if the force and displacement are harmonic functions of time, then we can apply complex representation. Let $u = Ue^{i\omega t}$. Then by differentiation with respect to t, we have $\dot{u} = i\omega Ue^{i\omega t} = i\omega u$. In this case a differentiation with respect to t is equivalent to a multiplication by $i\omega$. Applying this result to Eq. (1) of Sec. 2.11, we obtain

$$i\omega u = \frac{i\omega F}{\mu} + \frac{F}{\eta}. \tag{5}$$

This can be written in the form

$$F = G(i\omega)u, \tag{6}$$

which is the same as

$$Fe^{i\omega t} = G(i\omega)ue^{i\omega t}, \tag{6a}$$

where $G(i\omega)$ is called the *complex modulus of elasticity*. In the case of the Maxwell body,

$$G(i\omega) = i\omega\left(\frac{i\omega}{\mu} + \frac{1}{\eta}\right)^{-1}. \tag{7}$$

In a similar manner, Eqs. (3), (7), (19), (20), (22), and (23) of Sec. 2.11 can all be put into the form of Eq. (6), and the complex modulus of each model can be derived.

The complex modulus of elasticity of the Kelvin body (standard linear solid), corresponding to Eq. (7) of Sec. 2.11, is

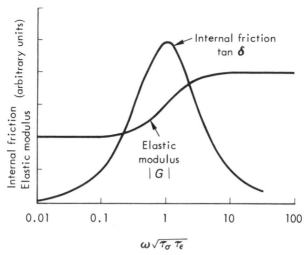

Figure 2.12:3 The dynamic modulus of elasticity $|G|$ and the internal damping $\tan\delta$ plotted as a function of the logarithm of frequency ω for a standard linear solid.

$$G(i\omega) = \frac{1 + i\omega\tau_\sigma}{1 + i\omega\tau_\varepsilon} E_R. \tag{8}$$

Writing

$$G(i\omega) = |G|e^{i\delta}, \tag{9}$$

where $|G|$ is the amplitude of the complex modulus and δ is the phase shift, we have

$$|G| = \left(\frac{1 + \omega^2\tau_\sigma^2}{1 + \omega^2\tau_\varepsilon^2}\right)^{1/2} E_R, \qquad \tan\delta = \frac{\omega(\tau_\sigma - \tau_\varepsilon)}{1 + \omega^2(\tau_\sigma\tau_\varepsilon)}. \tag{10}$$

The quantity $\tan\delta$ is a measure of "internal friction." When $|G|$ and $\tan\delta$ are plotted against the logarithm of ω, curves as shown in Fig. 2.12:3 are obtained. The internal friction reaches a peak when the frequency ω is equal to $(\tau_\sigma\tau_\varepsilon)^{-1/2}$. Correspondingly, the elastic modulus $|G|$ has the fastest rise for frequencies in the neighborhood of $(\tau_\sigma\tau_\varepsilon)^{-1/2}$.

2.13 Use of Viscoelastic Models

The viscoelastic models discussed in the previous sections are especially useful in biomechanics because biological tissues have viscoelastic features. In the laboratory, it is quite easy to determine the relaxation and creep curves. With suitable testing machines it is also easy to determine the complex modulus of a material subjected to a periodic forcing function. The amplitude and phase angle of the complex modulus can be plotted with respect to the frequency of oscillation. Having determined these curves experimentally, we try to fit them with a model. If a model (i.e., a differential equation such as Eqs. (1), (3), or (7) of Sec. 2.11, or a complex modulus such as that given in Eq. (8) of Sec. 2.12) can be found whose relaxation, creep, hysteresis, and complex modulus agree with the experimental data, then the material tested can be described by this model. Each model contains only a few material constants (such as η_1, μ_1, or τ_σ, τ_ε, E_R). The reams of experimental curves can then be replaced by a list of these material constants. Furthermore, the constitutive equation can then be used to analyze other problems concerned with the tissue. Many examples of such applications are presented in the chapters to follow.

I am not sure who first drew the model diagrams and wrote down Eqs. (1), (3), and (7) of Sec. 2.11, but the work of the people to whom these models are attributed is well known. James Clerk Maxwell (1831–1879), one of the greatest theoretical physicists of all times, was born in Edinburgh. When he was appointed to the newly founded professorship of experimental physics in Cambridge in 1871, he set about to establish a laboratory to study mechanical properties of matter. The first object of interest was air. He felt that air is viscoelastic. He built a testing machine consisting of two parallel disks rotating relative to each other to measure the viscosity of air. His mathematical description of air is the Maxwell model of viscoelasticity.

Woldemar Voigt (1850–1919) was a German physicist especially known for his work on crystallography. The Voigt model was named in his honor.

William Thomson (Lord Kelvin), 1824–1907, was born in Belfast, Ireland. He was educated at home by his father, James Thomson, a professor of mathematics at Glasgow. In 1846 he himself became a professor in the University of Glasgow. In the period 1848–1851 Thomson worked on the dynamic theory of heat and formulated the first and second laws of thermodynamics, which reconciled the work of Carnot, Rumford, Davy, Mayer, and Joule. In connection with the second law, he searched for supporting evidence in irreversible processes that are revealed in the mechanical properties of matter. He tested metals, rubber, cork, etc., in the form of a torsional pendulum (using these materials as the torsional spring). When the pendulum was set in motion, the amplitude of its oscillations decayed exponentially, and the number of cycles required for the amplitude to be reduced to one-half of the initial value could be taken as a measure of the "internal friction" of the material. Kelvin found that the material behavior can be explained by a model system possessing a spring and a dashpot, in parallel with another spring.

Thus, we see that these models were invented to correlate experimental data on real materials. Further generalization of these models, as is shown in Problems 2.12–2.16, may be necessary to correlate experimental data of a more complex nature.

The question of choosing the right model to fit the experimental data is an important one. Usually the first step is to compare the experimental curves of relaxation, creep, frequency response, and internal friction (complex modulus) with those of the theoretical models, such as those shown in Figs. 2.11:3, 2.11:4, and 2.12:3. If they look alike, then a curve fitting procedure can be used to determine the best constants. If the simple models cannot yield the desired features, then it would be necessary to consider generalized models, such as those listed in Problems 2.12–2.16 infra. Hints are often obtainable from the general character of the theoretical and experimental curves. For example, the generalized Kelvin or Voigt models of order n would have a relaxation function of the type

$$k(t) = a + c_1 e^{-t/\tau_1} + c_2 e^{-t/\tau_2} + \cdots + c_n e^{-t/\tau_n}, \tag{1}$$

which contains n exponential functions. The creep functions would also contain n exponential functions. The imaginary part of the corresponding complex modulus of elasticity, representing the internal friction, would have n peaks in the frequency spectrum. The real part of the complex modulus, plotted with respect to frequency, would look like a staircase with n steps. If such were the features of the experimental curve, then a generalized linear viscoelastic model of order n is suggested.

It often happens that many relaxation functions of the type of Eq. (1) can fit the relaxation data (because the determination of the constants $c_1, c_2 \cdots$ $\tau_1, \tau_2 \cdots$ is nonunique and multiple choices are possible). Similarly, many such functions can fit the creep data. Then a choice can be made by demanding that the chosen model be the one that fits *all* the experimental data, including relaxation, creep, hysteresis, amplitude of the complex modulus, and the internal friction spectrum. An example of this is given in Sec. 7.6.

It may happen that none of the models can fit *all* the experimental data. Then we may have to concede that the linearized viscoelasticity theory does not apply at all.

2.14 Methods of Testing

The ideal constitutive equations named above are abstractions of nature. The vast literature of fluid and solid mechanics is based principally on these idealized equations. Behavior of materials in the biological world usually deviates from these simple relationships. It is important to test the materials to determine how closely their mechanical behavior can be represented by simplified constitutive equations.

Testing the mechanical properties of biological materials does not differ in principle from testing industrial materials, except possibly in three aspects: (1) it is seldom possible to get large samples of biological materials; (2) strict attention must be given to keep the samples viable and in a condition as close as possible to that *in vivo*; and finally (3), many materials are nonhomo-

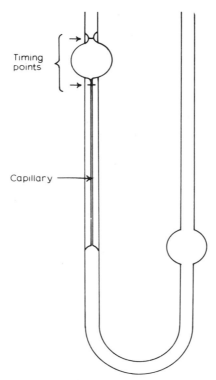

Timing points

Capillary

Figure 2.14:1 An Ostwald viscometer.

geneous. These special features usually call for special testing methods and equipment.

Biomechanical testing methods will be discussed throughout this book. Here, however, the classical topics of viscometry is discussed as an introduction.

One of the simplest methods for measuring fluid viscosity uses flow in a long circular cylindrical tube. A standard type, the Ostwald viscometer, is shown in Fig. 2.14:1. It is known (see Eq. (8) of Sec. 3.3) that for a Newtonian fluid in a laminar (nonturbulent) flow in such a tube the coefficient of viscosity η is related to the pressure gradient dp/dL, the volume rate of flow $\dot{Q}$, and the tube radius R by the equation

$$\eta = \frac{\pi R^4}{8\dot{Q}} \frac{dp}{dL}. \tag{1}$$

If pressure is given in dyn/cm², L and R in cm, and $\dot{Q}$ in cm³/s, then η is in dyn s/cm², or poise. If pressure is given in N/m², or Pascal, and length in m, then η is in N·s/m². One poise equals 0.1 Ns/m². To use Eq. (1) to calculate η, one must make sure that the flow is laminar, and that the "end effect" is corrected for, i.e., instead of using P/L for dp/dL, where P is the difference of

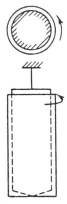

Figure 2.14:2 A Couette viscometer.

pressures at the ends and L is the length of the "capillary," one makes corrections for the anomaly at the ends, including the "kinetic energy correction" to allow for the effects of accelerations with the sudden change of radius of the vessel through which the fluid flows. An Ostwald viscometer may be used to test a non-Newtonian fluid (such as blood) to determine the "apparent" viscosity (see Sec. 5.2 for definition), but methods for correcting the end effects are often unknown for non-Newtonian fluids.

Another viscometer is the Couette type, using flow between two coaxial cylinders. An example is shown in Fig. 2.14:2. The outer cylinder (radius R_2), which contains the sample, is rotated at a constant angular speed ω. The inner cylinder (radius R_1) is suspended on a torsion wire centrally within the outer cylinder, so that the material to be tested fills the annular space between the cylinders. The viscosity of the liquid causes a torque to be transmitted to the inner cylinder which can be measured by a transducer. The shear rate is not constant across the gap, but the variation is small if the gap is small. The coefficient of viscosity for a Newtonian fluid in a Couette viscometer is given by the equation

$$\eta = \frac{M(R_2^2 - R_1^2)}{4\pi\omega h R_1^2 R_2^2},\tag{2}$$

where M is the measured torque, h is the height of the inner cylinder immersed in the liquid, and ω is the angular speed of the outer cylinder.

For higher accuracy and smaller samples, one may use the cone-plate viscometer illustrated in Fig. 2.14:3. The sample occupies the conical space. When the plate rotates, the linear velocity increases with the radial distance; but since the gap also increases linearly with the radius, the shear rate remains constant throughout if the cone remains stationary. One might suppose that the magnitude of the cone angle would not matter; but, for rather complex hydrodynamic reasons, this is not the case, and very flat cones must be used.

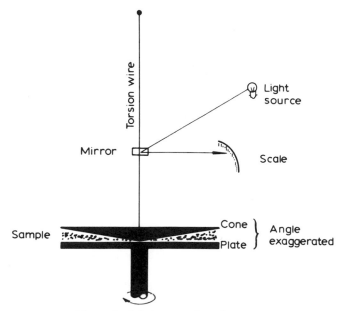

Figure 2.14:3 A cone-plate rheometer.

The Couette or cone-plate rheometers can be operated not only with a steady rotation at constant speed, but also in an oscillatory mode or in step changes. The lower cylinder, cone, or plate is caused to oscillate harmonically, and the torque response of the upper member can be recorded. The magnitude and phase relationship between the torque and shear rate will yield the complex modulus of a viscoelastic material (see Sec. 2.12).

A general instrument called Weissenberg's "rheogoniometer" can be used either for rotation or for sinusoidal oscillation, and can operate with either coaxial cylinders or cone-plates. It can measure not only the torque, but also the normal force (the "upthrust") on the upper member. The upthrust, known as the "Weissenberg effect," occurs in certain non-Newtonian fluids when they are sheared. If such a fluid is sheared between two cylinders (in rotation) as in a Couette flow, the fluid will climb up the inner cylinder if it is not totally immersed. The rheogoniometer measures such a thrust; hence, it measures forces "at all angles" (Greek, *gonia*, angle).

There are other types of rheometers that measure complex moduli. In one of these a vertical rod is oscillated sinusoidally within a hollow cylinder containing the material. In another the material is placed between a spherical bob and a concentric hemispherical cup. A third one uses a tuning fork to drive an oscillatory rod in the fluid, either in axial motion or in lateral motion.

Another popular method is that of a falling (or rising) sphere. In the simplest case, a metal or plastic sphere is timed as it falls or rises through a known distance in a liquid. If the liquid is opaque, the passage of the ball can be detected electromagnetically. If the fluid is Newtonian, and the container is much larger than the sphere, and if the Reynolds number is much smaller than 1, then the falling velocity is given by Stokes' formula

$$v = \frac{2r^2 \Delta \rho g}{9\eta},$$

where v is the velocity off all, r is the radius of the sphere, $\Delta \rho$ is the difference between the densities of the sphere and of the liquid, η is the coefficient of viscosity, and g is the gravitational constant. Unfortunately, this formula does not work if the Reynolds number $(vr\rho/\eta)$ is greater than 1, and if the container is not very much larger than the sphere. In the case of a large Reynolds number or a small container, more complex hydrodynamic considerations must be given. Sometimes a sphere rolling or sliding down an inclined plane is used to determine the viscosity of the fluid; in such cases, careful corrections must be made for inertia effects around the sphere. Sometimes a falling sphere apparatus can be modified so that the sphere can be dragged sideways by a magnet. On release, the elastic recovery can be measured.

In all cases of viscometry, it is absolutely necessary to have a thorough understanding of the hydrodynamics of the flow field. The distributions of velocity, pressure, and normal shear stresses must be known precisely. Otherwise one would not understand what is being measured. And this theoretical analysis has to be done for every hypothetical constitutive equation the fluid might be assumed to satisfy. This is why biomechanics is as much a theoretical subject as it is experimental.

So far we have mentioned instruments for measuring viscosity or viscoelasticity of fluids. For solid materials there are a number of commercial testing machines available. Generally, the specimen is clamped and stretched or shortened at a specific rate while the force and displacement are both recorded. For biological specimens, the usual problems of small size and the the necessity for keeping the specimen viable cause difficulty. Usually it is preferred that no measuring instrument should touch the specimen in the region where deformation is measured.

A noncontact method for three-dimensional analysis of blood vessels *in vivo* and *in vitro* is shown in Fig. 2.14:4. The vessel is soaked in a saline bath at 37°C. The dimensions between markers in the longitudinal and circumferential directions are measured with two closed-circuit television cameras (TV) and two video dimension analyzers (VDA). By rotating the camera, any scene may be selected, and the image can be magnified to any size by introducing different lenses and extension tubes in the camera optics. At a given optical magnification, the resolution of the obtainable measurements is limited primarily by the TV camera. With a quality camera, resolution exceeding 0.1% of full width of the screen is readily attainable. The VDA utilizes the video signal from the TV camera, and forms a DC voltage analog of the distance between two selected points in the scene. The optical density of the space between the two points must be sufficiently different from that of the surrounding space. Following acquisition, dimensional changes

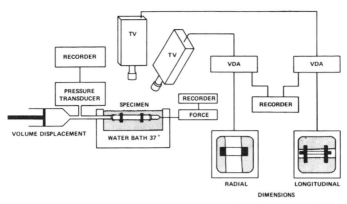

Figure 2.14:4 A noncontact method for three-dimensional analysis of vascular elasticity *in vivo* and *in vitro*. From Fronek, K., Schmid-Schoenbein, G., and Fung, Y. C. (1976) *J. Appl. Physiol.* 40, 634–637.

are automatically tracked. Since the image tube of a standard TV camera is scanned horizontally to form the video signal to be transmitted to the monitor, dimensions measured by an interposed VDA are synchronized along the horizontal axis. The resulting analog voltage, precalibrated in dimensional units of interest, can be displayed on a meter or strip-chart recorder. This noncontact type of optical dimension measurement makes it possible to select any portion of the specimen for study. Because the mechanical properties are influenced by the mounting of the specimen, it is an additional advantage of this approach that a middle portion less influenced by the stress disturbances due to clamped ends can be selected for measurement.

2.15 Mathematical Development of Constitutive Equations

The most famous constitutive equations are the simplest: those of the non-viscous ideal fluid, the Newtonian viscous fluid, and the Hookean elastic solid. Then the linear laws of viscoelasticity were introduced by Maxwell, Voigt, and Thomson (Lord Kelvin), culminating in Boltzmann's formulation (see Sec. 2.11). For finite deformation, the nonlinear relationship between strains and deformation gradients was recognized by Cauchy, Green, St. Venant, Almansi, and Hamel (see Sec. 2.4). Associated with finite deformation the strain rate tensor and stress rate tensors were defined, and nonlinear constitutive equations for elastic, viscoelastic, and viscoplastic materials were developed. Rivlin is certainly the one who provided the largest number of exact solutions to the nonlinear theory of elasticity of the Neo-Hookean and Mooney–Rivlin materials, which have some resemblance to rubber. Trusdell's books and papers are most influential in establishing the rigor, vigor, and broad horizon of continuum mechanics in the 1950s–1960s. Also of great importance are the books of Green and Zerna, Green and Adkins and Eringen. The author's book

Foundations of Solid Mechanics (Fung, 1965) is an introduction to the classical theory. It contains references to the authors named above.

For nonlinear viscoelastic behavior, Green and Rivlin (1957) and Green, Rivlin, and Spencer (1959) constructed a multiple integral constitutive equation, which may be arranged as a series in which the *n*th term is of degree *n* in the strain components. Pipkin and Rogers (1968) constructed a multiple integral constitutive equation which is arranged in a series whose first term represents the result of a one-step test (the stress due to a step change of strain), and whose *n*th term represents a correction due to the *n*th step. In the meantime, non-Newtonian fluid mechanics developed along with the development of polymer plastics industry. See Bird et al. (1977).

Until recently the constitutive equations of most biological tissues were unknown. Lack of this knowledge was the most important handicap to the development of biomechanics, because without constitutive equations boundary-value problems cannot be formulated, detailed analysis cannot be made, and predictions cannot be tested and evaluated. Borrowing from the existing continuum mechanics literature did not produce a good yield, because the best known materials like metals and rubbery materials seem to have few counterpart in living tissues as far as the mechanical properties are concerned. This is the reason for this book: to give an account of the mechanical properties of living tissues, consolidated in constitutive equations whenever possible.

Problems

2.1 Consider deformation of a body. Let the position of a point in the initial (undeformed) configuration of the body be denoted by the position vector **a**, with components a_1, a_2, a_3 when referred to a set of rectangular cartesian coordinates. When the body deforms, the position of the point **a** is moved to **x**, with components x_1, x_2, x_3. Consider an infinitesimal line element $d\mathbf{a}$ (a vector with components da_1, da_2, da_3) in the initial configuration. This line element is transformed into $d\mathbf{x}(dx_1, dx_2, dx_3)$ in the deformed configuration. Set down the definitions of strains as relationships between $d\mathbf{a}$ and $d\mathbf{x}$.

2.2 A set of rectangular cartesian coordinates x_1, x_2, x_3 is changed into another set x_1', x_2', x_3', through an orthogonal transformation:

$$x_1' = \beta_{11}x_1 + \beta_{12}x_2 + \beta_{13}x_3,$$
$$x_2' = \beta_{21}x_1 + \beta_{22}x_2 + \beta_{23}x_3,$$
$$x_3' = \beta_{31}x_1 + \beta_{32}x_2 + \beta_{33}x_3,$$

or

$$x_i' = \beta_{ij}x_j \tag{1}$$

for short. How do the strain components $e_{11}, e_{22}, e_{33}, e_{12}, e_{23}, e_{31}$ change when the deformation is referred to the new coordinates x_1', x_2', x_3'? Let the strain components referred to x_i' be denoted by e_{ij}'. Write down the relationship between e_{ij}' and e_{ij}.

Note: The simplest answer is obtained by recalling that strain is a tensor and thus obeys the tensor transformation law Eq. (2.2:7), i.e.,

$$e'_{km} = e_{ji}\beta_{kj}\beta_{mi}$$

2.3 Show that the following combinations of strain components $e_{11}, e_{22}, e_{33}, e_{12}, e_{23}, e_{31}$ do not change when the reference coordinates x_1, x_2, x_3 are changed to x'_1, x'_2, x'_3 through the orthogonal transformation

$$x'_i = \beta_{ij}x_j.$$

$$I_1 = e_{11} + e_{22} + e_{33},$$

$$I_2 = e_{11}e_{22} - e_{12}^2 + e_{22}e_{33} - e_{23}^2 + e_{33}e_{11} - e_{13}^2,$$

$$I_3 = \begin{vmatrix} e_{11} & e_{12} & e_{13} \\ e_{21} & e_{22} & e_{23} \\ e_{31} & e_{32} & e_{33} \end{vmatrix}.$$

Note: The corresponding relationship for the stress tensor is given in *The First Course of Continuum Mechanics*, Fung (1993), Sec. 4.5.

This result will be used in this book. So we give two examples of solution.

Solution 1

The simplest way is to identify I_1, I_2, I_3 as the coefficients of the characteristic equation which is invariant:

$$|e_{ij} - \lambda\delta_{ij}| = 0 \tag{2}$$

as is done in the *First Course*, See. 4.5.

Solution 2

We use the fact that for an orthogonal transformation given by Eq. (1) the β_{ij} matrix obeys the relation

$$(\beta_{ij})(\beta_{ij})^T = (\beta_{ij})(\beta_{ij})^{-1} = (\delta_{ij}) \tag{3}$$

where $(\ )^T$ denotes transpose, $(\ )^{-1}$ denotes inverse, and δ_{ij} denotes Kronecker delta. See *First Course*, Sec. 2.4.

Since Eq. (1) transforms e_{ij} into e'_{ij}, then

$$I'_1 = e'_{\alpha\alpha} = \beta_{\alpha k}\beta_{\alpha L}e_{kl} = \delta_{kl}e_{kL} = e_{kk} = I_1, \tag{4}$$

and I_1 is invariant. To show $I'_2 = I_2$, it is best to use the identity

$$J_2 = \tfrac{1}{3}e_{\alpha\alpha}^2 - I_2 = \tfrac{1}{3}I_1^2 - I_2, \tag{5}$$

where

$$J_2 = \tfrac{1}{2}e'_{ij}e'_{ij} \tag{6}$$

and e'_{ij} is the strain deviator tensor

$$e_{ij} = \tfrac{1}{3}e_{\alpha\alpha}\delta_{ij} + e'_{ij}. \tag{7}$$

Then, in the new coordinates,

$$J'_2 = \tfrac{1}{2}\beta_{ik}\beta_{jl}e'_{kl}\beta_{im}\beta_{jn}e'_{mn}$$

$$= \tfrac{1}{2}\delta_{km}\delta_{ln}e'_{kl}e'_{mn} = \tfrac{1}{2}e'_{mn}e'_{mn} = J_2. \tag{8}$$

Hence J_2 is invariant. Then, by Eqs. (4) and (5), I_2 is also invariant. *Alternatively,*
one notes that, by Eq. (2.3:19)

$$I_2 = \tfrac{1}{2}(e_{ii}e_{jj} - e_{ij}e_{ij}). \tag{9}$$

We have seen that e_{ii} is invariant. Hence it is sufficient to show that $e_{ij}e_{ij}$ is
invariant. This can be done as in the above. For I_3, note that in matrix notation:

$$\begin{bmatrix} e'_{11} & e'_{12} & e'_{13} \\ \cdot & \cdot & \cdot \\ \cdot & \cdot & \cdot \end{bmatrix} = (\beta_{ij})(e_{ij})(\beta_{ij})^T. \tag{10}$$

Hence, by taking the determinant of both sides,

$$I'_3 = |e'_{ij}| = |\beta_{ij}||e_{ij}||\beta_{ij}^T| = |e_{ij}| = I_3 \tag{11}$$

because $|\beta_{ij}| = 1$.

2.4 Give definitions of principal strains and principal axes of deformations. Let $e_1, e_2,$
e_3 be the principal strains. Show that the invariants I_1, I_2, I_3 named in Problem
2.3 have the meaning

$$I_1 = e_1 + e_2 + e_3,$$
$$I_2 = e_1e_2 + e_2e_3 + e_3e_1,$$
$$I_3 = e_1e_2e_3.$$

What are the physical meanings of these invariants? Discuss especially the
physical meaning of I_1, I_2, I_3 when the strains e_{ij} are infinitesimal. Cf. *First Course,*
Sec. 5.7.

2.5 Consider a deformation as described in Problem 2.1. Let a line element $d\mathbf{a}$ have
components $da_1 = ds_0,\ da_2 = 0,\ da_3 = 0$, i.e., with length ds_0 in the direction of
the coordinate axis x_1. When the body is deformed. the line element $d\mathbf{a}$ becomes
$d\mathbf{x}$ (with components ds, 0, 0) whose length is ds. The ratio ds/ds_0 is called the
stretch ratio, and is denoted by λ_1. Show that the Green's strain component
E_{11} is

$$E_{11} = \tfrac{1}{2}(\lambda_1^2 - 1).$$

Similarly, the other two normal strain components are

$$E_{22} = \tfrac{1}{2}(\lambda_2^2 - 1), \qquad E_{33} = \tfrac{1}{2}(\lambda_3^2 - 1),$$

where λ_2, λ_3 are stretch ratios referred to coordinate axes orthogonal to x_1.

If the coordinate axes x_1, x_2, x_3 were principal axes, then E_{11}, E_{22}, E_{33} are
principal strains, and $\lambda_1, \lambda_2, \lambda_3$ are principal stretch ratios. Thus, it is seen that the
three principal stretch ratios completely define the deformation of a body, provided
that the orientations of the principal axes are known. Cf. *First Course,* Sec. 5.6.

2.6 Review tensor transformation laws. By direct substitution, show that the tensors
$\delta_{ij}\delta_{kl}$ and $\delta_{ik}\delta_{jl} + \delta_{il}\delta_{jk}$ are isotropic.

It can be shown (see *First Course,* Fung 1993, Chapter 8), that any isotropic
tensor of rank 4 can be represented in the form

$$\lambda\delta_{ij}\delta_{kl} + \mu(\delta_{ik}\delta_{jl} + \delta_{il}\delta_{jk}) + \nu(\delta_{ik}\delta_{jl} - \delta_{il}\delta_{jk}),$$

where λ, μ, ν are scalars. Hence deduce that the generalized Hooke's law between stress and strain,

$$\sigma_{ij} = C_{ijkl}e_{kl},$$

can be reduced to the form

$$\sigma_{ij} = \lambda e_{\alpha\alpha}\delta_{ij} + 2\mu e_{ij}$$

if the material is isotropic.

2.7 Following the same reasoning as in Problem 2.6, show in detail that if stresses in a viscous fluid are a linear function of the strain rate, and if the fluid is isotropic, then

$$\sigma_{ij} = -p\delta_{ij} + \lambda V_{\alpha\alpha}\delta_{ij} + 2\mu V_{ij},$$

where V_{ij} is the strain rate tensor defined for a velocity field v_1, v_2, v_3:

$$V_{ij} = \frac{1}{2}\left(\frac{\partial v_i}{\partial x_j} + \frac{\partial v_j}{\partial x_i}\right),$$

and λ and μ are numerical constants.

Explain the meaning of the term p. Why is this term necessary for the constitutive equation of a fluid but is not required (see Problem 2.6) in the constitutive equation of an elastic solid?

2.8 Derive Navier–Stokes equations for a Newtonian viscous fluid such as water. Cf. *First Course*, Sec. 11.1.

2.9 Derive Navier–Stokes equations for a non-Newtonian viscous fluid such as blood. Cf. Sec. 3.2 *infra*.

2.10 Although the no-slip boundary condition on a solid wall is well accepted for homogeneous Newtonian viscous fluids such as air, water, and oil; for particulate flow of fluids such as the blood, which contains cellular bodies to the extent of about 45% by volume, one wonders if the no-slip condition applies at the solid–fluid interface. Invent a method to investigate this problem. Design and sketch the apparatus needed for your investigation. Explain in detail why your method would be decisive.

2.11 Under strain, an infinitesimal spherical region in the initial state is transformed into an ellipsoid in the deformed state. The principal axes of the ellipsoid are the principal axes of the strain. The ratios of the lengths of principal axes of the ellipsoid to the radius of the sphere are the *principal stretches* $\lambda_1, \lambda_2, \lambda_3$. The ratio of the volume of the infinitesimal body in the deformed state to that in the initial state is

$$\frac{dV}{dV_0} = \lambda_1\lambda_2\lambda_3.$$

Express the volume ratio in terms of the principal strains E_1, E_2, E_3, or e_1, e_2, e_3, or $\varepsilon_1, \varepsilon_2, \varepsilon_3$ defined in the senses of Green, Almansi, and Cauchy, respectively. Show that

$$\lambda_1^2\lambda_2^2\lambda_3^2 = 8I_3 + 4I_2 + 2I_1 + 1,$$

where the I's are the invariants given in Problems 2.3 and 2.4, with e standing for Green's strain.

2.12 Consider the generalized Kelvin model of a linear viscoelastic body as shown in Fig. P2.12:(a), where μ's are spring constants and η's are the viscosity coefficients of the dashpots. Derive (a) the differential equation relating the force F and displacement u, (b), the creep function, and (c), the relaxation function.

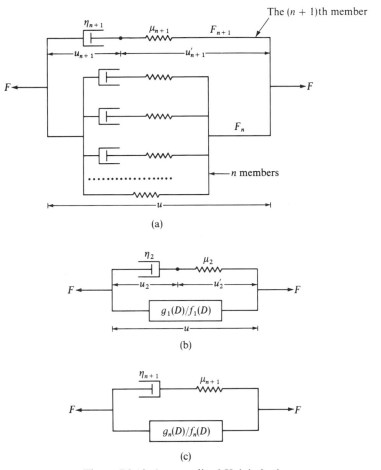

Figure P2.12 A generalized Kelvin body.

Solution. A Kelvin body may be regarded as adding a Maxwell body (see Fig. 2.11:1) in parallel with a spring. A *generalized Kelvin body of order n* is one that has n Maxwell bodies in parallel with a spring. See Fig. P2.12:(a). To derive the differential equation that governs the generalized Kelvin body, note first that if we write d/dt by the symbol D, then Eq. (7) of Sec. 2.11 can be written symbolically as

$$f_1(D)F = g_1(D)u, \tag{1}$$

where

$$f_1(D) = 1 + \frac{\eta_1}{\mu_1}D, \qquad g_1(D) = \mu_0\left(1 + \frac{\eta_1}{\mu_1}D\right) + \eta_1 D. \tag{2}$$

This is the equation for a Kelvin body, or a *generalized Kelvin body of order* 1. We can also write Eq. (1) symbolically as

$$F = \frac{g_1(D)}{f_1(D)} u. \tag{3}$$

Now consider a *generalized Kelvin body of order* 2. This may be represented symbolically as in Fig. P2.12:(b). Compare this with Fig. 2.11:1. We see that they are the same if μ_0 is replaced by $g_1(D)/f_1(D)$, and all subscripts 1 are replaced by subscript 2. Thus we can deduce its differential equation,

$$f_2(D)F = g_2(D)u, \tag{4}$$

with

$$f_2(D) = f_1(D)\left(1 + \frac{\eta_2}{\mu_2}D\right),$$

$$g_2(D) = g_1(D)\left(1 + \frac{\eta_2}{\mu_2}D\right) + \eta_2 f_1(D)D. \tag{5}$$

Finally, consider the *generalized Kelvin body of order* $n+1$ as shown in Fig. P2.12:(a). It is equivalent to the symbolic version shown in Fig. P2.12:(c). Hence, we can readily write down its differential equation,

$$f_{n+1}(D)F = g_{n+1}(D)u, \tag{6}$$

where

$$f_{n+1}(D) = f_n(D)\left(1 + \frac{\eta_{n+1}}{\mu_{n+1}}D\right),$$

$$g_{n+1}(D) = g_n(D)\left(1 + \frac{\eta_{n+1}}{\mu_{n+1}}D\right) + \eta_{n+1}f_n(D)D. \tag{7}$$

2.13 Determine the complex modulus of elasticity of the generalized Kelvin body of order n when it is subjected to a periodic force of circular frequency ω (rad/s).

2.14 A *generalized Maxwell body of order* n is one that is composed of a spring and n Voigt bodies connected in series, as illustrated in Fig. P2.14. Determine the differential equation relating the force and displacement of such a body, and the appropriate initial conditions. Derive the complex modulus when the body is subjected to a periodic force of circular frequency ω rad/s.

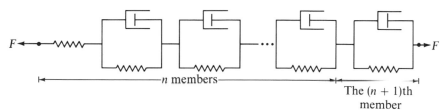

Figure P2.14 A generalized Maxwell body.

2.15 If Voigt models were put in series as shown in Fig. P2.15, then since for the ith unit, $F = \mu_i u_i + \eta_i \dot{u}_i$ so that $u_i = (\mu + \eta D)^{-1}F$, we have

$$u = u_1 + u_2 + \cdots + u_n = \sum_{i=1}^{n} \frac{1}{\mu_i + \eta_i D} F.$$

Show that the model retains the undesirable feature of the Voigt model: a step displacement calls for an infinitely large force at the instant of step application. This feature is corrected by adding a spring as in the model shown in Fig. P2.14.

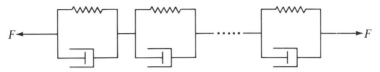

Figure P2.15 Voigt models in series.

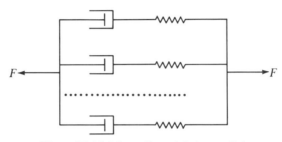

Figure P2.16 Maxwell models in parallel.

2.16 If n Maxwell models are put in parallel as shown in Fig. P2.16, show that

$$F = \sum_{i=1}^{n} \frac{D}{D/\mu_i + 1/\eta_i} u.$$

This system still has the feature of a Maxwell model: a constant force will produce an infinitely large displacement u when time $t \to \infty$. This difficulty can be corrected by adding a spring as in the Kelvin model. Compare Fig. P2.16 with Fig. P2.12.

2.17 A Newtonian fluid flows in a long tube of rectangular cross section, which has a width w and a height h. Formulate the mathematical problm in the form of a differential equation for the axial velocity and the associated boundary conditions. Suggest an appropriate method of solution when $w \gg h$.

   You may wish to use such a rectangular tube to measure the fluid viscosity instead of the circular tube illustrated in Fig. 2.14:1. Give a formula to compute the coefficient of viscosity.

   Note: You will have to solve a partial differential equation known as the Poisson's equation.

2.18 Consider a Couette viscometer (Fig. 2.14:2). Derive an equation to compute the coefficient of viscosity from the measured torque, the angular speed, and the dimensions of the viscometer. Suitable simplifying assumption based on the narrowness of the gap is acceptable.

2.19 Consider a cone–plate viscometer (Fig. 2.14:3). Do the same as in Problem 2.18. You may assume that the cone angle is small.

2.20 Whole blood is suspected to be viscoelastic. Design some experiments to reveal the viscoelastic properties of blood.

2.21 Consider a Maxwell body specified in Sec. 2.11, Fig. 2.11:1. Let it be subjected to a cycle of stretching and returning to the original position according to the history:

$$u = u_0 + ct \qquad \text{for} \qquad 0 \leq t \leq t_m,$$
$$= u_0 - c(t - t_m) \qquad \text{for} \qquad t_m \leq t \leq 2t_m.$$

Compute F as function of time. Find the area of the loop when F is plotted against u. The hysteresis is the ratio of the area of the loop divided by the area under the loading curve. What is the hysteresis of a Maxwell body?

2.22 What are the hysteresis characteristics of the Voigt and Kelvin bodies? See Problem 2.21.

References

Bird, R. B., Hassager, O., Armstrong, R. C., and Curtiss, C. F. (1977) *Dynamics of Polymeric Liquids*: Vol. 1. *Fluid Mechanics*; Vol. 2. *Kinetic Theory*. Wiley, New York.

Boltzmann, L. (1876) Zur theorie der elastischen Nachwickung. *Ann. Phys. Suppl.* **7**, 624–654.

Fung, Y. C. (1965) *Foundations of Solid Mechanics*. Prentice-Hall, Englewood Cliffs, NJ.

Fung, Y. C. (1993) *A First Course in Continuum Mechanics*, 3rd edition. Prentice-Hall, Englewood Cliffs, NJ.

Green, A. E. and Adkins, J. E. (1960) *Large Elastic Deformations*. Oxford University Press, London.

Green, A. E. and Rivlin, R. S. (1957) The mechanics of nonlinear materials with memory. *Arch. Rat. Mech. Anal.* **1**, 1–21.

Green, A. E., Rivlin, R. S., and Spencer, A. J. M. (1959) The mechanics of nonlinear materials with memory: Part 2. *Arch. Rat. Mech. Anal.* **3**, 82–90.

Pipkin, A. C. and Rogers, T. G. (1968) A nonlinear integral representation for visco-elastic behavior. *J. Mech. Phys. Solids* **16**, 59–74.

CHAPTER 3

The Flow Properties of Blood

3.1 Blood Rheology: An Outline

Blood is a marvelous fluid that nurtures life, contains many enzymes and hormones, and transports oxygen and carbon dioxide between the lungs and the cells of the tissues. We can leave the study of most of these important functions of blood to hematologists, biochemists, and pathological chemists. For biomechanics the most important information we need is the constitutive equation.

Human blood is a suspension of cells in an aqueous solution of electrolytes and nonelectrolytes. By centrifugation, the blood is separated into *plasma* and *cells*. The plasma is about 90% water by weight, 7% plasma protein, 1% inorganic substances, and 1% other organic substances. The cellular contents are essentially all *erythrocytes*, or *red cells*, with *white cells* of various categories making up less than 1/600th of the total cellular volume, and *platelets* less than 1/800th of the cellular volume. Normally, the red cells occupy about 50% of the blood volume. They are small, and number about 5 million/mm³. The normal white cell count is considered to be from 5000 to 8000/mm³, and platelets from 250 000 to 300 000/mm³. Human red cells are disk shaped, with a diameter of 7.6 μm and thickness 2.8 μm. White cells are more rounded and there are many types. Platelets are much smaller and have a diameter of about 2.5 μm.

If the blood is allowed to clot, a straw-colored fluid called *serum* appears in the plasma when the clot spontaneously contracts. Serum is similar to plasma in composition, but with one important colloidal protein, *fibrinogen*, removed while forming the clot. Most of the platelets are enmeshed in the clot.

The specific gravity of red cells is about 1.10; that of plasma is 1.03. When plasma was tested in a viscometer, it was found to behave like a Newtonian viscous fluid (Merrill et al., 1965), with a coefficient of viscosity about 1.2 cP

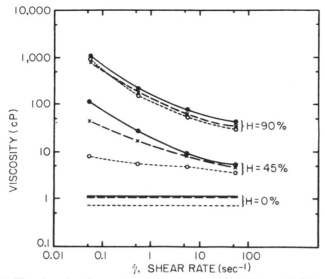

Figure 3.1:1 The viscosity-shear rate relations in whole blood (•—•), defibrinated blood (×–––×), and washed cells in Ringer soluion (○---○), at 45% and 90% red cell volume concentrations. From Chien et al. (1966), by permission.

(Gregersen et al., 1967; Chien et al., 1966, 1971). When whole blood was tested in a viscometer, its non-Newtonian character was revealed. Figure 3.1:1 shows the variation of the viscosity of blood with the strain rate when blood is tested in a Couette-flow viscometer whose gap width is much larger than the diameter of the individual red cells. For the curves in Fig. 3.1:1, the shear rate $\dot{\gamma}$ is defined as the relative velocity of the walls divided by the width of the gap. The viscosity of blood varies with the *hematocrit*, H, the percentage of the total volume of blood occupied by the cells. It varies also with temperature (see Fig. 3.1:2) and with disease state, if any.

There is a question about what happens to the blood viscosity when the strain rate is reduced to zero. Cokelet et al. (1963) insist that the blood has a finite yield stress. They say that at a vanishing shear rate the blood behaves like an elastic solid. They deduced this conclusion on the basis that their torque measuring device had a rapid response, and could be used to measure transient effects. They studied the time history of the torque after the rotating bob had been stopped suddenly, and compared the transient response for blood with that for a clay suspension, which is known to have a finite yield stress. They showed that the values deduced from this experiment agreed within a few percent with those obtained by extrapolation of the Casson plot as shown in Fig. 3.1:3. Merrill et al. (1965a) also used a capillary viscometer to see if blood in a capillary could maintain a pressure difference across the tube ends without any detectable fluid flow. Such a pressure difference was detected and found to agree with the yield stress determined by Cokelet's method.

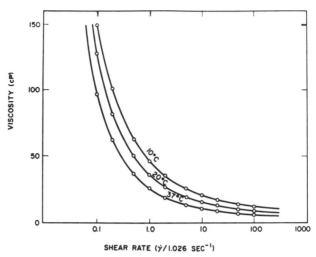

Figure 3.1:2 The variation of the viscosity of human blood with shear rate $\dot{\gamma}$ and temperature for a male donor, containing acid citrate dextrose, reconstituted from plasma and red cells to the original hematocrit of 44.8. From Merrill et al. (1963a, p. 206), by permission.

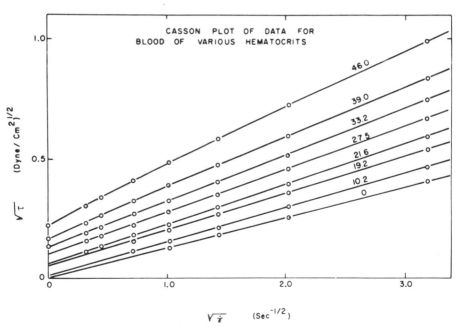

Figure 3.1:3 Casson plot of very low shear rate data for a sample of human blood at a temperature of 25°C. Range of linearity decreases as the hematocrit increases. Plasma is Newtonian. Data from Cokelet et al. (1963), by permission.

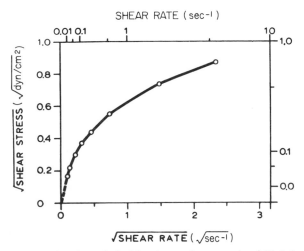

Figure 3.1:4 Casson plot for whole blood at a hematocrit of 51.7 (temp. = 37°C), according to Chien et al. (1966), by permission.

It must be understood that by "existence" of a yield stress is meant that no sensible flow can be detected in a fluid under a shearing stress in a finite interval of time (say, 15 min). The difficulty of determining the yield stress of blood as $\dot{\gamma} \to 0$ is compounded by the fact that an experiment for very small shear rate is necessarily a transient one if that experiment is to be executed in a finite interval of time. For blood the analysis is further complicated by the migration of red cells away from the walls of the viscometer when $\dot{\gamma}$ is smaller than about 1 s^{-1}. Cokelet's analysis takes these factors into account, and requires considerable manipulation of the raw data (see Cokelet, 1972). If the maximum transient shear stress reached after the start of an experiment at constant shear rate is plotted directly with respect to the nominal shear rate, the result appears as shown in Fig. 3.1:4 by Chien et al. (1966). The differences between the plots of Figs. 3.1:3 and 3.1:4 are caused mainly by the data analysis procedures, and partly reflect the dynamics of the instrument as well as that of the blood state.

The data of Cokelet et al. for a small shear rate, say $\dot{\gamma} < 10 \text{ s}^{-1}$, and for hematocrit less than 40%, can be described approximately by Casson's (1959) equation

$$\sqrt{\tau} = \sqrt{\tau_y} + \sqrt{\eta \dot{\gamma}}, \tag{1}$$

where τ is the shear stress, $\dot{\gamma}$ is the shear strain rate, τ_y is a constant that is interpreted as the yield stress in shear, and η is a constant. The fitting of this equation to experimental data can be seen in Fig. 3.1:3, from which it is clear that for hematocrit below 33% the experimental data points fall quite accurately on straight lines. For higher hematocrit (39% and above), deviations from straight lines are more evident. The yield stress τ_y given by Merrill

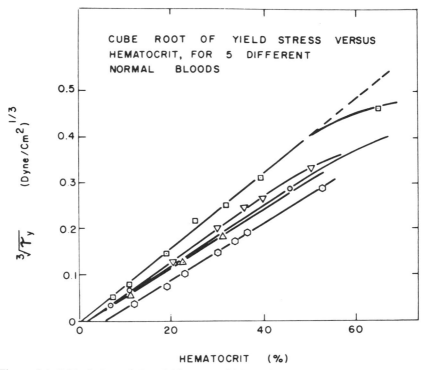

CUBE ROOT OF YIELD STRESS VERSUS HEMATOCRIT, FOR 5 DIFFERENT NORMAL BLOODS

HEMATOCRIT (%)

Figure 3.1:5 Variation of the yield stress of blood from five different donors with hematocrit. Tests on samples with hematocrits less than the applicable abscissa intercept showed no yield stress. From Merrill et al. (1963a), by permission. ACD was used as anticoagulant.

et al. (1963) is shown in Fig. 3.1:5. Note that τ_y is very small: of the order of 0.05 dyn/cm^2, and is almost independent of the temperature in the range 10–37°C. τ_y is markedly influenced by the macromolecular composition of the suspending fluid. A suspension of red cells in saline plus albumin has zero yield stress; a suspension of red cells in plasma containing fibrinogen has a finite yield stress.

At high shear rate, whole blood behaves like a Newtonian fluid with a constant coefficient of viscosity, as is seen in Fig. 3.1:1. In other words, for sufficiently large values of $\dot{\gamma}$ we have

$$\tau = \mu\dot{\gamma} \quad \text{or} \quad \sqrt{\tau} = \sqrt{\mu}\sqrt{\dot{\gamma}}, \qquad \mu = \text{const.} \tag{2}$$

As the shear rate $\dot{\gamma}$ increases from 0 to a high value, there is a transition region in which the stress–strain rate relation changes from that described by Eq. (1) to that described by Eq. (2). This is illustrated in Fig. 3.1:6, with data from Brooks et al. (1970). The part of the data to the right of the dashed curve belongs to the Newtonian region, described by Eq. (2); straight regression lines pass through the origin. The part to the left of the dashed curve

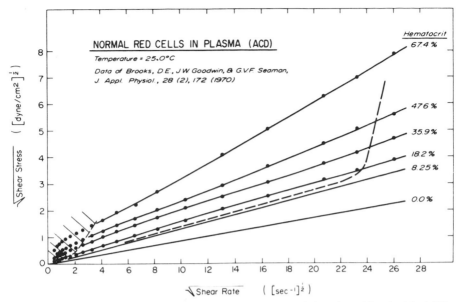

Figure 3.1:6 Casson plot of higher shear rate rheological data for a blood with ACD anticoagulant. The actual data points for plasma and the 8.25% hematocirt red cell suspension are not shown for clarity; these fluids appear to be Newtonian. Viscometer surfaces were smooth. From Cokelet (1972), by permission.

is the non-Newtonian region. The range of validity of Eq. (1) is smaller for higher hematocrit, and the transition region from Eq. (1) to Eq. (2) is larger for higher hematocrit. For hematocrit below 40% at 37°C, Eqs. (1) and (2) can be assumed to merge smoothly.

This outline shows the major features of blood flow in a viscometer of the Couette-flow type. The width of the channel in which blood flows in these testing machines is much larger than the diameter of the red blood cells. The blood is considered as a *homogeneous fluid*. This is a reasonable way to look at the blood when we analyze blood flow in large blood vessels. But all our blood vessels are not so large. In the human body there are about 10^{10} blood vessels whose diameter is about the same as that of the red blood cells, ranging from 4 to 10 μm. These are called the *capillary blood vessels*. When blood flows in capillary blood vessels, the red cells have to be squeezed and deformed, and move in single file. In this case, then, it would be more useful to consider blood as a *nonhomogeneous* fluid, of at least two phases, one phase being the blood cells, the other being the plasma. Between large arteries and veins and the capillaries there are blood vessel of varying diameters. At what level can we regard blood as homogeneous? This is a question whose answer depends on the fluid mechanical features that one wishes to examine. Flow of blood in microvessels has many unique features, which will be discussed in Chapter 5.

In the present chapter, we shall first put the experimental results in a form suitable for further analysis. The problem of blood flow in a circular cylindrical tube will be discussed. We shall then turn to the question why blood viscosity behaves the way it does. Through this kind of examination, we will gain some understanding of blood rheology in states of health or disease. We shall conclude this chapter with a discussion of blood coagulation and medical applications of blood rheology.

3.2 The Constitutive Equation of Blood Based on Viscometric Data and Casson's Equation

Blood is a mixture of blood plasma and blood cells. The mixture, when tested in viscometers whose characteristic dimension is much larger than the characteristic size of the blood cells, yields the data outlined in Sec. 3.1. In these viscometers, and, by implication, in large blood vessels whose diameters are much larger than the characteristic size of the blood cells, blood may be treated as a homogeneous fluid. The mechanical properties of the mixture as a homogeneous fluid can be described by a constitutive equation. In this section, such a constitutive equation based on the viscometric data outlined in Sec. 3.1 is formulated as an *isotropic* and *incompressible fluid*. The assumption of isotropy is motivated by the idea that when the shear stress and shear strain rate are zero, the blood cells have no preferred orientation. The assumption of incompressibility is based on the fact that, in the range of pressures concerned in physiology, the mass densities of the plasma, the cells, and the mixture as a whole are unaffected by the pressure.

The rheology of blood revealed in Sec. 3.1 differs from that of a Newtonian fluid only in the fact that the coefficient of viscosity is not a constant. The constitutive equation of an isotropic incompressible Newtonian fluid is

$$\sigma_{ij} = -p\delta_{ij} + 2\mu V_{ij}, \tag{1}$$

where

$$V_{ij} = \frac{1}{2}\left(\frac{\partial u_i}{\partial x_j} + \frac{\partial u_j}{\partial x_i}\right), \qquad V_{ii} = V_{11} + V_{22} + V_{33} = 0. \tag{2}$$

Here σ_{ij} is the stress tensor, V_{ij} is the strain-rate tensor, u_i is the velocity component, δ_{ij} is the isotropic tensor or Kronecker delta, p is the hydrostatic pressure, and μ is a constant called the *coefficient of viscosity*. The indices i, j range over $1, 2, 3$, and the components of the tensors and vectors are referred to a set of rectangular cartesian coordinates x_1, x_2, x_3. The summation convention is used, so that a repetition of an index means summation over the index, e.g., $V_{ii} = V_{11} + V_{22} + V_{33}$. Note the factor $\frac{1}{2}$ in our definition of strain rate in Eq. (2). This definition makes V_{ij} a tensor. In older books, shear rate is defined as twice this value. Thus $\dot{\gamma}$ of Fig. 3.1:1 is equal to $2V_{12}$ if the coordinate axes x_1, x_2 point at directions parallel and perpendicular to the wall, respectively.

Blood does not obey Eq. (1) because μ is not a constant. μ varies with the strain rate. Thus blood is said to be *non-Newtonian*. Similar experiments do verify, however, that *normal plasma alone is Newtonian*. Therefore the non-Newtonian feature of whole blood comes from the cellular bodies in the blood.

How can we generalize the Newtonian equation (1) to accommodate the non-Newtonian behavior of blood? What guidance do we have?

The most important principle we learned from continuum mechanics is that any equation must be tensorially correct, i.e., every term in an equation must be a tensor of the same rank. Thus, if we decide to try an equation of the type of Eq. (1) for blood, under the assumption that its mechanical behavior is isotropic, then μ is a scalar, i.e., μ must be a scalar function of the strain rate tensor V_{ij}. Now, V_{ij} is a symmetric tensor of rank 2 in three dimensions. It has three invariants which are scalars. Hence the coefficient of viscosity μ must be a function of the invariants of V_{ij}. These invariants are:

$$I_1 = V_{11} + V_{22} + V_{33},$$

$$I_2 = \begin{vmatrix} V_{11} & V_{12} \\ V_{21} & V_{22} \end{vmatrix} + \begin{vmatrix} V_{22} & V_{23} \\ V_{32} & V_{33} \end{vmatrix} + \begin{vmatrix} V_{33} & V_{31} \\ V_{13} & V_{11} \end{vmatrix},$$

$$(3)$$

$$I_3 = \begin{vmatrix} V_{11} & V_{12} & V_{13} \\ V_{21} & V_{22} & V_{23} \\ V_{31} & V_{32} & V_{33} \end{vmatrix}.$$

See the author's *First Course*, 3rd edn., Chapter 5. But we have assumed blood to be incompressible; hence I_1 vanishes. But when $I_1 = 0$, I_2 is negative valued, and it is more convenient to use a positive-valued invariant J_2 defined by

$$J_2 = \tfrac{1}{3}I_1^2 - I_2 = \tfrac{1}{2}V_{ij}V_{ij}. \tag{4}$$

Hence μ must be a function of J_2 and I_3. From Eq. (4) it is seen that J_2 is directly related to the shear strain rate. For example, if V_{12} is the only non-vanishing component of shear rate, then $J_2 = V_{12}^2$. From experiments described in Sec. 3.1 we know that the blood viscosity depends on the shear rate, hence on J_2. Whether it depends on I_3 or not is unknown because in most viscometric flows (e.g., Couette, Poiseuille, and cone-plate flows, see Sec. 2.14) I_3 is zero. Thus, we assume that μ is a function of J_2 and propose the following constitutive equation for blood when it flows:

$$\sigma_{ij} = -p\delta_{ij} + 2\mu(J_2)V_{ij}. \tag{5}$$

Let us now compare this proposal with experimental results. For the simple shear flow shown in Fig. 2.7:1 in Sec. 2.7, we have the shear rate

$$\dot{\gamma} = \frac{v}{h} = 2 \cdot \frac{1}{2}\left(\frac{\partial v_1}{\partial x_2} + \frac{\partial v_2}{\partial x_1}\right) = 2V_{12}. \tag{6}$$

(Note the factor of 2 due to the difference in the old and the tensorial definition of strain rate as remarked previously.) In this case, all other components V_{ij} vanish, and

$$J_2 = V_{12}^2,$$ (7)

so that from Eq. (6),

$$|\dot{\gamma}| = 2\sqrt{J_2},$$ (8)

whereas from Eq. (5), we have the shear stress

$$\sigma_{12} = 2\mu(J_2)V_{12} = \mu\dot{\gamma}.$$ (9)

It is shown in Sec. 3.1 that in steady flow in larger vessels, the experimental results may be expressed in Casson's equation, Eq. (1) of Sec. 3.1,

$$\sigma_{12} = [\sqrt{\tau_y} + \sqrt{\eta|\dot{\gamma}|}]^2$$ (10)

in a wide range of $\dot{\gamma}$. Comparing Eqs. (9) and (10), we see that

$$\mu = \frac{[\sqrt{\tau_y} + \sqrt{\eta|\dot{\gamma}|}]^2}{|\dot{\gamma}|}.$$ (11)

Thus, we conclude that the constitutive equation for blood is, in the range of $\dot{\gamma}$ when blood flows,

$$\sigma_{ij} = -p\delta_{ij} + 2\mu(J_2)V_{ij},$$ (12a)

where

$$\mu(J_2) = [(\eta^2 J_2)^{1/4} + 2^{-1/2}\tau_y^{1/2}]^2 J_2^{-1/2}.$$ (12b)

Equation (12) is valid when $J_2 \neq 0$ and sufficiently small. On the other hand, when J_2 is sufficiently large, the experimental results reduce to the simple statement that $\mu =$ constant, so that Eq. (1) applies. The point of transition from Eqs. (12a,b) to the Newtonian equation, $\mu =$ constant, depends on the hematocrit, H (the volume fraction of red blood cells in whole blood). For normal blood with a low hematocrit, $H = 8.25\%$, μ was found to be constant over the entire range of shear rate from 0.1 to 1000 s^{-1}. When $H = 18\%$, the blood appears to be Newtonian when $\dot{\gamma} > 600$ s^{-1}, but obeys Eq. (12) for smaller $\dot{\gamma}$. For higher H the transition point increases to $\dot{\gamma} \simeq 700$ s^{-1}. See Fig. 3.1:6.

The flow rule must be supplemented by another stress–strain relation when the blood is not flowing, i.e., when $V_{ij} = 0$. We know so little about the behavior of blood in this condition, however, that only a hypothetical constitutive equation can be proposed. A Hooke's law, for example, may be suggested when there is no flow, because the stress and strain are both very small. (The "yielding stress," τ_y, in Eq. (11) or (12), is only of the order of 0.05 dyn/cm^2. The weight of a layer of water 1 mm thick, spreading over 1 cm^2, produces a compressive stress of 100 dyn/cm^2!) For a complete formulation, we need another condition, the "yielding condition," to define at what stress level flow must occur. In the theory of plasticity, the yielding condition is often stated in terms of the second invariant of the stress deviation, defined as

$$J_2' = \tfrac{1}{2}\sigma_{ij}'\sigma_{ij}', \tag{13}$$

where

$$\sigma_{ij}' = \sigma_{ij} - \tfrac{1}{3}\sigma_{kk}\delta_{ij}. \tag{14}$$

Flow or yielding occurs when J_2' exceeds a certain number K. Thus:

$$V_{ij} = \begin{cases} 0 & \text{if } J_2' < K, \\ \dfrac{1}{2\mu}\sigma_{ij}' & \text{if } J_2' \geq K. \end{cases} \tag{15}$$

The stress–strain law when there is no flow, the yielding condition, and the flow rule together describe the mechanical behavior of the blood.

3.2.1 Other Aspects of Blood Rheology

Actually the rheological properties of blood are more complex than what is portrayed above. If blood is tested dynamically, it can be made to reveal all features of viscoelasticity. Furthermore, the viscoelastic characteristics of blood change with the level of strain and strain history; hence it is *thixotropic*. These complex properties are probably unimportant in normal circulatory physiology, but can be significant when one tries to use blood rheology as a basis of clinical applications to diagnosis of diseases, pathology, or biochemical studies.

3.2.2 Why Do We Need the General Constitutive Equation?

We have obtained, with some effort in theoretical reasoning, a constitutive equation for blood that is consistent with our experimental knowledge. Why is this complicated constitutive equation needed? The answer is that although the simple Casson equation, Eq. (1) of Sec. 3.1, suffices in the analysis of simple problems in which the strain rate tensor can be calculated a priori (without using the constitutive equation), in more complex problems it is insufficient. Casson's equation is all we need in analyzing Poiseuille and Couette flows, for which the shear stress and strain distributions are known from statics and kinematics. But if we wish to analyze the flow of blood at the point of bifurcation of an artery into two branches, or the flow through a stenosis, or the flow in aortic sinus, etc., the stress and strain rate distributions are not known. To analyze these problems, it is necessary to write down the general equations of motion based on an appropriate constitutive equation and solve them. For these problems the general constitutive equation is necessary.

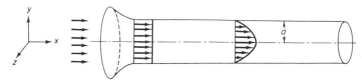

Figure 3.3:1 Velocity profiles in a steady laminar flow into a circular cylindrical tube.

3.3 Laminar Flow of Blood in a Tube

Let us consider the flow of blood in a circular cylindrical tube. We shall assume the flow to be *laminar*, that is, *not turbulent*. We shall also assume that the tube is long and the flow is steady, so that the conditions of flow change neither with the distance along the tube, nor with the time. Under these assumptions we can analyze the flow with a simple ad hoc approach, which is presented below.

We shall use polar coordinates for this problem. The polar axis coincides with the axis of the cylinder. See Fig. 3.3:1. The flow obeys Navier–Stokes equations of motion of an incompressible fluid. The boundary condition is that blood adheres to the tube wall (the so-called *no-slip* condition). Since the boundary condition is axisymmetric, the flow is also axisymmetric and the only nonvanishing component of velocity is $u(r)$ in the axial direction; $u(r)$ is a function of r alone, and not of x. Isolate a cylindrical body of fluid of radius r and unit length in the axial direction, as shown in Fig. 3.3:2(a). This body is subjected to a pressure p_1 on the left-hand end, p_2 on the right-hand end, and shear stress τ on the circumferential surface. Since $p_1 - p_2 = -1 \cdot (dp/dx)$ acts on an area πr^2, and τ acts on an area $1 \cdot 2\pi r$, we have, for equilibrium, the balance of forces

$$\tau \cdot 2\pi r = -\pi r^2 \frac{dp}{dx},$$

or

$$\tau = -\frac{r}{2}\frac{dp}{dx} \qquad \text{(Stokes, 1851).} \qquad (1)$$

This important result is shown in Fig. 3.3:2(b).

Now we must introduce a constitutive equation that relates the shear stress τ to the velocity gradient. Let us first consider a Newtonian fluid.

3.3.1 A Newtonian Fluid

By the definition of Newtonian viscosity, we have

$$\tau = -\mu \frac{du}{dr}. \qquad (2)$$

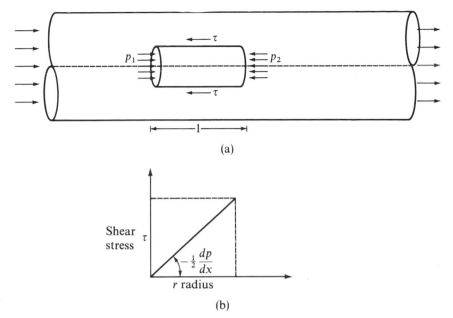

Figure 3.3:2 Steady flow in a long circular cylindrical pipe. (a) A free-body diagram of a centrally located element on which pressure and shear stresses act. (b) Relationship between the shear stress τ and the radial distance r from the axis of symmetry.

A substitution of Eq. (2) into Eq. (1) yields

$$\frac{du}{dr} = \frac{r}{2\mu} \frac{dp}{dx}. \tag{3}$$

Since the left-hand side is a function of r, the right-hand side must be also. Hence dp/dx cannot be a function of x. But since the fluid does not move in the radial direction, the pressures in the radial direction must be balanced, and p cannot vary with r. Hence the pressure gradient dp/dx must be a constant. Therefore, we can integrate Eq. (3) to obtain

$$u = \frac{r^2}{4\mu} \frac{dp}{dx} + B, \tag{4}$$

where B is an integration constant. B can be determined by the boundary condition of no-slip:

$$u = 0 \qquad \text{when} \quad r = a. \tag{5}$$

Combining Eqs. (4) and (5) yields the solution

$$u = -\frac{1}{4\mu} (a^2 - r^2) \frac{dp}{dx}, \tag{6}$$

which shows that the velocity profile is a parabola, as sketched in Fig. 3.3:1.

The *rate of flow* through the tube can be obtained by integrating velocity times area over the tube cross section:

$$\dot{Q} = 2\pi \int_0^a ur\,dr. \tag{7}$$

On substituting Eq. (6) into Eq. (7), we obtain the well-known formula

$$\dot{Q} = -\frac{\pi a^4}{8\mu}\frac{dp}{dx}. \tag{8}$$

A division of the rate of flow by the cross-sectional area of the tube yields the mean velocity of flow,

$$u_m = -\frac{a^2}{8\mu}\frac{dp}{dx}. \tag{9}$$

3.3.2 Blood, with Viscosity Described by Casson's Equation

If we compute the shear strain rate of the fluid at the tube wall, using the formulas obtained above for a Newtonian fluid, we obtain

$$\left.\frac{du}{dr}\right|_{r=a} = \frac{1}{2}\frac{a}{\mu}\frac{dp}{dx} \qquad \text{from Eq. (3) or (6),} \tag{10}$$

or

$$\left.\frac{du}{dr}\right|_{r=a} = -\frac{4u_m}{a} \qquad \text{from Eqs. (3) and (9).} \tag{11}$$

If physiological data on u_m, a, μ, and dp/dx are substituted into Eqs. (10) and (11), we find that $\dot{\gamma}$ ($= -du/dr$ at $r = a$) ranges from 100 to 2000 sec^{-1} in large and small arteries, and 20 to 200 sec^{-1} in large and small veins. Thus in the neighborhood of the blood vessel wall, the shear rate is high enough for the Newtonian assumption to be valid for normal blood. Toward the center of the tube, however, the shear gradient tends to zero, and the non-Newtonian feature of the blood will become more evident.

Let us assume that the blood is governed by Casson's equation, Eq. (1) of Sec. 3.1 or Eq. (12) of Sec. 3.2. We assume that the flow is laminar, uniaxial, axisymmetric, and without entrance effect (far away from the ends of the tube).

Equation (1) is valid for any fluid, hence for blood. The shear stress acting on any cylindrical surface ($r = $ const.) is proportional to r, as shown in Fig. 3.3:2 and reproduced in Fig. 3.3:3. On the wall, the shear stress is τ_w. At a certain point on the stress axis it is τ_y, the yield stress. This corresponds to a radius r_c, shown on the horizontal axis. If the shear stress is smaller than the yield stress, that is, $\tau < \tau_y$ or $r < r_c$, the blood will not flow. If it moves at all it would have to move like a rigid body. Therefore, the velocity profile depends on the relative magnitude of τ_y and τ_w. Now, by Eq. (1),

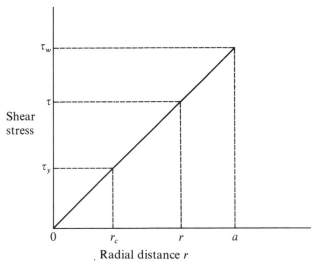

Figure 3.3:3 The relationship between shear stress and radial distance in a steady pipe flow. The shear stress reaches the yields stress of blood τ_y at radial distance r_c.

$$\tau_w = -\frac{a}{2}\frac{dp}{dx}, \qquad \tau_y = -\frac{r_c}{2}\frac{dp}{dx}. \tag{12}$$

Therefore, if $\tau_y > \tau_w$, i.e., $r_c > a$, then there is no flow:

$$u = 0 \qquad \text{when} \qquad -\frac{dp}{dx} < \frac{2\tau_y}{a}. \tag{13}$$

If $\tau_y < \tau_w$, i.e., $r_c < a$, or

$$-\frac{dp}{dx} > \frac{2\tau_y}{a}, \tag{14}$$

then the velocity profile of the flow will be like that sketched in Fig. 3.3:4. In the core, $r < r_c$, the profile is flat. Between r_c and a, Casson's equation applies:

$$\sqrt{\tau} = \sqrt{\eta}\sqrt{\dot{\gamma}} + \sqrt{\tau_y}, \tag{15}$$

η is called *Casson's coefficient of viscosity*. Taking the square root of Eq. (1) and using Eq. (15), we obtain

$$\sqrt{-\frac{r}{2}\frac{dp}{dx}} = \sqrt{\tau_y} + \sqrt{\eta}\sqrt{\dot{\gamma}}. \tag{16}$$

Solving for $\dot{\gamma}$, we have

$$-\frac{du}{dr} = \dot{\gamma} = \frac{1}{\eta}\left(\sqrt{-\frac{r}{2}\frac{dp}{dx}} - \sqrt{\tau_y}\right)^2. \tag{17}$$

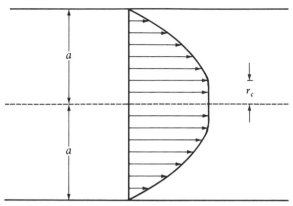

Figure 3.3:4 The velocity profile of laminar flow of blood in a long circular cylindrical pipe.

Integrating this equation and using the no-slip boundary condition $u = 0$ when $r = a$, we have

$$-\int_r^a \frac{du}{dr}\, dr = u|_r - u|_a = \frac{1}{\eta} \int_r^a \left(\sqrt{-\frac{r}{2}\frac{dp}{dx}} - \sqrt{\tau_y} \right)^2 dr,$$ (18)

or

$$u = -\frac{1}{4\eta}\frac{dp}{dx}\left[a^2 - r^2 - \tfrac{8}{3}r_c^{1/2}(a^{3/2} - r^{3/2}) + 2r_c(a - r)\right] \quad \text{for } (r_c \le r \le a).$$ (19)

At $r = r_c$, the velocity u becomes the core velocity u_c:

$$u_c = -\frac{1}{4\eta}\frac{dp}{dx}\left(a^2 - \tfrac{8}{3}r_c^{1/2}a^{3/2} + 2r_c a - \tfrac{1}{3}r_c^2\right)$$

$$= -\frac{1}{4\eta}\frac{dp}{dx}(\sqrt{a} - \sqrt{r_c})^3(\sqrt{a} + \tfrac{1}{3}\sqrt{r_c}).$$ (20)

And for all values of r between 0 and r_c,

$$u = u_c.$$ (21)

The velocity distribution is given by Eqs. (19) and (21) if $r_c < a$, i.e., if Eq. (14) applies; whereas the flow is zero if the inequality sign in Eq. (14) is reversed.

We can now find the rate of volume flow by an integration:

$$\dot{Q} = 2\pi \int_0^a ur\, dr.$$ (22)

Using Eqs. (13), (19), and (21) for appropriate ranges of pressure gradient and radius, we obtain

$$\dot{Q} = \frac{\pi a^4}{8\eta}\left[-\frac{dp}{dx} - \frac{16}{7}\left(\frac{2\tau_y}{a}\right)^{1/2}\left(-\frac{dp}{dx}\right)^{1/2} + \frac{4}{3}\left(\frac{2\tau_y}{a}\right) - \frac{1}{21}\left(\frac{2\tau_y}{a}\right)^4\left(-\frac{dp}{dx}\right)^{-3}\right]$$ (23)

if $dp/dx > (2\tau_y/a)$; whereas

$$\dot{Q} = 0 \qquad \text{if} \quad -\frac{dp}{dx} < \frac{2\tau_y}{a}. \tag{24}$$

If we introduce the notation

$$\xi = \left(\frac{2\tau_y}{a}\right)\left(-\frac{dp}{dx}\right)^{-1}, \tag{25}$$

then Eq. (23) can be written as

$$\dot{Q} = -\frac{\pi a^4}{8\eta}\frac{dp}{dx}F(\xi), \tag{26}$$

where

$$F(\xi) = 1 - \frac{16}{7}\xi^{1/2} + \frac{4}{3}\xi - \frac{1}{21}\xi^4. \tag{27}$$

Equation (26) is similar to Poiseuille's law, but with a modifying factor $F(\xi)$. These results were obtained by S. Oka (1965, 1974). Figure 3.3:5 gives Oka's graph of $F(\xi)$ vs. ξ. It is seen that the flow rate decreases rapidly with increasing ξ. For $\xi > 1$, there is no flow, and $\dot{Q} = 0$. Oka also obtained the interesting result shown in Fig. 3.3:6, that if one plots the square root of $\dot{Q}$ vs. the square root of the pressure gradient, one obtains a curve that resembles the flow vs. shear stress curve of a Bingham plastic material (see Problem 3.17, p. 97). Taking the square root on both sides of Eq. (26), expanding the square root of $F(\xi)$ in a power series of $\xi^{1/2}$, and retaining only the first power, one obtains the asymptotic equation

$$Q^{1/2} = \left(\frac{\pi a^4}{8\eta}\right)^{1/2}\left(-\frac{dp}{dx}\right)^{1/2}(1 - \tfrac{8}{7}\xi^{1/2})$$

$$= \left(\frac{\pi a^4}{8\eta}\right)^{1/2}\left[\left(-\frac{dp}{dx}\right)^{1/2} - \tfrac{8}{7}p_c^{1/2}\right], \tag{28}$$

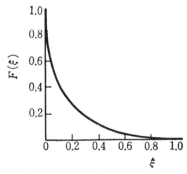

Figure 3.3:5 The function $F(\xi)$ from Eq. (37) of Sec. 3.5, which is the ratio of the flow rate in a tube of a fluid obeying Casson's equation to that of a Newtonian fluid with the same Casson viscosity. The variable ξ is inversely proportional to the pressure gradient; see Eq. (36) of Sec. 3.5. From Oka (1974), by permission.

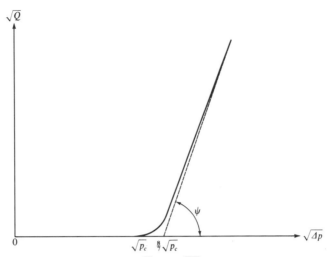

Figure 3.3:6 The relationship between $\sqrt{Q}$ and $\sqrt{\varDelta p}$; Q is the flow rate of a fluid obeying Casson's equation, and $\varDelta p$ is the pressure gradient. From Oka (1974), by permission.

where $p_c = 2\tau_y/a$. This is plotted as the dotted line in Fig. 3.3:6. The solid curve is the exact solution. The dotted line is an asymptote whose slope is

$$\tan \psi = \left(\frac{\pi a^4}{8\eta}\right)^{1/2}. \tag{29}$$

These results give complete information on the laminar flow of blood in circular cylindrical tubes.

3.4 Speculation on Why Blood Viscosity Is the Way It Is

Since normal plasma is Newtonian, there is no doubt that the non-Newtonian features of human blood come from blood cells. How the red blood cells move when blood flows is the central issue.

It has been known for a long time (Fahraeus, 1929) that human red blood cells can form aggregates known as rouleaux (Fig. 3.4:1), whose existence depends on the presence of the proteins fibrinogen and globulin in plasma. (Bovine blood does not form rouleaux.) The slower the blood flows, or rather, the smaller the shear rate, the more prevalent are the aggregates. When the shear rate tends to zero, it is speculated that human blood becomes one big aggregate, which then behaves like a solid. A solid may be viscoelastic or viscoplastic. If the blood aggregate behaves as a plastic solid, then a yield stress exists which can be (but does not have to be) identified with the constant τ_y in Casson's equation [Eq. (1) of Sec. 3.1].

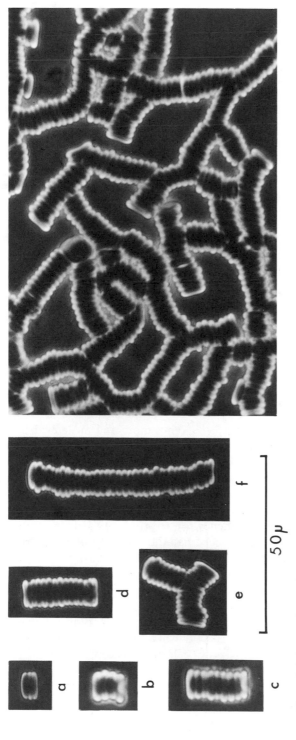

Figure 3.4:1 Rouleaux of human red cells photographed on a microscope slide showing single linear and branched aggregates (left part) and a network (right part). The number of cells in linear array are 2, 4, 9, 15, and 36 in a, b, c, d, and f, respectively. From Goldsmith (1972b), by permission.

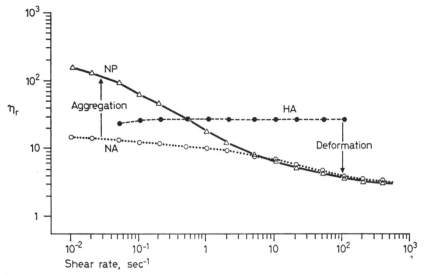

Figure 3.4:2 Logarithmic relation between relative apparent viscosity and shear rate in three types of suspensions, each containing 45% human RBC by volume. Suspending medium (plasma) viscosity = 1.2 cP ($= 1.2 \times 10^{-3}$ N s m^{-2}). NP = Normal RBC in plasma; NA = normal RBC in 11% albumin; HA = hardened RBC in 11% albumin. From Chien (1970), by permission.

When the shear rate increases, blood aggregates tend to be broken up, and the viscosity of blood is reduced. As the shear rate increases further, the deformation of the red cells becomes more and more evident. The cells tend to become elongated and line up with the streamlines. This further reduces the viscosity. The effects of aggregation and deformation of the red cells are well illustrated in Fig. 3.4:2, from Chien (1970). In this figure, the line NP refers to normal blood. The line NA refers to a suspension of normal red blood cells in an albumin-Ringer solution that does not contain fibrinogen and globulins. The tendency toward rouleaux formation was removed in the latter case, and it is seen that the viscosity of the suspension was reduced, even though the viscosity of the albumin-Ringer solution was adjusted to be the same as that of the plasma of the NP case. The third curve, HA, refers to a suspension of hardened red blood cells in the same albumin-Ringer solution. (Red cells can be hardened by adding a little fixing agent such as glutaraldehyde to the solution.) Hence the difference between the NP and NA curves indicates the effect of cell aggregation, whereas that between NA and HA indicates the effect of cell deformation.

The amazing fluidity of human blood is revealed in Fig. 3.4:3, in which the viscosity of blood at a shear rate $> 100 \, \text{s}^{-1}$ (at which particle aggregation ceased to be important) is compared with that of other suspensions and

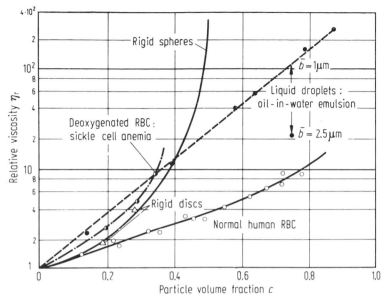

Figure 3.4:3 Relative viscosity of human blood at 25°C as a function of red cell volume fraction, compared to that of suspensions of rigid latex spheres, rigid disks, droplets, and sickled erythrocytes, which are virtually nondeformable. From Goldsmith (1972b), by permission.

emulsions. At 50% concentration, a suspension of rigid spheres will not be able to flow, whereas blood is fluid even at 98% concentration by volume. The much higher viscosity of blood with deoxygenated sickle cells is also shown: it tells us why sickle cell anemia is such a serious disease.

Since in a field of shear flow with a velocity gradient in the y direction the velocity is different at different values of y, any suspended particle of finite dimension in the y direction will tumble while flowing. The tumbling disturbs the flow and requires expenditure of energy, which is revealed in viscosity. If n red cells form a rouleaux, the tumbling of the rouleaux will cause more disturbance than the sum of the disturbances of the n individual red cells. Hence breaking up the rouleaux will reduce the viscosity. Further reduction can be obtained by deformation of the particle. If the particle is a liquid droplet, for example, it can elongate to reduce the dimension in the y direction, thus reducing the disturbance to the flow. A red cell behaves somewhat like a liquid droplet; it is a droplet wrapped in a membrane. These factors explain the reduction of viscosity with increasing shear rate.

Detailed studies of the tumbling and deformation of the red cells and rouleaux in shear flow have been made by Goldsmith and his associates by observing blood flow in tiny circular cylindrical glass tubes having diameters from 65 to 200 μm under a high powered microscope. The tubes lay on a vertically mounted platform, which was moved mechanically or hydraulically upward as the particles being tracked flowed downward from a syringe reservoir. The cell motion was photographed. Particle behavior at hematocrits $> 10\%$ was studied by using tracer red cells in a transparent suspension of hemolyzed, unpigmented red cells—so-called *ghost cells*. The ghosts were also biconcave in shape, although their mean diameter ($\sim 7.2\ \mu$) and volume ($7.4 \times 10^{-11}\ \text{cm}^3$) were somewhat smaller than those of the parent red cells.

Figure 3.4:4 shows the tumbling of an 11- and a 16-cell rouleau of red cells in Poiseuille flow at a shear rate $\dot{\gamma} < 10\ \text{sec}^{-1}$ (at shear stress <0.2 dyn cm^{-2}). This was observed at very low Reynolds numbers, and in a very dilute suspension.

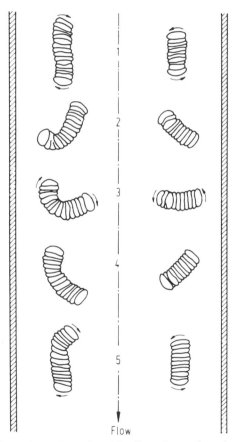

Figure 3.4:4 Rotation of an 11- and a 16-cell rouleau of erythrocytes in Poiseuille flow; $\dot{\gamma} < 10\ \text{sec}^{-1}$. Bending commenced at position 2 where the fluid stresses are compressive to the rouleaux. The longer particle did not straighten out in the succeeding quadrant, in which the stresses in the rouleaux are tensile. From Goldsmith and Marlow (1972), by permission.

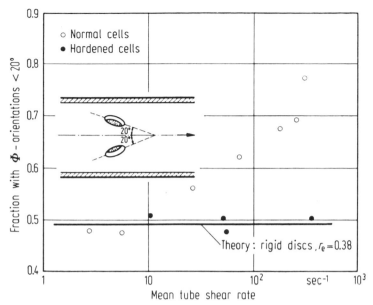

Figure 3.4:5 The increasing fraction of normal red cells (open circles) found in orientation within $\pm 20°$ with respect to the direction of flow as the shear rate is increased in a tube flow. The orientation distribution of hardened cells (closed circles) is independent of the flow speed and radial location in the tube, and agrees with that calculated for rigid disks having an equivalent thickness-to-diameter ratio of $r_e = 0.38$. From Goldsmith (1971), by permission.

The tendency of the deformable red cells to be aligned with the streamlines of flow at higher shear rates is illustrated in Fig. 3.4:5 which refers to observations made in very dilute suspensions in which particle interactions were negligible. It is seen that as the shear rate $\dot{\gamma}$ was increased, more and more cells were found in orientations in which their major axes were aligned with the flow. By contrast, rigid but still biconcave erythrocytes, produced by hardening with gluturaldehyde, continued to show orientations independent of shear rate.

The features shown in Figs. 3.4:4 and 3.4:5 are for isolated red cells or aggregates. Normal blood contains a high concentration of red blood cells,— with a hematocrit ratio (defined as cell volume/blood volume) of about 0.45 in large vessels, and 0.25 in small arterioles or venules. At such a high concentration, the cells crowd each other: no one cell can act alone. Goldsmith's (1972b) observations then show that

(1) The velocity profile in the tube is no longer parabolic as in Poiseuille flow;
(2) deformation of the erythrocytes in blood occurs to a degree that is not attributable to shear alone;
(3) the particle paths exhibit erratic displacements in a direction normal to the flow.

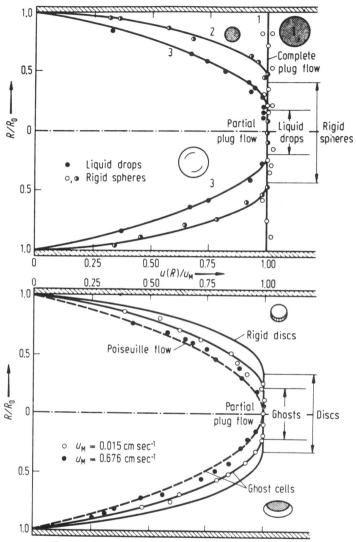

Figure 3.4:6 Comparison of dimensionless velocity profiles of tube flow of a fluid containing particles. b is the radius of the particles. R_0 is the inner radius of the tube. The upper panel shows the flow with 32% suspensions of rigid spheres ($b/R_0 = 0.112$ and 0.056 in curves 1 and 2, respectively) and liquid drops ($b/R_0 = 0.078$, curve 3). The lower panel shows the flow with 32% suspensions of rigid disks ($b/R_0 = 0.078$) and ghost cells ($b/R_0 = 0.105$). The lines drawn are the best fit of the experimental points; the dashed line is calculated from Eq. (6) of Sec. 3.3 in the form $U(R)/U(0) = 1 - R^2/R_0^2$. Note that by complete plug flow in curve 1 (upper part), we do not imply slip of fluid at the wall. Close to the boundary there must be a steep velocity gradient in the suspending fluid. From Goldsmith (1972b), by permission.

The velocity profiles measured in model systems with ghost cells, as well as rigid disks, liquid droplets, and rigid spheres, are shown in Fig. 3.4:6, and can be explained as follows. In a flow of a homogeneous fluid, the profile is parabolic (Poiseuille flow, shown by a dotted line in the lower panel). If there is a *single* isolated *rigid* sphere in the flow, and if the radius of the sphere b is much smaller than the radius of the tube R_0, and if the Reynolds number of the flow is smaller than 1, then the spherical particle will translate along a path parallel to the tube axis with a velocity which, except when the particle is very close to the wall, is equal to the velocity of the undisturbed fluid at the same radial distance. As the concentration of rigid spheres is increased, the particle velocity profile remains parabolic, provided that $b/R_0 < 0.04$ and the volume concentration $c < 0.2$. As the concentration c and the particle-tube-radius ratio b/R_0 are increased further, the velocity profile becomes blunted in the center of the tube. Flow in this central region may be called a *partial plug flow*. For a suspension of rigid spheres the velocity profile is independent of flow rate and suspending phase viscosity, and the pressure drop per unit length of the tube is directly proportional to the volume flow rate $\dot{Q}$. In contrast, flow of a concentrated, monodispersed oil-in-oil emulsion has a velocity profile which is appreciably less blunted than the rigid particle case, at the same values of c and b/R_0, and moreover, this profile is *dependent* on the flow rate and the suspending phase viscosity. In a given system, the lower the flow rate, and the greater the suspending phase viscosity, the greater was the degree of blunting.

Similar non-Newtonian behavior was exhibited by ghost cell plasma suspensions at concentrations from 20% to 70%. Figure 3.4:6 (lower panel) shows the results obtained at a concentration $c = 0.32$, a mean velocity of flow = 1.04 tube diameters per sec; ($U_M = 0.015$ cm sec^{-1}); and a tube radius of 36 μm. Upon increasing the mean velocity of flow to 47 tube diameters per sec ($U_M = 0.676$ cm sec^{-1}), the velocity distribution in the ghost cell suspension became almost parabolic. These features are consistent with the analytical results of Sec. 3.3.

The influence of particle crowding on cell deformation can be expected; but it becomes quite dramatic if we think of the meaning of the curves shown in Fig. 3.4:3. As seen in that figure, the relative viscosity of human blood is considerably lower than that of concentrated oil-in-water emulsions. Are the red blood cells more deformable than the liquid droplets? Are the red cells, in a crowded situation, able to squeeze and move past each other more readily than colliding liquid droplets? The answer is "yes." Direct observation has shown large distortion of red cells and rouleaux at very low shear rates $\dot{\gamma} < 5$ sec^{-1} or shear stress < 0.07 dyn cm^{-2}. An example is shown in Fig. 3.4:7, at a cell concentration (hematocrit) $c = 0.5$. The explanation is believed to lie in the biconcave shape of the red cells. In Chapter 4, Secs. 4.3 and 4.4, it is shown that because of the biconcave shape, the internal pressure of an isolated red cell must be the same as the external pressure if the bending rigidity of the cell membrane is negligibly small. Therefore, the cell membrane

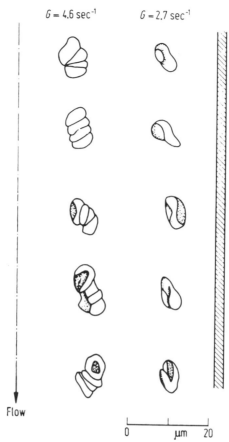

$G = 4.6\ \text{sec}^{-1}$ $G = 2.7\ \text{sec}^{-1}$

Flow

$\underset{0}{\vdash}\underset{\mu m}{\perp}\underset{20}{\dashv}$

Figure 3.4:7 Tracings from a cine film showing the continuous and irregular deformation of a single erythrocyte and a 4-cell rouleau in a ghost cell suspension, $H = 0.5$. The cell is shown at intervals of 0.4, 0.6, 2.0, and 3.2 s, respectively, in which time the isolated corpuscle would execute just over half a rotation. From Goldsmith (1972a), by permission.

is unstressed in the normal biconcave configuration. Futher, large deformation of the cell can take place without stretching the cell membrane, and hence needs little energy. On the other hand, a liquid droplet in emulsion is maintained by surface tension. In a static condition, a droplet must be spherical if the surface tension is uniform. In a shear flow, the droplets become ellipsoidal in shape; whereas in the crowded situation of a concentrated emulsion, large distortion of shape from that of a regular ellipsoid was noted. Such distortion from the spherical shape increases the surface area of the droplet, and hence the surface energy (surface tension is equal to surface energy per unit area). Thus it becomes plausible that the red cell, by packaging into the biconcave shape, is more deformable than a liquid droplet without a cell membrane. It also follows from this discussion that a detailed analysis of the rheology of blood or emulsion at high concentration needs data on the viscosities of the liquids and hemoglobin, as well as on the surface tension and cell membrane elasticity.

The third feature of concentrated particulate flow named above; the erratic sidewise movement of particles, has been recorded extensively by Goldsmith. Such erratic motion is expected because of frequent encounters of a particle with neighboring particles. The particle path therefore, must show features of a random walk. These observations offer a qualitative explanation of the way blood flows the way it does.

3.5 Fluid–Mechanical Interaction of Red Blood Cells with a Solid Wall

It has been pointed out by Thoma (1927) that in a tube flow there seems to be a tendency for the red cells to move toward the axis of the tube, leaving a marginal zone of plasma, whose width increases with increase in the shear rate. There is a layer close to the wall of a vessel that is relatively deficient of suspended material. In dilute suspensions, this "wall effect" has been measured by Goldsmith in the creeping flow regime (Reynolds number $\ll 1$). In emulsions the deformation of a liquid drop results in its migration across the streamlines away from the tube wall. Such lateral movement is not observed with rigid spherical particles; thus the deformability of the particle appears to be the reason for lateral migration.

A similar difference in flow behavior was found between normal red blood cells (RBC) and glutaraldehyde-hardened red cells (HBC), as illustrated in the upper panels of Fig. 3.5:1, which show the histograms of the number–concentration distributions of cells as a function of radial distance, at a section 1 cm downstream from the entry in a tube of 83 μm diameter at a Reynolds number about 0.03. Even more striking is the lateral migration in dextran solutions at Reynolds numbers $R_n = 3.7 \times 10^{-3}$ and 9.1×10^{-4}, shown in the lower panels. It is seen that very few cells are present in the outer half of the tube. Dextran solutions have a higher viscosity than Ringer or Ringer-plasma solutions. In dextran solutions the red cells are deformed into ellipsoidlike shapes. Similar observations of flow containing rouleaux of red cells show that rouleaux migrate to the tube axis faster than individual red cells.

Inward migration of deformable drops, fibers, and red cells away from the wall in both steady and oscillatory flows has been observed also at higher Reynolds numbers ($R_n > 1$), when the effects of fluid inertia are significant. At $R_n > 1$ nondeformable particles also exhibit lateral migration in dilute suspensions.

When the cell concentration is high, the crowding effect acts against migration away from the wall into the crowded center. Measurements made by Phibbs (1969) in quick-frozen rabbit femoral arteries, by Bugliarello and

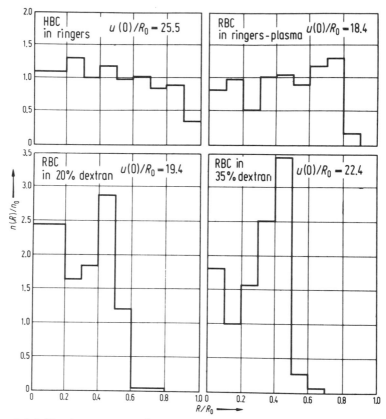

Figure 3.5:1 Number of cells/cm³ suspension, $n(R)$, divided by the syringe reservoir concentration, n_0, at intervals of 0.1 R_0, 1 cm downstream from the reservoir; $R_0 = 41.5$ μm, $c = 2 \times 10^{-3}$. The mean tube shear rates $= u(0)/R_0$ were approximately the same in each suspension. If the number distribution were uniform, then $n(R)/n_0 = 1$ at all R/R_0. From Goldsmith (1972b), by permission.

Sevilla (1971) on cine films of blood flow in glass tube, and by Blackshear et al. (1971) on ghost cell concentration, make it appear unlikely that at hematocrits of 40%–45% and normal flow rates, the plasma-rich zone can be much larger than 4 μm in vessels whose diameters exceed 100 μm.

This plasma-rich (or cell-rare) zone next to the solid wall, although very thin, has important effect on blood rheology. In the first place, measurement of blood viscosity by any instrument which has a solid wall must be affected by the wall layer. The change in cell concentration in the wall layer makes the blood viscosity data somewhat uncertain. Thus we are forced to speak of "apparent" viscosity (see Chapter 5, Sec. 5.1), rather than simply of the viscosity. It makes it necessary to specify how "smooth" or "serrated" the surface of viscometers must be. In the second place, the smaller the blood vessel is, the greater will be the proportion of area of the vessel occupied by the wall layer, and greater would be its influence on the flow. The low hematocrit in the wall layer lowers the average hematocrit in small blood vessels. And when a small blood vessel branches off from another vessel, it draws more fluid from the wall layer where the hematocrit is low. The result is a lower average hematocrit in the smaller blood vessel; and hence a lower apparent viscosity. This will be discussed in Chapter 5.

3.6 Thrombus Formation and Dissolution

Blood clots are formed on an injured inner wall of blood vessels and on contact with the surfaces of medical devices. When a circulating blood comes into contact with such a surface, the platelets in the blood adhere to the surface, release a number of chemicals, attract more platelets to form a larger aggregation, generate thrombin, and form fibrin, resulting in a thrombus. In time

TABLE 3.6:1 Properties of Human Clotting Factors*

Clotting factor	Synonym	Molecular weight (number of chains)	Normal plasma concentration (μg/ml)
Intrinsic system			
Factor XII	Hageman or contact factor	80 000 (1)	29
Prekallikrein	Fletcher factor	80 000 (1)	50
High-molecular-weight Kininogen	Fitzgerald factor	120 000 (1)	70
Factor XI	Plasma thrombo-plastin antecedent	160 000 (2 dimer)	4
Factor IX	Christmas factor	57 000 (1)	4
von Willebrand factor	vWF	1–2 000 000 (series of G-10 subunits)	7
Factor VIII:C	Antihemophilic factor	200 000–350	0.1
Extrinsic system			
Factor VII	Proconvertin	55 000 (1)	1
Tissue factor	Tissue thromboplastin	45 000 (1)	1
Common pathway			
Factor X	Stuart–Prower factor	59 000 (2)	5
Factor V	Proaccelerin	330 000 (1)	5–12
Prothrombin	Factor II	70 000 (1)	100
Fibrinogen	Factor I	340 000 (6: $A\alpha_2, B\beta_2, \gamma_2$)	2500 (250 mg/dl)
Factor XIII	Fibrin stabilizing factor	300 000 (4: a_2, b_2)	10

*Factor III is tissue thromboplastin. Factor IV is calcium. There is no factor VI.

plasmin is generated in the thrombus, and fibrinolysis begins, ending in the dissolution of the thrombus.

The blood clotting process is a cascade of chemical processes with many participants. The principal chemicals are listed in Table 3.6:1. In the table, those activating and coagulation factors reside in the blood plasma are called *intrinsic*, those residing in the cells (not in the plasma) are called *extrinsic*. A very brief sketch is given below. Details should be obtained from hematological, pharmacological, and medical books. An National Institute of Health (NIH) report edited by McIntire (1985) is very helpful.

3.6.1 Thrombogenesis

Figure 3.6:1 shows the major chemicals involved at the first stages of platelet adhesion and aggregation. The injured endothelium exposes collagen of the basement membrane which interacts with the glycoprotein on the platelet membrane and the von Willebrand factor which is synthesized by endothelial cells and is present in the plasma, and on the platelets. The adherent plate-

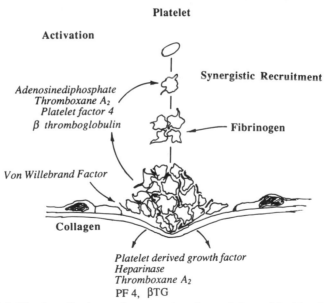

Figure 3.6:1 Platelet adhesion and aggregation after an injury of the blood vessel wall. Aggregation of platelets requires a rapid mobilization of fibrinogen receptors on the membranes of the platelets, and a calcium-dependent interplatelet bridging by fibrinogen. Various factors are shown. Figure was redrawn after an example in *Guidelines for Blood–Material Interactions*, which is a Report of the National Heart, Lung, and Blood Institute (NHLBI) Working Group of the U.S. Department of Health and Human Services, Public Health Service NIH Publ. No. 85–2185, 1985.

lets then release adenosine diphosphate (ADP), thromboxane A_2, fibrinogen, factor V, platelet factor 4, beta thromboglobulin, and a platelet-derived growth factor.

The released ADP and thromboxane A_2 act synergistically to recruit circulating platelets and enlarge the aggregate. Essential to this step is the mobilization of a platelet-membrane fibrinogen receptor and the calcium-dependent interplatelet bridging by fibrinogen.

Thrombus growth and stabilization depends on the formation of thrombin. Thrombin is generated by means of the intrinsic coagulation pathway through contact factor activation by subendothelial collagen, or when tissue thromboplastin from disrupted endothelial cells activates the extrinsic coagulation cascade. Once formed, thrombin stimulates the synthesis of thromboxane A_2 and the release of ADP, thus promoting platelet aggregation. Subsequently, thrombin generates fibrin from fibrinogen. Fibrin stabilizes the growing platelet mass to form a viscoelastic clot.

The contact factor (Hageman, or factor XII) activation begins by the adsorption of the contact factor to the negatively charged collagen surface. Factors XI, prekallikrein, and kininogen are also involved. This step is calcium independent. The adsorption of factor XII to collagen changes the molecular configuration of factor XII and exposes its hydrophobic active sites previously unavailable to the external environment, and the intrinsic system cascade begins. The fibrinolytic system is also activated by factor XII via plasminogen proactivator. Factors XI, V, and the von Willebrand factor may also be adsorbed to an artificial surface.

The activation of factor IX by XIa and the activation of factor X are calcium dependent. Factors IX, X, VII, and prothrombin are vitamin K dependent.

The *extrinsic system* is initiated by the activation of factor VII when it interacts with an intracellular tissue factor, or leukocytes. Tissue factor is present in large amounts in the brain and lung, and found in the intima of large blood vessels.

The *common pathway* begins as factor X is activated by factor VIIa or IXa.

The concentration of prothrombin in plasma is sufficient to allow a few molecules of activated initiator to generate a large burst of thrombin activity, which induces platelet aggregation, and converts fibrinogen into fibrin. Fibrin monomers polymerize nonenzymatically to gelate the fluid.

3.6.2 Thrombus Dissolution

Within the thrombus, plasmin digests fibrin to produce progressively smaller fragments to eventually dissolve the clot. The blood contains plasminogen (molecular weight 90 000, normally at the 120 μg/ml level) which is enmeshed in the thrombus. It can be activated intrinsically by the contact factor XII, etc.; or extrinsically with an activator originating from the blood vessel wall; or by drugs such as streptokinase or urokinase. Activation releases plasmin. Plasmin is an active serine protease, which hydrolyzes fibrin polymers.

TABLE 3.6:2 Antithrombotic Agents

Agent	Action
Fibrinolytic	
Streptokinase, urokinase	Plasminogen activators
Anticoagulant	
Heparin	Enhances inhibition of proteases by thrombin III
Warfarin	Vitamin K antagonist
Antiplatelet	
Aspirin	Decreases platelet aggregation and release
Sulfinpyrazone	Not established
Dipyridamole	Decreases platelet adhesion
Ticlopidine	Blocks ADP induced platelet interaction with fibrinogen and vWF
PGI_2	Activates platelet adenyl cyclase

In practice, a thrombus can be continuously formed and dissolved; so its composition is a result of chemical dynamics. Some well-known anti-thrombotic agents are listed in Table 3.6:2.

Thrombosis can be a threat to life. But it can stop internal bleeding if there were a break in the blood vessel, or external bleeding when there is an injury. Coagulation of blood seals the wound and saves lives. Contraction of the damaged blood vessels is another mechanism of life saving.

Blood coagulation is the result of a cascading activation of factors. An almost total absence of any one of these factors will make the coagulation process extremely slow. For example, if one takes the calcium away, blood will not clot. If one draws blood with a collecting vessel treated with oxalate or citrate, the blood will flow freely. Rheological studies of blood clotting are often made either for clinical reasons or for pharmacological development. A summary of some more popular instruments is given in Scott-Blair (1974). The instrument 'thrombelastograph" of Hartert (1962, 1975) needs blood sample of only 0.3 ml.

3.7 Medical Applications of Blood Rheology

The most obvious use of blood rheology in clinical medicine is to identify diseases with any change in blood viscosity. Data collected for this purpose have been presented by Dintenfass in two books (1971, 1976). For our purpose it suffices to cite a few examples. Figure 3.7:1 shows a comparison of

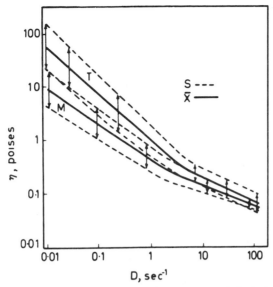

Figure 3.7:1 Arithmetic means (full lines) and one standard deviation (broken lines) of viscosities in normal men (M) and in patients with various thrombotic diseases (T). Viscosity, η, in poises, is plotted against the rate of shear, D, in sec^{-1}, on a log–log scale, where $\bar{x}$ is the arithmetic mean, and s is the standard deviation. Experimental data show log-normal distribution. From Dintenfass (1971), p. 11, by permission.

Dintenfass' data on the blood viscosity of normal healthy persons and patients with various thrombotic diseases. The diseased persons have higher viscosity. As we have seen in Sec. 3.4, elevation of blood viscosity at low shear rates indicates aggregation, whereas that at high shear rates suggests loss of deformability of the red blood cells. The viscosity change suggests some disease related changes in the blood.

Figure 3.7:2 suggests another use of lowered blood viscosity. It shows Langsjoen's (1973) result that a reduction of blood viscosity consequent to the infusion of dextran 40 solution in cases of acute myocardial infarction led to a significant improvement in both immediate and long-term survival. Dextran 40 (molecular weight about 40 000) solution dilutes the blood. The physiological effect of hemodilution is not simple, but the changed rheology must be a principal factor. For hemodilution, see Messmer and Schmid-Schoenbein (1972).

The fluid added to the bloodstream to make up the lost volume of blood due to hemorhage when a person suffers a wound is called *plasma expander*. Dextran solution is a good expander. Any blood substitute for long term use must have the right rheological property.

Another important rheological factor in clinical use is the coagulation characteristics of the blood. An obvious example is the hemophilic patient's difficulty with blood coagulation. On the other hand, hypertension (high

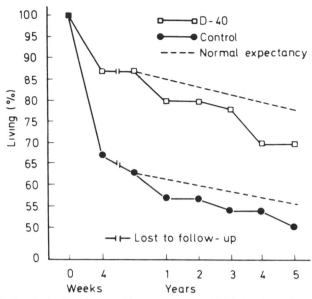

Figure 3.7:2 Survival of patients with acute myocardial infarction after conservative treatment (73 patients) and after treatment by infusion of dextran 40 (65 patients). Survival is recorded as the percentage of the original group. Note that there were more surviving patients in the dextran-treated group after 5 years than in the controls (conservative treatment) after 4 weeks. This gives cogent support to the premise that blood viscoscity is an important determinant of myocardial work. Reproduced from Langsjoen (1973), by permission.

blood pressure) and arteriosclerosis seem to correlate with an increase in the elasticity of the thrombi.

A third parameter in clinical use is the *erythrocyte sedimentation rate*. In anticoagulated blood, red cells may still clump together and cause a sedimentation. Fahraeus (1918) first studied this effect seriously. He noticed that blood from pregnant women sedimented faster than that of nonpregnant women, and believed that he might have found a convenient test for pregnancy. However, he soon afterwards found the same more rapid sedimentation in some male patients! It was then established that an increased erythrocyte sedimentation rate served as a good indication for both male and female patients that all was not well, and that quite a variety of conditions, many of them serious, increased the sedimentation rate. This is discussed in considerable detail in Dintenfass (1976).

It is quite clear that blood rheology as discussed in this chapter, referring to flow in vessels much larger than the diameter of red blood cells, reflects the interaction of red cells in bulk of whole blood. The "hyperviscosity" of blood in some disease states reflects the changes in hematocrit, plasma, and red cell deformability. On the other hand, the critical sites of interaction of red blood cells with blood vessels are in the capillary blood vessels. In the microcirculation bed, the deformability of the red cells is subjected to a really severe test. Hence it is expected that the pathological aspects of hyperviscosity will be seen in microcirculation. To pursue this matter, we shall consider the mechanical properties of red cells in the following chapter, and the flow properties of blood in microvessels in Chapter 5.

Problems

3.1 Show that, for an incompressible fluid, $I_1 = 0$, where I_1 is given in Eq. (3) of Sec. 3.2.

3.2 Thus far we have treated blood as a viscous fluid. This is undoubtedly permissible if blood flows in a steady-state condition. But since blood is a suspension of blood cells in plasma, and the cells are capable of interacting with each other, it is expected that blood will exhibit viscoelastic properties when conditions permit, just like many other polymer solutions. Speculate, on theoretical grounds, on what may be expected in the following situations:

(a) A volume of blood sits in a condition of stationary equilibrium in a Couette viscometer or in a circular cylindrical tube (Poiseuille viscometer). A harmonic oscillation of very small amplitude is imposed on the blood in the viscometer. What would be the relationship between force and velocity? How would one express the relationship through the method of a complex variable? What is the phase relationship? What do the real and imaginary parts of the complex modulus mean?

(b) Let the blood in the viscometer flow in a steady state and then superimpose a small harmonic oscillation on the flow in the viscometer. How would the complex modulus vary with the steady-state shear strain rate?

(c) Instead of harmonic oscillations of small amplitude as considered above, impose a small step function in velocity (Couette) or pressure gradient (Poiseuille), and discuss the expected response as a function of time.

(d) If the step function considered in (c) is finite in amplitude, could there be nonlinear effects which depend on the amplitude? The change in viscoelastic properties of a material with respect to time after a finite disturbance is called a *thixotropic* change. Thixotropy of blood may be described by a complex modulus of viscosity as a function of time after the initiation of disturbance (e.g., a step function); the modulus being obtained by a superposed small harmonic perturbation. Speculate on the possible thixotropic properties of blood.

Experimental results on the features named above as well as a mechanical and mathematical model of the viscoelasticity of blood expressed in terms of springs and dashpots are discussed by G. B. Thurston (1979).

3.3 *Entry flow of blood from a large reservoir into a pipe.* So far we have analyzed the condition of flow in a pipe in a region far away from the entry section. At the entry section ($x = 0$), where the pipe is connected to a large reservoir, the velocity profile is uniform, as shown in Fig. 3.3:1. As the distance from the entry section increases, the profile changes gradually to the steady-state profile of a parabola (if the fluid is Newtonian) or the flat-topped parabola of Fig. 3.3:4 (if the fluid is blood). Consider blood. We can trace the change as sketched in Fig. P3.3. At

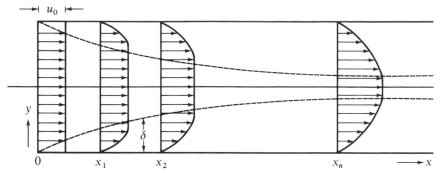

Figure P3.3 Entry flow of blood into a pipe. Dotted lines show the boundary layer of yielding.

$x = x_1$ the velocity has to change from u_0 at the center to 0 on the wall, where the no-slip condition applies. This creates a high shear strain rate at the wall, which has to decrease to zero at the center. At a certain distance δ from the wall is a point where the shear stress is equal to the yield stress of the blood. Beyond $y = \delta$, the velocity profile remains flat. As x increases, δ increases. At $x = x_n$ the velocity profile tends to the steady state, as shown in Fig. 3.3:4.

The surface $y = \delta(x)$ is the *yielding surface*, which divides the region (at the center) in which the blood is solid-like, from the region (next to the wall) in which the blood is a flowing fluid. It is a surface on which the shear stress is exactly equal to the yield stress. The rate at which $\delta(x)$ grows with x depends on the Reynolds number of flow N_R. Give a qualitative and mathematical analysis of $\delta(x)$ as a function of N_R.

3.4 With the constitutive equations of blood presented in Sec. 3.2, derive the equation of motion of blood in a form that is a generalization of the Navier–Stokes equation. Discuss the range of validity of the equations. Discuss any simplification that results if the shear strain rate is sufficiently high.

3.5 *Boundary Conditions.* The general field equations of continuity and motion are to be solved with boundary conditions which must be posed in such a way that the field equations have meaningful solutions, and must lead to solutions in agreement with experimental results. The condition at the interface between a fluid and a solid has raised the question of whether to allow relative slip between fluid and solid or not. Historically, for water, the question was resolved in favor of *no-slip* on the basis of precise experimental results of Poiseuille and Hagen on flow in circular cylindrical tubes. For Newtonian fluids the *no-slip* condition between fluid and solid has been established and has found no exception in the past 150 years.

Blood is a mixture of cells and a fluid. Consider the conditions at a blood–solid interface theoretically. What should the boundary conditions be?

3.6 Reduce the equations you obtained in Problem 3.4 to a dimensionless form. Introduce a characteristic length L, a characteristic velocity V, a characteristic viscosity η which is one of the constants in Eq. (12b) of Sec. 3.2, and dimensionless coordinates $x_i' = x_i/L$, velocities $u_i' = u_i/V$, and parameters

$$p' = \frac{p}{\rho V^2}, \quad t' = \frac{Vt}{L}, \quad N_R = \frac{VL\rho}{\eta},$$

where N_R is the Reynolds number.

3.7 If laboratory model testing is to be done for blood flow, on what basis would you design the models. A scaled model may be desired to facilitate observations on velocity, pressure, forces, etc. It may be convenient to use water or a polymer fluid as the working fluid in place of blood plasma. Discuss the principles of kinematic and dynamic similarities in model design.

3.8 Assume that Casson's equation [Eq. (11) of Sec. 3.2] describes the viscosity of blood. Both τ_y and η are functions of the hematocrit. Experimental data (see Fig. 3.1:5) on τ_y may be presented by an empirical formula,

$$\tau_y = (a_1 + a_2 H)^3.$$

For normal blood, we may take $a_1 = 0$, $a_2 = 0.625$, when τ_y is in dyn cm^{-2} and H, the hematocrit, is a fraction. Experimental data on η can be expressed in many ways. Let us assume that for large blood vessels (Cokelet et al., 1963)

$$\eta = \eta_0 \frac{1}{(1 - H)^{2.5}},$$

where η_0 is the viscosity of the plasma.

For the capillary blood vessels Casson's equation does not hold; but we may assume

$$\mu = \eta_0 \frac{1}{1 - CH}.$$

We may take an average value $C = 1.16$ for pulmonary capillaries (Yen-Fung, 1973) and H in capillaries to be 0.45 times that of the systemic hematocrit in large arteries. The peripheral resistance from capillary blood vessels may be assumed to be a constant fraction of the total peripheral resistance. For the lung this fraction may be as high as $\frac{1}{3}$. For other parts of the body, this fraction is perhaps 15%.

With these pieces of information, let us consider the question "What is the best hematocrit that minimizes the work of the heart while the total amount of oxygen delivered to the tissues of the body remains constant?" Such a question is important in surgery or hemodilution, in deciding the proper amount of plasma expander to use.

To answer this question we may consider blood flow as a flow in large and small vessels in series. The pressure drop in each segment is equal to the flow times the resistance which is proportional to blood viscosity. The total pressure difference the heart has to create is the sum of the pressure drop of the segments. The rate at which work is done by the heart is equal to the cardiac output (flow) times the pressure difference. The total amount of oxygen delivered to the tissues is proportional to the product of cardiac output and the hematocrit if the lung functions normally.

What would be the optimum hematocrit if the blood pressure is fixed and oxygen delivery is maximized?

3.9 The analysis of tube flow given in Sec. 3.3 ignores the radial migration of red cells away from the wall. As discussed in Sec. 3.5, for blood flowing in a tube, the immediate neighborhood of the solid wall is cell free. For human blood with a red cell concentration above 0.4 and tube diameter > 100 µm, the cell-free layer is a about 4 μm thick. For a dilute suspension, e.g., at red cell concentration 0.03, the thickness of the cell-free layer is larger (see Fig. 3.5:1). Use the information on the dependence of the constants τ_y and η of the Casson equation on the hematocrit as given in Problem 3.8 to revise the velocity profile and flow rate given in Eqs. (19) and (21) of Sec. 3.3. For this purpose assume the hematocrit to be constant away from the cell-free layer next to the wall. This assumption is quote good at normal hematocrit (0.2–0.5), but is rather poor for very dilute suspensions.

3.10 Consider the effect of non-Newtonian features of blood on the Couette flow of blood between concentric circular cylinders. Assume the outer cylinder to rotate, the inner cylinder stationary, Casson's equation to apply, cell-free layers next to solid walls with a thickness of a few (say, 4) µm, and absence of end effects.

3.11 Analyze blood flow in a cone-plate or a cone–cone viscometer analogous to Problem 3.10.

3.12 Blood exposed to air will form a surface layer at the interface, which has a specific surface viscosity and surface elasticity, depending on the plasma constitution. In viscometry using a Couette-flow or cone-plate arrangement, one must avoid reading the torque due to the surface layer. A *guard ring* was invented to shield the stationary cylinder from the torque transmitted through the surface layer. The ring can be held in place by an arm that is fixed to the laboratory floor, and unattached to the instrument. Propose a design for such a guard ring.

3.13 Consider the following thought experiment. Take normal red blood cells and suspend them in an isotonic dextran solution of viscosity η_0. η_0 can be varied by varying the concentration of dextran. Measure the viscosity at a sufficiently high shear rate $\dot{\gamma}$ so that the influence of cell aggregates is insignificant. Let the viscosity be $\mu = \eta_r \eta_0$, where η_r is the "relative viscosity." How would η_r vary with η_0? η_r reflects the effect of cell deformation. Would the red cells be more readily deformed in a more viscous suspending medium? Or less so? Predict curves of η_r vs. $\dot{\gamma}$ for various values of η_0.

  Note. The effect of cell deformation on the apparent viscosity depends on the flow condition. Consider Couette flow which is a case investigated by Chien (1972) experimentally. In this case η_r decreases monotonically when the shear strain rate increases.

3.14 Discuss the pros and cons for several types of viscometers from the point of view of practical laboratory applications (convenience of operation, accuracy of data, amount of fluid samples needed, data reduction procedures, and procedures required for correction of errors).

3.15 Consider the flow of a Newtonian viscous fluid in a straight tube of infinite length and a rectangular cross section of width a and thickness b. Derive the equation that governs the velocity distribution in the tube. Express the volume flow rate as a function of the pressure gradient and the dimensions of the cross section. Obtain results for the specially simple case in which $b \gg a$. Then repeat the analysis if the fluid obeys Casson's equation, at least for the case in which $b \gg a$.

3.16 A fluid is said to obey a *power law* viscosity if the shear stress τ in a Couette flow is related to the shear rate $\dot{\gamma}$ by the equation

$$\dot{\gamma} = \frac{1}{k}\tau^n \qquad (n > 0). \tag{1}$$

This relationship is illustrated in Fig. P3.16.

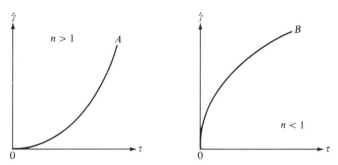

Figure P3.16 Power-law viscosity. Curve A, $n > 1$; curve B, $n < 1$.

For a steady flow of an incompressible fluid obeying the power law (1) in a circular cylindrical tube of radius R, show that the axial velocity is

$$u = \frac{1}{2^n(n+1)k}\left(\frac{\Delta p}{L}\right)^n (R^{n+1} - r^{n+1}), \tag{2}$$

where Δp is the difference of pressures at the ends of a tube of length L. The rate of flow (volume) is

$$Q = \frac{\pi R^{n+3}}{2^n(n+3)k}\left(\frac{\Delta p}{L}\right)^n. \tag{3}$$

Let the average velocity be $U = Q/\pi R^2$, then

$$\frac{u}{U} = \frac{n+3}{n+1}\left[1 - \left(\frac{r}{R}\right)^{n+1}\right]. \tag{4}$$

3.17 A material is said to be a *Bingham plastic* if the shear stress τ is related to the shear rate $\dot{\gamma}$ by the equation

$$\dot{\gamma} = \frac{1}{\eta_B}(\tau - f_B) \quad \text{when} \quad \tau > f_B; \quad \text{and} \quad \dot{\gamma} = 0 \quad \text{when} \quad \tau < f_B.$$

The stress f_B is called the *yield stress*. The constant η_B is called *Bingham viscosity*. The relation is illustrated in Fig. P3.17.

The steady flow of a Bingham plastic in a circular cylindrical tube of length L subjected to a pressure difference Δp can be analyzed in a manner similar to that presented in Sec. 3.3. Show that:

(a) If $\dfrac{\Delta p}{L} > \dfrac{2f_B}{R}$ and $r_B = \dfrac{2f_B L}{\Delta p}$, then

$$u = \frac{\Delta p}{4\eta_B L}(R - r_B)^2 \qquad \text{when} \quad r < r_B,$$

$$= \frac{\Delta p}{4\eta_B L}[R^2 - r^2 - 2r_B(R - r)] \qquad \text{when} \quad r > r_B.$$

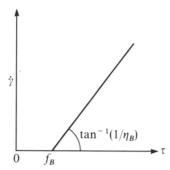

Figure P3.17 Flow curve of a Bingham plastic.

(b) If $\dfrac{\Delta p}{L} < \dfrac{2f_B}{R}$, then there is no flow: $u = 0$.

The flow rate is zero if $\Delta p < p_B \equiv 2Lf_B/R$, and is

$$Q = \frac{\pi R^4}{8\eta_B L}\left[\Delta p - \frac{4}{3}p_B + \frac{1}{3}\frac{p_B^4}{(\Delta p)^3}\right] \qquad \text{if} \quad \Delta p > p_B.$$

3.18 A plastic material obeying the following relation between the shear stress τ and shear rate $\dot\gamma$ is called a plastic of *Herschel–Bulkley* type:

$$\dot\gamma = \frac{1}{k}(\tau - f_H)^n \quad \text{if} \quad \tau > f_H, \qquad \text{and} \qquad \dot\gamma = 0 \quad \text{if} \quad \tau < f_H,$$

where k, f_H, and $n > 0$ are constants. The Bingham plastic of Problem 3.17 is a special case in which $n = 1$, whereas the power law of Problem 3.16 is a special case when $f_H = 0$. For the general case, find the velocity distribution in a steady laminar flow of an incompressible Herschel–Bulkley fluid in a circular cylindrical tube.

3.19 Discuss the turbulent flow of water in a circular cylindrical tube. Consider the question of the relationship of mean velocity of flow with pressure gradient. Does it remain linear? How is the flow rate related to pressure gradient? Discuss resistance to flow in a turbulent regime.

   Note. Many books deal with this subject. See references listed in Chapter 11 of Fung, *A First Course in Continuum Mechanics*.

3.20 Prove that blood viscosity as we know it cannot be represented by the Boltzmann equation

$$\tau(t) = \int_{-\infty}^{t} G(t - \tau)\frac{d\gamma(\tau)}{d\tau}\,d\tau,$$

where $\tau(t)$ is the shear stress, $\gamma(t)$ is the shear strain rate, and $G(t - \tau)$ is a continuous function of the variable $t - \tau$.

3.21 A Newtonian fluid flows steadily through a capillary tube whose radius R is

$$R(x) = a + \varepsilon \sin\frac{\pi x}{L}$$

where a, L, and ε are constants, with $\varepsilon \ll a$, $L \gg a$. The Reynolds number of flow is small, $\ll 1$. What is the pressure distribution in the tube as a function of x?

Hint. Under the assumption that $\varepsilon/a \ll 1$, $a/L \ll 1$, you may treat the Poiseuille equation as a differential equation for the pressure p:

$$\frac{dp}{dx} = -\frac{8\mu Q}{\pi R^4}$$

Q is constant, R is variable. Solve the equation above as a power series in ε/a.

3.22 Consider an elastic tube whose radius R varies with the pressure according to the following law:

$$R(x) = a(x)[1 + \alpha p(x)],$$

where p is the local transmural pressure, α is the flexibility coefficient, and $a(x)$ is the radius of the tube when $p = 0$. $a(x)$ is a function of x and is not a constant. For a constant blood viscosity μ and a one-dimensional flow in a tube of length L, what is the volumetric flow rate $\dot{Q}$ in the tube as a function of the pressures at the entry (p_a) and exit (p_v) of the tube? α is a constant. Use the hint given in Prob. 3.21.

3.23 Derive the following equations for the laminar steady flow of a non-Newtonian incompressible fluid in a circular cylindrical tube; under the assumption that the shear stress τ_w at the wall is a single-valued function of the local rate of deformation:

$$\tau_w = \frac{D(\Delta p)}{4L},$$

$$\dot{\gamma}_w = \frac{\tau_w}{4}\frac{d}{d\tau_w}\left(\frac{8\overline{V}}{D}\right) + \frac{3}{4}\left(\frac{8\overline{V}}{D}\right),$$

where $\dot{\gamma}_w$ = shear rate at the wall,
  D = tube diameter,
  ΔP = axial pressure difference for two points at a distance L apart,
  L = axial distance,
  $\overline{V}$ = bulk average velocity.
Ref. Markovitz, 1968. *Physics Today*, **21**, 23–30.

References

Barbee, J. H. and Cokelet, G. R. (1971) The Fahraeus effect. *Microvasc. Res.* **3**, 1–21.

Biggs, R. and MacFarlane, R. G. (1966) *Human Blood Coagulation and Its Disorders*, 3rd edition. Blackwell, Oxford.

Blackshear, P. L., Forstrom, R. J., Dorman, F. D., and Voss, G. O. (1971) Effect of flow on cells near walls. *Fed. Proc.* **30**, 1600–1609.

Brooks, D. E., Goodwin, J. W., and Seaman, G. V. F. (1970) Interactions among erythrocytes under shear. *J. Appl. Physiol.* **28**, 172–177.

Bugliarello, G. and Sevilla, J. (1971) Velocity distribution and other characteristics of steady and pulsatile blood flow in fine glass tubes. *Biorheology* **7**, 85–107.

Casson, M. (1959) A flow equation for pigment-oil suspensions of the printing ink type. In *Rheology of Disperse Systems*, C. C. Mills (ed.) Pergamon, Oxford, pp. 84–104.

Charm, S. E. and Kurland, G. S. (1974) *Blood Flow and Micro Circulation*. Wiley, New York.

Chien, S. (1970) Shear dependence of effective cell volume as a determinant of blood viscosity. *Science* **168**, 977–979.

Chien, S. (1972) Present state of blood rheology. In *Hemodilution. Theoretical Basis and Clinical Application*, K. Messmer and H. Schmid-Schoenbein (eds.) Int. Symp. Rottach-Ergern, 1971, S. Karger, Basel, pp. 1–45.

Chien, S., Usami, S., Taylor, M., Lundberg, J. L., and Gregersen, M. I. (1966) Effects of hematocrit and plasma proteins of human blood rheology at low shear rates. *J. Appl. Physiol.* **21**, 81–87,

Chien, S., Usami, S., and Dellenbeck, R. J. (1967) Blood viscosity: Influence of erythrocyte deformation. *Science* **157**, 827–831.

Chien, S., Usami, S., Dellenbeck, R. J., and Gregersen, M. (1970) Shear dependent deformation of erythrocytes in rheology of human blood. *Am. J. Physiol.* **219**, 136–142.

Chien, S., Luse, S. A., and Bryant, C. A. (1971) Hemolysis during filtration through micropores: A scanning electron microscopic and hemorheologic correlation. *Microvasc. Res.* **3**, 183–203.

Chien, S., Usami, S., and Skalak, R. (1984) Blood flow in small tubes. In E. M. Renkin, and C. C. Michel (eds.) *Handbook of Physiology, Sec. 2, The Cardiovascular System*, Vol. IV, Part 1. American Physiological Society, Bethesda, MD, pp. 217–249.

Cokelet, G. R. (1972) The rheology of human blood. In *Biomechanics: Its Foundation and Objectives*, Fung, Perrone, and Anliker (eds.) Prentice-Hall, Englewood Cliffs, NJ, pp. 63–103.

Cokelet, G. R., Merrill, E. W., Gilliland, E. R., Shin, H., Britten, A., and Wells, R. E. (1963) The rheology of human blood measurement near and at zero shear rate. *Trans. Soc. Rheol.* **7**, 303–317.

Dintenfass, L. (1971) *Blood Microrheology*. Butterworths, London.

Dintenfass, L. (1976) *Rheology of Blood in Diagnostic and Preventive Medicine*. Butterworths, London.

Fung, Y. C. (1965) *Foundations of Solid Mechanics*. Prentice-Hall, Englewood Cliffs, NJ.

Fung, Y. C. (1993) *A First Course in Continuum Mechanics*, 3rd edition. Prentice-Hall, Englewood Cliffs, NJ.

Goldsmith, H. L. (1971) Deformation of human red cells in tube flow. *Biorheology* **7**, 235–242.

Goldsmith, H. L. (1972a) The flow of model particles and blood cells and its relation to thrombogenesis. In *Progress in Hemostasis and Thrombosis*, Vol. 1, T. H. Spaet (ed.) Grunte & Stratton, New York, pp. 97–139.

Goldsmith, H. L. (1972b) The microrheology of human erythrocyte suspensions. In *Theoretical and Applied Mechanics Proc. 13th IUTAM Congress*, E. Becker and G. K. Mikhailov (eds.) Springer, New York.

Goldsmith, H. L. and Marlow, J. (1972) Flow behavior of erythrocytes. I. Rotation and deformation in dilute suspensions. *Proc. Roy. Soc. London B* **182**, 351–384.

Gregersen, M. I., Bryant, C. A., Hammerle, W. E., Usami, S., and Chien, S. (1967) Flow characteristics of human erythrocytes throughy polycarbonate sieves. *Science* **157**, 825–827.

Hartert, H. and Schaeder, J. A. (1962) The physical and biological constants of thrombelastography. *Biorheology* **1**, 31–40.

Hartert, H. (1975) Clotting layers in the rheo-simulator. *Biorheology* **12**, 249–252.

Haynes, R. H., (1962) The viscosity of erythrocyte suspensions. *Biophys. J.* **2**, 95–103.

Langsjoen, P. H. (1973) The value of reducing blood viscisity in acute myocardial infarction. No. **11**. Karger, Basel, pp. 180–184.

Larcan, A. and Stoltz, J. F. (1970) *Microcirculation et Hemorheologie.* Masson, Paris.

McIntire, L. V. (ed.) (1985) *Guidelines for Blood–Material Interactions.* Report of a National Heart, Lung, and Blood Institute Working Group. U. S. Dept. of HHS, PHS, and NIH. NIH Publication No. 85-2185.

McMillan, D. E. and Utterback, N. (1980) Maxwell fluid behavior of blood at low shear rate. *Biorheology* **17**, 343–354.

McMillan, D. E., Utterback, N. G., and Stocki, J. (1980) Low shear rate blood viscosity in diabetes. *Biorheology* **17**, 355–362.

Merrill, E. W., Cokelet, G. C., Britten, A., and Wells, R. E. (1963) Non-Newtonian rheology of human blood effect of fibrinogen deduced by "subtraction." *Circulation Res.* **13**, 48–55.

Merrill, E. W., Gilliland, E. R., Cokelet, G. R., Shin, H., Britten, A., and Wells, R. E. (1963) Rheology of human blood, near and at zero flow. *Biophys. J.* **3**, 199–213.

Merrill, W. E., Margetts, W. G., Cokelet, G. R., and Gilliland, E. W. (1965) The Casson equation and rheology of blood near zero shear. In *Symposium on Biorheology*, A Copley (ed.) Interscience Publishers, New York, pp. 135–143.

Merrill, E. W., Benis, A. M., Gilliland, E. R. Sherwood, R. K., and Salzman, E. W. (1965) Pressure-flow relations of human blood in hallow fibers at low flow rates. *J. Appl. Physiol.* **20**, 954–967.

Messmer, K. and Schmid-Schoenbein, H. (eds.) (1972) *Hemodilation: Theoretical Basis and Clinical Application.* Karger, Basel.

Oka, S. (1965) Theoretical considerations on the flow of blood through a capillary. In *Symposium on Biorheology*, A. L. Copley (ed.) Interscience, New York, pp. 89–102.

Oka, S. (1974) *Rheology–Biorheology.* Syokabo, Tokyo (in Japanese).

Phibbs, R. H. (1969) Orientation and distribution of erythrocytes in blood flowing through medium-sized arteries. In *Hemorheology*, A. C. Copley (ed.) Pergamon Press, New York, pp. 617–630.

Rand, R. P. and Burton, A. C. (1964) Mechanical properties of the red cell membrane. *Biophys. J.* **4**, 115–136.

Rand, P. W., Barker, N., and Lacombe, E. (1970) Effects of plasma viscosity and aggregation on whole blood viscosity. *Am. J. Physiol.* **218**, 681–688.

Rowlands, S., Groom, A. C., and Thomas, H. W. (1965) The difference in circulation times between erythrocyte and plasma in vivo. In *Symposium on Biorheology*, A. Copley, (cd.) Interscience Publishers, New York, pp. 371–379.

Scott-Blair, G. W. (1974) *An Introduction to Biorheology.* Elsevier, New York.

Thurston, G. B. (1972) Viscoelasticity of human blood. *Biophys. J.* **12**, 1205–1217.

Thurston, G. B. (1973) Frequency and shear rate dependence of viscoelasticity, of human blood. *Biorheology* **10**, 375–381; (1976) **13**, 191–199; (1978) **15**, 239–249; (1979) **16**, 149–162.

Thurston, G. B. (1976) The viscosity and viscoelasticity of blood in small diameter tubes. *Microvasc. Res.* **11**, 133–146.

Vadas, E. B., Goldsmith, H. L., and Mason, S. G. (1973) The microrheology of colloidal dispersions. I. The microtube technique. *J. Colloid Interface Sci.* **43**, 630–648.

Whitmore, R. L. (1968) *Rheology of the Circulation.* Pergamon Press, New York.

Yen, R. T. and Fung, Y. C. (1973) Model experiments on apparent blood viscosity and hematocrit in pulmonary alveoli. *J. Appl. Physiol.* **35**, 510–517.

CHAPTER 4

Mechanics of Erythrocytes, Leukocytes, and Other Cells

4.1 Introduction

In the previous chapter, we studied the flow properties of blood. In this chapter, we turn our attention to the blood cells. We give most of the space to the red blood cells, but treat the white blood cells and other cells toward the end of the chpater.

Red blood cells are the gas exchange units of animals. They deliver oxygen to the tissues of all organs, exchange with CO_2, and return to the lung to unload CO_2 and soak up O_2 again. This is, of course, what circulation is all about. The heart is the pump, the blood vessels are the conduits, and the capillary blood vessels are the sites where gas exchange between blood and tissue or atmosphere takes place.

If one observes human red blood cells suspended in isotonic solution under a microscope, their beautiful geometric shape cannot escape attention (see Fig. 4.1:1; which shows two views of a red blood cell, one a plane view, and one a side view). The cell is disk-shaped. It has an almost perfect symmetry with respect to the axis perpendicular to the disk. The question is often asked: Why are human red cells so regular? Why are they shaped the way they are? When red cells grow in the bone marrow, they have nuclei, and their shape is irregular. Then as they mature, they expel their nuclei and enter the blood. They circulate in the body for 120 days or so, then swell into spherical shape and become hemolyzed by macrophages in the spleen.

In circulating blood, however, the red blood cells are severely deformed. Figure 4.1:2 shows photographs of blood flow in the capillary blood vessels in the mesentery of the rabbit and the dog. Note how different the cell shapes are compared with the isolated floating cells shown in Fig. 4.1:1! Some

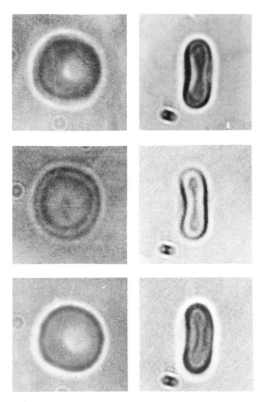

Figure 4.1:1 Two views of a normal human red blood cell suspended in isotonic solution as seen under a conventional optical microscope. Left: plane views. Right: side views. The three sets of images differ only in slight changes in focusing. These photographs illustrate the difficulty in accurately measuring the cell geometry with an optical microscope because of light diffraction. The isotonic Eagle-ablumin solution is made up of 6.2 g NaCl, 0.36 g KCl, 0.13 g NaH_2PO_4 H_2O, 1.0 g $NaHCO_3$, 0.18 g $CaCl_2$, 0.15 g $MgCl_2$ 6 H_2O, 0.45 g dextrose, and 1.25 g bovine serum albumin in 800 ml distilled water, buffered at pH 7.4. From Fung, Y.C. (1968) Microcirculation dynamics. *Biomed. Instrumentation* **4**, 310–320.

authors describe the red cells circulating in capillary blood vessels as parachute-shaped, others describe them as shaped like slippers or bullets.

The study of red blood cell geometry and deformability will throw light on the mechanical properties of cells and cell membranes, and thus is of basic importance to biology and rheology. It is also of value clinically, because change of shape and size and strength of red cells may be indicative of disease.

The term *leukocyte* or *white blood cell*, stands for a class of five morphologically distinct cells. Leukocytes play a key role in the immune system, i.e., in the protective mechanisms of the body against diseases, and in tissue inflammation, in wound healing, and in other physiological and pathological pro-

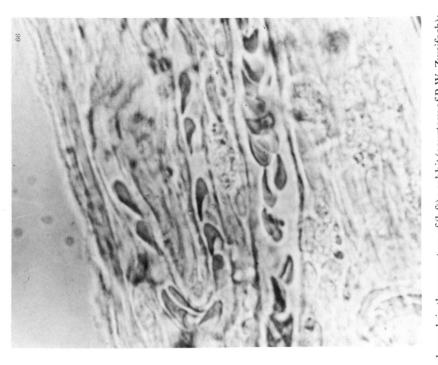

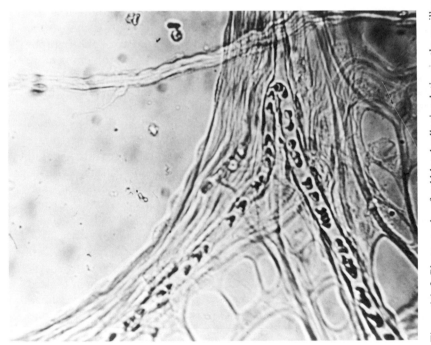

Figure 4.1:2 Photographs of red blood cells circulating in the capillary blood vessels in the mesentery of (left) a rabbit (courtesy of B.W. Zweifach); and of (right) a dog (courtesy of Ted Bond). From Fung, Y.C. (1969) Blood flow in the capillary bed. *J. Biomechanics* **2**, 353–372.

cesses. In human blood, the ratio of the number of leukocytes to that of erythrocytes is 1 to 1000. They are such a minority in blood cell population that they have little influence on the coefficient of viscosity of blood in viscometers. In microcirculation, however, leukocytes may adhere to the blood vessel wall and obstruct the flow. The obstruction is related to the deformability of the leukocytes. The study of leukocyte mechanics is, therefore, of great interest.

4.2 Human Red Cell Dimensions and Shape

All evidence shows that red blood cells are extremely deformable. They take on all kinds of odd shapes in flowing blood in response to hydrodynamic stresses acting on them. Yet when flow stops the red cells become biconcave disks. In isotonic Eagle-albumin solution the shape of the red cell is remarkably regular and uniform. We speak of red cell dimensions and shape in this static condition of equilibrium.

An accurate determination of red cell geometry is not easy. Photographs of a red blood cell in a microscope are shown in Fig. 4.1:1, from which one would like to determine the cell's diameter and thickness. The border of the cell appears as thick lines in the photographs. How can one determine where the real border of the cell is? In other words, how can we choose points in the thick image of the border in order to measure the diameter and thickness and other characteristics of red cell geometry?

The thickness of the cell border that appears in the photograph is not due to the cell being "out of focus" in the microscope. If we change the focus up and down slightly, the border moves, and a sequence of photographs as shown in Fig. 4.1:1 is obtained. Thus it is clear that the problem is more basic. It is related to the interaction of light waves with the red cells, because the thickness of the red cell is comparable with the wavelength of the visible light.

If we do not want to be arbitrary in making up a rule to read the photograph of images in a microscope for the determination of dimensions, we should have a rational procedure. Such a procedure must take into account the wave characteristics of light (physical optics). One approach is to describe the red cell geometry by an analytic expression with undetermined coefficients, calculate the image of such a body under a microscope according to the principles of physical optics, then compare the calculated image with the photographed image in order to determine the unknown coefficients. Such a program has been carried out by the author in collaboration with E. Evans (Evans and Fung, 1972). We used an interference microscope to obtain a photograph of the phase shift of light that passes through the red cell in a microscope. Figure 4.2:1 shows such a photograph. We assumed the following formula to describe the thickness distribution of the red blood cell:

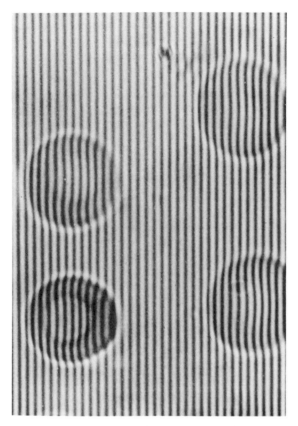

Figure 4.2:1 Picture of red blood cells taken with an interference microscope. Phase shift of light waves through the red cell leads to the deviations from the base lines.

$$D(r) = [1 - (r/R_0)^2]^{1/2}[C_0 + C_2(r/R_0)^2 + C_4(r/R_0)^4], \qquad (1)$$

where R_0 is the cell radius, r is the distance from the axis of symmetry, and C_0, C_2, C_4, and R_0 are numerical coefficients to be determined. The phase image corresponding to Eq. (1) is determined according to the principle of microscopic holography. The parameters C_0, C_2, C_4, and R_0 are determined by a numerical process of minimization of errors between the calculated and photographed images. It was possible in this way to determine the geometric dimensions of the red cell to within 0.02 μm. Hence a resolution of 0.5%–1% for the cell radius and 1% for the cell thickness is achieved. The resolution of the cell surface area and cell volume are 2% and 3%, respectively.

Results obtained by Evans and Fung (1972) for the red cells of a man are shown in Tables 4.2:1 through 4.2:4. The cells were suspended in Eagle-albumin solution (see the legend of Fig. 4.1:1) at three different tonicities (osmolarity). At 300 mosmol the solution is considered isotonic. At 217 mosmol it is hypotonic. At 131 mosmol the red cells became spheres. In

TABLE 4.2:1 Statistics of 50 RBC in Solution at 300 mosmol at pH 7.4. From Evans and Fung (1972)

	Diameter	Minimum thickness	Maximum thickness	Surface area	Volume
Average	7.82 μm	0.81 μm	2.58 μm	135 μm^2	94 μm^3
Standard deviation σ	± 0.62 μm	± 0.35 μm	± 0.27 μm	± 16 μm^2	± 14 μm^3
2nd mom. M_2	3.77×10^{-1}	1.20×10^{-1}	7.13×10^{-2}	2.46×10^2	2.02×10^2
3rd mom. M_3	2.19×10^{-1}	-2.51×10^{-3}	3.26×10^{-4}	3.58×10^3	2.11×10^3
4th mom. M_4	4.81×10^{-1}	3.22×10^{-2}	1.73×10^{-2}	2.16×10^5	1.59×10^5
Skewness G_1	0.97	-0.063	0.018	0.96	0.76
Kurtosis G_2	0.57	-0.70	0.58	0.75	1.11
χ^2 for 10 groups	35.1	10.2	8.2	22.1	18.7

TABLE 4.2:2 Statistics of 55 RBC at 217 mosmol. From Evans and Fung (1972)

	Diameter	Minimum thickness	Maximum thickness	Surface area	Volume
Average	7.59 μm	2.10 μm	3.30 μm	135 μm^2	116 μm^3
Standard deviation σ	± 0.52 μm	± 0.39 μm	± 0.39 μm	± 13 μm^2	± 16 μm^3
2nd mom. M_2	2.66×10^{-1}	1.56×10^{-1}	1.50×10^{-1}	1.57×10^2	2.40×10^2
3rd mom. M_3	5.54×10^{-1}	1.02×10^{-2}	1.45×10^{-2}	-5.76×10^1	-1.25×10^3
4th mom. M_4	3.96×10^{-1}	8.82×10^{-2}	5.26×10^{-2}	1.21×10^5	2.07×10^5
Skewness G_1	0.41	0.17	0.26	-0.03	-0.35
Kurtosis G_2	2.98	0.80	-0.62	2.21	0.78
χ^2 for 10 groups	18.6	4.0	7.2	10.0	6.37

TABLE 4.2:3 Statistics of 50 RBC at 131 mosmol. From Evans and Fung (1972)

	Diameter	Surface area	Volume	Index of refraction difference
Average	6.78 μm	145 μm^2	164 μm^3	0.0447
Standard deviation σ	± 0.32 μm	± 14 μm^2	± 23 μm^3	± 0.0043
2nd mom. M_2	1.01×10^{-1}	1.84×10^2	5.34×10^2	1.82×10^{-5}
3rd mom. M_3	2.83×10^{-3}	4.42×10^2	3.29×10^3	8.83×10^{-9}
4th mom. M_4	2.31×10^{-2}	7.83×10^4	6.77×10^5	9.60×10^{-10}
Skewness G_1	0.09	0.18	0.28	0.12
Kurtosis G_2	-0.68	-0.63	-0.56	0.02
χ^2 for 10 groups	18.5	15.9	14.2	10.5

TABLE 4.2:4 Shape Coefficients for the Average RBC. From Evans and Fung (1972)

Tonicity (mosmol)	R_0 (μm)	C_0 (μm)	C_2 (μm)	C_4 (μm)
300	3.91	0.81	7.83	−4.39
217	3.80	2.10	7.58	−5.59
131	3.39	6.78	0.0	0.0

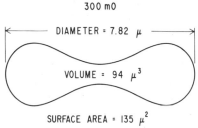

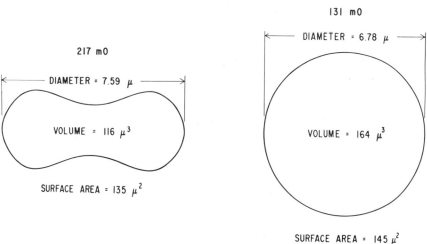

Figure 4.2:2 Scaled cross-sectional shape of the average RBC and other geometrical data of Evans. From Evans and Fung (1972).

these tables the standard deviations of the samples are listed in order to show the spread of the statistical sample. The sample size was 50.

Figure 4.2:2 shows the cross-sectional shape and other geometric data of the average RBC from Evans and Fung (1972). It is seen that the thickness is most sensitive to environmental tonicity changes; the surface area is least sensitive. It is remarkable that the average surface area of the cell remained constant during the initial stages of swelling (300–217 mosmol).

The volume of the cell increases as the tonicity decreases. The χ^2 statistics data in Tables 4.2:1–4 show that the distribution of the maximum and minimum thickness and the volume are the closest to being normal for each tonicity. Since these geometric properties are functionally related, it is not possible for all of them to have a normal distribution. For example, if the volume of a spherical cell is normally distributed, then the radius is not.

A more extensive collection of data based on the interferometric method named above was done by the author and W. C. O. Tsang (see Tsang, 1975). Table 4.2:5 shows the data from blood samples of 14 healthy men and women, with a total of 1581 cells. Figure 4.2:3 shows the average cell shape. Classification according to races, sexes, and ages showed no significant difference.

The red cell dimensions listed in Tables 4.2:1–5 may be compared with those published previously, but it must be realized that older data based on photographs through optical microscopes cannot discriminate any dimension to a sensitivity of 0.25 μm on account of the diffractive uncertainties mentioned in Sec. 4.1, and illustrated in Fig. 4.1:1. Data published before 1948 are summarized in Ponder (1948). Other important sources are those of Houchin, Munn, and Parnell (1958), Canham and Burton (1968), and Chien et al. (1971).

TABLE 4.2:5 The Geometric Parameters of Human Red Blood Cells. Statistics of Pooled Data from 14 Subjects; Sample Size $N = 1581$. 300 mosmol. From Tsang (1975)

	Diameter (μm)	Minimum thickness (μm)	Maximum thickness (μm)	Surface area (μm^2)	Volume (μm^3)	Sphericity index
Mean	7.65	1.44	2.84	129.95	97.91	0.792
Std. error of mean	± 0.02	± 0.01	± 0.01	± 0.40	± 0.41	± 0.001
Std. dev., σ	0.67	0.47	0.46	15.86	16.16	0.055
Min. value	5.77	0.01	1.49	86.32	47.82	0.505
Max. value	10.09	3.89	4.54	205.42	167.69	0.909
Skewness G_1	0.26	0.46	0.52	0.53	0.30	-1.13
Kurtosis G_2	1.95	1.26	0.24	0.90	0.30	3.27

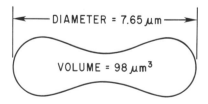

Figure 4.2:3 The average human red cell. From Tsang (1975); 14 subjects, $N = 1581$.

The sphericity index in Table 4.2:5 and Table 4.3:1 (p. 119) is equal to $4.84 \, (\text{cell volume})^{2/3} \cdot (\text{cell surface area})^{-1}$.

4.3 The Extreme-Value Distribution

How large is the largest red blood cell in a blood sample? The answer obviously depends on the size of the sample, but it cannot be obtained reliably by the ordinary procedure of adding a multiple of standard deviation to the mean. To assess the extreme value in a large sample, one has to use the statistical distribution of extreme values. The theory of statistics of extremes is a well-developed subject. E. J. Gumbel (1958) has consolidated the theory in a book. He made many applications of the theory to such problems as the prediction of flood and drought, rain and snow, fatigue strength of metals, gust loading on aircraft, quality control in industry, oldest age in a population, etc. His method has been used by the author and P. Chen (Chen and Fung, 1973) to study red blood cells.

The method may be briefly described as follows. Take a random sample of blood consisting of, say, 100 red cells. By observation under a microscope, select and measure the diameter of the largest red cell in the sample. Throw away this sample; take another sample of 100 red cells, and repeat the process. In the end we obtain a set of data on the largest diameter in every 100. From this set of data we can predict the probable largest diameter in a large sample of, say, 10^9 cells.

It is fortunate that the asymptotic formula for the probability distribution of the largest among n independent observations turns out to be the same for a variety of initial distributions of the *exponential type*, which includes the exponential, Laplace, Poisson, normal, chi-square, logistic, and logarithmically transformed normal distributions. In these distributions the variates are unlimited toward the right, and the probability functions $F(x)$ converge with increasing x toward unity at least as quickly as an exponential function. The distribution of the diameter of red cells is most likely of the exponential type (past publications usually claim it to be approximately normal; see, for example, Ponder, 1948; Houchin, Munn, and Parnell, 1958; Canham and Burton, 1968), and hence we have great confidence in the validity of the extreme distribution. This was found to be the case by Chen and Fung (1973).

According to Gumbel (1954), the probability that the largest value in a sample of size n will be equal to or less than a certain value x is given by

$$F(x) = \exp[-\exp(-y)], \tag{1}$$

where

$$y = \alpha(x - u), \tag{2}$$

and α, u, are parameters depending on n. The parameter u is the *mode*, and is the most probable value of x. The inverse of the parameter α is a measure of dispersion, called the *Gumbel slope*. The parameters are determined by the following formulas:

$$\frac{1}{\alpha} = \frac{S\sqrt{6}}{\pi}, \qquad u = \bar{x} - \frac{0.57722}{\alpha}, \tag{3}$$

where $\bar{x}$ is the mean and S is the standard deviation of the extreme variate. The variable y is dimensionless and is called the *reduced largest value*. For a continuous variate there is a probability $1 - F(x)$ for an extreme value to be equal to or exceeded by x. Its reciprocal,

$$T(x) = \frac{1}{1 - F(x)}, \tag{4}$$

is called the *return period*, and is the number of observations required so that, on the average, there is one observation equalling or exceeding x.

Gumbel (1954, 1958) has reduced the procedure of testing the goodness of fit of the mathematical formula to any specific set of observed data, as well as the evaluation of parameters α and u, to a simple graphical method. He constructed a probability paper for extreme values in which the reduced largest value y, the cumulative probability $F(x)$, and the return period $T(x)$, are labeled on the abscissa, while the variate x is labeled linearly on the ordinate. This graph paper is given in King (1971). An example is given in Fig. 4.3:1. The entire set of n observed largest values is plotted onto this paper

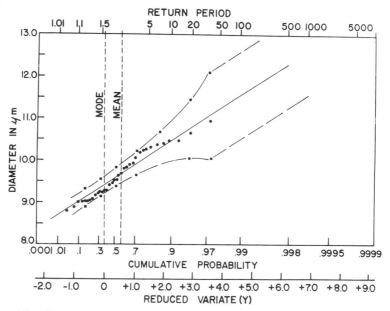

Figure 4.3:1 Extreme value distribution of red blood cells plotted on Gumbel's probability paper. The sample batch size is 100 cells. The set of data on the largest diameter in every 100 cells is plotted. The least-squares fit straight line is represented by $x = 9.386 + 0.4568y$. Circles represent experimental data. Squares are computed at 95% confidence limits. From Chen and Fung (1973).

in the following manner. List all the data points in order of their magnitude, x_1 being the smallest, x_m the mth, and x_n the largest. Plot x_m against a cumulative probability

$$\bar{F}_m = \frac{m}{n+1}. \tag{5}$$

If the theoretical formula (1) applies, the data plotted should be dispersed about a straight line:

$$x = u + \frac{y}{\alpha}. \tag{6}$$

The intercept and the slope of the fitted straight line give us the parameters u and α. Having this straight line, we can find the expected largest value corresponding to any desired large return period T from the reduced variate:

$$y = \log_e T(x). \tag{7}$$

Chen and Fung (1973) used the method of resolution of microscopic holography (Evans and Fung, 1972) discussed in Sec. 4.2, to obtain the dimensions of the red cells, and then used the Gumbel method to determine the extreme-value distribution.

The experimental results of four subjects in the age range 22–29 are summarized in Table 4.3:1. A typical extreme value plot for a subject is shown in Figure 4.3:1. Within the 95% confidence limit, it can be seen that the dis-

TABLE 4.3:1 Distribution of Diameter, Area, Volume, Maximum Thickness, Minimum Thickness, and Sphericity Index of Cells with the Largest Diameter in Samples of 100 Cells Each. 300 mosmol. From Chen and Fung (1973)

Subject	PC	JP	DV	MY
Sex	M	M	M	M
No. of Samples	55	36	35	37
Mode (μm) of largest diameter	9.083	9.372	9.386	9.168
Gumbel slope ($1/\alpha$) of largest diameter	0.5204	0.5255	0.4548	0.5314
Area	175.34	184.42	182.00	178.01
(μm^2) $\pm$ SD	$\pm$20.21	$\pm$20.29	$\pm$14.25	$\pm$20.37
Volume	119.87	129.08	132.38	130.03
(μm^3) $\pm$ SD	$\pm$17.45	$\pm$17.37	$\pm$16.81	$\pm$19.64
Maximum thickness	2.367	2.449	2.456	2.510
(μm) $\pm$ SD	$\pm$0.281	$\pm$0.217	$+$0.249	$\pm$0.229
Minimum thickness	0.751	1.086	0.9716	1.0281
(μm) $\pm$ SD	$\pm$0.358	$\pm$0.318	$\pm$0.2815	$\pm$0.314
Sphericity index	0.6711·	0.6703	0.6907	0.6979

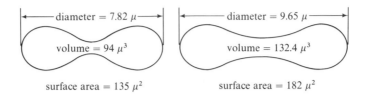

A NORMAL CELL   A LARGE CELL

Figure 4.3:2 Comparison between a normal red blood cell from Evans and Fung (1972) and an extreme cell from Chen and Fung (1973).

tribution of the largest diameter follows the theoretical asymptote for the exponential type of initial distribution. The Gumbel slope is relatively constant for all the subjects. When compared to the study of normal blood samples, the large cells have a larger surface area and volume, the minimum thickness is higher, the maximum thickness remains approximately the same, and the sphericity is lower. A correlation study shows that for the most probable largest cell in a sample of size n, the cell surface area and the minimum thickness are proportional to the cell diameter. The maximum thickness seems to be independent of the cell diameter. The sphericity index decreases with increasing cell diameter. There is no correlation between cell volume and cell diameter for the largest cell.

Figure 4.3:2 shows a comparison between a mean cell from Evans and Fung (1972) and an extreme cell from Chen and Fung (1973). The difference in cell shape is quite evident. If we predict the size of the largest cell in a population of 10^8 cells, then for subject PC we obtain an impressive value of 15.66–17.06 μm for its diameter, as compared with the mean value of 7.82 μm; i.e., the largest cell's diameter is more than twice that of the average.

The same technique can be used to study the smallest red cell in a given sample, but there seem to be no reason to do so. However, a study of the smallest cross section of a blood vessel may have meaning. Sobin et al. (1978) have used this method to determine the elasticity of pulmonary arterioles and venules. We can think of many biological variables whose extreme values are of interest. The method deserves to be widely known.

4.4 The Deformability of Red Blood Cells (RBC)

In static equilibrium a red cell is a biconcave disk. In flowing through capillary blood vessels, red cells are highly deformed. An illustration of deformed red blood cells in flowing blood in the mesentery of a dog is shown in Fig. 4.1:2 in Sec. 4.1. The order of magnitude of the stresses that correspond to such a large deformation may be estimated as follows. Assume that a capillary blood vessel 500 um long contains 25 RBC and has a pressure drop of 2 cm

H_2O. Then the pressure drop is about 0.08 g/cm^2 per RBC. Imagine that the RBC is deformed into the shape of a cylindrical plug with an end area of 25 μm^2 and a lateral area of 90 μm^2. The axial thrust 0.08 × 25 × 10^{-8} g is resisted by the shear force γ × 90 × 10^{-8} acting on the lateral area. Then the shear stress γ is of the order of 0.08 × $\frac{25}{90}$ = 0.02 g/cm^2, or 20 dyn/cm^2. Such a small stress field induces a deformation with a stretch ratio of the order of 200% in some places on the cell membrane! This stress level may be compared with the "critical shear stress" of about 420 dyn/cm^2, which Fry (1968, 1969) has shown to cause severe changes in the endothelial cells of the arteries. It may also be compared with the "shearing" stress of about 50–1020 dyn/cm^2 acting at the interface between a leukocyte and an endothelium when the leukocyte is adhering to or rolling on the endothelium of a venule (Schmid-Schoenbein, Fung, and Zweifach, 1975).

Red cell deformation can also be demonstrated in Couette flow and in Poiseuille flow (Schmid-Schonbein and Wells, 1969; Hochmuth, Marple, and Sutera, 1970). In fact, one of the successful calculations of the viscosity of the hemoglobin solution inside the RBC was made by Dintenfass (1968) under the assumption that the cell deforms like a liquid droplet in a Couette flow. Based on an analytical result on the motion of liquid droplets obtained earlier by G. I. Taylor, Dintenfass calculated the viscosity of the RBC contents to be about 6 cP. Experiments by Cokelet and Meiselman (1968) and Schmidt-Nielsen and Taylor (1968) on red cell contents, obtained by fracturing the cells by freezing and thawing and removing the cell membranes by centrifugation, showed that the hemoglobin solution in the red cell behaves like a viscous fluid with a coefficient of viscosity of about 6 cP. Thus one infers that the red cell is a liquid droplet wrapped in a membrane.

The question why red blood cells are biconcave in shape has been debated for many years. Ponder (1948, p. 22) thinks that the biconcave disk shape is the optimum geometry for oxygen transfer. The mathematics is correct but the reasoning is doubtful, because red cells assume a biconcave disk shape only when they are in static equilibrium: in a flowing condition they are deformed into the shape of a bullet or slipper in the capillary blood vessels, where oxygen transfer takes place. Canham (1970) thinks that the red cell membrane is flat if it is unstressed, that it behaves like a flat plate, and has a finite bending rigidity and negligible resistance to stretching. Then he shows that if such a flat bag is filled with fluid, the biconcave disk shape is the one that minimizes the potential energy. Such an analysis is interesting but somewhat futile. because it is known (see Sec. 4.6) that the red cell membrane has a finite resistance to stretching but a very low bending rigidity; and the flat bag has a very specific edge which cannot be identified on the red blood cell. Perhaps it is more meaningful to accept the fact that the shape of the red cell at static equilibrium is biconcave and ask *what special consequences are implied by the biconcavity*. In fact, the answer to such a question can yield information about the stresses in the cell membrane, the pressure in the interior of the cell, and the deformability and strength of the cell. Biconcavity is a geometric property.

A deduction made from geometry and the first principles of mechanics will be simple and general. Uncluttered by special chemical and biological hypotheses, the theory will be as transparent as a proposition in geometry. We shall present such a simple analysis in the following section.

4.5 Theoretical Considerations of the Elasticity of Red Cells*

Let us make the following hypothesis about the red blood cell: (1) The cell membrane is elastic and has a finite strength. (2) Inside the cell is an incompressible Newtonian viscous fluid. (3) In static equilibrium floating in a homogeneous Newtonian fluid it is an axisymmetric biconcave disk. (4) In this state, there is no bending moment in the cell membrane. Then we shall derive theoretically the following conclusions:

C(1) The biconcave state is the zero-stress state.
C(2) At the biconcave state, the transmembrane pressure difference is zero.
C(3) There exists an infinite number of large deformations of the cell which preserves the volume of the cell, surface area of the cell membrane, and without stretching and tearing of the membrane anywhere. Bending is concentrated along certain lines.

As a proof, let us consider an axially symmetric biconcave shell of revolution in static equilibrium under internal pressure. Because of the axial symmetry, it is necessary only to consider a meridional cross section as shown in Fig. 4.5:1. A point P on the shell can be identified by a pair of cylindrical polar coordinates (r, z) or by r and the angle ϕ between a normal to the shell and the axis of revolution. At each point on the shell, there are two principal radii of curvature: One is that of the meridional section, shown as r_1 in Fig. 4.5:1. The other is equal to r_2, a distance measured on a normal to the meridian between its intersection with the axis of revolution and the middle surface of the shell. The lines of principal curvature are the meridians and parallels of the shell.

The stresses in the shell may be described by stress resultants (membrane stresses) and bending moments. The stress resultants are the mean stress in the membrane multiplied by the thickness. The bending moments arise from the variation of the stresses within the thickness of the membrane, and is associated with the change of curvature of the membrane. We assume that at the natural state there is no bending moment in the shell. Hence we need only to examine the membrane stresses.

The stress resultants acting on an element of the shell surface bounded between two meridional planes, $\theta = \theta_0$ and $\theta = \theta_0 + d\theta$, and two parallel planes, $\phi = \phi_0$ and $\phi = \phi_0 + d\phi$, are shown in Fig. 4.5:2. The symbol N_ϕ denotes the normal stress resultant (force per unit length, positive if in tension)

---

* The material in this section is taken from Fung (1966).

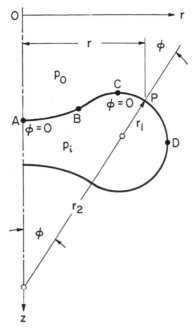

Figure 4.5:1 Meridional cross-section of the red blood cell. Angle ϕ, the radial distance r, or the arc length s may be chosen as the curvilinear coordinates for a point on the cell membrane. Note the principal radii of curvature r_1, r_2 at point P. Point B is a point of inflection.

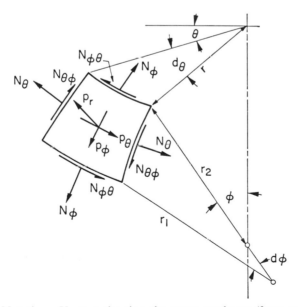

Figure 4.5:2 Notations. Vectors showing the stress resultants (force per unit length in the cell membrane) N_ϕ, N_θ, $N_{\phi\theta}$ and the external loads (force per unit area) p_r, p_θ, p_ϕ.

acting on sections ϕ = const., N_θ denotes the normal stress resultant (tension) per unit length acting on sections θ = const. The symbol $N_{\theta\phi}$ is the shear resultant per unit length acting on sections θ = const., and $N_{\phi\theta}$ is the shear acting on ϕ = const. For axisymmetric loading, $N_{\theta\phi} = N_{\phi\theta} = 0$. The equations of equilibrium are (see Flügge, 1960, p. 23), for a shell subjected to a uniform internal pressure p_i and a uniform external pressure p_o:

$$\frac{d}{d\phi}(rN_\phi) - r_1 N_\theta \cos\phi = 0, \tag{1}$$

$$\frac{N_\phi}{r_1} + \frac{N_\theta}{r_2} = p_i - p_o. \tag{2}$$

Solve Eq. (2) for N_θ, substitute into Eq. (1), and multiply by $\sin\phi$ to obtain

$$\frac{d}{d\phi}(r_2 N_\phi \sin^2\phi) = r_1 r_2 (p_i - p_o)\cos\phi \sin\phi. \tag{3}$$

An integration gives the result

$$N_\phi = \frac{1}{r_2 \sin^2\phi}\left[\int r_1 r_2 (p_i - p_o)\cos\phi \sin\phi \, d\phi + c\right], \tag{4}$$

where c is a constant. From Fig. 4.5:1, we can read off the geometric relations

$$r = r_2 \sin\phi, \qquad \frac{dr}{d\phi} = r_1 \cos\phi, \qquad \frac{dz}{d\phi} = r_1 \sin\phi. \tag{5}$$

Thus Eq. (4) can be reduced to the form

$$N_\phi = \frac{1}{r \sin\phi}\int_0^r (p_i - p_o)r \, dr. \tag{6a}$$

This solution yields the alarming result that at the top point C in Fig. 4.5:1 where $\phi = 0$ and r is finite, we must have $N_\phi \to \infty$ if $p_i - p_o \neq 0$. This result, that a finite pressure differential implies an infinitely large stress at the top (point C), is so significant that a special derivation may be desired. Consider the equilibrium of a circular membrane that consists of a portion of the shell near the axis, say, inside the point B in Fig. 4.5:1. The isolated circular membrane is shown in Fig. 4.5:3(a). The total vertical load acting on the membrane is

$$(p_i - p_o)\pi r^2.$$

This load must be resisted by the vertical component of the stress resultant acting on the edges, namely, $N_\phi \sin\phi$, multiplied by the circumference $2\pi r$. Hence

$$2\pi r N_\phi \sin\phi = (p_i - p_o)\pi r^2,$$

or

$$N_\phi = \frac{p_i - p_o}{2}\frac{r}{\sin\phi}. \tag{6b}$$

At C, $r \neq 0$, $\phi = 0$, and N_ϕ must become infinitely large if $p_i \neq p_o$. Figure

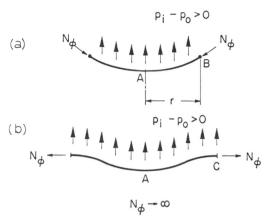

Figure 4.5:3 Equilibrium of a polar segment of the cell membrane. Pressure load is balanced by the membrane stress resultant N_ϕ. Transverse shear and bending stresses in the thin cell membrane are ignored. (a) At B the slope of the membrane $\neq 0$. (b) At C the slope $= 0$.

4.5:3(b) shows this very clearly. Here the slope at C is horizontal. No matter how large N_ϕ is, the stress resultant on the edges cannot balance the vertical load. Hence $N_\phi \to \infty$ if the vertical load is finite.

Perhaps no biological membrane can sustain such a large stress. Therefore, if we insist that $p_i - p_o \neq 0$, then we shall have a major difficulty. This difficulty is removed if we admit the natural conclusion that the pressure differential across the cell wall vanishes:

$$p_i - p_o = 0. \tag{7}$$

Then it follows that for a normal biconcave red cell, the membrane stresses N_ϕ, N_θ vanish throughout the shell:

$$N_\phi = N_\theta = 0. \tag{8}$$

Thus, indeed, a floating biconcave red cell is at zero-stress state. This proves the conclusions C(1), C(2) listed on p. 122.

Our conclusion is based on the biconcave geometry of the red cell. Nucleated cells and some pathological erythrocytes do not assume biconcave geometry; consequently, our conclusion does not apply. Sickle cells have a crystalline hemoglobin structure inside the cell. Leukocytes have gelated pseudo pods. To them our analysis does not apply because in their interior the stress field cannot be described by pressure alone. Hence we do not know the zero-stress state of leukocytes.

4.5.1 Stretching and Bending Rigidities of the Red Cell Membrane

If a deformation changes the curvature of a shell, then the bending rigidity must be considered. If the bending moment is not uniform, then there must be transverse shear (in sections perpendicular to the membrane). With bending and transverse shear, a pressure difference $p_i - p_o$ can be tolerated.

As it is well known in the theory of plates and shells (see, e.g., Flügge, 1960), the bending rigidity of a thin isotropic plate or a thin shell is proportional to the cubic power of the wall thickness:

$$\text{bending rigidity } D = \frac{Eh^3}{12(1 - v^2)}, \tag{9}$$

where E is Young's modulus, h is the wall thickness, and v is Poisson's ratio. In contrast, the extensional rigidity is Eh, proportional to the first power of the wall thickness. When the thickness h is very small, the bending rigidity is much smaller than the extensional rigidity.

Large deformation of very thin shells often occur in such a way that the extensional deformation is small everywhere (because of the high extensional rigidity), while large curvature change is concentrated in some small areas (so that the product of curvature change and bending rigidity is significant in these areas). In the limit we are lead to the consideration of deformations in which significant bending occurs along a system of lines on the shell. Deformations of this type is discussed in the following paragraphs.

4.5.2 Isochoric Applicable Deformations

We shall now consider a red cell as a deformable shell filled with an incompressible fluid. Let the shell be impermeable to the fluid. Any deformation of the shell must preserve the volume of the incompressible fluid. We call such a deformation *isochoric*. The deformation of an erythrocyte without mass transport across its membrane is isochoric.

In the theory of differential geometry (see, for example, Graustein, 1935, or Struik, 1950), a surface is said to be *applicable* to another surface if it can be deformed into the other by a continuous bending without tearing and without stretching. A deformation of a surface into an applicable surface involves no change in the intrinsic metric tensor on the surface. No change in the membrane strain is involved in such a deformation. Accordingly, no change in membrane stresses will occur in such a deformation.

A deformation of a shell by continuous bending without tearing or stretching and with the enclosed volume kept constant is called an *isochoric applicable deformation*. From what has been said above, we conclude that an isochoric applicable deformation of an erythrocyte will induce no change in the membrane stresses in the cell wall.

Consider a spherical shell made of an elastic material. Let the shell be filled with an incompressible liquid and be placed in another fluid. Now suppose the external pressure is reduced. If the internal pressure does not change, the shell would expand. But such an expansion is impossible because the volume is fixed. Hence the internal pressure must be reduced to equal the reduced external pressure so that the stress in the shell will remain unchanged. Hence an elastic spherical shell transmits pressure changes perfectly, so that the changes in internal pressure are exactly equal to the changes in external pressure.

Next consider an arbitrary elastic shell containing an incompressible fluid. Although no longer so obvious, the same argument as above applies, and we conclude that neither a change in the membrane stress in the shell, nor a change in the pressure differential across the shell, can be achieved by manipulating the external pressure alone, as long as the deformation is isochoric and applicable.

It is known in differential geometry that if two surfaces are applicable to each other, their Gaussian curvature (also called the total curvature, which is the product of the two principal curvatures) must be the same at corresponding points. Thus a developable surface (whose Gaussian curvature is zero) is applicable to another developable surface. A spherical surface is applicable to another spherical surface of the same radius. Some examples are shown in Fig. 4.5:4. In Figs. 4.5:4(a) and (b) it is seen that cylindrical

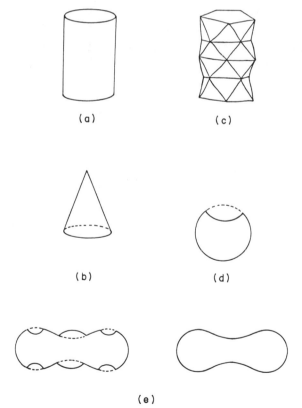

(a)

(c)

(b)

(d)

(e)

Figure 4.5:4 Examples of applicable surfaces. In (e) an isochoric applicable deformation of a red blood cell is shown.

and conical shells are applicable to each other and to flat sheets if they can be cut open. In Fig. 4.5:4(c) a diamond-patterned surface composed of flat triangles can be shown to be applicable to a cylinder. In Fig. 4.5:4(d) a dimple with reversed curvature on a sphere is applicable to the original sphere. Although deformations shown in Figs. 4.5:4(c) and (d) are not isochoric, they are important modes of buckling of cylindrical and spherical shells. In Fig. 4.5:4(e) a biconcave shell similar to an erythrocyte is shown. In this case an infinite variety of isochoric applicable surfaces exists.

This proves the conclusion C(3) listed on p. 122. The great liberty with which a biconcave erythrocyte can deform isochorically into an infinite variety of applicable surfaces without inducing any membrane stresses in its cell membrane, and without any change in the pressure differential between the interior and the exterior offers a disarming simplicity to the mechanics of erythrocytes.

4.6 Cell Membrane Experiments

In the preceding section, we considered the consequences of the biconcave disk shape of the red blood cell and concluded that the geometry makes the cell flaccid so that it can deform into a great variety of shapes *without* inducing any membrane stresses (i.e., without stretching the membrane in any way), provided that the small bending stress (and rigidity) of the cell membrane is ignored. This great flexibility is, of course, a great asset to the red cell, which has to circulate all the time and to go through very narrow tubes in every cycle. If the cell can deform and squeeze through the narrow tubes without inducing any stress, then wear and tear is minimized.

It follows that in order to investigate the stress–strain relationship of the red cell membrane, one must induce "off design" deformations in the cell, deformations that are not obtainable by *isochoric* (without changing volume) and *applicable* (without stretching the cell membrane in any way) transformations. The following are several popular types of experiments:

1. Osmotic swelling in hypotonic solution (Figs. 4.2:2 and 4.6:1).
2. Compression between two flat plates [Fig. 4.6:2(b)].
3. Aspiration by a micropipet [Figs. 4.6:2(a) and 4.6:6].
4. Deflection of the surface by a rigid spherical particle [Fig. 4.6:2(c)].
5. Fluid shear on a cell tethered to a flat plate at one point (Fig. 4.6:5).
6. Transient recovery of a deformed cell (Fig. 4.6:7).
7. Plastic flow and other details of tethers (Fig. 4.6:8).
8. Forced flow through polycarbonate sieves.
9. Thermal effect of elastic response.
10. Chemical manipulation of the cell membrane.

Grouped according to their objectives, these experiments can be classified as follows. *Strain types* are varied in the following experiments:

a. Membrane surface area changes (Exps. 2, 3, 4, and the latter stages of Exp. 1).
b. Area does not change, stretching different in different directions (earlier stages of Exp. 1 and parts of membrane in Exps. 2–8).
c. Bending energy not negligible compared with stretching energy (earlier stages of Exp. 1).

With respect to variation in *time*, there are two kinds of experiments.

a. Static equilibrium (Exps. 1–5);
b. dynamic process (Exps. 6 and 7).

With respect to *material properties*, there are three kinds:

a. Elastic (Exps. 1–5);
b. viscoelastic (Exp. 6);
c. viscoplastic (Exp. 7).

Through these experiments, information about the elasticity, viscoelasticity, and viscoplasticity of the red blood cell membrane is gathered. Some comments follow.

4.6.1 Osmotic Swelling

Osmotic swelling of a red cell is one of the most interesting experiments in biomechanics. When a red cell is placed in a hypotonic solution, it swells to the extent that the osmotic pressure is balanced by the elastic stress and surface tension in the cell membrane. Returning the cell to an isotonic solution recovers the geometry, so the transformation is reversible. Results of such experiments are given in Tables 4.2:1–4.2:4. Eagle-albumin solution at 300 mosmol is considered isotonic. In a hypotonic solution of 217 mosmol, the cell volume is increased 23%, but the surface area is unchanged. At 131 mosmol the volume is increased 74% but the surface area is increased only 7%. At 131 mosmol the red cell is spherical and it is on the verge of hemolysis. At a tonicity smaller than 131 mosmol the cells are hemolyzed. This shows that the red cell membrane area cannot be stretched very much before it breaks. On the other hand, a look at the red cells in other experiments, especially in tethering (Exp. 5) and aspiration (Exp. 3), convinces us that the red cell membrane is capable of very large deformation provided that its area does not have to change (locally as well as globally). A deformation called "pure shear," stretching in one direction and shrinking in the perpendicular direction, is sustainable. Furthermore, it was found that the elastic modulus for cell membrane area change [tensile stress $\div$ ($\Delta A/A$), A being the membrane area] is three to four orders of magnitude larger than the shear modulus (shear stress $\div$ shear strain). For these reasons, experiments are classified according to whether cell membrane area is changed or not.

In the initial stages of osmotic swelling (say, from 300 to 217 mosmol), the cell geometry is changed considerably without noticeable change in surface area. In this stage the bending rigidity of the cell membrane, though very small, is very important: without taking bending into account, the sequence of swelling configurations cannot be explained. Conversely, the swelling experiment provides information about cell membrane bending. See Zarda, et al. (1977).

A model experiment on a thin-walled rubber model of a red blood cell reveals some details which are quite instructive. Such a model with a diameter of about 4 cm and a wall thickness of about 40 μm (so that the radius-to-thickness ratio is about 500) is shown in Fig. 4.6:1. When this rubber cell was immersed in a water tank and its internal volume was controlled by water injection, the sequence of geometric changes was photographed and

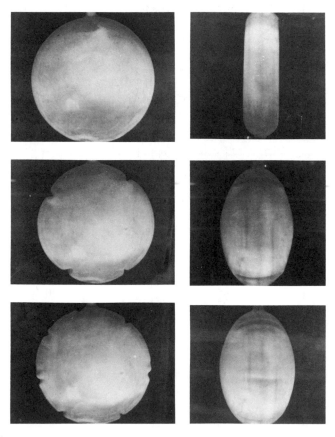

Figure 4.6:1 Thin-walled rubber model of red blood cell filled with water and immersed in a water tank. Figure shows the change of cell geometry with internal pressure. Top: natural shape. Second row: polar and equatorial regions buckled. Third row: the number of equatorial buckles increases with increasing pressure. From Fung (1966).

is shown in the figure. The top two pictures show the initial shape without pressure difference. The following pictures show the change in shape under increasing internal pressure. In the initial stage, a sudden bulge of the dimples (i.e., buckling in the polar region of the cell) can be easily detected. The cell shown in the second row of Fig. 4.6:1 has already buckled. Note that the central, polar region bulges out. However, while the poles bulge, the equator of the cell contracts. In the successive pictures it is seen that the equator contracted so much that wrinkles parallel to the axis of symmetry were formed. The number of equatorial wrinkles increased with increasing internal pressure. At large internal pressure the cell swelled so much that the equator became smooth again.

This picture sequence is interesting in revealing a type of buckling in a thin-walled elastic shell when it is subjected to an increasing internal pressure. Engineers are familiar with elastic buckling of thin shell structures (e.g., spheres, cylinders, airplanes, submarines) subjected to external pressure. But here we have a type of the opposite kind. The principle is simple enough: on increasing internal pressure and volume, the red cell swells into a more rounded shape. In doing so the perimeter at the equator is shortened, and buckling occurs.

The method of making the thin-walled rubber model of the RBC may be recorded here because it can be used to make other models of biomechanical interest. To make such a rubber shell model, a solid metal model of the RBC was machined according to the geometry shown in Fig. 4.2:1, with a diameter of about 4 cm. A female mold of plaster of Paris was then made from the solid metal model. From the female mold solid RBC models were cast with a material called "cerrolow" (Peck-Lewis Co., Long Beach, Calif.), which has a melting point of about 60°C. At room temperature this material appears metallic and can be machined and polished. A stem was left on each mold. Holding the stem, we dipped the mold into a cup of latex rubber in which a catalyst was added. Upon drying in air, a rubber membrane covered the mold. Each dip yielded a membrane of thickness of about 20 μm. In this way, a cell model with a wall thickness of 20–40 μm can be made satisfactorily. The solidified rubber model was then rinsed in hot water, in which the cerrolow melted and could be squeezed out, leaving a thin-walled red cell model.

4.6.2 Area Dilatation Experiments

The methods illustrated in Fig. 4.6:2 are aimed at changing the surface area of the cell. The *micropipet aspiration* method sucks a portion of the cell membrane into the pipet. The *compression* method flattens a cell. The *particle* method requires implantation of a magnetic particle into the cell, followed by observing the cell's distortion when it is placed in a magnetic field. See brief description in the legend of Fig. 4.6:2. These methods are

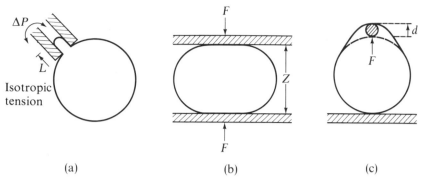

Figure 4.6:2 Three examples of mechanical experiments in which the dominant stress resultants are circumferential and longitudinal tensions. If the membrane encapsulated system is spherical or nearly spherical initially, and the interior volume is fixed, then area dilation is the primary deformation produced in these experiments. The examples above schematically illustrate (a) micropipette aspiration, (b) compression between two flat surfaces, and (c) deflection of the surface by a rigid spherical particle. In (a), ΔP is the pressure difference between the pipette interior and the outside medium, and L is the length of the aspirated projection of the membrane. In (b), F is the force on each plate where the plates are separated from eah other by a distance z. In (c), F is the force which displaces the magnetic particle and the surface by a distance d. From Evans and Skalak (1979), by permission.

best suited for spherical cells (e.g., sphered red cells or sea urchin eggs). They can yield the modulus of elasticity with respect to area change of the membrane. But their analysis is by no means simple. See Evans and Skalak (1979) for detailed discussions of experimental results and their interpretations and analysis.

Experimenting on sphered red blood cells, Evans and Hochmuth (1976) found that to aspire cell membrane into a pipet a pressure of the order of 10^5 dyn/cm^2 is needed. Assuming that that pressure is about equal to $p_i - p_o$, we can compute the membrane tension (stress resultant), N_ϕ, according to Laplace's formula given in Problem 1.4 in Sec 1.6 of Chapter 1. If N_ϕ is related to the area dilatation by the equation

$$N_\phi = K \frac{\Delta A}{A_0}, \tag{1}$$

where ΔA is the change of membrane area and A_0 is the initial area, and K is a constant of proportionality, which may be called the *areal modulus of elasticity*, Evans and Hochmuth found that K is about 450 dyn/cm at 25°C. Analytical details are given in Sec. 4.6.7.

4.6.3 Membrane Shear Experiments

Distortion of a membrane without change of area is called *pure shear*. If a membrane is stretched in the x direction by a stretch ratio λ and shrunk

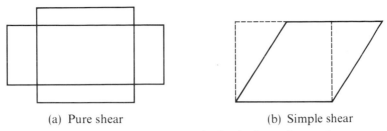

(a) Pure shear   (b) Simple shear

Figure 4.6:3 (a) Pure shear and (b) simple shear of a membrane.

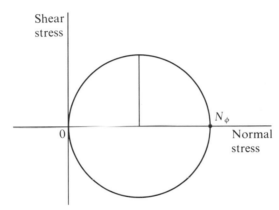

Figure 4.6:4 Mohr's circle of stress for the uniaxial stress state.

in the perpendicular, y direction by a stretch ratio λ^{-1}, the deformation is pure shear; see Fig. 4.6:3(a). In contrast, the deformation shown in Fig. 4.6:3(b) is called *simple shear*. A general deformation of a membrane can be considered as composed of a pure shear plus an areal change. Since the red cell membrane is so peculiar that it can sustain a large shear deformation but only a small area change (the cell membrane ruptures if the area is changed by more than a few percent), it is important to experiment on these two types of deformation separately.

Stresses acting on any small cross-sectional area in the membrane can be resolved into a normal stress and a shear stress. If a membrane is subjected to a tensile stress resultant N_ϕ in the x direction and 0 in the y direction, then on a cross section inclined at 45° to the x axis, the shear and normal stress resultants are both $N_\phi/2$. This is called a *uniaxial stress state*: it is an effective way to impose pure shear. The Mohr's circle of stress resultants for this state is shown in Fig. 4.6:4.

The tethering experiment illustrated in Fig. 4.6:5 is designed to induce a uniaxial tension state of stress into the cell membrane. If red blood cells are suspended in an isotonic solution and the suspension is left in contact with a glass surface, some cells will become attached to the glass. Then if the solution is made to flow relative to the glass, the solution will sweep past the tethered red cell. This is analogous to the situation in which we

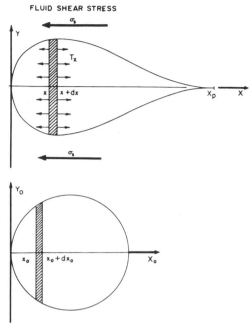

Figure 4.6:5 Schematic illustration of the uniaxial extension at constant area of a membrane element defined by $(x_0, x_0 + dx_0)$ into the element $(x, x + dx)$ under the action of a uniform fluid shear stress, σ_s. x_p is the location of the attachment point for the disk model. From Evans and Skalak (1979), by permission.

hold a balloon tied to a string and stand in a strong wind. The wind will deform the balloon. In the case of the red cell the Reynolds number is very small and the viscous force dominates. Figure 4.6:5 shows the stress field in a membrane element of the red cell in this situation. By photographing the deformed red cell, one can approximately assess the modulus of elasticity of the cell membrane in a uniaxial state of stress. The shear modulus found by Hochmuth et al. (1973) by this method is of the order of 10^{-2} dyn/cm.

A different experiment for the same purpose is illustrated in Fig. 4.6:6. Suction is used to draw into a micropipette a portion of the cell membrane in the region of the poles of a flaccid red cell. Large deformation of the cell membrane occurs in the vicinity of the pipette mouth. If one assumes that the area of the membrane does not change, and that the suction is so localized that the cell remains flaccid and biconcave away from the pipette's mouth, then an approximate analysis of the membrane deformation can be made. The deformation is pure shear. With bending rigidity neglected, the equations of equilibrium can be set up. Combined with a constitutive equation (such as those proposed in Sec. 4.7), the deformation can be related to the sucking pressure in the pipette. Comparison of the experimental results with the calculated results will enable us to determine the elastic constants

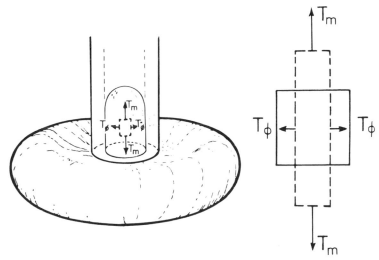

Figure 4.6:6 Illustration of membrane extension and deviation between force resultants produced by micropipette aspiration of a flaccid red blood cell. The principal force resultants (tensions) can be decomposed into isotropic and deviatoric (shear) contributions. From Evans and Skalak (1979), by permission.

in the constitutive equation. If the membrane stress resultants are N_m, N_ϕ, as shown in Fig. 4.6:6, and the stretch ratio in the direction of N_m is λ_m, then the maximum shear stress resultant is

$$S_{\max} = \frac{N_m - N_\phi}{2}, \tag{2}$$

and the stress–strain relationship, Eq. (3) of Sec. 2.8, or Eq. (10) of Sec. 4.7, is reduced to the form

$$S_{\max} = 2\mu \cdot (\text{max. shear strain}). \tag{3}$$

Waugh (1977) found the value of the shear modulus μ to be 6.6×10^{-3} dyn/cm at 25°C, with an 18% standard deviation for 30 samples. This is more than four orders of magnitude smaller than the areal modulus. Further analysis is given in 4.6.6.

4.6.4 Viscoelasticity of the Cell Membrane

The experiments considered above may be performed in a transient manner to measure response of the cell to step loading or step deformation, or to oscillatory loading and deformation, thus obtaining the viscoelastic characteristics of the cell membrane.

Hochmuth et al. (1979) experimented on the viscoelastic recovery of red cells by pulling a flaccid red cell disk at diametrically opposite locations on the rim of the cell: the cell is attached to glass on one side and pulled

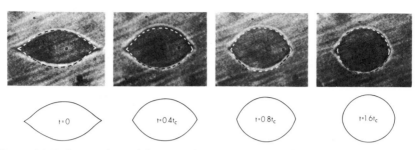

Figure 4.6:7 Comparison of the extensional recovery predicted for a viscoelastic disk model to that of a series of photographs taken of an actual red blood cell during recovery. The correlation provides a time constant, t_c, for the membrane recovery of 0.10 s in this case. The dashed lines are an overlay of the predicted model behavior onto the outline of the cell. From Evans and Skalak (1979), by permission.

with a micropipet from the opposite side. The cell is then released and the recovery is observed with high speed photography. (See Fig. 4.6:7.)

A related case is suddenly stopping the flow in the experiment illustrated in Fig. 4.6:5 and observing the recovery of the red cell. Observation of cell response to suddenly applied pressure and release in micropipet aspiration [Fig. 4.6:2(a)] was done by Chien et al. (1978).

4.6.5 Viscoplastic Flow

When a flaccid red cell held with a micropipette for a certain length of time is suddenly expelled from the pipette, a residual "bump" remains in the membrane, which is interpreted as due to plastic deformation. At a given sucking pressure, the height of the bump was found to be proportional to the length of time the cell is held in the pipette.

Figure 4.6:8 illustrates another experiment by Evans and Hochmuth (1976). They observed that irrecoverable flow of material commences when the membrane shear resultant exceeds a threshold, or "yield" value. The membrane of a red cell tethered to a glass plate will flow plastically if the shear flow exceeds a certain limit. Analysis of the plastic flow is given in Sec. 4.6.8.

4.6.6 Mechanics of Micropipet Experiments

The micropipet experiments illustrated in Fig. 4.6:2 have yielded many pieces of information about erythrocytes, leukocytes, endothelial cells, etc. In this method, a cell to be studied is put in a suitable liquid medium. A micropipette with a suitable diameter at the mouth is used to suck on the wall of the cell. The suction pressure can be controlled (e.g., by lowering the reservoir of the fluid filling the pipet) to within 1 dyn/cm^2 (0.1 Pa). For a pipet with a 2 μm

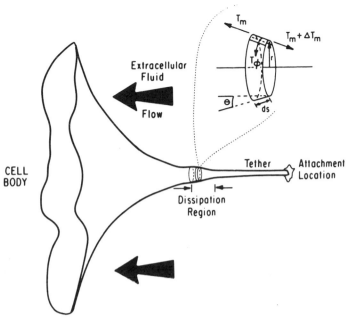

Figure 4.6:8 Schematic illustration of the microtether and the region of viscous dissipation in the "necking" region, where the plastic flow occurs. The geometry of a membrane flow element is shown in the enlarged view. From Evans and Hochmuth (1976), by permission.

tip diameter, a pressure difference of 1 dyn/cm^2 would produce a tiny suction force of 0.3 pN. The shape of the cell outside the pipet and the length of the lip sucked into the pipet can be photographed and measured as a function of the aspiration pressure.

From the pipet aspiration pressure and cell deformation relationship, one should be able to deduce the material constants of the constitutive equations of the materials of the cell provided that (1) the structure and distribution of the materials of the cell is known, (2) the forms of the constitutive equations of the materials are known so that only a set of unknown constants are to be determined, and (3) the stresses and strains in the cell can be computed. These three requirements are nontrivial. The importance of (3) is evident. To do (3), one must know (1) and (2). The difficulty of (3) is well known in the theories of elasticity, viscoelasticity, and thin shells.

Under drastic simplifying assumptions, the stress–strain distribution in the cell body and cell membrane can be computed. For example, if the cell membrane is homogeneous (i.e., of uniform thickness and uniform mechanical property, if the cell contents is a liquid, if the condition is static, and if the cell body outside the pipet and the part at the tip of the lip sucked into the pipet are spherical in shape, then the stress resultant N_ϕ in the cell membrane outside the pipet, and the longitudinal stress resultant N_z of the cell membrane in the pipet (Figs. 4.5:2 and 4.6:2a) are given by the Laplace equations (see Chapter 1, Sec. 1.9)

$$p_i - p_p = \frac{2N_z}{R_p} \tag{4}$$

$$p_i - p_o = \frac{2N_\phi}{R_o}, \tag{5}$$

where p_i is the pressure in the cell, p_o is the pressure outside the cell, p_p is the pressure in the pipet, and R_p is the radius of the pipet. R_o is the radius of the cell outside the pipet. These formulas are derived from the assumptions named above and the equations of equilibrium, and are valid for any constitutive equation. If, in addition, one assumes that the frictional force between the glass pipet and cell membrane is negligible, then the equation of equilibrium of a strip of the cell membrane around the tip of the pipet yields

$$N_z = N_\phi. \tag{6}$$

Combining Eqs. (4)–(6), we obtain

$$p_o - p_p = 2N_\phi \left(\frac{1}{R_p} - \frac{1}{R_o} \right). \tag{7}$$

These formulas have been used by Evans and Rawicz (1990) to calculate the tension required to smooth out the thermal undulations or "Brownian movement" of the membrane of 10–20 μm phospholipid vesicles. This tension was found to be on the order of 0.01 to 0.1 dyn/cm. Rand and Burton (1964) used these formulas to estimate the failure strength of the red cell membrane, which was found to be of the order of 10 dyn/cm (10^{-4} n/cm). Needham and Nunn (1990) used them to calculate the tension at failure of various phospholipid bilayer membranes and found it to be also in the order of 10 dyn/cm.

4.6.7 Modulus of Elasticity of Areal Changes of Cell Membrane

The micropipet experiment can be used to determine the modulus of elasticity of areal changes of cell membrane. This is based on an idea developed in the preceding section that the red cell membrane has a modulus of elasticity for areal changes about four orders of magnitude higher than its shear modulus of elasticity. The cell content is incompressible. The membrane is semipermeable to various solutes and the cell volume can be changed by mass transfer across the cell membrane. Referring to Fig. 4.6:2(a), we consider a red cell swollen into a sphere in a medium having a suitable osmotic pressure. A micropipet with a tip inner radius of R_p sucks on the cell membrane with a pressure p_p and creates a lip of length L_p. Originally, the volume of the cell is $V_0 = (4/3)\pi R_o^3$. With pipet aspiration, the cell radius is changes to $R_o - \Delta R_o$ and the cell volume remains to be V_0:

$$V_0 = \tfrac{4}{3}\pi(R_o - \Delta R_o)^3 + \pi R_p^2 L_p + \tfrac{2}{3}\pi R_p^3, \tag{8}$$

whereas the cell membrane area is

$$A = 4\pi R_o^2 + 2\pi R_p L_p + 2\pi R_p^2. \tag{9}$$

In aspiration experiments, ΔR_o is a function of L_p while both ΔR_o and L_p depend on p_p. Differentiating Eqs. (8) and (9), writing the differential of ΔR_o as dR_o, and remembering that V_0 is a constant, we have

$$O = 4R_o^2 \, dR_o + R_p^2 \, dL_p, \tag{10}$$

$$dA = 8\pi R_o \, dR_o + 2\pi R_p \, dL_p. \tag{11}$$

Using Eqs. (10) and (11), we obtain

$$dA = 2\pi R_p^2 \left(\frac{1}{R_p} - \frac{1}{R_o} \right) dL_p. \tag{12}$$

Equation (12) shows that a small change of cell surface area, dA, can be converted to a measurable change of the length of the lip in the pipet, dL_p. Using Eq. (12) in Eq. (1), one can determine the modulus of elasticity of areal changes of cell membrane, K. By this method, Waugh and Evans (1979) obtained a value of K of about 500 dyn/cm for the red blood cell membrane, whereas Needham and Nunn (1990) showed that K is 1700 dyn/cm for certain cholesterol–lipid mixtures.

4.6.8 Other Measurements by Micropipet

Evans and Hochmuth (1976) used a small glass fiber to touch the membrane of a red blood cell sucked by a micropipet and showed that the cell membrane can be pulled out as a thin tube or "tether" as shown in Fig. 4.6:8. The ability of cell membrane to do so is an illustration of the small shear modulus of elasticity vs. the large areal modulus of elasticity of the red cell membrane. Evans and Hochmuth (1976) observed that the flow of membrane material commences when the shear stress exceeds a threshold, suggesting a visco-plastic behavior of the red cell membrane. Hochmuth and Evans (1983) considered the relationship between the diameter of the tether and the aspiration pressure, under the hypotheses of constant cell volume and constant cell membrane area, carrying to the limit of $N_\phi \to \infty$, and obtained a limiting tether radius of 5.5 nm. If the limiting tether radius is the mean mass radius, then they obtained a cell membrane thickness of 7.8 nm.

Hochmuth (1987), using the method shown in Fig. 4.6:6, showed that when a red cell is released from the pipet, the membrane and cell quickly recover a normal, biconcave shape in about 0.1 s. Thus, the membrane has a characteristic surface viscosity η of about 10^{-3} poise·cm. The surface viscosity is equal to the product of the coefficient of viscosity and the membrane thickness. Since the membrane thickness is about 10 nm, the coefficient of viscosity of the membrane is of the order of 1000 poise, much higher than that of the hemoglobin solution in the red cell. Hence the shape recovery is dominated by the viscosity of the membrane, and not of the cytoplasm.

Evans (1983) and Bo and Waugh (1989) used micropipet to study the buckling of red cell membrane to determine its bending rigidity. They also studied the diameter of the tether (Fig. 4.6:8) which depends on the bending rigidity. They found that the ratio of the bending moment per unit length (dyn. cm/cm) to the membrane curvature (cm^{-1}) has a value on the order of 10^{-12} dyn. cm.

4.6.9 Forced Flow of Red Cells through Polycarbonate Sieves

Gregersen et al.'s (1965) experiment on the flexibility of red cells remains a classic. They used a polycarbonate paper that has been bombarded by neutrons from a nuclear reactor (General Electric) and etched full of holes. These holes have the geometry of circular cylindrical tunnels. Their diameter can be controlled to within a relatively narrow limit. Gregersen et al. used this paper as a filter; supported it with a grid at the bottom of a small cup. Several ml of blood was put in the cup. A suction pressure was applied underneath the filter, and the period of time for the blood to be emptied was measured. During this time all the red cells in the cup flowed through the tunnels in the polycarbonate paper.

Many things were discovered by this experiment. For example, it was found that 2.3 μm diameter seems to be the lower limit of the tunnels through which human blood can be forced to flow through without hemolysis. Lingard (1974) showed that the white blood cells in the blood tend to plug off these tunnels and cause nonlinearity in the results. After removing the leukocytes with appropriate methods, the polycarbonate sieve offers a practical means to measure the flow properties of red cells.

4.6.10 Thermal and Chemical Experiments

The experiments mentioned above can be repeated at different temperatures to obtain the dependence of elastic and viscoelastic responses on temperature. The results provide some thermoelastic relations of the cell membrane. See Evans and Skalak (1979) and Waugh and Evans (1979).

Chemical manipulations to change spectrin or surface tension are associated with change of the shape of the red cell. See Steck (1974), Bennett and Branton (1977), and discussions in Sec. 4.8.

4.7 Elasticity of the Red Cell Membrane

An outstanding feature of the red cell membrane is that it is capable of large deformation with little change in surface area. This fact has been observed for a long time and was elaborated by Ponder (1948). The data shown in

Sec. 4.2 tell us that the red cell membrane is not unstretchable, but that even when the cell is sphered in a hypotonic solution and on the verge of hemolysis the increase of surface area is only 7% of the normal.

A second remarkable feature is that the shape of the red cell at equilibrium in a hydrostatic field is extremely regular. The stress and strain history experienced by the red cell does not affect its regular shape in equilibrium. This suggests that the cell membrane may be considered elastic.

Based on these observations, several stress–strain relationships for the red cell membrane have been proposed. It was shown by Fung and Tong (1968) that for a two-dimensional generalized plane-stress field in an isotropic material, under the general assumption that the stress is an analytic function of the strain, the most general stress–strain relationship may be put in the form

$$\sigma_1 = \frac{Y}{1 - v^2}(E_1 + vE_2), \qquad \sigma_2 = \frac{Y}{1 - v^2}(E_2 + vE_1), \tag{1a}$$

or

$$E_1 = \frac{1}{Y}(\sigma_1 - v\sigma_2), \qquad E_2 = \frac{1}{Y}(\sigma_2 - v\sigma_1), \tag{1b}$$

where Y and v are elastic constants that are functions of the strain invariants,

$$E_1 + E_2, \qquad E_1 E_2, \tag{2}$$

σ_1, σ_2 are the principal stresses referred to the principal axes x_1, x_2, and E_1, E_2 are the principal strains. Although Eq. (1) has the appearance of Hooke's law, it is actually nonlinear if Y and v do not remain constant.

To apply Eq. (1) to the membrane, we may integrate the stresses through the thickness of the membrane and denote the stress resultants by N_1, N_2; then use Eq. (1) with σ_1, σ_2 replaced by N_1, N_2. Y is then the Young's modulus multiplied by the membrane thickness. The strains E_1, E_2 are measured in the plane of the membrane, and can be defined either in the sense of Green (Lagrangian) or in the sense of Almansi (Eulerian). If Green's strains are used, then the principal strains E_1, E_2 can be expressed in terms of the principal stretch ratios λ_1, λ_2:

$$E_1 = \tfrac{1}{2}(\lambda_1^2 - 1), \qquad E_2 = \tfrac{1}{2}(\lambda_2^2 - 1), \tag{3}$$

where λ_1 is the quotient of the changed length of an element in the principal direction (corresponding to the principal strain E_1) divided by the original length of that element in the unstressed state, and λ_2 is the stretch ratio corresponding to E_2.

If the area remains constant, then $\lambda_1 \lambda_2 = 1$. If the membrane material is incompressible, then $\lambda_1 \lambda_2 \lambda_3 = 1$, where λ_3 denotes the stretch ratio in the direction perpendicular to the membrane. Hence if the area of an incompressible membrane remains constant, then the thickness of the membrane cannot change ($\lambda_3 = 1$). The reverse is also true. Since the red cell membrane area

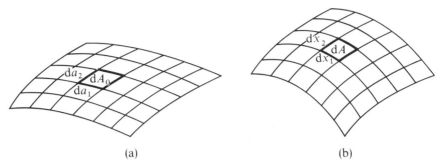

Figure 4.7:1 Deformation of a membrane. (a) Unstressed; (b) deformed.

remains almost constant, we conclude that its membrane thickness does not change much when the cell deforms.

The nonlinear Eq. (1) does not exhibit directly the special feature that it is easy to distort the cell membrane, but hard to change its area. This feature requires $v \to 1$. To exhibit this Skalak et al. (1973) and Evans and Skalak (1979) introduced the following strain measures:

$$(\lambda_1\lambda_2 - 1) \quad \text{for area and} \quad \tfrac{1}{2}(\lambda_1^2 - \lambda_2^2) \quad \text{for shear.} \tag{4}$$

The physical meaning of the quantity $(\lambda_1\lambda_2 - 1)$ is *the fractional change of the area of the membrane*. This can be seen by reference to Fig. 4.7:1, in which an element $da_1 \times da_2$ of area dA_0 in the unstressed membrane is deformed into an element $dx_1 \times dx_2$ of area dA; and the ratio is

$$\frac{dA}{dA_0} - 1 = \frac{dx_1}{da_1}\frac{dx_2}{da_2} - 1 = \lambda_1\lambda_2 - 1. \tag{5}$$

On the other hand, according to Eq. (3),

$$E_1 - E_2 = \tfrac{1}{2}(\lambda_1^2 - \lambda_2^2). \tag{6}$$

By Mohr's circle, one-half of $|E_1 - E_2|$ is the maximum shear strain, γ_{max} (Fig. 4.7:2). With these strain measures, Skalak et al. and Evans and Skalak wrote the following strain–stress resultant relationship:

$$N_1 = K(\lambda_1\lambda_2 - 1) + \mu\frac{\lambda_1^2 - \lambda_2^2}{2},$$

$$N_2 = K(\lambda_1\lambda_2 - 1) - \mu\frac{\lambda_1^2 - \lambda_2^2}{2}. \tag{7}$$

N_1, N_2 are the principal stress resultants. One-half of their sum is the mean stress resultant:

$$N_{\text{mean}} = \tfrac{1}{2}(N_1 + N_2) = K(\lambda_1\lambda_2 - 1). \tag{8}$$

One-half of their difference is the maximum shear stress resultant:

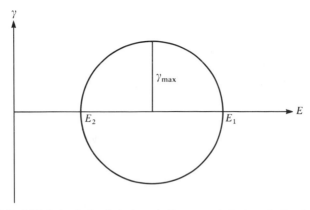

Figure 4.7:2 A Mohr's circle of strain. e is the normal strain; γ is the shear strain.

$$S_{\max} = \tfrac{1}{2}|N_1 - N_2| = \mu \frac{|\lambda_1^2 - \lambda_2^2|}{2}. \tag{9}$$

Equation (8) shows that the mean stress resultant is proportional to the areal change. K is an *elastic modulus for area dilatation* in units of dyn/cm. If we divide K by the cell membrane thickness, then we obtain the elastic modulus with respect to areal dilatation in units of newton/cm^2. On the other hand, Eq. (9) can be written as

$$S_{\max} = 2\mu\gamma_{\max}. \tag{10}$$

Equation (10) is of the same form as the Hooke's law, and the parameter μ can be called the *membrane shear modulus*. μ divided by the cell membrane thickness yields the usual shear modulus in units of newton/cm^2.

The moduli K and μ are functions of the strain invariants. But as a first approximation they may be treated as constants. If the constant K is much larger than μ, then the resistance of the membrane to change of area is much greater than that for distortion without change of area.

The first estimate of the elastic modulus of the red cell membrane was given by Katchalsky et al. (1960) based on sphering experiments in a hypotonic solution. The estimated value of the elastic modulus during the spherical phase of the cell and just before hemolysis is 3.1×10^7 dyn/cm^2 (as corrected by Skalak et al., 1973). The general range of this estimate was confirmed by Rand and Burton (1964) based on experiments in which red cells were sucked into micropipettes whose diameter was of the order of 2 μm in diameter. Rand and Burton gave the range of the moduli as 7.3×10^6 to 3.0×10^8 dyn/cm^2. Later, Hochmuth and Mohandas (1972) reported experiments in which red blood cells adhering to a glass surface were elongated due to shearing stress applied by the flow of the suspending fluid over the cells; they estimated that the modulus of elasticity is of the order of 10^4 dyn/cm^2. In another type of test in which the deformed cells were allowed to recover their natural shape, Hoeber and Hochmuth (1970) gave a value 7.2×10^5 dyn/cm^2. Skalak et al. (1973) observed that the elongation in the sphering and pipette experiments that give the higher estimates is about 8%; that in the uniaxial tests giving the lower modulus is 40%–60%. They therefore suggest that the lower values are appropriate for the constant μ, and the higher values are appropriate for the constant K. If the thickness of the cell membrane is assumed to be 50 Å, then $\mu = 10^4$ dyn/cm^2 and $h = 50$ Å gives a value $h\mu = 0.005$ dyn/cm; whereas $K = 10^8$ dyn/cm^2 and $h = 50$ Å gives $hK = 50$ dyn/cm.

4.8 The Red Cell Membrane Model

What kind of material makes up a membrane that has the elasticity, visco-elasticity, and viscoplasticity discussed above? The exact ultrastructure of the cell membrane of the blood cell is still unknown. A conceptual model advocated by Evans and Skalak (1979) is shown in Fig. 4.8:1. It shows the basic structure of a *unit membrane* consisting of two layers of phospholipid molecules (*bilayer*) with their hydrophilic heads facing outward and hydro-phobic tails locking into each other in the interior by hydrophobic forces. Globular proteins are partially embedded in the membrane, and partially protude from it. These are called *integral proteins*. Other proteins forming a network lining the endoface parallel to the membrane are called the *skeletal proteins*. *Linking proteins* connect the integral and skeletal proteins. From a

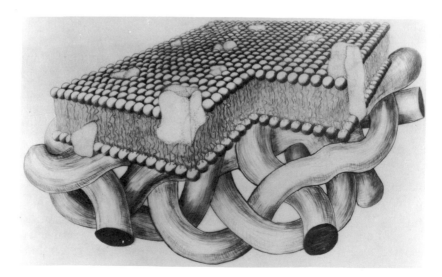

Figure 4.8:1 An idealized view of the red blood cell membrane composite. The under-neath spectrin network provides structural rigidity and support for the fluid lipid-protein layer of the membrane. From Evans and Skalak (1979), by permission.

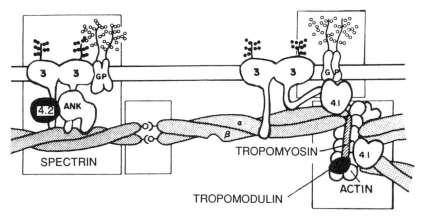

Figure 4.8:2 Molecular organization of red cell membrane skeleton. Ank = ankyrin; GP = glycophorin. Provided by L. A. Sung; modified from Lux and Becker (1989).

point of view of mechanics, the lipid layers and the proteins are one composite structure. Figure 4.8:2, by Amy Sung, shows these proteins in greater detail.

The lipid bilayer was demonstrated by electron microscopy. The existence of globular protein in the membrane was demonstrated later (Seifriz, 1927; Norris, 1939). By studying the location of antibodies in the cell membrane through electron microscopic photographs, Singer and Nicolson (1972) concluded that the cell membrane behaves like a fluid; the globular protein (the antibody in their case) can move about in the lipid bilayer in a way similar to the way icebergs move about on the surface of the ocean. The lipid bilayer has a characteristic thickness that is invariable. The mobility of the protein and lipid molecules is restricted to the plane of the membrane. This model is therefore called the *fluid mosaic* model. Experiments also showed, however, that cell membranes exhibit properties of a solid, e.g., elasticity. Since the lipids are in a fluid state, the solid characteristics of the membrane must be attributed to connections of proteins and other molecules associated with the membrane. Marchesi et al. (1969, 1970) identified spectrin on the red cell membrane. Steck (1974) and Singer (1974) suggested that the spectrin supports the lipid bilayer.

The proteins contribute approximately one-half of the weight of the red cell membrane (Chien and Sung, 1990a). They are generally identified by treating the red cell ghost membrane with sodium dodecyl sulfate and applying it onto polyacrylamide gel for electrophoresis. The proteins are numbered starting from the top of the gel, where the high molecular weight species are located, giving rise to protein bands 1 and 2 (spectrin), 2.1 (ankyrin), 3 (anion exchange protein), 4.1, 4.2, 5 (Actin), 6 (glyceraldehyde-3-phosphate dehydrogenase), 7, 8, and globin. Glycophrins and anion exchange protein are integrated in the membrane. Spectrin and actin are skeletal proteins. Ankyrin, and proteins 4.1, 4.2 are linking proteins. The molecular composition of all these proteins are known (see Chien and Sung, 1990a, for references).

4.9 The Effects of Red Cell Deformability on Turbulence in Blood Flow

Turbulence is an important feature of flow. Turbulence in blood flow affects the resistance to flow, the shear stress acting on the blood vessel wall, the tensile stress in the endothelial cell membrane, and the mass transport characteristics from the blood to the vessel wall. Turbulence has implications on the initiation of atherosclerosis, and the formation of blood clot. In a pulsatile blood flow, whether turbulence sets in or not depends on the values of two dimensionless parameters: the peak Reynolds number, and a frequency parameter called the Womersley number. See Fung (1990), Chapter 5, Secs. 5.12 and 5.13. Naturally, it is also affected by blood rheology, which in turn depends on the deformability of red blood cells.

Stein and Sabbah (1974) used a constant temperature hot-film anemometer to study blood flow in arteries. They showed that turbulence augments the sickling process of sickle-hemoglobin-containing erythrocytes, and the formation of thrombi. They made two arteriovenous shunts in each of eight dogs under anesthesia, one from each femeral artery to the contralateral femoral vein. One shunt contained a turbulence-producing device, the other did not. Both shunts are tubing with an internal diameter of 4.8 mm. Turbulence is produced in one by creating a converging–diverging section with an orifice diameter of 1.6 mm, and a divergent angle of about 45°. Flow in both shunts was kept at comparable levels in both shunts for 7 min. The shunts were then removed and thrombi were taken from them and immediately weighed. The total results show that virtually all of the thrombi were collected from the turbulent shunts, and that the weight can be expressed by the regression line:

thrombus weight in grams $= 3 \times 10^{-4}$ (Reynolds No.) $- 0.02$.

The experimental Reynolds number range was 200 to 900, based on a diameter of 4.8 mm for laminar flow and 1.6 mm for turbulent flow.

The effect of hematocrit (volume fraction of red cells in blood) on he intensity of turbulence was studied by Stein et al. (1975) *in vitro* by allowing the blood to flow through a tube with a turbulence producing stenosis which has a conical divergent section with a divergent angle of about 45°. The instantaneous velocity of flow was measured with a hot film anemometer. The longitudinal velocity $u(x, t)$ is written as a sum of a steady mean value U and a fluctuation $u'(x, t)$:

$$u(x, t) = U + u'(x, t).$$

If the average value of a quantity over time and over the space of the anemometer probe is indicated by a bar over the quantity, then the turbulent fluctuation u' is defined by the equations

$$\overline{u'(x, t)} = 0,$$

$$\overline{u(x, t)} = U,$$

$$\overline{[u'(x, t)]^2} = \overline{u'^2}.$$

The root mean square $(\overline{u'^2})^{1/2}$ is called the *absolute intensity of turbulence*, and the ratio of the root mean square of turbulence to U is called *the relative intensity of turbulence*:

$$\text{relative intensity of turbulence} = \frac{(\overline{u'^2})^{1/2}}{U}.$$

Stein et al. compared the results of flow with blood of various hematocrit with flows of a mixture of plasma and dextrose of nearly identical viscosity and density. They showed that at hematocrits between 20% and 30%, the intensity of turbulence of the blood was over twice that of the equally viscous and dense plasma. The addition of more red cells to reach a hematocrit of 40% caused a smaller difference between blood and comparable plasma. Further, Sabbah and Stein (1976) showed that hardening the red cells with glutaraldehyde caused and increase of the tendency for the flow to become turbulent, and in the turbulent regime increased the relative intensity of turbulence. The least-squares regression lines of the experimental results are of the form

$$\text{relative intensity of turbulence} = a\,(\text{Reynolds No.}) + b$$

with a ranging from 0.00027 to 0.0004 and b ranging from 0.27 to 0.38 for the blood of four patients.

4.10 Passive Deformation of Leukocytes

When leukocytes are put in an isotonic solution containing EDTA, active spontaneous movement stops and the cells respond passively to external loads. Sung et al. (1982, 1988a) performed micropipet sucking and release experiments on these passive cells and obtained the records of the shape change of human neutrophils in aspiration and recovery. Dong et al. (1988) analyzed these records with a leukocyte model which assumes the cell to be an elastic shell containing a Maxwell viscoelastic fluid. The wall of the shell consists of a layer of cortex wrapped in a lipid membrane. The cortex represents an actin-rich layer near the cell surface, and is assumed to have a residual tensile stress which keeps the cell in spherical shape at the no-load state. The lipid-bilayer membrane is assumed to have an excess area which folds on the cortex. The cell is in a passive state. All materials are assumed to be incompressible.

Under the assumptions that the combined thickness of the cell membrane and the cortex layer is much smaller than the cell radius, that the bending of the cortex can be ignored, and that only axisymmetric deformation is considered, the equations of equilibrium of the elastic cortex are the same as Eqs. (1) and (2) of Sec. 4.5. The solution is given by Eqs. (3)–(6) of Sec. 4.5, except that $p_i - p_o$, the radial loading per unit area, is a function of the polar angle ϕ in the present case, whereas $p_i - p_o$ was a constant for the red blood cell in Sec. 4.5. Dong et al. assumed the existence of a prestress which is a uniform hoop tension in a spherical shell. This initial tension is carried as a parameter in the analysis.

For the interior of the cell Dong et al. used a rectangular Cartesian frame of reference and denoted the stress tensor by σ_{ij}, which is split into a hydrostatic pressure p and a stress deviation tensor τ_{ij}:

$$\sigma_{ij} = -p\delta_{ij} + \tau_{ij} \qquad (i,j = 1, 2, 3), \tag{1}$$

where δ_{ij} is the Kronecker delta. A Maxwell fluid is assumed. Let v_i ($i = 1, 2, 3$) be the velocity inside the leukocyte, and $\dot{\gamma}_{ij}$ be the strain rate tensor [Eq. (2.4:3)]

$$\dot{\gamma}_{ij} = \frac{1}{2}\left(\frac{\partial v_i}{\partial x_j} + \frac{\partial v_j}{\partial x_i}\right), \tag{2}$$

then the constitutive equation of a Maxwell fluid is [Sec. 2.11, Eq. (1)]

$$\tau_{ij} + \frac{\mu}{k}\dot{\tau}_{ij} = 2\mu\dot{\gamma}_{ij} \tag{3}$$

in which each dot represents a differentiation with respect to time, k (with units of N/m^2) is an elastic constant, and μ (with units $N \cdot s/m^2$) is a coefficient of viscosity. On neglecting the gravitational and inertial forces, the equation of equilibrium is (Chapter 2)

$$\frac{\partial \sigma_{ij}}{\partial x_j} = 0. \tag{4}$$

The equation of continuity is

$$\frac{\partial v_j}{\partial x_j} = 0. \tag{5}$$

With Eqs. (1)–(3), Eq. (4) can be written as

$$\mu\frac{\partial^2 v_i}{\partial x_j \partial x_j} = \frac{\partial p}{\partial x_i} + \frac{\mu}{k}\frac{\partial}{\partial_i}\frac{\partial p}{\partial t}. \tag{6}$$

In the case of slow axisymmetric viscous flow associated with a spherical boundary, a general solution of Eq. (6) expressed in terms of spherical harmonics has been given by H. Lamb in his *Hydrodynamics*, 6th ed., 1932, Cambridge Univ Press, reprinted by Dover Publications 1945, Chapter XI, Secs. 336 and 352. Lamb's solution can be easily generalized to a linear viscoelastic solid. In spherical coordinates (r, ϕ), where ϕ is the meridional angle, the solution is expressed in terms of the Legendre polynomials, $L_n(\eta)$, where $\eta = \cos \phi$, and n is the order of the polynomial. Dong et al. (1988) gave

$$v = \sum_{n=1}^{\infty} \left\{ \frac{(n + 3)r^2}{2(n + 1)(2n + 3)}\left(\frac{1}{\mu}\nabla P_n + \frac{1}{k}\nabla\frac{\partial P_n}{\partial t}\right) + \nabla\Phi_n \right.$$
$$\left. - \frac{n\mathbf{r}}{(n + 1)(2n + 3)}\left(\frac{1}{\mu}P_n + \frac{1}{k}\frac{\partial P_n}{\partial t}\right)\right\}, \tag{7}$$

where $\mathbf{r}$ is the radial position vector, and ∇ is the gradient operator.

The components of the velocity vector are, in the axisymmetric case,

$$v_r = \sum_{n=1}^{\infty} \left\{ \frac{nr}{2(2n+3)} \left(\frac{1}{\mu} P_n + \frac{1}{k} \frac{\partial P_n}{\partial t} \right) + \frac{n}{r} \Phi_n \right\}, \tag{8}$$

$$v_\phi = \sum_{n=1}^{\infty} \left\{ \frac{1}{r} \frac{\partial \Phi_n}{\partial \phi} + \frac{(n+3)r}{2(n+1)(2n+3)} \left(\frac{1}{\mu} \frac{\partial P_n}{\partial \phi} + \frac{1}{k} \frac{\partial^2 P_n}{\partial \phi \partial t} \right) \right\}, \tag{9}$$

in which P_n and Φ_n are defined as

$$P_n = \alpha_n(t) \left(\frac{r}{a_0} \right)^n L_n(\eta), \tag{10}$$

$$\Phi_n = \beta_n(t) \left(\frac{r}{a_0} \right)^n L_n(\eta) \tag{11}$$

and

$$a_0 = \text{original radius of the leukocyte},$$

$$\alpha_n, \beta_n = \text{coefficients to be determined}, \tag{12}$$

$$\eta = \cos \phi.$$

The corresponding radial stress is

$$\sigma_r = \tau_{r0} e^{-(k/\mu)t} + \sum_{n=1}^{\infty} \left\{ -P_n + 2k \int_0^t e^{-(k/\mu)(t-t')} \left[\frac{n(n+1)}{2(2n+3)} \right. \right.$$

$$\left. \left. \times \left(\frac{1}{\mu} P_n + \frac{1}{k} \frac{\partial P_n}{\partial t} \right) + \frac{n(n-1)}{r^2} \Phi_n \right] dt' \right\}, \tag{13}$$

where τ_{r0} is the radial stress at $t = 0^+$.

With these analytic solutions, Dong et al. (1988) analyzed the experimental results of Sung et al. (1988a). They also developed a finite element method to compute the cell motion. For the human neutrophil, they obtained the following parameters that best fit the experimental data:

$$k = 28.5 \text{ N/m}^2,$$

$$\mu = 30.0 \text{ N} \cdot \text{s/m}^2, \tag{14}$$

$$T_0 = 3.1 \times 10^{-5} \text{ N/m} = 0.031 \text{ dyn/cm}.$$

In a later paper, Dong, Skalak, and Sung (1991) used micropipet to study the cytoplasmic rheology of passive neutrophils. They showed that the internal structure of the neutrophil is inhomogeneous. The nucleus is much stiffer and more viscous than the cytoplasm. As a result the mechanical properties of the cell as a whole depends on the degree of deformation. If a passive white blood cell is modeled as a viscoelastic body bounded by a cortical shell with a persistent tension, then the various viscoelastic coefficients depend on the degree of cell deformation. Dong et al. (1991) used Pipkin's (1964) constitutive equation for a viscoelastic body in curvilinear coordinates:

$$\sigma^{ij}(t) = -p(t)G^{ij} + \frac{\partial z^i \partial z^j}{\partial x^a dx^b} g^{am} g^{bn} \left\{ 2k\gamma_{mn}(t) \right.$$

$$\left. - \frac{2k^2}{\mu} \int_0^t \dot{\gamma}_{mn}(t - t') \exp\left(-\frac{k}{\mu} t'\right) dt' \right\}, \tag{15}$$

where σ^{ij} is the contravariant components of Cauchy stress (or spatial stress) referred to the deformed state in the z^i coordinate system ($i, j = 1, 2, 3$) at time t. k and μ are the elastic and viscous coefficients, respectively. p is an arbitrary hydrostatic pressure in the viscoelastic body. g_{ab} are the covariant components of the metric tensor referred to the initial state in the x^a coordinates ($a, b = 1, 2, 3$). g^{ab} and G^{ij} are the contravariant components of the metric tensor referred to x^a and z^i coordinates, respectively. γ_{mn} is the Green strain tensor whose covariant components in the x^a configuration are defined as ($m, n = 1, 2, 3$):

$$\gamma_{mn} = \frac{1}{2}\left(\frac{\partial z^i}{\partial x^m}\frac{\partial z^i}{\partial x^n} - g_{mn}\right). \tag{16}$$

A dot is a differentiation with respect to time.

The cortical layer of the cell is characterized by two principal membrane tensions, N_1 and N_2 (Fig. 4.5:2). It is assumed that ($i = 1, 2$)

$$N_i = T_0 + E_a A(\lambda_1, \lambda_2) + E_s B_i(\lambda_1, \lambda_2), \tag{17}$$

in which λ_1 and λ_2 are two stretch ratios in the meridional and circumferential directions, respectively. E_a, and E_s are two elastic moduli. A and B_i are functions of the principal stretch ratios that characterize the isotropic and anisotropic elastic deformations, respectively, which are suggested as

$$A(\lambda_1, \lambda_2) = (\lambda_1 \lambda_2 - 1)^p, \tag{18}$$

$$B_i(\lambda_1, \lambda_2) = \frac{1}{2\lambda_1 \lambda_2}\left(\lambda_i^2 - \frac{1}{\lambda_1^2}\right), \tag{19}$$

where ρ is a positive finite number, $\rho > 1$.

A sliding boundary condition is imposed on the cortical layer surface where the cell contacts the pipet wall. A finite element model is used for calculation. Dong et al. (1988), is a linearized version. See also Schmid-Schönbein et al. (1983) and Tözeren et al. (1984, 1989).

Evans and Yeung (1989) and Needham and Hochmuth (1990) also studied human white blood cells with the micropipet method. They showed, however, that the cell behaves under a steady suction pressure as a Newtonian liquid drop with a cortical tension as long as it is in a resting state. They found that the cytoplasmic viscosity of resting neutrophils and other white cells to be of the order of 10^3 poise which is 10^5 times that of water. Further work is needed to resolve this controversy.

4.11 Cell Adhesion: Multipipets Experiments

An example of the use of more than one pipet to study the adhesion between cells is shown in Fig. 4.11.1 Sung et al. (1986) studied the conjugation and separation of a cytotoxic T cell (human clone FI, with specificity of HLA-DRw6) and a target cell (JY: HLA-A2, -B7, -DR4, w6) prior to delivery of the lethal hit. The figure shows a sequence of events detailed in the legends. Note the deformation of the cells when pulled by the forces imposed by the pipets, and the formation of tether when the cells separated. The tether seems to behave as an elastic band. A critical tensile stress S_c in the conjugated area of the cells must be exceeded in order to separate the two cells. Sung et al. found that the S_c for the FI–JY pair is $(1.529 \pm .045$ SD$) \times 10^4$ dyn/cm^2. This junction avidity for the FI–JY pair is 6 times larger than the critical stress for separating an FI–FI pair, and 13 times larger than the critical stress to separate a JY–JY pair.

Similar experiments have been done by Sung et al. (1988b) on the adhesion of cells to glass, plastic, lipid membranes, and bilayer membranes containing various molecules.

4.12 Topics of Cell Mechanics

The study of the mechanics of erythrocytes and leukocytes throws the door open to cell mechanics in general. Some significant new topics and references are listed below, in Secs. 4.12.1–4.12.5.

The mechanics of single cells and cells in a continuum will undoubtedly occupy an important position in Biomechanics of the future. But the subject has not been advanced far enough today to explain all the mechanical properties of the tissues and organs discussed in this volume. A planned volume by Skalak et al., entitled *Biomechanics of Cells*, to be published by Springer-Verlag, contains an excellent exposition of the subject.

4.12.1 Active Amoeboid Movement

The machinery, mechanism, and mechanics of active movement and adhesion of leukocytes, amoebae, and other cells are extremely important to physiology, pathology, and bioengineering. For example, activated leukocytes can adhere to the endothelial wall, increase resistance to blood flow, or, in capillaries, block the flow (Schmid-Schönbein 1987). Sometimes they are seen to penetrate the endothelial cells and move from the blood to the interstitial space. But the mechanical properties of the cell constituents in the active state are largely unknown. See Holberton (1977).

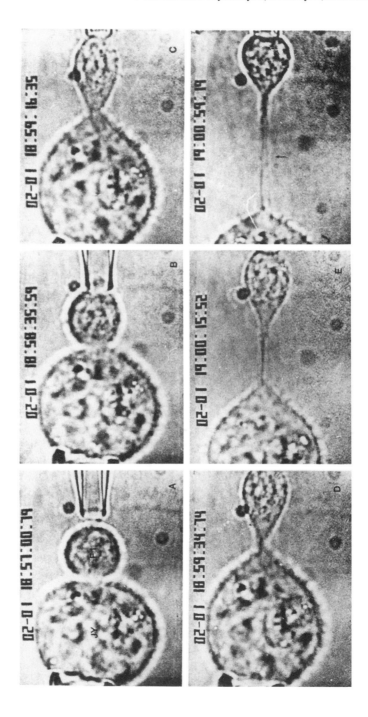

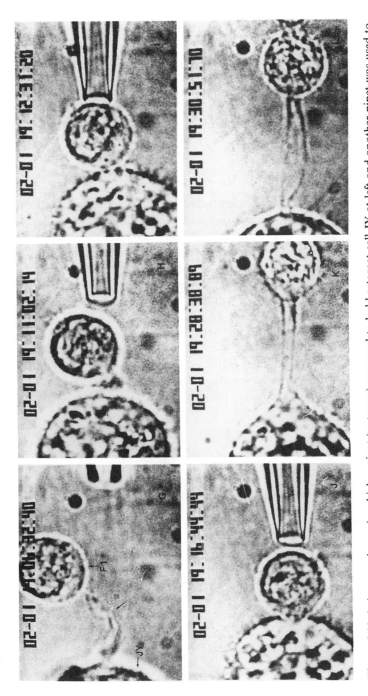

Figure 4.11:1 An experiment in which a pipet (not seen) was used to hold a target cell JY at left and another pipet was used to hold a T cell FI on the right-hand side. (A) Conjugation of JY and FI cell was released from its holding pipette. (B) Reaspiration of FI and alignment of the two pipettes. (C and D) FI was pulled away from JY, and both cells showed deformation. (E and F) The FI cell was increasingly separated from the JY cell as the FI-holding pipette was pulled away. The conjugated area was reduced to membrane tethers. (G) The pressure in the FI-holding pipet was released, and FI was free in medium, only with some membrane tether (arrow) attached to JY. (H and I) FI was gradually drawn close to JY as the membrane tether disappeared. (J) FI reconjugated with JY. (K and L) FI was again pulled away from JY; the force requirement was comparable to that of the first trial; the experiments are reproducible. From K.-L. P. Sung, L. A. Sung, M. Crimmins, S. J. Burakoff, and S. Chien (1986) *Science* **234**, 1405–1408. Reproduced by permission.

4.12.2 Stretch-Activated Channels and Cell Volume Regulation

The application of patch-clamp technique has lead to the discovery of a class of "gated" channels in cell membrane. These channels are activated by simply stretching the membrane and appear to be triggered by tension or stress development in cytoskeletal elements closely associated with the cell membrane.

Stretch-activated channels have been identified in the plasma membrane of E. Coli, yeast, plant, invertebrate, and vertebrate cells. The majority prefer monovalent cations and are mildly selective for K^+ over Na^+, but there are several reports of anion-selective stretch activated channels, and Ca^+ channels (Sachs, 1990).

Schultz (1989) suggests that the stretch-activated channels regulates the cell volume of all cells. When the osmotic pressure outside a cell is decreased, the cell swells in response and stretches its membrane, and activates the stretch-activated channels. Then K^+ (and anion) inside of the cell will pass out through the stretch-activated channels, decreasing the osmotic pressure inside the cell, and achieving some degree of volume regulation and tension control.

4.12.3 Asymmetry of the Lipids in Cell Membranes

Op den Kamp (1979) has shown that in platelet and red cell membranes the two layers of lipids in the bi-layer structure of a cell membrane are asymmetric, Fig. 4.12:1. The outer layer is composed of phosphatidylcholine and sphingo-

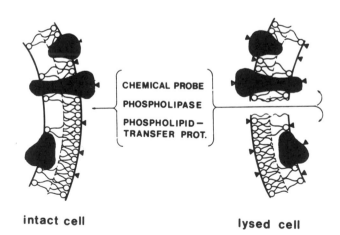

CHEMICAL PROBE

PHOSPHOLIPASE

PHOSPHOLIPID–
TRANSFER PROT.

intact cell   **lysed cell**

Figure 4.12:1 Principle of localization experiment of cell membrane. *Circles with two tails* represent phospholipids. Although not depicted, equal amount of phospholipids are present in inner and outer layers. Large *black shapes* are membrane proteins. Some of the reagents used are listed in the middle of the figure. *Arrowheads* indicate reactions with reagents. Adapted from Op den Kamp (1979) by Zwaal (1988). Reproduced by permission.

myelin, both of which have a phosphocholine head group. The inner layer has phosphatidylethanolamine and phosphatidylserine, both of which have amino group in their head group. This discovery was made by an experiment whose principle is shown in Fig. 4.12:1. Intact and lysed cells are treated with reagents (chemical probe, see Fig. 4.12:1) that cannot permeate the intact cells. In experiments with intact cells, the reagents react with the outer layer only. In experiments with lysed cells, the reagents react with both the inner and outer layers. After the reactions are completed, the lipids can be extracted and those that were susceptible to the probes can be identified. By comparison of the two results, information is obtained about the inner layer alone.

The tails are also asymmetric. The choline-phospholipids are more saturated than the amino-phospholipids. The amount of cholesterol in the outer layer is about twice that in the inner layer.

Zwaal (1978, 1988) has done experiments which show that the activation of platelets causes a redistribution of the phospholipids in the two layers of the platelets cell membrane. Cell membrane phospholipids react with activating factor X; hence offering a local control of blood clotting.

4.12.4 Biochemical Transduction of Mechanical Strain

Lanyon et al. (1982, 1984) and Rubin and Lanyon (1985) first measured the amount of deformation in bone during normal loading in animals and humans. Strains $< 5 \times 10^{-4}$ were not found to stimulate, those between 5×10^{-4} and 1.5×10^{-3} appeared to maintain bone mass, and strains $> 1.5 \times 10^{-3}$ increased bone mass. Peak strains during strenuous exercises are in the range of 3×10^{-3} and the strain rate about 5×10^{-3} per second. Microcrack failure occurs at a strain of 3×10^{-2}. Jones et al. (1991) tested osteoblast-like cells and skin fibroblasts and found that only periostal (bone surface) osteoblasts are sensitive to strains biochemically within the physiological range. Osteoblasts derived from the haversian system and skin fibroblasts do not respond biochemically except at higher strains. Jones et al. (1991) found that the transduction mechanism is located in the cytoskeleton and activates the membrane phospholipase C. Application of strains $> 10^{-2}$ results in a change of morphology of osteoblasts to become fibroblast-like.

4.12.5 Effect of Mechanical Forces in Organogenesis

Mechanical forces play important growth regulatory roles in bone (DeWitt et al., 1984), cartilage (Klein-Nulend et al., 1987), connective tissue (Curtis and Seehar, 1978), epithelial tissue (Odell et al., 1981), lung (Rannels, 1989), cardiac muscle (Morgan et al., 1987), smooth muscle (Leung et al., 1976), skeletal muscle (Stewart, 1972; Strohman et al., 1990; Vandenburgh, 1988), nerve (Bray 1984), and blood vessels (Leung et al., 1976, Folkman and

Handenschild, 1980; Ingber and Folkman, 1989). Vandenburgh and Karlisch (1989) have used computerized equipment to impose mechanical forces during cell culture, and demonstrated the organogenesis from a single cell (skeletal muscle myoblast) in vitro. Tissue transformation from muscle into bone in vivo is reported by Khouri et al. (1991). An interesting mathematical description of morphogenesis is given by Odell et al. (1981).

Problems

4.1 Show that a circular cylindrical surface and a circular cone are both *applicable* to a plane, i.e., they can be deformed into a plane without stretching and tearing. See Sec. 4.5.2. The same is true for cylinders and cones of arbitrary cross section. Note that the statement is true only in the sense of differential geometry; and not in a global sense. To deform a whole cylinder into a plane, a cut has to be made. The cut cylindrical surface can then be spread out into a plane. Verify that the deformations shown in Fig. 4.5:4 are applicable deformations.

4.2 According to Schmid-Schoenbein and Wells (1969), a red blood cell can be seen "tank-treading" in a shear flow; that is, under the microscope its wall seems to rotate about itself like the belt of a tank (Fig. P4.2). Different points on the belt take turns being at the front, bottom, rear, and on top. It follows that the material of the red cell membrane at the dimples and the equator is not fixed.

Show that, in accordance with the analysis given in Sec. 4.5, the biconcave disk is applicable to itself when it "tank-treads" (Fig. P4.2). Would the membrane stress in the cell membrane be changed in such a motion?

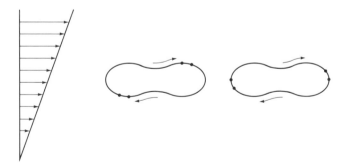

Figure P4.2 "Tank treading" of red blood cells in shear flow.

4.3 Imagine an arbitrary continuous curve in a three-dimensional space. A tangent line can be drawn at every point on the curve. As one moves continuously along the curve, the tangent lines of successive points form a continuum and sweep out a surface that is called a *tangent surface*. Show that a tangent surface generated by an arbitrary space curve is *developable*, i.e., it is *applicable* to a plane.

4.4 Prove that, for two surfaces to be applicable to each other as defined in Sec. 4.5.2, the total curvature (product of the principal curvatures) must remain the same at the corresponding points.

4.5 Draw Mohr's circles for the stress states in the cell membrane corresponding to uniaxial tension, biaxial tension, and various stages of osmotic swelling illustrated in Fig. 4.6:1. Draw Mohr's circles for the corresponding strain states; thus verify the discussions of experiments presented in Sec. 4.6.

4.6 Study an original paper in which one of the experiments discussed in Sec. 4.6 was first published (see Evans and Skalak, 1979, Chapter V, for detailed history and references). Discuss the analysis carefully from the point of view of continuum mechanics. Are the equations of equilibrium, the constitutive equation, the equation of continuity, and the boundary conditions satisfied rigorously? What are the approximations introduced? How good are these approximations? In which way can the approximations seriously affect the accuracy of the result?

4.7 Why is a red cell so deformable but a white cell is less so?

4.8 What are the sources of bending rigidity of a red cell membrane?

4.9 What is the evidence that the bending rigidity of a red cell membrane exists? In what circumstances is the bending rigidity important?

4.10 What is the evidence that the hydrostatic pressure in the red cell is about the same as that outside the cell?

4.11 What are the factors that determine the cell volume of a red blood cell? Most cells have mechanisms regulating their volume, see Sec. 4.12.2. Is red cell an exception? Explain the apparent contradiction.

4.12 How can a cell deform without changing its surface area and volume? Can a sphere do it? Formulate a mathematical theory for such a deformation.

4.13 What is the evidence that the hemoglobin in the red blood cell is in a liquid state? When a hemoglobin solution is examined under an x ray, a definite diffraction pattern exists. Why does the existence of such a crystalline pattern not in conflict with the idea that the solution is in a liquid state? Why is it so difficult to assign a cell membrane thickness to a red cell? The values of red cell membrane thickness given in the literature vary over a wide range.

4.14 A diver dives 30 m underwater. What is the internal pressure in the diver's red blood cells? Give a theoretical proof of your conclusion.

4.15 A red blood cell moves in a small, tightly fitting capillary blood vessel. What are the equations governing the flow of plasma around the red cell in the vessel? What are the boundary conditions? How could you prove the "lubrication" effect that arises in this situation as pointed out by Lighthill (Sec. 5.8)?

4.16 A spherocyte (a spherical cell) of radius a is squeezed between two flat plates. When the force is F, the area of contact between the cell and the plate is a small circle of radius b, $b \ll a$. What is the internal pressure in the cell? What is the stress resultant in the cell membrane?

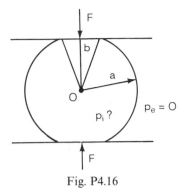

Fig. P4.16

4.17 The membrane of an erythrocyte carries electric charge. How would the cell deform in a high-frequency electric field? How is the cell deformation related to the frequency and intensity of the electric field? How can this deformation be measured? See Kage et al. (1990).

References to Erythrocytes

Bennett, V. and Branton, D. (1977) Selective association of spectrin with the cytoplasmic surface of human erythrocyte plasma membranes. Quantitative determination with purified (32 p) spectrin. *J. Biol. Chem.* **252**, 2753–2763.

Bessis, M. (1956) *Cytology of the Blood and Blood-Forming Organs.* Grune and Stratton, New York.

Blackshear, P. L., Jr. (1972) Mechanical hemolysis in flowing blood. In *Biomechanics: Its Foundations and Objectives.* Fung, Perrone, and Anliker (eds.) Prentice-Hall, Englewood Cliffs, NJ.

Bo, L. and Waugh, R. E. (1989) Determination of bilayer membrane bending stiffness by tether formation from giant, thin-walled vesicles. *Biophys. J.* **55**, 509–517.

Braasch, D. and Jennett, W. (1968) Erythrozyten flexibilität, Hämokonzentration und Reibungswiderstand in Glascapillaren mid Durchmessern zwischen 6 bis 50 μ. *Pflügers Arch. Physiol.* **302**, 245–254.

Bränemark, P.-I. (1971) *Intravascular Anatomy of Blood Cells in Man.* Monograph. Karger, Basel.

Brailsford, J. D. and Bull, B. S. (1973) The red cell—A macromodel simulating the hypotonic-sphere isotonic disk transformation. *J. Theor. Biol.* **39**, 325–332.

Bull, B. S. and Brailsford, J. D. (1975) The relative importance of bending and shear in stabilizing the shape of the red blood cell. *Blood Cells* **1**, 323–331.

Canham, P. B. (1970) The minimum energy of bending as a possible explanation of the biconcave shape of the human red blood cell. *J. Theor. Biol.* **26**, 61–81.

Canham, P. B. and Burton, A. C. (1968) Distribution of size and shape in populations of normal human red cells. *Circulation Res.* **22**, 405–422.

Chen, P. and Fung, Y. C. (1973) Extreme-value statistics of human red blood cells. *Microvasc. Res.* **6**, 32–43.

Chien, S. (1972) Present state of blood rheology. In *Hemodilution: Theoretical Basis and Clinical Application*, K. Messmer and H. Schmid-Schoenbein (eds.) Karger, Basel.

Chien, S., Usami, S., Dellenback, R. T., and Gregersen, M. I. (1967) Blood viscosity: Influence of erythrocyte deformation. *Science* **157**, 827–829.

Chien, S., Usami, S., Dellenback, R. J., and Bryant, C. A. (1971) Comparative home-orheology—Hematological implications of species differences in blood viscosity. *Biorheology* **8**, 35–57.

Chien, S., Sung, K. L. P., Skalak, R., Usami, S., and Tözeren, A. (1978) Theoretical and experimental studies on viscoelastic properties of erythrocyte membrane. *Biophys. J.* **24**, 463–487.

Chien, S. and Sung, L. A. (1990a) Molecular basis of red cell membrane rheology. *Biorheology* **27**, 327–344.

Chien, S., Feng, S.-S., Vayo, M., Sung, L. A., Usami, S., and Skalak, R. (1990b) The dynamics of shear disaggregation of red blood cells in a flow channel. *Biorheology* **27**, 135–147.

Cohen, W. D. (1978) Observations on the marginal band system of nucleated erythrocytes. *J. Cell Biol.* **78**, 260–273.

Cohen, W. D., Bartlet, D., Jaeger, R., Langford, G., and Nemhauser, I. (1982) The cytoskeletal system of nucleated erythrocytes. I. Composition and function of major elements. *J. Cell Biol.* **93**, 828–838.

Cokelet, G. R. and Meiselman, H. J. (1968) Rheological comparison of hemoglobin solutions and erythrocyte suspensions. *Science* **162**, 275–277.

Cokelet, G. R., Meiselman, J. H., and Brooks, D. E. (eds.) (1980) *Erythrocyte Mechanics and Blood Flow*. Alan Liss, New York.

Dick, D. A. T. and Lowenstein, L. M. (1958) Osmotic equilibria in human erythrocytes by immersion refractometry. *Proc. Roy. Soc. London B* **148**, 241–256.

Dintenfass, L. (1968) Internal viscosity of the red cell and a blood viscosity equation. *Nature* **219**, 956–958.

Evans, E. A. (1983) Bending elastic modulus of red blood cell membrane derived from buckling instability in micropipet aspiration tests. *Biophys. J.* **43**, 27–30.

Evans, E. and Fung, Y. C. (1972) Improved measurements of the erythrocyte geometry. *Microvasc. Res.* **4**, 335–347.

Evans, E. A. and Hochmuth, R. M. (1976) Membrane viscoelastocity. *Biophys. J.* **16**, 13–26.

Evans, E. A., Waugh, R., and Melnik, L. (1976) Elastic area compressibility modulus of red cell membrane. *Biophys. J.* **16**, 585–595.

Evans, E. A. and Skalak, R. (1979) *Mechanics and Thermodynamics of Biomembranes*. CRC Press, Boca Raton, FL.

Evans, E. A. and Rawicz, W. (1990) Entropy-driven tension and bending elasticity in condensed-fluid membranes. *Phys. Rev. Lett.* **64**, 2094–2097.

Flügge, W. (1960) *Stresses in Shells*. Springer-Verlag, Berlin.

Fry D. L. (1968) Acute vascular endothelial changes associated with increased blood velocity gradients. *Circulation Res.* **22**, 165–197.

Fung, Y. C. (1966) Theoretical considerations of the elasticity of red cells and small blood vessels. *Fed. Proc. Symp. Microcirc.* **25**, Part I, 1761–1772.

Fung Y. C. (1968) Microcirculation dynamics. In: *Biomedical Sciences Instrumentation*. Instrument Society of America. Plenum Press, New York, Vol. 4, pp. 310–320.

Fung, Y. C. and Tong, P. (1968) Theory of the sphering of red blood cells. *J. Biophys.* **8**, 175–198.

Graustein, W. C. (1935) *Differential Geometry*. Macmillan, New York.

Gregersen, M. I., Bryant, C. A., Hammerle, W. E., Usami, S., and Chien, S. (1967) Flow characteristics of human erythrocytes through polycarbonate sieves. *Science* **157**, 825–827.

Gumbel, E. J. (1954) *Statistical Theory of Extreme Value and Somne Practical Applications.* National Bureau of Standards, Applied Math. Ser. 33. Superintendent of Documents, U.S. Government Printing Office, Washington, D.C., pp. 1–51.

Gumbel, E. J. (1958) *Statistics of Extremes.* Columbia University Press, New York.

Hochmuth, R. M. (1987) Properties of red blood cells. In *Handbook of Bioengineering,* R. Skalak and S. Chien (eds.) McGraw-Hill, New York, Chapter 12.

Hochmuth, R. M., Marple, R. N., and Sutera, S. P. (1970) Capillary blood flow. I. Erythrocyte deformation in glass capillaries. *Microvasc. Res.* **2**, 409–419.

Hochmuth, R. M. and Mohandas, N. (1972) Uniaxial loading of the red cell membrane. *J. Biomech.* **5**, 501–509.

Hochmuth, R. M., Mohandas, N., and Blackshear, Jr., P. L. (1973) Measurement of the elastic modulus for red cell membrane using a fluid mechanical technique. *Biophys. J.* **13**, 747–762.

Hochmuth, R. M., Worthy, P. R., and Evans, E. A. (1979) Red cell extensional recovery and the determination of membrane viscosity. *Biophys. J.* **26**, 101–114.

Hochmuth, R. M., Evans, E. A., Wiles, H. C., and McCown, J. T. (1983) Mechanical measurement of red cell membrane thickness. *Science* **220**, 101–102.

Hoeber, T. W. and Hochmuth, R. M. (1970) Measurement of red blood cell modulus of elasticity by in vitro and model cell experiments. *Trans. ASME Ser. D,* **92**, 604.

Houchin, D. W., Munn, J. I., and Parnell, B. L. (1958) A method for the measurement of red cell dimensions and calculation of mean corpuscular volume and surface area. *Blood* **13**, 1185–1191.

Kage, H. S., Engelhardt, H., and Sackman, E. (1990) A precision method to measure average viscoelastic parameters of erythrocyte populations. *Biorheology* **27**, 67–78.

Katchalsky, A., Kedem, D., Klibansky, C., and DeVries, A. (1960) Rheological considerations of the haemolysing red blood cell. In *Flow Properties of Blood and Other Biological Systems,* A. L. Copley and G. Stainsby (eds.) Pergamon, New York, pp. 155–171.

King, J. R. (1971) *Probability Chart for Decision Making.* Industrial Press, New York.

Lingard, P. S. (1974 et seq) Capillary pore rheology of erythrocytes. I. Hydroelastic behavior of human erythrocytes. *Microvasc. Res.* **8**, 53–63. II. Preparation of leucocyte-poor suspension. *ibid,* **8**, 181–191 (1974). III. Behavior in narrow capillary pores. *ibid.,* **13**, 29–58 (1977). IV. Effect of pore diameter and hematocrit. *ibid.,* **13**, 59–77 (1977). V. Glass capillary array. *ibid.,* **17**, 272–289 (1979).

Lingard, P. S. and Whitmore, R. L. (1974) The deformation of disk-shaped particles by a shearing fluid with application to the red blood cell. *J. Colloid Interface Sci.* **49**, 119–127.

Lipowsky, R. (1991) The conformation of membranes. *Nature* **349**, 475–481.

Lux, S. E. and Becker, P. S. (1989) Disorders of the red cell membrane skeleton: Hereditary spherocytosis and hereditary elliptocytosis. In *The Metabolic Basis of Inherited Disease,* 6th ed., C. R. Scriver, A. L. Beaudet, W. S. Sly, and D. Valle, McGraw-Hill, New York, Vol. 2, pp. 2367–2408.

Marchesi, V. T., Steers, E., Tillack, T. W., and Marchesi, S. L. (1969) Properties of spectrin: A fibrous protein isolated from red cell membranes. In *Red Cell Membrane,* G. A. Jamieson and T. J. Greenwalt (eds.) Lippincott, Philadelphia, p. 117.

Marchesi, S. L., Steers, E., Marchesi, V. T., and Tillack, T. W. (1970) Physical and chemical properties of a protein isolated from red cell membranes. *Biochemistry* **9**, 50–57.

Needham, D. and Nunn, R. S. (1990) Elastic deformation and failure of lipid bilayer membranes containing cholesteral. *Biophys. J.* **58**, 997–1009.

Norris, C. H. (1939) The tension at the surface and other physical properties of the nucleated erythrocyte. *J. Cell. Comp. Physiol.* **14**, 117–133.

Op den Kamp, J. A. F. (1979) Lipid asymmetry in membranes. *Ann. Rev. Biochem.* **48**, 47–71.

Ponder, E. (1948) *Hemolysis and Related Phenomena*. Grune and Stratton, New York.

Rand, R. H. and Burton, A. C. (1964) Mechanical properties of the red cell membrane. I. Membrane stiffness and intracellular pressure. II. Viscoelastic breakdown of the membrane. *Biophys. J.* **4**, 115–135; 303–316.

Sabbah, H. N. and Stein, P. D. (1976) Effect of erythrocytic deformability upon turbulent blood flow. *Biorheology* **13**, 309–314.

Schmid-Schoenbein, H. and Wells, R. E. (1969) Fluid drop-like transition of erythrocytes under shear. *Science* **165**, 288–291.

Schmidt-Nielsen, K. and Taylor, C. R. (1968) Red blood cells: Why or why not? *Science* **162**, 274–275.

Secomb, T. W., Skalak, R., Özkaya, N., and Gross, J. F. (1986) Flow of axisymmetric red blood cells in narrow capillaries. *J. Fluid Mech.* **163**, 405–423.

Seifriz, W. (1927) The physical properties of erythrocytes. *Protoplasma* **1**, 345–365.

Singer, S. J. (1974) The molecular organization of membranes. *Ann. Rev. Biochem.* **43**, 805–833.

Singer, S. J. and Nicolson, G. L. (1972) The fluid mosaic model of the structure of cell membranes. *Science* **175**, 720–731.

Skalak, R. (1973) Modeling the mechanical behavior of red blood cells. *Biorheology* **10**, 229–238.

Skalak, R., Chen, P. H., and Chien, S. (1972) Effect of hematocrit and rouleaux on apparent viscosity in capillaries. *Biorheology* **9**, 67–82.

Skalak, R., Tözeren, A., Zarda, R. P., and Chien, S. (1973) Strain energy function of red blood cell membranes. *Biophys. J.* **13**, 245–264.

Skalak, R. and Zhu, C. (1990) Rheological aspects of red blood cell aggregation. *Biorheology* **27**, 309–325.

Steck, T. L. (1974) The organization of proteins in the human red cell membrane. *J. Cell. Biol.* **62**, 1–19.

Stein, P. D. and Sabbah, H. N. (1974) Measured turbulence and its effect on thrombus formation. *Circulation Res.* **35**, 608–614.

Stein, P. D., Sabbah, H. N., and Blick, E. F. (1975) Contribution of erythrocytes to turbulent blood flow. *Biorheology* **12**, 293–299.

Stokke, B. T. (1984) The role of spectrin in determining mechanical properties, shapes, and shape transformations of human erythrocytes. Ph.D. Thesis. University of Trandheim, Norway.

Struik, D. J. (1950) *Lectures on Classical Differential Geometry*. Addison-Wesley, Cambridge, MA.

Sugihara-Seki, M. and Skalak, R. (1988) Numerical study of asymmetric flows of red blood cells in capillaries. *Microvasc. Res.* **36**, 64–74.

Sugihara-Seki, M. and Skalak, R. (1989) Stability of particle motions in a narrow channel flow. *Biorheology* **26**, 261–277.

Tözeren, H. and Skalak, R. (1979) Flow of elastic compressible spheres in tubes. *J. Fluid Mech.* **95**, 743–760.

Tözeren, A., Skalak, R., Fedorciw, B., Sung, K. L. P., and Chien, S. (1984) Constitutive equations of erythrocyte membrane incorporating evolving preferred configuration. *Biophys. J.* **45**, 541–549.

Tözeren, A., Sung, K. L. P., and Chien, S. (1989) Theoretical and experimental studies on cross-bridge migration during cell disaggregation. *Biophys. J.* **50**, 479–487.

Tsang, W. C. O. (1975) The size and shape of human red blood cells. M. S. Thesis. University of California, San Diego, La Jolla, California.

Wang, H. and Skalak, R. (1969) Viscous flow in a cylindrical tube containing a line of spherical particles. *J. Fluid Mech.* **38**, 75–96.

Waugh, R. and Evans, E. A. (1979) Temperature dependence of the elastic moduli of red blood cell membrane. *Biophys. J.* **26**, 115–132.

Waugh, R. E., Erwin, G., and Bouzid, A. (1986) Measurement of the extensional and flexural rigidities of a subcellular structure: Marginal bands isolated from erythrocytes of the newt. *J. Biomech. Eng.* **108**, 201–207.

Zarda, P. R., Chien, S., and Skalak, R. (1977) Elastic deformations of red blood cells. *J. Biomech.* **10**, 211–221.

References to Leukocytes and Other Cells

Atherton, A. and Born, G. V. R. (1972) Quantitative investigations of the adhesiveness of circulating polymorphonuclear leukocytes to blood vessel walls. *J. Physiol. (London)* **222**, 447–474.

Bray, C. (1984) Axonal growth in response to experimentally applied tension. *Dev. Biol.* **102**, 379–389.

Chien, S., Schmid-Schönbein, G. W., Sung, K. L. P., Schmalzer, E. A., and Skalak, R. (1984) Viscoelastic properties of leukocytes. In *White Blood Cell Mechanics: Basic Science and Clinical Aspects.* H. L. Meiselman and M. A. Lichtman (eds.) Plenum Press, New York, pp. 19–51.

Curtis, A. S. G. and Seehar, G. M. (1978) The control of cell division by tension or diffusion. *Nature (London)* **274**, 52–53.

DeWitt, M. T., Handley, C. J., Oakes, B. W., and Lowther, D. A. (1984) In vitro response of chondrocytes to mechanical loading. The effects of short term mechanical tension. *Connective Tissue Res.* **12**, 97–109.

Dong, C., Skalak, R., Sung, K.-L. P., Schmid-Schönbein, G. W., and Chien, S. (1988) Passive deformation analysis of human leukocytes. *J. Biomech. Eng.* **110**, 27–36.

Dong. C., Skalak, R., and Sung, K.-L. P. (1991) Cytoplasmic rheology of passive neutrophils. *Biorheology* **28**, 557–567.

Evans, E. A. (1984) Structural model for passive granulocyte behavior based on mechanical deformation and recovery after deformation tests. In *White Cell Mechanics* (H. J. Meiselman, M. A. Lichtman, and P. L. LaCelle (eds.) Alan Liss, New York.

Evans, E. A. and Yeung, A. (1989) Apparent viscosity and cortical tension of blood granulocytes determined by micropipet aspiration. *Biophys. J.* **43**, 27–30.

Fenton, B. M., Wilson, D. W., and Cokelet, G. R. (1985) Analysis of the effects of measured white blood cell entrance times on hemodynamics in a computer model of a microvascular bed. *Pflügers Arch.* **403**, 396–401.

Folkman, J. and Handenschild, C. (1980) Angiogenesis in vitro. *Nature (London)* **288**, 551–556.

Holberton, D. V. (1977) Locomotion of protozoa and single cells. In *Mechanics and Energetics of Animal Locomotion*. R. McN. Alexander and G. Goldspink (eds.), Chapman and Hall, London, Chapter 11, pp. 279–326.

Hurley, J. V. (1963) An electron microscopic study of leukocytic emigration and vascular permeability in rat skin. *Austral. J. Exp. Biol.* **41**, 171–186.

Huxley, H. E., Bray, D., and Weeds, A. G. (eds.) (1982) Molecular biology of cell locomotion. *Phil. Trans. Roy. Soc. London* **B.299**, 145–327.

Ingber, D. E. and Folkman, J. (1989) How does extracellular matrix control capillary morphogenesis? *Cell* **58**, 803–805.

Jones, D. B., Nolte, H., Scholübbers, J.-G., Turner, E., and Veltel, D. (1991) Biochemical signal transduction of mechanical strain in osteoblastlike cells. *Biomaterials*, **12**; 101–110.

Khouri, R. K., Koudsi, B., and Reddi, H. (1991) Tissue transformation into bone in vivo, a potential practical application. *JAMA* **266**, 1953–1955.

Klein-Nulend, J., Veldhuijzen, J. P., van de Stadt, R. J., Jos van Kampen, G. P., Keujer, R., and Burger, E. H. (1987) Influence of intermittent compressive force on proteoglycan content in calcifying growth plate cartilage in vitro. *J. Biol. Chem.* **262**, 15,490–15,495.

Lanyon, L. E., Goodship, A. E., Pye, C. J., and MacFie, J. H. (1982) Mechanically Adaptive bone remodeling. *J. Biomechanics* **15**, 141–154.

Leung, D. Y. M., Glagov, S., and Mathews, M. B. (1976) Cyclic stretching stimulates synthesis of matrix components by arterial smooth muscle cells in vitro. *Science* **191**, 475–477.

Lichtman, M. A. (1970) Cellular deformability during maturation of the myeloblast: Possible role of marrow egress. *New England J. Med.* **283**, 943–948.

Lipowsky, R. (1991) The conformation of membrane. *Nature* **349**, 475–481.

Lanyon, L. E. (1984) Functional strain as a determinant for bone remodeling. *Calcif. Tiss. Res.* **36**, 556–561.

Morgan, H. E., Gorden, E. E., Kira, Y., Chua, B. H. L., Russo, L. A., Peterson, C. L., McDermott, P. J., and Watson, P. A. (1987) Biochemical mechanisms of cardiac hypertrophy. *Annu. Rev. Physiol.* **49**, 533–543.

Needham, D. and Hochmuth, R. M. (1990) Rapid flow of passive neutrophils into a 4 μm pipet and measurement of cytoplasmic viscosity. *J. Biomech. Eng.* **112**, 269–276,

Odell, G. M., Oster, G., Alberch, P., and Burnside, B. (1981) The mechanical basis of morphogenesis. I. Epithelial folding and invagination. *Devel. Biol.* **85**, 446–462.

Op den Kamp, J. A. F. (1979) Lipid asymmetry in membranes. *Ann. Rev. Biochem.* **48**, 47–71.

Pipkin, A. C. (1964) Small finite deformations of viscoelastic solids. *Rev. Mod. Phys.* **36**, 1034–1041.

Rannels, D. E. (1989) Role of physical forces in compensatory growth of the lung. *Am. J. Physiol.* **257**, L179–L189.

Rubin, C. T. and Lanyon, L. E. (1985) Regulation of bone mass by mechanical strain magnitude. *Calcif. Tiss. Res.* **37**, 411–417.

Sachs, F. (1990) Mechanical transduction in biological systems. In *CRC Critical Reviews in Biomedical Engineering*. CRC Press, Orlando, FL.

Schmid-Schönbein, G. W., Fung, Y. C., and Zweifach, B. W. (1975) Vascular endothelium–leukocyte interaction. *Circulation Res.* **36**, 173–184.

Schmidt-Schönbein, G. W., Sung, K.-L. P., Tözeren, H., Skalak, R., and Chien, S. (1981) Passive mechanical properties of human leukocytes. *Biophys. J.* **36**, 243–256.

Schmid-Schönbein, G. W., Skalak, R., Sung, K. L.-P., and Chien, S. (1983) Human leukocytes in the active state. In *White Blood Cells, Morphology and Rheology as Related to Function*, U. Bagge, G. B. R. Bom, and P. Gaehtgens (eds.) Martinus Mijhoff, The Hague, pp. 21–31.

Schmid-Schönbein, G. W. (1987) Capillary plugging by granulocytes and the no-reflow phenomenon in the microcirculation. *Fed. Proc.* **46**, 2397–2401.

Schultz, S. G. (1989) Volume preservation: Then and now. *News in Physiol. Sci.* **4**, 169–172.

Stewart, D. M. (1972) The role of tension in muscle growth. In *Regulation of Organ and Tissue Growth*, R. J. Goss (ed.) Academic Press, New York, pp. 77–100.

Stossel, T. P. (1982) The structure of cortical cytoplasm. *Phil. Trans. Roy. Soc. London B* **299**, 275–289.

Strohman, R. C., Byne, E., Spector, D., Obinata, T., Micou-Eastwood, J., and Maniotis, A. (1990) Myogenesis and histogenesis of skeletal muscle on flexible membranes in vitro. *In Vitro Cell Dev. Biol.* **26**, 201–208.

Sung, K.-L. P., Schmid-Schönbein, G. W., Skalak, R., Schuessler, G. B., Usami, S., and Chien, S. (1982) Influence of physicochemical factors on rheology of human neutrophils. *Biophys. J.* **39**, 101–106.

Sung, K.-L. P., Sung, L. A., Crimmins, M., Burakoff, S. J., and Chien, S. (1986) Determination of junction avidity of cytolytic T cell and target cell. *Science* **234**, 1405–1408.

Sung, K.-L. P., Dong, C., Schmid-Schönbein, G. W., Chien, S., and Skalak, R. (1988a) Leukocyte relaxation properties. *Biophys. J.* **54**, 331–336.

Sung, K.-L. P., Sung, L. A., Crimmins, M., Burakoff, S. J., and Chien, S. (1988b) Biophysical basis of cell killing by cytotoxic T Lymphocytes. *J. Cell Sci.* **91**, 179–189.

Vandenburgh, H. H. (1988) A computerized mechanical cell stimulator for tissue culture: Effects on skeletal muscle organogenesis. *In Vitro Cell Dev. Biol.* **24**, 609–619.

Vandenburgh, H. H. and Karlisch, P. (1989) Longitudinal growth of skeletal myotubes in vitro in a new horizontal mechanical cell stimulator. *In Vitro Cell Dev. Biol.* **25**, 607–616.

Vandenburgh, H. H., Swasdison, S., and Karlisch, P. (1991) Computer-aided mechanogenesis of skeletal muscle organs from single cells in vitro. *FASEB J.* **5**, 2860–2867.

Zhu, C. and Skalak, R. (1988) A continuum model of protrusion of pseudopod in leukocytes. *Biophys. J.* **54**, 1115–1137.

Zhu, C., Skalak, R., and Schmid-Schönbein, G. W. (1989) One-dimensional steady continuum model of retraction of pseudopod in leukocytes. *J. Biomech. Eng.* **111**, 69–77.

Zwaal, R. F. A. (1978) Membrane and lipid involvement in blood coagulation. *Biochim. Biophys. Acta* **515**, 163–205.

Zwaal, R. F. A. (1988) Scrambling membrane phospholipids and local control of blood clotting. *News in Physiol. Sci.* **3**, 57–61.

CHAPTER 5

Interaction of Red Cells with Vessel Wall, and Wall Shear with Endothelium

5.1 Introduction

The sizes of the viscometers cited in Chapter 3 are so large that blood can be treated as a homogeneous fluid in them. The size of the individual red cells is many orders of magnitude smaller than the dimensions of the viscometers. The same condition holds in large blood vessels. The diameters of the capillary blood vessels, however, are comparable with the dimensions of the red cells. Hence in the capillaries, red blood cells must be treated as individuals. Blood must be regarded as a two-phase fluid: a liquid plasma phase and a deformable solid phase of the blood cells.

The necessity to consider the tight interaction between red cells and the walls of capillary blood vessels is very clearly shown by the photograph reproduced in Fig. 4.1:2. It is through this interaction that the red cells are severely deformed (compare the shape of cells shown in Fig. 4.1:2 to that shown in Fig. 4.1:1). The influence of the vessel wall on the red cells in blood, however, is not limited to the capillaries. In arterioles and venules, whose diameters range from 1 to 10 or 20 times the diameter of the red cell, the distribution of red cells in the vessel is also affected by the blood vessel wall. The cell distribution is rarified in the neighborhood of the vessel wall (Sec. 3.5); and the apparent viscosity of the blood is reduced (Sec. 5.2). In Secs. 5.3 and 5.4, we discuss the flow of red cells in narrow tubes. In Sec. 5.5, we discuss the flow of red cells in capillary blood vessels. In Secs. 5.6 and 5.7, we consider very narrow tubes whose diameters are smaller than the diameter of the red cell.

On the wall of the blood vessel there is a layer of endothelial cells. These cells cannot move, but can deform. They response to the shear stress imposed on the vessel wall by the flowing blood. They form a continuous layer through which any exchange of matter between the tissue and the blood must take place. Therefore they are believed to play a significant role in the genesis of such disease as atherosclerosis. We devote Secs. 5.9–5.15 to discuss how does the endothelium respond to the shear stress.

5.2 Apparent Viscosity and Relative Viscosity

The need to consider cell–vessel interaction makes the rheology of blood in micro-blood vessels very different from ordinary macroscopic rheology. To link these topics together, let us consider two intermediate terms that are useful in organizing experimental data, namely, the *apparent viscosity* and the *relative viscosity*. To explain their meaning, consider a flow through a circular cylindrical tube. If the fluid is Newtonian and the flow is laminar, we have the Hagen–Poiseuille formula [Eq. (8) of Sec. 3.3]:

$$\frac{\Delta p}{\Delta L} = \frac{8\mu}{\pi a^4} \dot{Q}, \tag{1}$$

where Δp is the pressure drop in length ΔL, μ is the coefficient of viscosity of the fluid, a is the radius of the tube, and $\dot{Q}$ is the volume rate of flow. If the fluid is blood, this equation does not apply; but we can still measure $\Delta p/\Delta L$ and $\dot{Q}$, and use Eq. (1) to calculate a coefficient μ:

$$\mu = \frac{\pi a^4}{8} \frac{1}{\dot{Q}} \frac{\Delta p}{\Delta L}. \tag{2}$$

The μ so computed is defined as the *apparent coefficient of viscosity for the circular cylindrical tube*, and is denoted by μ_{app}. If μ_0 denotes the viscosity of plasma, which is Newtonian, then the ratio μ_{app}/μ_0 is defined as the *relative viscosity*, and is denoted by μ_r. Note that the unit of apparent viscosity is (force·time/area), or the poise (P), whereas the relative viscosity is dimensionless.

The concept of relative viscosity can be generalized to an organ system. To measure the relative viscosity of blood in such a system we perfuse it with a homogeneous fluid and measure the pressure drop Δp corresponding to a certain flow $\dot{Q}$. We then perfuse the same system with whole blood and again measure the pressure drop at the same flow. The ratio $\Delta p/\dot{Q}$ for whole blood divided by $\Delta p/\dot{Q}$ for the specific fluid is the relative viscosity, μ_r, of blood relative to that fluid.

The concept of apparent viscosity can be extended to any flow regime, including turbulent flow, as long as we can compute it from a formula that is known to work for a homogeneous Newtonian fluid. The concept of relative viscosity can be extended to any flow system, even if we do not know its structural geometry and elasticity, as long as flow and pressure can be measured. Neither μ_{app} nor μ_r needs to be constant. They are functions of all the dimensionless parameters defining the kinematic and dynamic similarities; and if the system is nonlinear, they are functions of pressure p and flow $\dot{Q}$.

Apparent and relative viscosities are not intrinsic properties of the blood; they are properties of the blood and blood–vessel interaction, and depend on the data reduction procedure. There are as many definitions for apparent viscosities as there are good formulas for well-defined problems. Examples are: Stokes flow around a falling sphere, channel flow, flow through an orifice, and flow in a cylindrical tube. But if a vessel system has a geometry such that the theoretical problem for homogeneous fluid flow has not been solved, then we cannot derive an apparent viscosity for flow in such a system. But we can determine a relative viscosity if we can perform flow experiments in the system both with blood and with a homogeneous fluid.

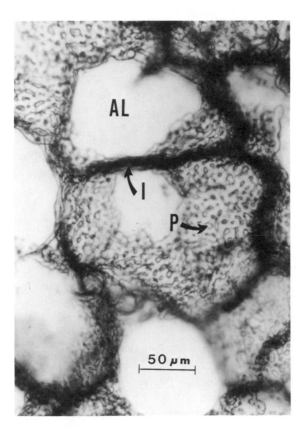

Figure 5.2:1 A view of the pulmonary capillary blood vessels of the cat photographed when the focal plane of the microscope is parallel to the plane of several interalveolar septa. The large white spaces are the alveolar air spaces (AL). The capillaries are the interconnected small channels seen on the septa. The islandlike spaces between the capillaries are called "posts" (*P*). "*I*" is an interalveolar septum perpendicular to the page. Courtesy of Dr. Sidney Sobin.

5.2.1 An Example

Let us consider the flow of blood in the capillary blood vessels of the lung. As an introduction, a few words about the anatomy of these vessels is necessary. The lung is an organ whose function is to oxygenate the blood. The venous blood of peripheral circulation is pumped by the right ventricle of the heart into the pulmonary artery. The artery divides and subdivides again and again into smaller and smaller blood vessels. The smallest blood vessels are the capillaries. Pulmonary capillaries have very thin walls (about 1 μm thick) across which gas exchange takes place. The oxygenated blood then flows into the veins and the heart. The pulmonary capillaries form a closely knit network in the interalveolar septa. Figure 5.2:1 shows a plan view of the network. Figure 5.2:2 shows a cross-sectional view of the network. It is clear that the network is two-dimensional. From a fluid dynamical point of view, one may regard the vascular space as a thin sheet of fluid bounded by two membranes that are connected by an array of *posts*. The thickness of the sheet depends on the transmural pressure (blood pressure minus alveolar gas pressure). If the transmural pressure is small and tending to zero, the sheet thickness tends to 4.3 μm in the cat, 2.5 μm in the dog, and 3.5 μm in man. In this condition the sheet thickness is smaller than one-half the diameter of the undeformed red blood cells of the respective species. Hence the red cells have to flow either sideways or become severely deformed. The sheet thickness increases almost linearly with increasing transmural pressure: at a rate of 2.23 μm/kPa (0.219 μm per cm H_2O) in the cat, 1.24 μm/kPa in the dog, and 1.29 μm/kPa in man.

For blood flow through such a complicated system we would want to know how the resistance depends on the following factors: the hematocrit, the geometry of the network (sheet), the thickness of the sheet relative to the size of the red blood cells, the rigidity of the red cells, the Reynolds number, the size of the posts, the distance between posts, the pattern of the posts, the direction of flow relative to the pattern of the posts, and so on. If we know how the resistance depends on these factors, then we can make model experiments, discuss the effects of abnormalities, compare the lungs of different animal species, etc.

To deal with the problem, we first make a dimensional analysis. Let p be the pressure, μ be the apparent coefficient of viscosity, U be the mean velocity of flow, h be the sheet thickness, w be the width of the sheet, ε be the diameter of the post, a be the distance between the posts, and θ be the angle between the mean direction of flow relative to a reference line defining the postal pattern. The postal pattern is further described by the ratio of the volume occupied by the blood vessels divided by the total volume of the sheet, called the *vascular-space-tissue ratio* and denoted by S. Sobin et al. (1970, 1972, 1979) have experimentally measured S in the lungs of the cat, rabbit, and man. Let us consider first the flow of a homogeneous Newtonian viscous fluid in this system. We want to know the relationship between the pressure gradient

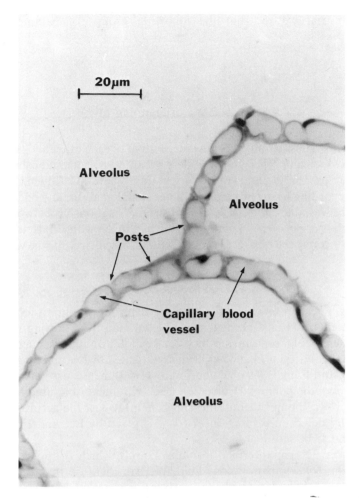

Figure 5.2:2 Another view of pulmonary capillary blood vessels of the cat photographed when the focal plane of the microscope cuts three interalveolar septa nearly perpendicularly. Courtesy of Dr. Sidney Sobin.

(rate of change of pressure with distance, a vector with components $\delta p/\delta x$, $\delta p/\delta y$ and denoted by ∇p), and the velocity, U, the viscosity coefficient, μ, the sheet thickness, h, and the other factors, w, ε, a, θ, S, and ρ. The physical dimensions of these quantities are, with F denoting force, L denoting length, and T, the time:

$$
\begin{aligned}
&\text{pressure: } FL^{-2}, && \text{pressure gradient: } FL^{-3}, \\
&\text{velocity: } LT^{-1}, && \text{sheet thickness: } L, \\
&\text{coefficient of viscosity} = \text{stress/shear rate: } FL^{-2}T, \\
&\text{density of fluid,} && \rho = \text{mass/vol: } FT^2L^{-4}, \\
&h, w, \varepsilon, a: L, && \theta, S: \text{dimensionless.}
\end{aligned}
$$

By simple trial we see that the following groups of parameters are dimensionless:

$$\frac{w}{h}, \quad S, \quad \frac{h}{\varepsilon}, \quad \frac{\varepsilon}{a}, \quad \theta,$$

$$\frac{h^2}{\mu U} \nabla p, \quad \frac{Uh\rho}{\mu} \equiv N_R \quad \text{(Reynolds' number)}. \tag{3}$$

These dimensionless parameters are independent of each other and they form a set from which any other dimensionless parameters that can be formed by these variables can be obtained by proper combinations of them.

Now, according to the principle of dimensional analysis, any relationship between the variables p, U, μ, h, etc. must be a relationship between these dimensionless parameters. In particular, the parameter we are most interested in, that connecting the pressure gradient with flow velocity, may be written as

$$\frac{h^2}{\mu U} \nabla p = F\left(\frac{w}{h}, \frac{h}{\varepsilon}, \frac{\varepsilon}{a}, S, \theta, N_R\right), \tag{4}$$

where F is a certain function that must be determined either theoretically or experimentally. For a homogeneous Newtonian fluid flowing in a pulmonary interalveolar septum, in which the Reynolds number is much smaller than 1, the theoretical problem has been investigated by Lee and Fung (1969), Lee (1969), and Fung (1969). Yen and Fung (1973) then constructed a simulated model of the pulmonary alveolar sheet with lucite, and used a silicone fluid (which was homogeneous and Newtonian) to test the accuracy of the theoretically determined function $F(w/h, \ldots)$ given by Lee and Fung; and it was found to be satisfactory.

We do have, then, a verified formula, Eq. (4), which is applicable to a homogeneous Newtonian viscous fluid. We can then use this formula for experimental determination of the apparent viscosity of blood flowing in pulmonary alveoli. From the experimental values of ∇p, U, h^2, etc. we can compute the apparent viscosity μ_{app} from the following formula:

$$\nabla p = -\frac{U}{h^2} \mu_{app} F\left(\frac{w}{h}, \frac{h}{\varepsilon}, \ldots\right). \tag{5}$$

The apparent viscosity is influenced by the plasma viscosity, the concentration of the red blood cells, the size of the red cells relative to the blood vessel, the elasticity of the red cells, etc. Let H be the hematocrit (the fraction of the red cell volume in whole blood), let D_c be the diameter of the red cell, and let μ_{plasma} be the viscosity of the plasma; then the apparent viscosity of blood in pulmonary alveoli must be a function of μ_{plasma}, H, D_c/h, etc. Hence, again on the basis of dimensional analysis, we can write

$$\mu_{app} = \mu_{plasma} f\left(H, \frac{D_c}{h}, \ldots\right), \tag{6}$$

where f is the function to be determined, and the dots indicate the cell elasticity and other parameters not explicitly shown.

Dimensional analysis is the basis for model experiments. For mechanical simulation the model must be geometrically and dynamically similar to the prototype. For geometric similarity the kinematic parameters listed in the first line of Eq. (3) must be the same for the model and prototype. For dynamic similarity the parameters in the second line of Eq. (3) must be the same for

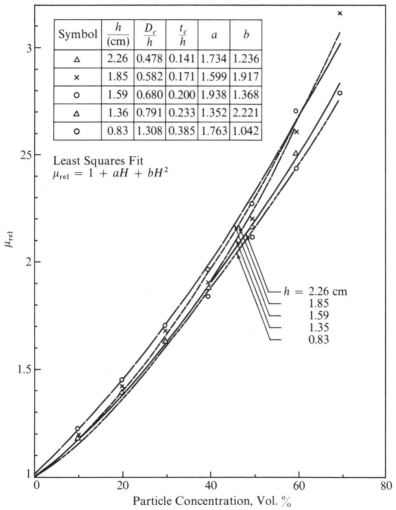

Symbol	$\dfrac{h}{(cm)}$	$\dfrac{D_c}{h}$	$\dfrac{t_c}{h}$	a	b
△	2.26	0.478	0.141	1.734	1.236
×	1.85	0.582	0.171	1.599	1.917
○	1.59	0.680	0.200	1.938	1.368
△	1.36	0.791	0.233	1.352	2.221
○	0.83	1.308	0.385	1.763	1.042

Least Squares Fit
$\mu_{rel} = 1 + aH + bH^2$

$h = 2.26$ cm
1.85
1.59
1.35
0.83

Figure 5.2:3 The relative viscosity of simulated whole blood (gelatin pellets in silicone oil) in a pulmonary alveolar model. Here D_c is the diameter of the red cell, h is the thickness of the alveolar sheet, t_c is the thickness of the red blood cell, and a, b are constants in Eq. (8). The model is so large that h is in centimeters, although in a real lung h is only a few microns. From Yen and Fung (1973).

them. When these parameters are simulated, any relationship obtained from the model is applicable to the prototype. Yen and Fung (1973) experimented on a scale model of the pulmonary alveolar sheet, with the red blood cells simulated by soft gelatin pellets and with the plasma simulated by a silicone fluid. Their results show that the pressure-flow relationship is quite linear, and that for $h/\varepsilon < 4$, the relative viscosity of blood with respect to plasma depends on the hematocrit H in the following manner:

$$\mu_{\text{relative}} = 1 + aH + bH^2. \tag{7}$$

Thus, the apparent viscosity of blood in pulmonary capillaries is

$$\mu_{\text{app}} = \mu_{\text{plasma}}(1 + aH + bH^2). \tag{8}$$

Since Eq. (4) is verified for the plasma, Eq. (8) can be used in Eq. (6). These relations are illustrated in Fig. 5.2:3, where the values of the constants a and b are listed.

This rather complex example is quoted here to show that (a) the definition of apparent viscosity is not unique, and (b) that it may not be simple. The selection of definition is guided only by its usefulness.

5.3 Effect of Size of the Blood Vessel on the Apparent Viscosity of Blood: The Fahraeus–Lindqvist Effect

For blood flow in cylindrical vessels, the apparent viscosity decreases with decreasing blood vessel diameter. This is shown in Fig. 5.3:1. This was first pointed out by Fahraeus and Lindqvist (1931), who tested blood flow in

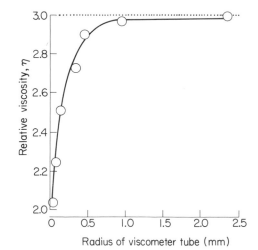

Figure 5.3:1 The change of relative viscosity of blood with the size of blood vessel. The data are Kümin's, analyzed by Haynes (*Am. J. Physiol.* **198**, 1193, 1960).

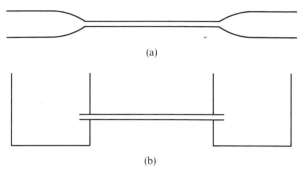

Figure 5.3:2 The capillary tubes used by (a) Fahraeus and Lindqvist (1931) and (b) by Barbee and Cokelet (1971) to measure the dependence of the apparent viscosity of blood on the diameter of circular cylindrical tubes.

glass tubes connected to a feed reservoir in an arrangement illustrated in Fig. 5.3:2. They showed this trend in tubes of diameter in the range 500–50 μm. Barbee and Cokelet (1971) extended this experiment and showed that the trend continues at least to tubes of diameter 29 μm. (Human blood was used. Remember that the average human red cell diameter is 7.6 μm.) These experiments were done at rates of flow so high that the apparent viscosity does not vary significantly with the flow rate.

An explanation of the Fahraeus–Lindqvist effect was provided by Barbee and Cokelet (1971) and is based on an observation made by Fahraeus himself. Fahraeus (1929) found that when blood of a constant hematocrit is allowed to flow from a large feed reservoir into a small tube, the hematocrit in the tube decreases as the tube diameter decreases. This trend was shown by Barbee and Cokelet (1971) to continue down to a tube diameter of 29 μm (cf. Fig. 5.3:2). Barbee and Cokelet then showed that if one measures the apparent viscosity of blood in a large tube (say, of a diameter about 1 mm) as a function of the hematocrit, and use the data to compute the apparent viscosity of the same blood in the smaller tube at the actual hematocrit in that tube, a complete agreement with the experimental data can be obtained. This is an important finding, because it not only extends the usefulness of the apparent viscosity measurements, but also furnishes insight into the mechanism of flow resistance in microcirculation.

Verification of these statements is shown in Figs. 5.3:3 and 5.3:4. Figure 5.3:3 shows Barbee and Cokelet's result on how the average hematocrit of the blood in a tube (H_T) varies with the tube size and the feed reservoir blood hematocrit (H_F). The "relative hematocrit" H_R, plotted on the ordinate, is the ratio H_T/H_F. This figure demonstrates the Fahraeus effect. Figure 5.3:4 shows how the Fahraeus effect can be used to explain the Fahraeus–Lindqvist effect. The experimental data on the wall shear stress (τ_w) are plotted vs. $\bar{U}$, the bulk average velocity of flow divided by the tube diameter. The data are plotted this way since it can be shown that for a fluid for which the rate of

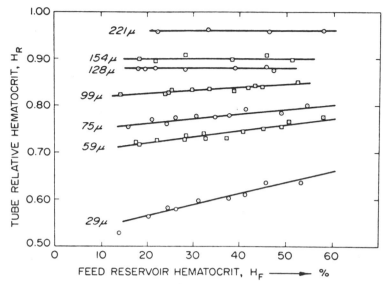

Figure 5.3:3 The Fahraeus effect: when blood flows from a reservoir into and through a small diameter tube, the average hematocrit in the tube is less than that in the reservoir. The tube relative hematocrit is defined as the average hematocrit of the blood flowing through a tube divided by the hematocrit of the blood in the reservoir feeding the tube. Numbers to the left of the lines are the tube diameters. From Barbee and Cokelet (1971), by permission.

deformation is just a function of the shear stress, the same function applying at all points in the tube, the data should give one universal curve, even when the data are obtained with tubes of different diameters. The top curve in the figure represents data obtained with an 811 μm tube with $H_F = 0.559$. On the next curve, the circles represent experimental data obtained with a 29 μm tube with $H_F = 0.559$. These two sets of data should coincide if the shear stress-shear rate relationships are the same in these two tubes. But they do not agree, because the hematocrits in these two tubes are different. The true tube hematocrit is $H_T = 0.358$. If we obtain flow data in an 811 μm tube with $H_F = 0.358 (= H_T)$, we find that these data are represented by the solid curve, which happens to pass through the circles. Other data points corrected for hematocrit show similar good agreement. Thus, in spite of the nonuniform distribution of red cells in the tubes, the Fahraeus–Lindqvist effect appears to be due entirely to the Fahraeus effect for tubes larger than 29 μm.

Why does the hematocrit decrease in small blood vessels? One of the explanations is that a *cell-free layer* exists at the wall. See Chapter 3, Fig. 3.5:1. Another explanation is that the red cells are elongated and oriented in a shear flow. See Chapter 3, Fig. 3.4:5. The cell-free layer reduces the hematocrit. The smaller the vessel, the larger the fraction of volume occupied by the cell-free layer, and the lower the hematocrit. Furthermore, if a small side

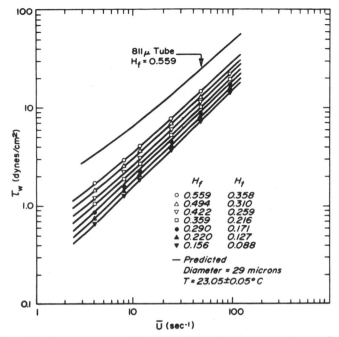

Figure 5.3:4 The flow behavior of blood in a 29 μ diameter tube. The symbols are the actual flow data, recorded as the wall shear stress τ_ω and the bulk average velocity divided by the tube diameter, U. The solid curves through the points represent the behavior of the blood predicted from the data obtained in an 811 μ diameter tube when the average tube hematocrit is the same as that experimentally found in the 29 μ tube. In an 811 μ tube, H_f, the feed reservoir hematocrit, and H_t, the average tube hematocrit, are equal. From Barbee and Cokelet (1971).

branch of a vessel draws blood from the vessel mainly from the cell-free layer, the hematocrit in the side branch will be smaller. This is usually referred to as *plasma skimming*.

The cell orientation effect reinforces the effect of plasma skimming. If a small side branch having a diameter about the same size as that of the red cell draws blood from a larger tube (such as a capillary branching from an arteriole) the *entry condition* into the small branch is affected by the orientation of the red cells. If the entry section is aligned with the red cells it will be easier for the cells to enter. If the entry is perpendicular to the cells, some cells skim over the small branch and do not enter, and the hematocrit in the small branch will be decreased.

With a cell-free layer, the viscosity at the wall is that of the plasma and is smaller than that of whole blood. We have seen (Fig. 3.1:1) that the viscosity of blood increases with hematocrit. In a tube the hematocrit is higher at the core. Here the viscosity of blood in the blood vessel is higher at the core and lower at the wall. This affects the velocity profile. The apparent viscosity is a result of both of these factors.

5.4 The Distribution of Suspended Particles in Fairly Narrow Rigid Tubes

Mason and Goldsmith (1969) conducted an exhaustive series of experiments with suspensions of variously shaped particles, which may be rigid, deformable, or merely viscous immiscible droplets flowing through straight rigid pipes. Their results provide the best insight into the cell-free layer mentioned in the preceding section.

For a suspension of neutrally buoyant rigid spheres, the explanation of the cell-free layer is primarily geometrical, in that the center of the spheres must be at least a radius away from the wall. At a very low Reynolds number, whether there is any dynamical effect tending to force spheres away from the wall is still uncertain. Isolated rigid spheres in Poiseuille flow can be shown to experience no net radial force; they rotate, but continue to travel in straight lines; thus they cannot create a cell-free layer.

When inertia force is not negligible, rigid spheres in a Poiseuille flow do, in general, experience a radial force. They exhibit a *tubular pinch effect* demonstrated experimentally by Segre and Silberberg (1962), in which particles near the wall move toward the axis, and particles near the axis move toward the wall. This effect is probably unimportant in arterioles or capillaries, where the Reynolds number is less than 1. Similar results are obtained for suspensions of rigid rods and disks. There is no overall radial motion unless inertial effect is important.

When the particles in suspension are deformable, Mason and Goldsmith found that they do experience a net radial hydrodynamic force even at very low Reynolds numbers, and tend to migrate towards the tube axis. The mechanism of this phenomenon is obscure, and a full theoretical analysis is not yet available.

5.5 The Motion of Red Cells in Tightly Fitting Tubes

Many capillary blood vessels in a number of organs have diameters smaller than the diameter of the resting red cell. It is extremely difficult to make *in vivo* measurements of velocity and pressure fields in these small blood vessels. To obtain some details, two alternative approaches may be taken: mathematical modeling and physical modeling. In this section we shall consider larger-scale physical model testing.

In Chapter 4, Sec. 4.6, we described how thin-walled rubber models of red blood cells can be made. These cells have a wall thickness to cell diameter ratio ranging from 1/100–1/150. Lee and Fung (1969) used these rubber models as simulated red cells in testing their interaction with tightly fitting tubes. An independent, similar testing was done by Sutera et al. (1970). Figure 5.5:1 shows a schematic diagram of the apparatus. The test section

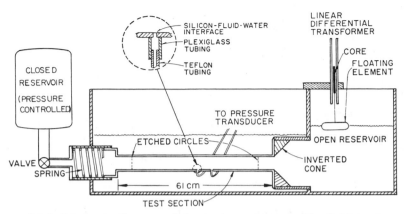

Figure 5.5:1 Schematic diagram of the testing apparatus. The fluid in the open reservoir at the right enters the test tube through the conical entry section, flows to the left, and exits into the closed reservoir. The wavy lines in the reservoirs indicate the fluid levels.

was made of interchangeable lucite tubes, 61 cm long, and with inner diameters 2.54, 3.15, 3.81, 4,37, and 5.03 (± 0.03) cm. A conical section with a 60° inclination from the tube axis was used to guide the flow into the test section. The flow was controlled by a pressure reservoir. A constant mean velocity ranging from 0.03 to 3 cm/sec could be maintained throughout each experiment. A silicone fluid with a viscosity 295 P at room temperature and a specific gravity 0.97 was used to simulate the plasma. The cell has a diameter of 4.29 ± 0.05 cm, a volume of 16.5 cm³, a wall thickness of 0.042 ± 0.01 cm in one model, and 0.03 in another, and is filled with the same silicone fluid. A nipple on the cell marks the pole location. The Reynolds number based on the cell diameter was in the range 0.0004–0.04.

If the tube diameter is larger than the cell diameter by more than 15%, the cell will flow in the stream without significant deformation. If the tube diameter is about equal to the resting cell diameter, then significant cell deformation occurs. Figure 5.5:2 shows such a case. Here V_M denotes the mean flow velocity, and V_c denotes the cell velocity. If the tube diameter is considerably smaller than the cell diameter, then large deformation of the cell occurs; there is severe buckling of the cell membrane; the leading edge of the cell bulges out, the trailing end caves in; and the natural tendency of the cell is to enter the tube sideways, or edge-on (with its equatorial plane parallel to the cylinder axis). Forcing the cell to the other orientation (with the axis of the cell and the tube parallel) was not successful. In the case in which the tube diameter is equal to the resting cell diameter, the edge-on configuration is more stable. It is possible in this case for the cell to assume an axisymmetric configuration, as is shown in frame D in Fig. 5.5:2, but this seems to be an unstable situation and is rarely seen.

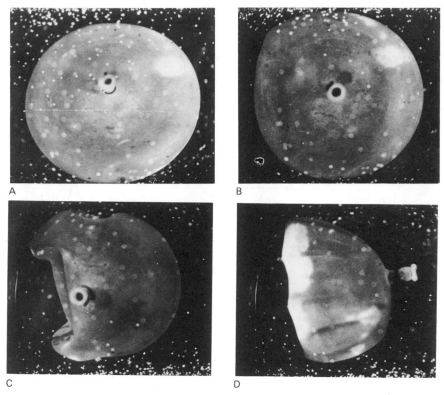

Figure 5.5:2 Change of shape of the 4.29 cm thin-walled cell in the 4.37 cm tube as the velocity increases: (A) Cell velocity $V_c = 0.12$, mean flow velocity $V_m = 0.10$; (B) $V_c = 0.36$, $V_m = 0.28$; (C) $V_c = 2.49$, $V_m = 1.95$; (D) $V_c = 2.59$, $V_m = 2.16$. All velocities are in cm/sec. Direction of flow: from left to right. From Lee and Fung (1969).

5.5.1 The Streamline Pattern and Cell Motion

The streamlines around the cell were determined in the 3.15 cm tube by photographing with short-duration exposure and are shown in the upper panel of Fig. 5.5:3. When the streamlines are redrawn to coordinates moving with the cell, they appear as *bolus* flow as shown in the lower panel of Fig. 5.5:3.

The cell velocity is related to the mean flow velocity. The experimental data can be fitted by the equation

$$V_c = k(V_M - \alpha) \tag{1}$$

for V_M greater than α. The constants determined are:

Cell/tube diameter ratio	k	α (cm/sec)
1.69	1.00	0.0
1.36	1.10	0.02
1.13	1.17	0.10
0.98	1.26	0.015

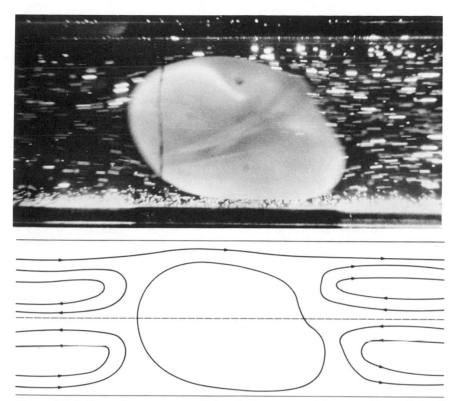

Figure 5.5:3 Short time exposures (0.5 sec) of tracer particles as a thin-walled cell of diameter 4.29 cm moves to the left in a 3.15 cm tube with a cell velocity $V_c =$ 0.93 cm/sec. The lower figure shows the streamlines relative to the cell. In the upper panel the cell is moving to the left. In the lower panel the streamlines are drawn with the cell held stationary and the tube moves to the right. From Lee and Fung (1969).

This is consistent with the observation that as the tube becomes larger, the cell will be more centrally located, where the velocity is higher than in the wall region.

5.5.2 Pressure Measurement

Figure 5.5:4 shows a typical record of pressure at a point midway in the 3.15 cm tube (cell diameter 4.29 cm). The pressure dropped at once as the fluid was sucked into the tube. When the cell approached the entrance, the pressure decreased further. After the cell passed the tap, the pressure returned approximately to the level in the tube before the entry of the cell. The difference between the pressure just upstream of the cell, p_u, and that just

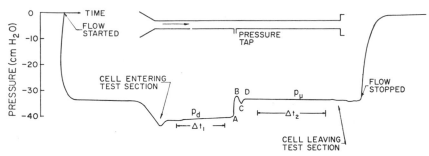

Figure 5.5:4 A typical record of the history of the pressure at the tap about midway in the test section and the change of fluid level in the open reservoir. The 4.29 cm cell moved at a velocity of 1.23 cm/sec in a 3.15 cm tube. The mean velocity was 1.18 cm/sec. The position of the cell at any given instant of time can be found roughly by drawing a vertical line from the time scale to the tube sketched above the graph. Note the change of pressure at the cell's entering the test section, passing the pressure tap, and leaving the test section. In the time intervals Δt_1 and Δt_2, the cell was more than one cell diameter away from the ends of the test section and the pressure tap. The pressure in Δt_1 is the downstream pressure p_d, that in Δt_2 is the upstream pressure p_u. From Lee and Fung (1969).

downstream of the cell, p_d, is a measure of the resistance to the motion of the red cell. It is a function of the cell velocity and the ratio of the diameters of the cell and tube, as is shown in Fig. 5.5:5.

The details of the pressure change as the cell passes over the tap are very interesting. At the leading edge of the cell (point A in Fig. 5.5:4), the pressure rises rapidly. It reaches a peak at point B (on the front portion of the cell). Then comes a valley at point C (near the trailing edge of the cell), where the pressure is lower than that at the rear of the cell. This kind of pressure distribution is consistent with the predictions of lubrication layer theory (Sec. 5.8). The higher pressure at B than C will tend to push fluid into the gap between the cell and the wall toward the rear of the cell, thus reducing the velocity gradient at the wall, and decreasing the wall friction and apparent viscosity of the blood.

5.5.3 Critique of the Model

One unsatisfactory feature of the red cell model made of rubber is that the rubber elasticity is quite different from that of the red cell membrane. For the rubber membrane, the shear modulus and areal modulus are of the same order of magnitude. For the red cell membrane, the two moduli differ by a factor of 10^4 (see Chapter 4, Sec. 4.6). Hence the cell membrane elasticity was not simulated. Furthermore, the bending rigidity of the red cell membrane is probably derived from electric charge or from fibrous protein attached to the membrane (see Chapter 4, Sec. 4.8), whereas that of the rubber

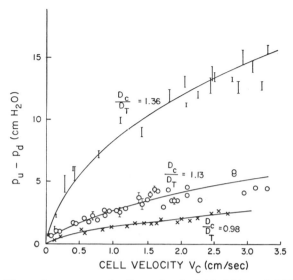

Figure 5.5:5 The resistance to red cell motion in 3.15, 3.81, and 4.37 cm tubes expressed in terms of the loss of pressure over and above that of the Poiseuille flow head, $p_u - p_d$, as a function of the cell velocity. The variation in the pressure difference was a result of minor shape changes of the red cell model and corresponds to the fluctuations of the downstream pressure. From Lee and Fung (1969).

membrane can be obtained only from the bending strain of the solid rubber. This failure to simulate the material properties could have effects which, however, are still unclear.

5.5.4 Observation of Red Cell Flow in Glass Capillaries

Figure 5.5:6 (from Hochmuth et al., 1970) shows data obtained from motion pictures of red cells traversing glass capillaries. The apparent plasma-layer thickness was calculated by taking the difference between the known capillary diameter and the maximum transverse dimension of the deformed cell. The plasma layer thickness seems to reach an asymptotic value at a velocity of about 2 mm/sec. The scatter of the data is due in part to the random orientation of the cells relative to the plane of focus.

The increased clearance between cell and vessel wall reduces shear stress at the capillary wall due to the passing of the red cell. Hence according to Fig. 5.5:6, the apparent viscosity of blood in the capillary should decrease as the velocity of flow increases; and by physical reasoning, the clearance will increase if the cell is more flexible, so the more flexible the cell is, the smaller the additional pressure drop due to the red cell.

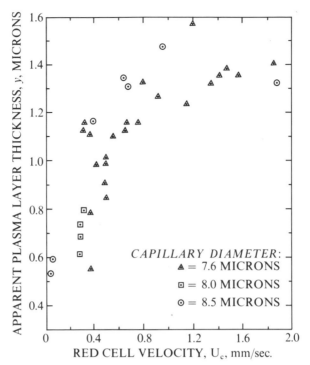

Figure 5.5:6 Variation of apparent plasma-layer thickness with red cell velocity in glass capillaries. Human red cells were used. From Hochmuth et al. (1970), by permission.

5.6 Inversion of the Fahraeus–Lindqvist Effect in Very Narrow Tubes

According to the result shown in Fig. 5.5:5 the additional pressure drop due to the motion of a single red blood cell increases greatly when the tube diameter becomes smaller than that of the resting red cell; and the smaller the tube, the higher is the resistance. Hence unless there is an extraordinary decrease in the number of red cells in the tube (which is not the case, as we shall show in Sec. 5.7), the resistance can be expected to increase with decreasing tube diameter when the tube is smaller than the red cell. This is the reverse of the Fahraeus–Lindqvist effect.

To calculate the resistance of red cells in a tightly fitting tube, we must know the interaction between the neighboring cells. Sutera et al. (1970) found that the additional pressure drop caused by a single cell is unaffected by a neighboring cell if the two are separated by a distance equal to or greater than one tube diameter. This short interaction distance is expected in low Reynolds number (N_R) flow. (In low Reynolds number tube flow the development length is of the order of one tube diameter, with a lower limit of 0.65 D_T in the limit $N_R \to 0$; see Lew and Fung (1969b, 1970).) Therefore, when the cells are separated by a space equal to or greater than one tube diameter, we can calculate the total pressure drop in the tube by adding the additional pressure drop due to individual cells to the Poiseuillean value. Since the pressure gradient in a Poiseuille flow is $-32\mu_0 V_M/D_T^2$ (using notations of the previous section, with V_M denoting the mean flow velocity, and D_T the tube diameter), we can write the total pressure drop per unit length of the tube as

$$-\frac{\partial p}{\partial x} = \frac{32\mu_0 V_M}{D_T^2} + n\Delta p^*, \tag{1}$$

where n is the number of cells per unit length, and Δp^* is the additional pressure drop per single cell. μ_0 is the coefficient of viscosity of the suspending fluid (plasma). If H_T is the hematocrit in the capillary tube, then n is equal to the volume of the tube per unit length divided by the volume of one red blood cell:

$$n = \frac{H_T \pi D_T^2}{4(\text{RBC vol.})}. \tag{2}$$

Δp^*, as presented in Fig. 5.5:6, can be made dimensionless with respect to the characteristic shear stress $\mu_0 V_M/D_T$. It is a function of D_c, D_T, V_M, μ_0, and the elasticity of the cell. Let the elastic modulus of the cell membrane for pure shear be denoted by E_C (see Chapter 4, Sec. 4.7); then a dimensionless parameter is $\mu_0 V_M/(E_c D_T)$. This parameter is the ratio of a typical shear stress $\mu_0 V_M/D_T$ to the membrane elasticity modulus, and thus can be called a *characteristic membrane strain*.

Figure 5.6:1 shows the normalized additional pressure drop of a single cell given by Sutera et al. (1970). We can then combine Eqs. (1) and (2) as

$$-\frac{\partial p}{\partial x} = \frac{32\mu_0 V_M}{D_T^2} + \frac{\pi H_T D_T^2}{4(\text{RBC vol.})}\left(\frac{\Delta p^*}{\mu_0 V_M/D_T}\right)\frac{\mu_0 V_M}{D_T}. \tag{3}$$

But if we invoke the concept of *apparent viscosity*, we can also write

$$-\frac{\partial p}{\partial x} = \frac{32\mu_{\text{app}} V_M}{D_T^2}. \tag{4}$$

Comparing Eqs. (4) and (5), we obtain the *relative apparent viscosity* μ_r:

$$\mu_r = \frac{\mu_{\text{app}}}{\mu_0} = 1 + \frac{\pi H_T}{128}\left(\frac{D_T^3}{\text{RBC vol.}}\right)\left(\frac{\Delta p^*}{\mu_0 V_M/D_T}\right). \tag{5}$$

If the data of Fig. 5.6:1 are examined in light of this equation, it will be found that μ_r increases with decreasing D_T/D_c when $D_T/D_c < 1$, for any fixed values of H_T. This is opposite to the Fahraeus–Lindqvist effect. If a simplified expression relating the hematocrit in the tube, H_T, to that in the reservoir, H_F, as given in Eqs. (1) and (2) of Sec. 5.7, is used in Eq. (5) above,

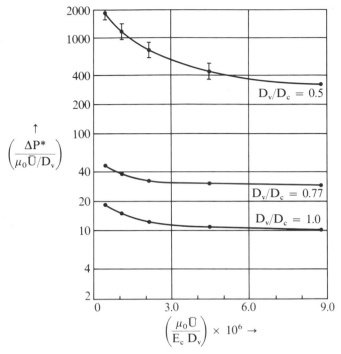

Figure 5.6:1 Additional pressure drop due to a single cell as a function of strain parameter, from large-scale model. In this figure, D_v is used in place of D_T, the vessel or tube diameter, and $\bar{U}$ is used in place of V_M the mean flow velocity. From Sutera et al. (1970), by permission.

then the results shown in Fig. 5.6:2 are obtained at a reservoir hematocrit of 40%.

Equations (1)–(5) are derived under the assumption that the red cells are sufficiently apart from each other in the tube. If the cells are more closely packed, the situation is more complex. Sutera (1978), citing his model experiments (1970), believes that these equations are applicable irrespective of the spacing between the red cells, so that the resistance is linearly proportional to the hematocrit without limitation on the hematocrit value. A similar conclusion is reached by Lighthill's (1968, 1969) lubrication layer theory, because the interaction between the red cell and the endothelium of the capillary blood vessel is localized. Experiments by Jay et al. (1972), however, showed that the relative viscosity of blood tends to be independent of the hematocrit for human blood flowing in glass tubes of 4 to 15 μm in diameter. Earlier, such a phenomenon was also described by Prothero and Burton (1961, 1962). Jay et al. believed that this is a direct contradiction to the Lighthill–Fitz-Gerald theory. Chien (1972) points out, however, that the theoretical analysis by Skalak et al. (1972) showed that when the spacing

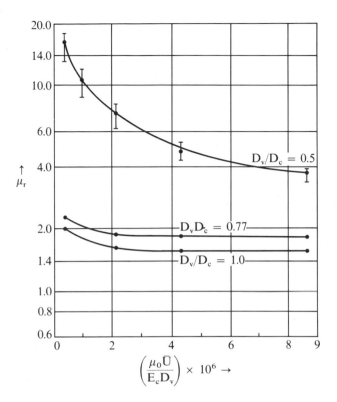

Figure 5.6:2 Apparent relative viscosity of blood in capillaries according to model experiment results of Sutera et al. (1970) and calculated for the "reservoir" or "feed" hematocrit $H_F = 40\%$. From Sutera (1978), by permission.

between the red cells is small, the plasma trapped between the cells moves with the cell. Hence when the hematocrit is sufficiently high, the cell-plasma core moves almost like a rigid body, and the apparent viscosity will be independent of the hematocrit. Furthermore, the size of the core depends on the deformation of the red cell. One can estimate the size of the red cell core from the experimental results shown in Fig. 5.5:6. It is seen that even in very narrow capillaries there is a sizable gap between the cell and endothelium if the blood flows at a speed close to the *in vivo* value. The gap becomes very small only if the flow velocity becomes very small, i.e., in a near stasis condition.

Summarizing, we conclude that the relative viscosity of single-file flow of red cells in very narrow capillaries is proportional to the tube hematocrit if the spacing between the red cells is of the order of a tube diameter or larger. If the spacing is smaller, and the flow velocity approaches 1 mm/sec or larger, then the cell-plasma core moves like a rigid body, and the relative viscosity will tend to be independent of the tube hematocrit. If the spacing between red cells is smaller than the tube diameter and the flow velocity is smaller than, say 0.4 mm/sec, then the gap between the cell and endothelium decreases, the resistance of each cell increases, and the relative viscosity will be proportional to the number of red cells per unit length of the vessel.

5.7 Hematocrit in Very Narrow Tubes

Since the apparent viscosity of blood depends on hematocrit [see, for example, Eq. (8) of Sec. 5.2], we should know the way hematocrit changes from one blood vessel to another in a microvascular bed. When one examines microcirculation in a living preparation under a microscope, one is impressed by the extreme nonuniformity in hematocrit distribution among the capillary blood vessels, and by the unsteadiness of the flow. The red blood cells are not uniformly distributed. In a sheet of mesentery or omentum, or a sheet of muscle, sometimes we see a long segment of the capillary without any red cells; at another instant we see cells tightly packed together. In any given vessel, the velocity of flow fluctuates in a random manner (Johnson and Wayland, 1967). Similarly, in the capillaries of the pulmonary alveoli, the velocity of flow fluctuates (Kot, 1971). The distribution of red cells appears "patchy" at any given instant of time (Warrell et al., 1972), i.e., the distribution is nonuniform in the alveolar sheet: capillaries in certain areas have a high density of red cells, while neighboring areas have few red cells.

These dynamic features have many causes. Figure 5.7:1 shows some examples. Consider a balanced circuit connecting a vessel at pressure p_1 to another at pressure p_0 (Fig. 5.7:1). Let us assume that all branches, A, B, C, D, E, are of equal length and diameter. Let us also assume that at the beginning all the red cells are identical and are uniformly distributed in all vessels. Then it is clear that there will be uniform flow in the branches A, B, C, D, but no flow in E. Now suppose that branch B receives one red cell more than A. The balance is then upset. The pressure drop in B will be increased, and a flow is created in the branch E. The same will happen if branch B, instead of getting an extra cell, gets a red cell that is larger than those in A, or a leukocyte. The flow in E, thus created, will continue unless the resistance is balanced again. Thus, because of the statistical spread of the red cell sizes, and the existence of leukocytes, continued fluctuation in branch E is expected. Finally, active control due to sphincter action of vascular smooth muscle will have the same effect, as is shown in the lower right panel in Fig. 5.7:1.

Another cause of flow fluctuation is the basic particulate nature of the blood. To clarify the idea, consider the situation shown in Fig. 5.7:2, in which rigid spherical particles flow down a tube which divides into two equal branches. Let the tube diameter be almost equal to the diameter of the spheres. Consider a sphere that has just reached the point of bifurcation. We shall see that the ball has a tendency to move into the branch in which the

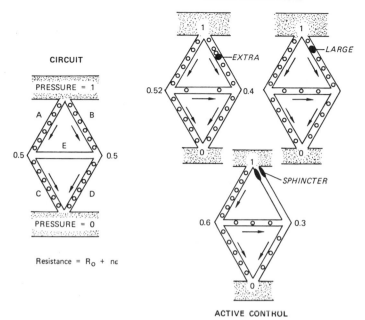

Figure 5.7:1 An idealized capillary blood flow circuit. (Left), balanced. (Right), balance upset by (a), an extra cell in branch B, (b) an extremely large cell in branch B, and (c) a sphincter contraction. From Fung (1973).

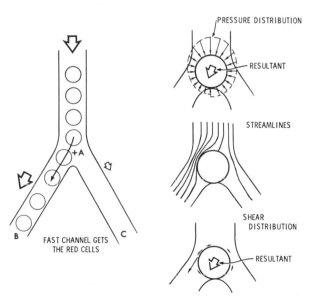

Figure 5.7:2 At a branch point of a capillary blood vessel, the branch with the faster stream gets the red blood cells. (Left), spherical balls flow in a cylindrical tube which bifurcates into two equal branches. (Right), the forces acting on the sphere at the moment when it is located at the point of bifurcation of the vessel. The resultants of the pressure forces (top) and shear stress (bottom) both point to the faster stream. From Fung (1973).

187

flow is faster. Let us assume that the flow in the branch to the left, AB, is faster than that in the branch to the right, AC (see the left-hand drawing in Fig. 5.7:2). The reason for the flow in AB to be faster must be that the pressure gradient is larger in AB than in AC. Now consider all the forces acting on the ball, which is situated centrally at the junction as shown on the right-hand side of Fig. 5.7:2. These are pressure forces and shear forces. The pressure on the top side of the ball is nearly symmetric with respect to right and left. The pressure on the underside is lower on the left as we have just explained. Therefore, the net pressure force will pull the ball into the left branch. Next consider the shear stress. Assume that the ball comes down the vertical tube without rotation. At the junction, the ball is temporarily stopped at the fork. The fluid will tend to stream past the ball at this position. This streaming is more rapid on the left-hand side, because we have assumed that the branch at the left has a faster flow. Therefore, the shear strain rate and the shear stress are larger on the left. The net shear force is a vector pointing into the left channel. Therefore, in this configuration both the pressure force and the shear force tend to pull the ball into the faster stream.

When the next ball reaches the junction the same situation prevails. Therefore, under the assumption made above the more rapid stream will get all the balls. If the velocity in AB is twice that in AC, the theoretical ratio of the density of red cells in AB to that in AC is not 2 to 1, but infinity.

In capillary blood flow, each blood cell would have to make the same decision when it reaches a crossroad: which branch should it go into? The decision making is influenced by the size of the blood cell relative to the blood vessel, the flexibility of the blood cell, and the ratio of the velocities in the branches. A detailed investigation of these factors has been made by the author and R. T. Yen (Yen and Fung, 1978). We showed that when a capillary blood vessel bifurcates into two daughter vessels of equal diameter, the red cells will be distributed nonuniformly to the two daughter branches if the velocities in the two branches are unequal. The faster side gets more cells. If the cell-to-tube diameter ratio D_c/D_t is of the order of 1 or larger, and the velocity ratio of the flow in the two branches is less than a certain critical value, then the hematocrit ratio in the two branches is proportional to the velocity ratio. If the velocity ratio exceeds a critical value, then the faster branch gets virtually *all* the red cells. In that case the hematocrit in the slower branch is zero. The critical velocity ratio is of the order of 2.5, with the exact value depending on the cell-to-tube diameter ratio D_c/D_t and the elasticity of the red cell.

A way to produce a large velocity ratio in capillary branches is to occlude the downstream end of a vessel. Svanes and Zweifach (1968), using a micro-occlusion technique, have shown that it is possible to clear any capillary vessels of its red cells.

Our discussion so far has stressed the importance of the entry condition of flow into branches of equal diameter on the hematocrit distribution. How

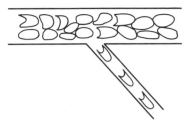

Figure 5.7:3 A small vessel branching out from a larger one.

about flow into branches of unequal diameter? Or flow from a reservoir into a small tube (as in Fahraeus' experiment, Fig. 5.3:2)? Or from a larger tube into a smaller side branch as in the case of a capillary blood vessel issuing from an arteriole (Fig. 5.7:3)? In these cases the hematocrit in the smaller branch may be reduced for three reasons:

(1) Due to the entry condition into the capillary (affected by the geometry of the entry section of the tube, and the flow just outside of the entrance section).
(2) Due to a decrease of hematocrit at the wall of the larger tube, where the capillary siphons off the blood (the *plasma skimming* effect discussed in Chapter 3, Sec. 3.5, and Chapter 5, Sec. 5.3).
(3) Due to the fact that the red cell moves faster than the plasma in the capillary blood vessel.

If the effects (1) and (2) are ignored, then we can calculate the effect of (3) as follows. Let the mean velocity of blood in a capillary be V_m, and that of the red cell be V_c. In a unit interval of time the capillary draws a volume of blood equal to $V_m A$ from the reservoir, where A is the cross-sectional area of the capillary. If the hematocrit of the reservoir is H_F, the volume of red cells flowing in is $V_m A H_F$. At the same time, the volume occupied by the red cells (which are arranged in single file in very narrow capillaries) is $V_c A H_T$, where H_T denotes the hematocrit in the capillary. Thus, obviously,

$$V_m A H_F = V_c A H_T,$$

or

$$H_T = H_F \frac{V_m}{V_c}. \tag{1}$$

This is a statement of the conservation of red cell volume. Since $V_m / V_c < 1$ (see Sec. 5.5), we have $H_T < H_F$, the Fahraeus effect.

The effect of the two factors (1) and (2) ignored above, however, may render Eq. (1) nonvalid. The actual hematocrit of the blood drawn in by the capillary from the arteriole (or reservoir) may not be equal to H_F, the average hematocrit in the larger vessel. The hematocrit in the reservoir in the neighborhood of the entry section can be smaller or larger than H_F by a factor F. Thus

$$H_T = FH_F \frac{V_m}{V_c}. \tag{2}$$

To clarify the entry effect and determine F, Yen and Fung (1977) made a model experiment in which gelatin particles (circular disks) were suspended in a silicone fluid to simulate the blood. When the diameter of the undeformed cell was equal to or greater than the tube diameter, the volume fraction of the cells in the tube was found to increase to a value equal to or greater than that in the reservoir. This is the reverse of the Fahraeus effect.

Additional experiments showed that the hematocrit in the tube could be greatly influenced by the flow condition outside the entrance to the tube. If the tube was nearly perpendicular to the main direction of flow in the reservoir (as is the case in most arteriole-capillary junctions), it was found that the velocity of flow in the reservoir just outside the entrance to the tube could affect the hematocrit in the tube. Some details are given in the following.

Two models were tested. See Fig. 5.7:4. The first model employs a tube with a smooth entry section opening into a still reservoir. For this model an "entry cone" is inserted in front of the tube, similar to the configuration tested by Fahraeus and Lindqvist [Fig. 5.3:2(a)]. The second model uses a tube with an abrupt entry section opening perpendicularly into a stream, which has a definite shear gradient. The end of the tube is cut squarely and sharply, similar to the configuration tested by Barbee and Cokelet [Fig. 5.3:2(b)]. The reservoir of the second model is a cylindrical tank containing a rotating inner core which imparts a steady shear flow to the fluid. This is considered to be the mainstream flow. The tube is placed perpendicular to the direction of main flow in the reservoir in order to simulate the flow condition at an arteriole-capillary blood vessel junction.

A suspension of gelatin pellets in a silicone fluid was used to simulate blood. The pellets were circular cylindrical disks, with a diameter of 1.08 cm and a thickness of 0.32 cm in the first model, and a diameter of 0.32 cm and a thickness of 0.1 cm in the second model. The buoyancy of the pellets was controlled by mixing a suitable amount of alcohol with water in making them. The viscosity of the fluid was 100 P, and the Reynolds number based on the tube diameter was 10^{-2} to 10^{-4}, similar to *in vivo* values in capillaries. In each experiment, the pellets were carefully mixed with the silicone fluid; the mixture was stirred until uniform and then poured into the reservoir. To determine the particle concentration at any flow condition, the flow was suddenly stopped, the number of particles present in the tube was counted, and their volume computed.

Figure 5.7:5(a) shows the ratio of hematocrit in the "capillary" tube to that in the reservoir of Model I, plotted against the ratio of particle to tube diameters. The H_T/H_F ratio increases with increasing ratios of D_c/D_T. Thus

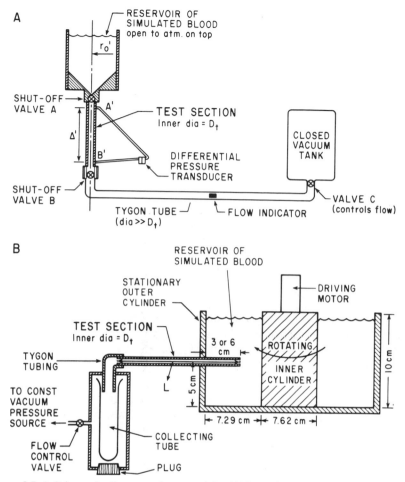

Figure 5.7:4 Schematic diagram of test models: (A) Model I, (B) Model II. For Model II, the length of the tygon tubing attached to the outflow end of the test tubes was 6.4 cm. The inner diameter of the tygon tubing was equal to the outer diameter of the test tube. The three test tubes have inner diameters $D_t = 0.67$, 0.45, and 0.32 cm; whereas the simulated red cells (gelatin pellets) have a diameter of 0.32 cm and a thickness of 0.1 cm. From Yen and Fung (1977).

for a given cell diameter, the tube hematocrit increases when the tube diameter decreases (reverse Fahraeus effect). This is more evident at lower reservoir concentration.

Figure 5.7:5(b) shows the variation of the hematocrit ratio H_T/H_F as a function reservoir hematocrit H_F for fixed values of D_c/D_t. It shows a tendency toward increased relative hematocrit in the capillary tube when the reservoir hematocrit is lowered.

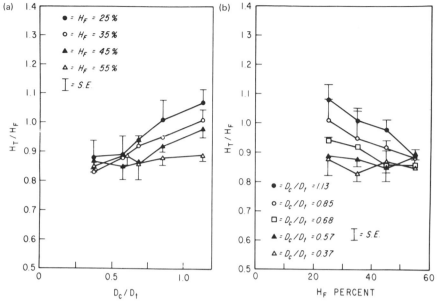

Figure 5.7:5 Experimental results from Model I, showing mean and standard errors of the mean. (a) Plot of ratio of *tube hematocrit* (H_T) to *feed hematocrit* (H_F) as a function of ratio of cell diameter (D_c) to tube diameter (D_t). (b) Tube hematocrit to feed hematocrit ratio (H_T/H_F) as a function of feed hematocrit (H_F). From Yen and Fung (1977).

Typical results from Model II are shown in Fig. 5.7:6. H_D is the "discharge" hematocrit of the outflow from the tube. H_F is the reservoir hematocrit. The ratio H_D/H_F is the factor F of Eq. (2), because by the principle of conservation of matter (red cell volume), we obtain, by reasoning identical to the derivation of Eq. (1),

$$\frac{H_D}{H_T} = \frac{\text{discharge hematocrit}}{\text{tube hematocrit}} = \frac{\text{mean speed of pellets}}{\text{mean speed of tube flow}} = \frac{V_c}{V_m}. \qquad (3)$$

Substituting Eq. (3) into Eq. (2), we obtain

$$F = \frac{H_D}{H_F}. \qquad (4)$$

Figure 5.7:6 shows that H_D/H_F, hence the factor F, varies with the ratio of the mainstream velocity at the entrance section of the tube, U_F, to the mean velocity of flow in the tube, U_T. The curves are seen to be bell-shaped. From a certain initial value of H_D/H_F, the discharge hematocrit rises to a peak value at a ratio of U_F/U_T in the range 1 to 4. For higher velocity ratios (U_F/U_T from 4 to 17), the discharge hematocrit declines.

In Fig. 5.7:7 the ratio of the hematocrit in the "capillary" tube to that in the reservoir is plotted against the ratio of the mainstream velocity in the reservoir at the tube entrance to the mean velocity of flow in the tube, for a

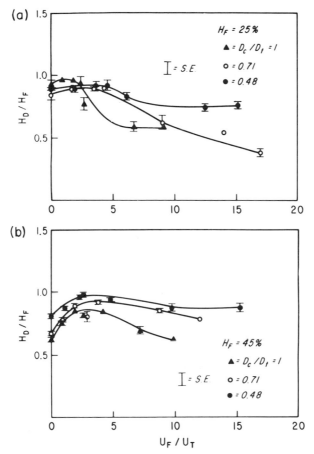

Figure 5.7:6 Experimental results from Model II, showing mean and standard errors of the mean. (a) Relative *discharge hematocrit* as a function of the ratio of tangential velocity in the reservoir at tube entrance and mean tube flow velocity for feed hematocrit $H_F = 25\%$. (b) The same for feed hematocrit $H_F = 45\%$. From Yen and Fung (1977).

reservoir hematocrit of 25% in Model II. A comparison of Figs. 5.7:7 and 5.7:6(a) shows that the capillary tube hematocrit and the discharge hematocrit are not equal. Both H_T and H_D vary with the mainstream velocity and are influenced by the entrance condition.

Motion pictures of the flow in tubes with D_c/D_t equal to 1 or larger show that most of the pellets (disks) enter the tube "edge-on." Thus the force of interaction between a pellet and the entry section of the tube is concentrated on two areas at the edge of the cell. In general, buckling of the cell membrane occurs at these points.

It is clear that the seemingly simple motion of suspended flexible particles from a reservoir into a small tube is a complicated phenomenon to analyze.

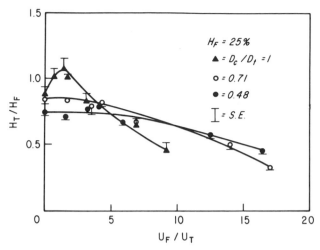

Figure 5.7:7 Relative *tube hematocrit* in Model II as a function of the ratio of tangential velocity in the reservoir at tube entrance to mean tube flow velocity for feed hematocrit $H_F = 25\%$. From Yen and Fung (1978).

5.8 Theoretical Investigations

Even in very narrow capillary blood vessels, moving red cells never come to solid-to-solid contact with the endothelium of the blood vessel. There is a thin fluid layer in between. The thicker the fluid layer, the smaller would be the shear strain rate, and the smaller the viscous stress. On the other hand, for any given thickness of the fluid layer, the shear stress depends on the velocity profile in the gap. Is there any way the shear stress on the vessel wall (endothelium) can be reduced, thereby reducing the resistance of the blood? The answer is "yes": by forcing the fluid into the gap in such a way as to reduce the slope of the velocity profile on the vessel wall. This is the hydrodynamic lubrication used in engineering, in journal bearings. Lighthill (1968, 1969) pointed out that such an effect can be expected from a red cell squeezing through a very narrow capillary vessel. The effect is due to a "leak-back" in the gap between the cell and the vessel wall.

The physical picture is illustrated in Fig. 5.8:1, from Caro et al. (1978). In the figure the velocity vectors are drawn with respect to an "observer" moving with the red cell. The cell then appears to be at rest, and the capillary wall moving backward with speed U. In Fig. 5.8:1(a) the velocity profile is assumed to be linear. Under this hypothesis the rate of fluid (volume) flow in the gap, which is equal to the area of the velocity profile multiplied by the length (in the direction perpendicular to the paper), will be nonuniform: smaller at the point x in the figure than that at y. This is impossible in an incompressible fluid; hence the hypothesis is untenable. It follows that the velocity profile cannot be linear everywhere.

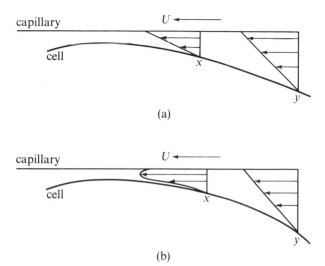

(a)

(b)

Figure 5.8:1 Diagram showing the need for leak-back past a red cell in a capillary (cell taken to be at rest, capillary wall moving). (a) Linear profile everywhere leads to nonuniform flow rate, which is not permitted. (b) Continuity upheld by the super-position of a parabolic component, requiring a pressure gradient. From Caro et al. (1978), p. 404, by permission.

In Fig. 5.8:1(b), a nonlinear velocity profile is shown at x. This profile has a zero velocity on the cell wall and U on the capillary, as the boundary conditions demand. In the gap the profile bulges to the left so that the area under the profile at x, where the gap is thinner, is equal to the area of the triangle at y. Then the principle of conservation of mass is satisfied. The slope of the nonlinear velocity profile at the capillary wall at point x is smaller than that at y. The shear stress, which is proportional to the slope, is therefore smaller at x then at y. In the special case sketched in Fig. 5.8:1(b), this slope is negative, so the shear stress acts in a direction to propel the blood!

In general, the more the velocity profile deviates from the linear profile, the more reduction of shear stress on the capillary wall is obtained. Correspondingly, the negative pressure gradient in the layer is reduced. Let Q be the volume flow rate associated with this departure from the linear profile (proportional to the area under the velocity profile minus $\frac{1}{2}Uh$, the area of a hypothetical linear profile). Let x be the distance along the capillary wall measured from left to right, and let h be the thickness of the layer. Then, according to a detailed analysis of force balance, Lighthill shows that the pressure gradient (dp/dx) is the sum of two terms, one positive and proportional to Q/h^3, and the other negative and proportional to $-U/h^2$. Thus

$$\frac{dp}{dx} = \frac{12\mu Q}{h^3} - \frac{6\mu U}{h^2}, \tag{1}$$

where μ is the viscosity of the plasma, and Q is the leak-back. The lubrication quality is derived from Q.

So far the analysis is indisputable. To complete it, we must compute the thickness h as a function of x. This cannot be done without specifying the elasticity of the red cell and the endothelium. Lighthill (1968) and Fitz-Gerald (1969a,b) assumed both of these behave like an elastic foundation, with local deflection directly proportional to local pressure. This assumption is questionable, especially with regard to the red cell which, according to what is described in Chapter 4, should behave more like a very thin-walled shell with a very small internal pressure. For this reason the many interesting predictions made by these authors remain to be evaluated.

The concept of a lubrication layer is pertinent to red cells moving in capillary blood vessels so narrow that severe deformation of the cells is necessary for their passage. In larger capillaries whose diameters are larger than that of the red cells, the deformation of the red cells is minor, and the mathematical analysis can be based on Stoke's equation for the plasma, which is a Newtonian fluid. Skalak and his students (see reviews in Skalak, 1972, and Goldsmith and Skalak, 1975) have carried out a sequence of investigations of flows of this type: the flow of a string of equally spaced rigid spheres in a circular cylindrical tube, the flow of oblate and prolate spheroids in the tube, the flow of spherical and deformable liquid droplets in the tube, and finally the flow of red blood cells with membrane properties specified by Eq. (7) of Sec. 4.7 in the tube. In the last case they also included very narrow tubes, which call for large deformation of the cells under the assumption that the radius of the cell at any point is linearly dependent on the pressure at that point. Among other things, they showed that, in the case of the flow of rigid spheres down the center of a tube, the pressure drop required for a given volume flow rate increases when the ratio of the diameters of the sphere, b, and of the tube, a, increases. But even when $b/a = 0.9$ and the spheres are touching, the pressure drop is only about 2.0 times that required by the same flow with plasma only, without spheres. But the apparent viscosity of whole blood is at least 2.5 times that of plasma, so the lubrication layer effect is operating even when b/a is as small as 0.9.

Secomb et al. (1986) have summarized the mathematical theories of axisymmetric red blood cells in narrow capillaries.

Skalak and Chien (1983) presented a theoretical model of rouleau formation and disaggregation. Over the years from 1966 to the present, Skalak and his students and associates have published a long series of papers on red cell flow in capillaries, introducing refinements and generalizations step by step. See Skalak (1990) for a summary.

5.9 The Vascular Endothelium

The vascular endothelium is a continuum of endothelial cells lining the blood vessels, Fig. 5.9:1. Studies by electron microscopy, molecular biochemistry,

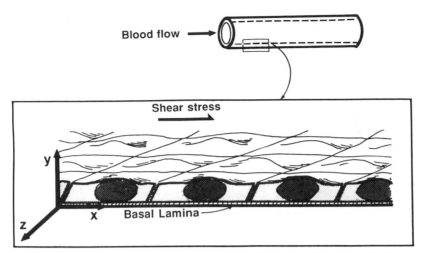

Figure 5.9:1 A schematic drawing of an arterial wall showing blood flow in the upper figure, and the enlarged view of the endothelium in the lower figure. The coordinates system shown here is used throughout this chapter. From Fung and Liu (1993), by permission of ASME.

gene expression, membrane technology, immunochemistry, etc., have yielded a wealth of information about the endothelium (Repin et al., 1984; Rhodin, 1980; Simionescu et al., 1975, 1976). The endothelial cell morphology, metabolism, and ultrastructure have been found to vary with the shear strain rate of the blood (Caro et al., 1971; Dewey et al., 1981; Flaherty et al., 1972; Fry, 1968; Gau et al., 1980; Helmlinger et al., 1991; Kim et al., 1989; Levesque et al., 1985; Nerem et al., 1990; Nollert et al., 1991; Sato et al., 1987, 1990; Sprague et al., 1987; Theret et al., 1988; Zarins et al., 1987). We would like to study the stresses in the endothelial cells. However, little is known about the mechanical properties of its internal parts, so we are not ready to make a full stress analysis. In this Section, we explore those aspects of the system which do not need the details of the cell model and the constitutive equations. We recognize only that the endothelial cells form a continuous layer, and that the cell membrane is an indispensible part of every cell. We assume that the endothelial cell membrane is solid-like because the cells can maintain their shape while they are subjected to a life-long shear force from the flowing blood. It reacts to the shear force with a deformation, not with a flow, and hence is a solid. The content of the endothelial cell can be represented as a composite mixture. Some components, such as the nucleus and actin fibers, may be expected to be solid-like, but the mixture as a whole may be fluid-like. At this time, the rheological properties of the content of endothelial cell are unknown. Hence, we shall make two mutually exclusive but together exhaustive hypotheses: either (1) the content is fluid-like, so that the cell membrane plays a dominant role in maintaining the shape of the endothelium; or (2) the content is solid-like, so that the cell membrane and the cell content together maintain the endothelium geometry. We investigate the consequences of these alternative hypotheses.

Under the hypothesis that the content of the endothelial cells is fluid-like, then in the steady state there is no internal flow and no shear stress in the cell content. Consequently, the cell membrane is the structure that resists the shear load from the blood flow at steady state.

The cell membrane is very thin. A very thin membrane is very easy to bucke (see Sec. 4.5) and cannot sustain a significant amount of compressive stress in its own plane. Hence we assume that (1) the cell membrane is so thin that it buckles easily and cannot support compression in its plane; and (2) a situation exists in which one of the principal strains in the deformed membrane is positive while the other one is negative or negligible. A stress analysis based on these assumptions is called a *tension field* theory. The analysis in Sec. 5.11–5.13 is based on this theory.

5.10 Blood Shear Load Acting on the Endothelium

In 1968, Fry called attention to the existence of a relationship between the shape of the endothelial cells and their nuclei and the shear stress in the blood. In 1969 Caro et al. called attention to a possible connection between arteriosclerossis and the shear stress imparted by the flowing blood. In human coronary arteries, Giddens et al. (1990) have shown that the axial shear stress varies in the range of 1 to 2 N/m^2, with a mean value around 1.6 Pa. Rodbard (1970) and Kamiya et al. (1980, 1984) have observed that the shear stress in flowing blood at the endothelial surface is of the same order of magnitude in all generations of arteries, large and small, including the aorta and capillaries. This, then, is the order of magnitude of the shear load acting on the endothelial cell membrane of arteries in contact with the flowing blood. The exact value will depend on the local condition: entry, exit, branching, flow separation, secondary flow, etc. The shear stress of the flow acting on the venules and veins is smaller because the volume flow rate is similar but the diameter of veins of any generation is larger than that of the arteries of the same generation. At a given flow, the wall shear scales inversely as the third power of the diameter. If the diameter is larger by 26%, the shear stress at the wall will be smaller by a factor of 2 at the same flow.

The stresses in the media and adventitia are much larger. See Chapter 8. In normal physiological conditions, the circumferential and longitudinal tensile stresses in rabbit arteries are in the order of 60 to 110 kPa. The radial stress is compressive. The maximum shear stress at the inner wall is equal to one-half of the difference between the max principal stress and the min principal stress, acting on a plane which is inclined at $45°$ to the principal axes. This is four orders of magnitude larger than the shear stress acting on the surface of the endothelium due to blood flow. The maximum shear stress elsewhere in the media is similarly several thousand times larger than the shear stress of the blood acting on the endothelial cell surface.

Test equipment to impose shear stress on cultured cells has been described by Dewey et al. (1981), Strong et al. (1982), Sakariassen et al. (1983), Koslow et al. (1986), and Viggers et al. (1986).

5.11 Tension Field in Endothelial Cell Membranes Under the Fluid Interior Hypothesis

Figure 5.9:1 shows a flow of blood which causes a shear stress to act on the blood vessel wall. A coordinate system is attached to the vessel wall. Below the blood vessel is shown a magnified schematic drawing of the endothelium, with a longitudinal cross section in the front.

Due to the special geometry of the endothelium, different parts of the cell membrane of each endothelial cell are subjected to different forces. The *upper* cell membrane is in contact with the blood. The *lower* cell membrane is adhered to the basal lamina. The *side* cell membranes connect the upper and lower membranes. Each endothelial cell membrane has a system of internal stress and strain. The components of stress in an upper cell membrane can be seen on a free-body diagram of a small rectangular element of the cell membrane as shown in Fig. 5.11:1. With reference to a rectangular Cartesian frame of reference with coordinates x, y, z, with the x axis pointing in the direction of the blood flow and the y axis normal to the membrane, the stress tensor

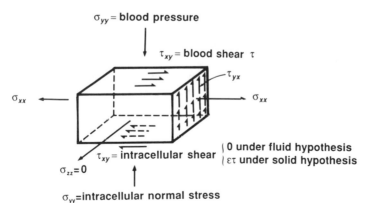

Figure 5.11:1 The components of stress acting on a small element of the upper endothelial membrane. The tensile stress σ_{xx} is usually much larger than all the other stresses. σ_{zz} is zero under the tension field assumption. On the top side of the membrane acts the blood pressure and shear. On the underside of the membrane acts the normal and shear stress of the cell content at the interface. Two alternative hypotheses are made with regard to the static stresses in the cell content. From Fung and Liu (1993), by permission of ASME.

has nine components:

$$\begin{pmatrix} \sigma_{xx} & \tau_{xy} & \tau_{xz} \\ \tau_{yx} & \sigma_{yy} & \tau_{yz} \\ \tau_{zx} & \tau_{zy} & \sigma_{zz} \end{pmatrix} \tag{1}$$

of which six are independent, because $\tau_{xy} = \tau_{yx}$, $\tau_{yz} = \tau_{zy}$, $\tau_{zx} = \tau_{xz}$. In the upper cell membrane, the magnitude of some component of stress is much larger than that of the others. The normal stress σ_{yy} is equal to the blood pressure on the top side, and to the intracellular normal stress on the under side. The shear stresses τ_{xz}, τ_{yz} are most likely to be very small at steady state if the fluid-mosaic concept of the lipid bilayer promulgated by Singer et al. (1972) is valid. If we accept this concept, then

$$\tau_{zx} = \tau_{xz} = \tau_{yz} = \tau_{zy} = 0. \tag{2}$$

The shear stress τ_{yx} on the upper surface of the upper endothelial cell membrane which is in contact with the flowing blood must be equal to the viscous shear stress of the blood, τ. The shear stress τ_{yx} on the bottom of the upper cell membrane depends on the rheological behavior of the cell content. Under the hypothesis that the cell content is fluid-like, then τ_{yx} would be zero if the plasma streaming inside of the cell can be ignored. Plasma streaming is believed to be small. Hence, in the upper cell membrane,

$$\tau_{xy} = \tau_{yx} = \tau \qquad \text{on the surface in contact with blood flow,}$$

$$\tau_{xy} = \tau_{yx} = 0 \qquad \begin{array}{l}\text{on the surface facing cell interior under the} \\ \text{fluid interior hypothesis}\end{array} \tag{3}$$

There are two normal stresses, σ_{xx} and σ_{zz}, left to be considered. We shall show that σ_{xx} is much larger than τ, and σ_{zz} is nearly zero. To explain the reasoning leading to this conclusion, we follow the conventional theory of plates and shells by introducing the *stress resultants* or *membrane tensions* as working variables. The stree resultants or membrane tensions N_x, N_z are defined by the integrals of σ_{xx}, σ_{zz} throughout the thickness of the membrane, h:

$$\int_0^h \sigma_{xx}\, dy = N_x, \qquad \int_0^h \sigma_{zz}\, dy = N_z. \tag{4}$$

The units of stress being Newton/m^2, those of N_x and N_z are Newton/m. Now, the equation of equilibrium of forces in the x direction acting on an element of the membrane is (Chapter 2)

$$\frac{\partial \sigma_{xx}}{\partial x} + \frac{\partial \tau_{xy}}{\partial y} + \frac{\partial \tau_{xz}}{\partial z} = 0. \tag{5}$$

Multiplying every term with dy and integrating all terms with respect to y from $y = 0$ to $y = h$, and noting Eqs. (2) and (3), we obtain

$$\frac{\partial N_x}{\partial x} + \tau = 0, \tag{6}$$

where τ is the shear stress in the blood at the endothelium. Assuming τ to be a constant in the length of an endothelial cell and integrating Eq. (6) with respect to x, we obtain

$$N_x = -\tau x + N_0, \tag{7}$$

where N_0 is an integration constant, which is independent of x, but can be a function of z. If $N_x = 0$ when $x = 0$, then the integration constant is zero, and N_x is seen to increase linearly with $-x$. If the endothelium is flat and is L m long, then at $x = -L$,

$$N_x = \tau L \text{ Newton/m}. \tag{8}$$

The stress σ_{xx} in the cell membrane is essentially uniform throughout the thickness because bending stress is negligible in a very thin membrane. Hence $N_x = \sigma_{xx} h$ and at the left edge $x = -L$,

$$\sigma_{xx} = N_x/h = \tau L/h. \tag{9}$$

The value of τ is between 1 to 2 N/m^2, the length of an endothelial cell is 10 to 60 μm. If we take τ to be 1 N/m^2, L to be 1 cm, and the thickness of the endothelial cell membrane h to be 10 nm, then the tensile stress σ_{xx} is equal to 10^6 N/m^2. If the length L is taken to be that of a single cell, of order 10 μm, then $\sigma_{xx} = 10^3$ N/m^2. If the upper endothelial cell membrane is wavy but the curvature is small, Eq. (9) remains a good approximation.

On the other hand, the equation of equilibrium in the z direction is

$$\frac{\partial \tau_{zx}}{\partial x} + \frac{\partial \tau_{zy}}{\partial y} + \frac{\partial \sigma_{zz}}{\partial z} = 0. \tag{10}$$

Noting Eq. (2) and integrating Eq. (10) first with respect to y from $y = 0$ to $y = t$, then with respect to z from 0 to x, we obtain

$$N_z = \text{const}. \tag{11}$$

If N_z is zero somewhere, then it is zero everywhere.

Thus, if we assume the cell content to be fluid-like and introduce the *tension field* hypothesis named in Sec. 5.9, then $N_z = 0$ and the stress distribution in the endothelial membrane is extremely simple: the circumferential tension is zero everywhere, the longitudinal tension increases linearly in the direction opposite to the flow.

5.12 The Shape of Endothelial Cell Nucleus Under the Fluid Interior Hypothesis

So far we have assumed the upper cell membrane to be a plane. In reality, that membrane is not a plane because of the existence of th nucleus (Fig. 5.9:1). To clarify the mechanical interaction between the nucleus, the cell membrane, and the blood flow in the vessel, let us assume, as a first step, that the nucleus

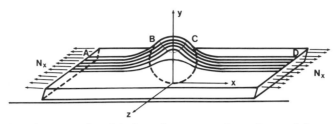

Figure 5.12:1 A perspective drawing of a rectangular cell containing a spherical nucleus. The upper cell membrane has tension in the x direction, and is pushed up by the nucleus. From Fung and Liu (1993), by permission.

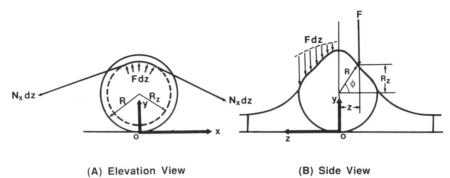

(A) Elevation View   (B) Side View

Figure 5.12:2 The elevation and side views of the nucleus, showing some geometric relations. The x axis points in the direction of blood flow. (A) View in line of the z axis. (B) View in line of the x axis, showing the deformation of the nucleus. From Fung and Liu (1993), by permission.

is a sphere whose diameter, $2R$, is somewhat larger than the average thickness of the endothelium. An idealized perspective sketch is given in Fig. 5.12:1. The lower cell membrane is attached to the basal lamina. The upper cell membrane is in contact with the nucleus. In the area of contact the vertical displacements of the membrane and nucleus must be the same. A coordinate system with x parallel to the direction of flow, y perpendicular to the basal lamina and passing through the center of the sphere, and a z axis normal to x, y is used. A plane normal to the z axis and at a distance to the origin equal to $z\,(< R)$ will intersect the sphere in a circle whose radius is [see Figs. 5.12:2(A) and 5.12:2(B)]

$$R_z = R \sin \varphi, \tag{1}$$

where φ is the polar angle defined by

$$z = R \cos \varphi. \tag{2}$$

Solving Eq. (2) for $\cos \varphi$ and substituting into Eq. (1) yields

$$R_z = R(1 - \cos^2 \varphi)^{1/2} = R(1 - z^2/R^2)^{1/2}. \tag{3}$$

Now, let us choose a series of planes normal to the z axis. These planes intersect the upper cell membrane and the spherical nucleus. The intercepts are sketched in Fig. 5.12:1. They are plane continuous curves whose radius of curvature in the area where the upper cell membrane and the nucleus are in contact is exactly R_z given in Eq. (3). If we consider each intercept as a thin strip of upper cell membrane with a width dz, then, since the upper cell membrane has a tension per unit width equal to N_x, the tensile force in the strip is $N_x dz$. This force has a radius of curvature R_z over the nucleus, as can be seen in Fig. 5.12:2(A). For equilibrium, there must be a distributed lateral force which can be computed by Laplace's formula (Chapter 1, Sec. 1.9). This lateral force, F, is the force of interaction between the upper cell membrane and the nucleus:

$$F = \frac{N_x}{R_z}. \tag{4}$$

This force F is vertical (parallel to the y axis) because the arc BC lies in a plane $z = $ const. Thus the force acting on the nucleus due to the endothelial membrane is, as shown in Fig. 5.12:2(B), nonuniform with respect to z (because R_z varies with z).

If the material property of the nucleus is either isotropic or is axisymmetric with respect to the vertical central axis, then such a load will produce a deformation which is sketched in Fig. 5.12:2(B). The nucleus appears narrowed at the shoulder and bulging at the bottom in the $y - z$ cross section. Thus the part of the nucleus protruding above the average height of the cell will appear somewhat elongated. Endothelial cell elongation was reported by Fry (1968), Kim et al. (1989), Dewey et al. (1981), Flaherty et al. (1972), Gau et al. (1980), Helmlinger et al. (1991), and Levesque and Nerem (1985). Although they did not present data on the nuclei, such elongation can be seen in some of their photographs.

5.13 Transmission of the Tension in the Upper Endothelial Cell Membrane to the Basal Lamina through the Sidewalls

Let us consider a confluent layer of endothelial cells whose profiles are schematically drawn in Fig. 5.13:1. The upper cell membranes are in contact with the flowing blood. The lower cell membranes are attached to the basal lamina. The side membranes of the neighboring cells are apposed to each other and together they are called *sidewalls*. These sidewalls may or may not transmit some of the tension in the upper cell membrane to the basel lamina. If they do, then the growth of the membrane tension in the upper membrane in the direction opposite to the blood flow will be reduced periodically by the sidewalls.

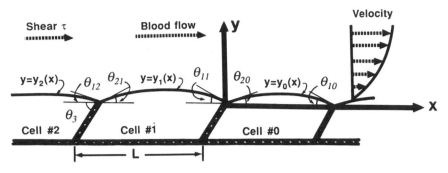

Figure 5.13:1 A longitudinal profile of the endothelium showing endothelial cells adhered to a basal lamina. The nomenclature of the cell numbers, dimensions, angles at the junctions, and the equations describing the curve of the upper cell membrane relative to the $x-y$ coordinates ($y = y_1(x)$, etc.) are given. From Fung and Liu (1993), by permission.

To analyze the force transmission problem, consider the equilibrium of forces in the membranes shown in Fig. 5.13:1. With a frame of reference as shown, assume that the condition is two-dimensional and independent of the coordinate z. The upper membrane of cell No. 1 is described by an equation $y = y_1(x)$, that of cell No. 2 by $y = y_2(x)$, etc. The tensile stress resultant in the upper membrane is denoted by N_x. The value of N_x in cell No. 1 at the right-hand end (junction with cell No. 0) is denoted by T_{11}, that at the left-hand end (junction with cell No. 2) is denoted by T_{21}. The slope of the upper membrane of cell No. 1 at the right-hand is denoted by $\tan \theta_{11}$, that at the left-hand end is denoted by $\tan \theta_{21}$. The first subscript denotes the end, "1" for the right, and "2" for the left. The second subscript denotes the cell number. The double subscripts system is used because successive cells may have different shapes and tensions. The equation of equilibrium of the upper membrane of cell 1 is [see Fig. 5.13:2(A)]

$$T_{21} \cos \theta_{21} = T_{11} \cos \theta_{11} + \tau L_1, \tag{1}$$

where L_1 is the length of the cell No. 1. Now consider the balance of forces at the junction of the upper membranes of cell No. 1 and cell No. 2 [Fig. 5.13:2(B)]. At this junction, the tension in the upper membrane to the left is T_{12}, that to the right is T_{21}, and that in the sidewall is T_3. The conditions of equilibrium of forces acting at the junction are

$$T_{21} \cos \theta_{21} - T_{12} \cos \theta_{12} = T_3 \cos \theta_3, \tag{2}$$

$$T_{21} \sin \theta_{21} + T_{12} \sin \theta_{12} = T_3 \sin \theta_3, \tag{3}$$

where the θ's are the angles of inclination indicated in Fig. 5.13:1. These equations can be used to compute the angle of inclination of the sidewall, θ_3 and the membrane tension, T_3. Dividing Eq. (2) by Eq. (3), we obtain

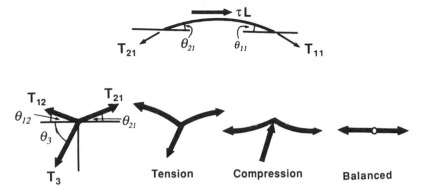

Figure 5.13:2 The shape of the upper cell membrane of cell No. 1 is shown in the upper drawing with the nomenclature of the angles and tensions indicated. The forces acting at the upper junction of cells No. 1 and No. 2 are shown in the lower drawings for several values of the angles θ_{12}, θ_{21} defined in the first figure. θ_{12}, θ_{21} both positive leads to tension in the side wall ($T_3 > 0$). θ_{12}, θ_{21} both negative leads to compression in the side wall ($T_3 < 0$). θ_{12}, θ_{21} both zero implies $T_3 = 0$. From Fung and Liu (1993), by permission.

$$\cot \theta_3 = (T_{21} \cos \theta_{21} - T_{12} \cos \theta_{12})/(T_{21} \sin \theta_{21} + T_{12} \sin \theta_{12}). \qquad (4)$$

Squaring both sides of Eqs. (2) and (3), adding, and simplifying, we obtain

$$T_3^2 = T_{12}^2 + T_{21}^2 - 2T_{12} T_{21} \cos(\theta_{12} + \theta_{21}). \qquad (5)$$

Since T_{12}, T_{21} are positive, Eq. (3) shows that the membrane tension T_3 in the sidewall is positive when θ_{12}, θ_{21} are positive, whereas T_3 becomes negative when θ_{12}, θ_{21} are negative.

According to our tension field hypothesis, T_3 cannot be negative (compressive). The smallest value T_3 can have is zero. Setting $T_3 = 0$ in Eqs. (2) and (3), we can deduce that

$$T_{12} \sin(\theta_{21} - \theta_{12}) = 0,$$
$$T_{21} - T_{12} \cos(\theta_{21} + \theta_{12}) = 0. \qquad (6)$$

Since T_{12}, T_{21} are positive, the unique solution of Eqs. (5) and (6) is

$$T_{12} = T_{21}, \qquad \theta_{12} = \theta_{21} = 0, \qquad \theta_3 = \pi/2, \qquad (7)$$

which says that the upper membrane must be flat at the junction and the sidewall must be vertical. This is obviously reasonable because if the sidewall has no tension, then the two tensile forces in the membranes must pull each other in a single straight line, see the last diagram in Fig. 5.13:2.

Hence the sidewall transmits tension in the upper cell membrane to the basement membrane if and only if θ_{12} and θ_{21} are positive, i.e., if and only if the upper cell membrane bulges into the blood stream. The membrane will bulge outward if there is a static pressure pushing it out. The static pressure in a cell is controlled by Starling's law for fluid movement across the cell membrane, which states that the rate of outward movement of fluid across the cell membrane, $\dot{m}$, is equal to the product of the coefficient of permeability, k, and the differences of the static pressure p and the osmotic pressure π inside the cell (subscript "i") and outside the cell (subscript "o"). Thus [see Fung (1990)]

$$\dot{m} = k[(p_i - p_o) - (\pi_i - \pi_o)]. \tag{8}$$

At equilibrium, the fluid movement is zero ($\dot{m}$ vanishes), the static pressure difference balances the osmotic pressure difference:

$$p_i - p_o = \pi_i - \pi_o. \tag{9}$$

The static pressure difference deflects the cell membrane according to Laplace's formula (Chapter 1, Sec. 1.9)

$$N_x \cdot \text{curvature} = p_i - p_o. \tag{10}$$

The curvature of the upper membrane of cell No. 1 shown in Figs. 5.13:1 and 5.13:2 is equal to the negative of the second derivative of the cell membrane surface given by the equaton $y = y_1(x)$, if the slope of the surface is sufficiently small. Hence, for small deflection, the differential equation for the cell membrane of the cell No. 1 is, on account of Eqs. (5.11:7) and (10),

$$-(T_{11} - \tau x)\frac{d^2 y_1}{dx^2} = p_i - p_o. \tag{11}$$

Introducing the dimensionless variables

$$y' = y_1/L_1, \qquad x' = x/L_1, \qquad T'_{11} = T_{11}/(L_1 \tau), \qquad p' = (p_i - p_o)/\tau, \tag{12}$$

we have

$$(T'_{11} - x')\frac{d^2 y'}{dx'^2} = -p'. \tag{13}$$

The boundary conditions are that the deflection y_1 must vanish at the two ends of the upper membrane (Fig. 5.13:1):

$$y' = 0 \qquad \text{when } x' = 0 \quad \text{and} \quad x' = -1. \tag{14}$$

The solution of Eqs. (13) and (14) is

$$\frac{dy'}{dx'} = p' \log(T'_{11} - x') + c_1,$$

$$y' = -p'(T'_{11} - x')[\log(T'_{11} - x') - 1] + c_1 x' + c_2,$$

$$c_2 = p' T'_{11}(\log T'_{11} - 1), \tag{15}$$

$$c_1 = -p'(T'_{11} + 1)[\log(T'_{11} + 1) - 1] + c_2.$$

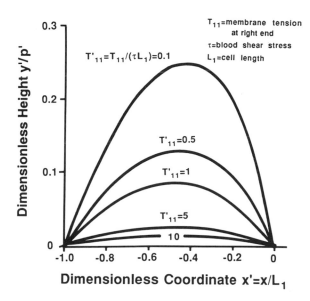

Figure 5.13:3 Theoretical shape of the upper cell membrane as a function of the parameters T'_{11} which is equal to the tension per unit width at the right-hand end of the cell membrane of cell No. 1 divided by the blood shear force per unit width, τL. The ordinate is dimensionless and is equal to the actual height divided by the product of the length of the cell and the dimensionless pressure parameter p', which is equal to the static pressure difference (also equal to the osmotic pressure difference) divided by the blood shear stress τ. The abscissa is the longitudinal coordinate.

It is seen that y' is linearly proportional to p' and y'/p' depends only on one variable, T'_{11}. Figure 5.13:3 shows the curves of y'/p' vs. x' with T'_{11} as a parameter.

5.13.1 The Case of Zero Static Pressure Difference $p_i - p_o = 0$

If $p_i - p_o = 0$, then Eq. (9) requires $\pi_i - \pi_o = 0$, so there will be no osmotic pressure difference between the content of the endothelial cell and the blood. In this case, Eq. (10) shows that the upper cell membrane must be flat; Eq. (15) shows $y' = 0$. Furthermore, the case $p_i < p_o$ is unattenable, because this will cause the cell membrane to bulge inward, compressing the sidewall [see the third figure of Fig. 5.13:2(B)], causing it to buckle, and returning the cell membrane to flat configuration, $p_i = p_o$.

When the upper cell membranes of successive endothelial cells are all flat at the cell junctions, then, in order to bear the shear stress of the flowing blood, the membrane tension N_x will increase linearly with the distance in a direction opposite to the blood flow. Figure 5.13:4 shows the consequence of this conclusion. The sketch at left shows a flow in a blood vessel with a side branch.

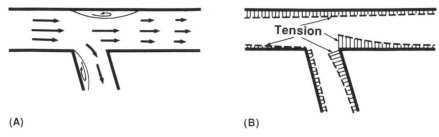

(A)   (B)

Figure 5.13:4 An illustration of the possible major difference between the distribution of blood shear stress acting on the vessel wall and the distribution of the tensile stress in the cell membrane of the endothelial cells in contact with the blood. The static pressure inside the cell is assume to be equal to the static pressure of the blood. Under this assumption, the tension in the cell membrane of one cell can be transmitted to the next cell and become accumulated. In the figure at left (A), the velocity profile in a vessel with a branch is shown with two separation regions having secondary flow. The shear stress is proportional to the velocity gradient (not shown). In the figure at right (B), the tensile stress in the upper cell membrane of the endothelial cells is plotted by tick marks perpendicular to the vessel wall. The higher the marker, the larger is the tensile stress. The dotted profile is that of the tensile stress distribution. From Fung and Liu (1993), by permission.

Two separation zones are shown in which the shear stress is reversed locally. In spite of the change of local shear stress on the vessel wall due to blood flow, in the present case the tensile stress in the upper endothelial cell membrane is transmitted from one cell to the next as shown in the diagram at right in Fig. 5.13:4, which indicates the magnitude of the tensile stress in the upper cell membrane by tick marks perpendicular to the vessel wall. The higher the marks are, the larger the tensile stress. This sketch is drawn for the case $p_i - p_o = 0$. In this case, the cumulative growth of membrane tension is somewhat mitigated by the reversed flow, but the tension remains high in these regions. At the apex of the flow divider, the tensile stress reaches a peak.

If the osmotic or static pressure difference were positive, $p_i - p_o > 0$, then the tensile stress could increase, decrease, or fluctuate depending on the cell geometry. See Fung and Liu (1993).

5.13.2 An Additional Principle is Needed to Determine the Value of $T_3' \cos \theta_3$

On a detailed examination, it is found that:

(a) When $p_i - p_o = 0$: $T_3' \cos \theta_3 = 0$, there is no transmission to basal lamina.
(b) When $p_i - p_o > 0$:

$T_3' \cos \theta_3 = 1$, all incremental tension is transmitted off.

$0 < T_3' \cos \theta_3 < 1$, tension increases progressively.

$T_3' \cos \theta_3 > 1$, tension is gradually reduced.

The case $p_i - p_o < 0$, if it existed, will be reduced to $p_i - p_o = 0$ by buckling of the sidewalls. We see that $T_3' \cos \theta_3$ is a key parameter which determines the tension in the cell membrane. This is a crucial number for experimenters to determine.

Theoretically, Eqs. (1), (2), and (3), with θ_{ij} as implicit functions of T_{ij}' and p', are three equations for four unknowns: T_{11}', T_{12}', T_{21}', and $T_3 \cos \theta_3$ if p' is regarded as a physiological variable. An additinal equation must be introduced before the solution can be definite.

I propose to invoke *the complementary energy theorem*, which states that of all stress fields that satisfy the equation of equilibrium and boundary conditions where stresses are prescribed, the "actual" one is distinguished by a stationary value of the complementary energy which is the sum of the strain energy expressed in terms of the stress components and the negative of an integral of the product of the force and displacement on that part of the boundary where displacements are specified. A derivation is given in Fung (1965), p. 293.

Under the fluid-like cell content hypothesis, the strain energy is concentrated in the cell membranes. We need the stress–strain relationship of the cell membrane. If that stress–strain relationship is as given in Sec. 4.7,

$$N_1 = 4\mu E_{11}, \tag{16}$$

where N_1 is the stress resultant, E_{11} is the strain, and μ is the shear modulus, then the strain energy density per unit area of the membrane expressed in terms of stress is

$$W_c = N_1 E_{11}/2 = N_1^2/(8\mu). \tag{17}$$

The complementary energy for the endothelium in tension field condition is

$$V_m = \frac{1}{8\mu} \int N_1^2 \, dS + \frac{1}{8\mu'} \int T_3^2 \, dS, \tag{18}$$

where the first integral is taken over the entire area of the upper cell membranes in a specified area of the endothelium, the second integral is taken over all the sidewalls. T_3 is tension in the sidewall, μ' is the sidewall's shear modulus. Membranes over the basal lamina contribute nothing because their displacement is zero.

Now, N_1 and T_3 are functions of the shear stress due to blood flow, τ, the osmotic pressure difference $\pi_i - \pi_o$, the static pressure difference $p_i - p_o$, the sidewall parameter $T_3 \cos \theta_3$, and T_{11}', T_{21}', T_{12}', as discussed in the preceding sections. The integration limits involve the height and length of the cells. The theorem supplies the desired additional equation:

$$\frac{\partial V_m}{\partial (T_3 \cos \theta_3)} = 0. \tag{19}$$

In an obvious way the calculation can be generalized to a three-dimensional pattern of cells in an endothelium. Then the morphometric data may be predicted, and compared with experiments.

5.14 The Hypothesis of a Solid-Like Cell Content

Let us now consider the alternative hypothesis that rheologically the content of the endothelial cell is a solid, i.e., it resists a static stress by a deformation from a zero stress state. Under the shear load of the blood, a system of stress and strain is induced in the cell through the following boundary conditions on the cell content-cell membrane interface (see Fig. 5.11:1):

$$\text{normal stress } \sigma_{yy} \text{ of cell content} = \text{that of cell membrane,} \tag{1}$$

$$\text{shear stress } \tau_{xy} \text{ of cell content} = \text{that of cell membrane.} \tag{2}$$

The cell content is loaded through these boundary conditions. Assume that a part of the shear load τ acting on the top side of the uppper cell membrane is transmitted through the membrane to its lower side, i.e., at the interface

$$\tau_{xy} \text{ of cell membrane} = \varepsilon\tau, \quad 0 \leqslant \varepsilon < 1, \tag{3}$$

then it follows that, at the interface,

$$\tau_{xy} \text{ of cell content} = \varepsilon\tau. \tag{4}$$

As long as the cell membrane is recognized as an important mechanical part of the cell, $\varepsilon \neq 1$, and the role played by the cell membrane can be examined through these boundary conditions.

For the cell membrane, the shear stress is τ on the surface in contact with the blood, and is $\varepsilon\tau$ on the surface in contact with the cell content. The equation of equilibrium of the cell membrane now becomes

$$\frac{\partial N_x}{\partial x} + (1 - \varepsilon)\tau = 0 \tag{5}$$

under the solid-content hypothesis. Hence, by integration,

$$N_x = -\int_0^x (1 - \varepsilon)\tau \, dx + N_0. \tag{6}$$

If ε were a constant, then

$$N_x = -(1 - \varepsilon)\tau x + N_0. \tag{7}$$

Hence, with a modification of replacing τ by $(1 - \varepsilon)\tau$, all the conclusions reached in the preceding sections remain valid. But ε may depend on x.

A similar argument applies to the sidewalls of the cells. If we denote the shear stress acting on the sidewall due to the solid content by $\varepsilon'\tau$, where ε' is a constant which may differ from ε, then under the solid-content hypothesis the equations of equilibrium of the forces in the cell membranes meeting at a junction of cells remain valid except that T_{12}, T_{21} should be reduced by a factor of $(1 - \varepsilon)$, and T_3 should be reduced by a factor of $(1 - \varepsilon')$.

To evaluate ε and ε', we have to know the constitutive equations of the internal structure of the endothelial cell, and solve the problem of stress and strain distribution. This is a big problem for the future, a problem which is full of exciting opportunities for discovery and clarification.

With regard to the tension field theory, we note that with a solid interior that supports the cell membrane elastically, the critical buckling stress of the cell membrane may be so high that the tension field theory may not apply. Then we have to analyze the cell as a solid. However, Kim et al. (1989) have shown that the F-actin stress fibers in endothelial cells are either bundled about the cell periphery, or are aligned with the direction of blood flow. This suggests that the elastic support given to the cell membrane by the stress fibers lies in the flow direction, and not in the direction perpendicular to the flow. Hence the possible compressive stress in the direction perpendicular to the flow could be very small compared with the tensile stress in the direction of flow, and the simplifying assumption of a tension field in the direction of the flow may remain valid.

5.15 The Effect of Turbulent Flow on Cell Stress

Experiments by Davies et al. (1986) have shown that turbulent flow with a mean correlation length of about five times the cell length can cause large increase in cell division and surface cell loss. So a look at the effect of turbulence on cell stress is of interest. Because of their small height, the endothelial cells lie totally in the laminar sublayer of the turbulent flow in Davies et al.'s experiment, and also in blood flow in larger arteries. The pressure and shear stress fluctuations of the turbulent flow act on the endothelium. Let the blood pressure acting on the upper endothelial cell membrane be separated into a mean part and a perturbation:

$$p_o = \bar{p}_o + p_o'(x, t), \tag{1}$$

where $\bar{p}_o$ is the mean value of p_o, and $p_o'(x, \tau)$ is a function of space and time whose average value over a sufficiently long period of time and a sufficiently large area vanishes:

$$\overline{p_o'(x, t)} = 0. \tag{2}$$

Similarly, the surface shear stress from the blood is split into a mean and a perturbation about the mean:

$$\tau = \bar{\tau} + \tau'(x, t), \tag{3}$$

where $\bar{\tau}$ is the mean value of τ and is a constant, whereas τ' is the perturbation about the mean, so that its mean value vanishes:

$$\overline{\tau'(x, t)} = 0. \tag{4}$$

In response to the external load p_o and τ, internal stresses are induced in the cell membranes and cell contents. Since there is motion in the cells, there is shear stress in the cell content under both fluid-like and solid-like hypotheses. Hence we use Eq. (5.14:4) to describe the shear stress on the inner wall of the cell membrane. The equation of motion of the upper cell membrane is obtained by adding an inertial force term (mass times acceleration) to the equation of equilibrium, Eq. (5.14:5). Under tension field theory and using Eq. (3), we have

$$\frac{\partial N_x}{\partial x} + (1 - \varepsilon)[\bar{\tau} + \tau'(x, t)] = \rho h \frac{\partial^2 u}{\partial t^2}, \tag{5}$$

where ρ is the mass density of the cell membrane, h is the membrane thickness, u is the displacement of the mass particles of the membrane, and t is time. A conservative estimate of the order of magnitude of the inertial force term is 10^{-5} N/m^2, if we assume $\rho \sim 10^3$ kg/m^3, $h \sim 10^{-8}$ m, $u \sim 10^{-6}$ m, $t \sim 10^{-3}$ sec. It is five orders smaller than the shear load $\bar{\tau}$ which is of the order of 1 N/m^2, and can be neglected in Eq. (5). The equation of motion (5) is reduced to Eq. (5.14:5), and the solution is given by Eq. (5.14:6). The cell membrane's stress response to the turbulent flow is instantaneous and quasi-static.

Thus, in a turbulent flow, the tensile stress in the upper cell membrane reacts to the pressure $\bar{p}_o + p'(x, t)$ and shear $\bar{\tau} + \tau'(x, t)$ in the same way as described in Secs. 5.13 and 5.14.

A major effect of turbulence on the tensile stress in the cell membrane is revealed by this analysis. Suppose that the static pressure difference $p_i - p_o$ fluctuates around a mean value of zero, i.e., $\bar{p}_i - \bar{p}_o = 0$. The instantaneous pressure difference $p_i - p_o$ takes on positive and negative values. When $p_i - p_o$ is instantaneously positive, the tension in the upper cell membranes can be transmitted to the basal lamina. When $p_i - p_o$ is instantaneously negative, the transmission in the sidewalls ceases and the tensile stress from one cell is transmitted to the next and the stress accumulates. The larger the turbulence scale the more severe is the accumulation. This is a kind of off-and-on chain reaction, whose interval, duration, and severity are statistical.

Thus in a turbulent flow, the tension in the upper cell membranes fluctuates, transiently tending to pull neighboring cells apart. Separation is possible if the adhesion of the sidewalls is not perfect at the junction at all times. The sidewall tension also tends to pull the cell away from the basal laminar. In the meantime, as described in Sec. 5.12, the upper cell membrane tension will compress the nuclei. When the cell membrane tension oscillates, the dynamic action will induce an oscillatory motion in the nuclei. These transient events are more severe whenever $p_i - p_o$ can be negative, irrespective of the mean value of $\bar{p}_i - \bar{p}_o$. This suggests that these fluctuations may contribute to the surface cell loss observed by Davies et al. (1986).

In conclusion, we see that when cells form a continuum, the stress in one cell is affected by stresses in all cells.

Nollert et al. (1991) have shown that mass transport through endothelium is affected by shear stress. They did not look into the tensile stress in the cell membrane. The classical Pappenheimer's (1953) "pore" concept of mass transport through a membrane suggests the importance of the tensile stress in the membrane because the pore dimension must be sensitive to the tensile stress.

Other mechanisms of mass transport, channels, pumps, etc., may be affected by membrane tension. Markin and Martinac (1991) have proposed such a theory.

Problems

5.1 Show that, theoretically, in man the average circulation time for red blood cells is different from the average circuation time of blood plasma. Do you feel uneasy about this concept?

   Note. A flowing river carries sand in its current; the average speed of the water does not have to be equal to that of the sand.

5.2 Discuss the hematocrit variation in the large and small arteries and veins, arterioles, venules, and capillaries.

5.3 Discuss the ways in which an entry section can influence the hematocrit distribution.

5.4 Since red blood cells and albumin or other plasma proteins can be selectively labeled with radioactive substances, one should be able to measure the circulation time of red blood cells and plasma by the indicator dilution technique. Look into the literature on indicator dilution for evidence of the different circulation time of red cells and plasma.

5.5 Formulate a mathematical theory of flow of a fluid in a pipe containing particles (spherical or otherwise) whose diameters are comparable to that of the pipe. Write

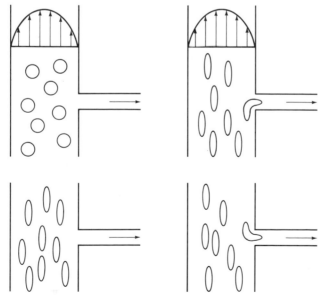

Figure P5.6 Flow into a small branch from a larger vessel. The tendency of a particle to enter the branch depends on the resultant force acting on the particle when it is in the neighborhood of the entry section.

down all the field equations and boundary conditions. Show that the problem is mathematically well defined (theoretically solvable and has a unique solution). Outline a precedure to determine the apparent viscosity of the fluid and particle mixture as a function of the size and shape of the particles and the vessel.

5.6 A small capillary vessel leaves a larger vessel at a right angle (Fig. P5.6). The fluid contains particles whose diameter is comparable to that of the small branch. Consider first spherical particles. Analyze the tendency for the sphere to enter the small branch as a function of the ratio of the average velocity in the larger vessel to that in the small branch. Then consider red cells. In shear flow the red cells tend to be elongated and lined up with the flow. What is the chance for a red cell to be sucked into the capillary branch? Use the result to discuss the expected hematocrit ratio in the two branches.

5.7 Blood flow in arteries with severe stenosis often gives out bruit (vibrational noise of the arterial wall) due to flow separation (Fig. P5.7). Flow separation can also occur at arterial bifurcation points if the velocities in the two daughter branches are very different. Describe the change of apparent viscosity of flow in these vessels due to flow separation as a function of the flow velocities.

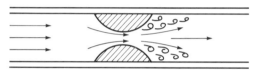

Figure P5.7 Flow separation in divergent flow. Divergence can be seen from the pattern of the streamlines.

5.8 Consider the bifurcating circular cylindrical tube shown in Fig. 5.7:2. Let the spheres shown in the figure be replaced by flexible disks (such as red blood cells in small capillaries). Analyze the new situation and determine the direction of motion of the pellets.

Use this result to analyze single-file flow of red cells in a bifurcating capillary blood vessel and derive a relationship between the hematocrits in the two daughter branches and the velocity ratio v_1/v_2 (cf. Yen and Fung, 1978).

5.9 Consider turbulent flow in a blood vessel. Turbulences consist of eddies of various sizes. The statistical features of these eddies can be stated in the form of velocity correlation functions in space and time. From these correlation functions, scales of turbulence can be defined. Now let the fluid be blood, and consider two related questions: (a) the effect of red cells on turbulence, especially on the transition from laminar to turbulent flow, and (b) the effect of turbulence on the stresses induced in the membranes of red cells and platelets. These interactions obviously depend on the ratio of the red cell diameter to the scale of turbulence, or the spatial correlation of the velocity field. Present a rational discussion of these questions.

5.10 The question (a) of the preceding problem is relevant to the shear stress imparted by the fluid on the blood vessel wall, and hence may be significant with respect to the genesis of artherosclerosis. The question (b) is relevant to hemolysis and thrombosis. Discuss the significance of the solutions of Problem 5.9 with regard to the situation described in Problem 5.7.

5.11 Consider the flow of water in a circular cylindrical pipe. As the flow velocity increases, the flow becomes turbulent. How does the apparent viscosity change with velocity of flow?

Note. The theory of turbulence is complex. See Kuethe and Chow (1986) for an introductory exposition.

References to Blood Cells in Microcirculation

Barbee, J. H. and Cokelet, G. R. (1971) The Fahreus effect. *Microvasc. Res.* **34**, 6–21.

Braasch, D. and Jennett, W. (1986) Erythrozytenflexibilitat, Hämokonzen tration und Reibung swiderstand in Glascapillarem mit Durchmessern Zwischen 6 bis 50 μ. *Pfügers Archiv. Physiol.* **302**, 245–254.

Caro, C. G., Pedley, T. J., Schroter, R. C., and Seed, W. A. (1978) *The Mechanics of Circulation.* Oxford University Press, New York.

Chien, S. (1972) Present status of blood rheology. In *Hemodilution: Theoretical Basis and Clinical Application,* K. Messmer and H. Schmid-Schönbein (eds.) Karger, Basel, pp. 1–45.

Fahraeus, R. (1929) The suspension stability of blood. *Physiol. Rev.* **9**, 241–274.

Fahraeus, R. and Lindqvist, T. (1931) Viscosity of blood in narrow capillary tubes. *Am. J. Physiol.* **96**, 562–568.

Fitz-Gerald, J. M. (1969a) Mechanics of red cell motion through very narrow capillaries. *Proc. Roy. Soc. London, B* **174**, 193–227.

Fitz-Gerald, J. M. (1969b) Implications of a theory of erythrocyte motion in narrow capillaries. *J. Appl. Physiol.* **27**, 912–918.

Fitz-Gerald, J. M. (1972) In *Cardiovascular Fluid Dynamics,* D. H. Bergel (ed.) Academic, New York, Vol. 2, Chapter 16, pp. 205–241.

Fung, Y. C. (1969a) Blood flow in the capillary bed. *J. Biomechan.* **2**, 353–372.

Fung, Y. C. (1969b) Studies on the blood flow in the lung. *Proc. Can. Congr. Appl. Mech.,* May, 1969. University of Waterloo, Canada, pp. 433–454,

Fung, Y. C. (1973) Stochastic flow in capillary blood vessels. *Microvasc. Res.* **5**, 34–48.

Goldsmith, H. L. (1971) Deformation of human red cells in tube flow. *Biorheology* **7**, 235–242.

Goldsmith, H. L. and Skalak, R. (1975) Hemodynamics. *Ann. Rev. Fluid Mech.* **7**, 213–247.

Gross, J. F. and Aroesty, J. (1972) Mathematical models of capillary flow: A critical review. *Biorheology* **9**, 255–264.

Haynes, R. H. (1960) Physical basis of the dependence of blood viscosity on tube radius. *Am. J. Physiol.* **198**, 1193–1200.

Hochmuth, R. M., Marple, R. N., and Sutera, S. P. (1970) Capillary blood flow. 1. Erythrocyte deformation in glass capillaries. *Microvasc. Res.* **2**, 409–419.

Jay, A. W. C., Rowlands, S., and Skibo, L. (1972) *Cand. J. Physiol. Pharmacol.* **5**, 1007–1013.

Johnson, P. C. and Wayland, H. (1967) Regulation of blood flow in single capillaries. *Am. J. Physiol.* **212**, 1405–1415.

Kot, P. (1971) Motion picture shown at the Annual Meeting of the Microcirculatory Society, Atlantic City, N. J., April 1971.

Kuethe, A. M. and Chow, C. Y. (1986). *Foundations of Aerodynamics* 4th Edn. Wiley, N.Y.

Lee, J. S. (1969) Slow viscous flow in a lung alveoli model. *J. Biomech.* **2**, 187–198.

Lee, J. S. and Fung Y. C. (1969). Modeling experiments of a single red blood cell moving in a capillary blood vessel. *Microvasc. Res.* **1**, 221–243.

Lew, H. S. and Fung, Y. C. (1969a) The motion of the plasma between the red cells in the bolus flow. *Biorheology*, **6**, 109–119.

Lew, H. S. and Fung, Y. C. (1969b) On the low-Reynolds-number entry flow into a circular cylindrical tube. *J. Biomech.* **2**, 105–119.

Lew, H. S. and Fung, Y. C. (1969c) Flow in an occluded circularly cylindrical tube with permeable wall. *Zeit. angew. Math.* Physik **20**, 750–766.

Lew, H. S. and Fung, Y. C. (1970a) Entry flow into blood vessels at arbitrary Reynolds number. *J. Biomech.* **3**, 23–38.

Lew, H. S. and Fung, Y. C. (1970b) Plug effect of erythrocytes in capillary blood vessels. *Biophys. J.* **10**, 80–99.

Lighthill, M. J. (1968) Pressure forcing of tightly fitting pellets along fluid-filled elastic tubes. *J. Fluid Mech.* **34**, 113–143.

Mason, S. G. and Goldsmith, H. L. (1969) The flow behavior of particulate suspensions. In *Circulatory and Respiratory Mass Transport*. A Ciba Foundation Symposium, G. E. W. Wolstenholme and J. Knight (eds.) Churchill, London, p. 105.

Prothero, J. and Burton, A. C. (1961) The physics of blood flow in capillaries. *Biophys. J.* **1**, 565–579; **2**, 199–212; **2**, 213–222 (1962).

Secomb, T. W., Skalak, R., Özkaya, N., and Gross, J. F. (1986) Flow of axisymmetric red blood cells in narrow capillaries. *J. Fluid Mech.* **163**, 405–423.

Segre, G. and Silberberg, A. (1962) Behavior of macroscopic rigid spheres in Poiseuille flow. *J. Fluid Mech.* **14**, 136–157.

Seshadri, V., Hochmuth, R. M., Croce, P. A., and Sutera, A. P. (1970) Capillary blood flow. III. Deformable model cells compared to erythrocytes in vito. *Microvasc. Res.* **2**, 434–442.

Skalak, R. (1972) Mechanics of the microcirculation. In *Biomechanics: Its Foundations and Objectives*, Y. C., Fung, N. Perrone, and M. Anliker (eds.) Prentice-Hall, Englewood Cliffs, NJ, pp. 457–500.

Skalak, R. (1973) Modeling the mechanical behavior of red blood cells. *Biorheology* **10**, 229–238.

Skalak, R., Chen, P. H., and Chien, S. (1972) Effect of hematocrit and rouleau on apparent viscosity in capillaries. *Biorheology* **9**, 67–82.

Skalak, R. and Chien, S. (1983) Theoretical models of rouleau formation and dis-aggregation. *Ann. New York Academy of Sciences*. Part 1, Vol 416, pp. 138–148.

Skalak, R. (1990) Capillary flow: history, experiments, and theory *Biorheology* **27**, 277–293.

Sobin, S. S., Tremer, H. M., and Fung, Y. C. (1970) Morphometric basis of the sheet-flow concept of the pulmonary alveolar microcircumation in the cat. *Circulation Res.* **26**, 397–414.

Sobin, S. S., Fung, Y. C., Tremer, H. M., and Rosenquist, T. H. (1972) Elasticity of the pulmonary alveolar microvascular sheet in the cat. *Circulation Res.* **30**, 440–450.

Sutera, S. P. (1978) Red cell motion and deformation in the microcirculation. *J. Biomech. Eng.* **100**, 139–148.

Sutera, S. P. and Hochmuth, R. M. (1968) Large scale modeling of blood flow in the capillaries. *Biorheology* **5**, 45–78.

Sutera, S. P., Seshadri, V., Croce, P. A., and Hochmuth, R. M. (1970) Capillary blood flow. II. Deformable model cells in tube flow. *Microvasc. Res.* **2**, 420–433.

Svanes, K. and Zweifach, B. W. (1968) Variations in small blood vessel hematocrits produced in hypothermic rats by micro-occlusion. *Microvasc. Res.* **1**, 210–220.

Tong, P. and Fung, Y. C. (1971) Slow particulate viscous flow in channels and tubes— Application to biomechanics. *J. Appl. Mech.* **38**, 721–728.

Warrell, D. A., Evans, J. W., Clarke, R. O., Kingaby, G. P., and West, J. B. (1972) Patterns of filling in the pulmonary capillary bed. *J. Appl. Physiol.* **32**, 346–356.

Yen, R. T. and Fung, Y. C. (1973) Model experiments on apparent blood viscosity and hematocrit in pulmonary alveoli. *J. Appl. Physiol.* **35**, 510–517.

Yen, R. T. and Fung, Y. C. (1977) Inversion of Fahraeus effect and effect of mainstream flow on capillary hematocrit. *J. Appl. Physiol.* **42**(4), 578–586.

Yen, R. T. and Fung, Y. C. (1978) Effect of velocity distribution on red cell distribution in capillary blood vessels. *Am. J. Physiol.* **235**(2), H251–H257.

References to Endothelial Cells

Caro, C. G., Fitz-Gerald, J. M., and Schroter, R. C. (1971) Atheroma and arterial wall shear—observation, correlation, and proposal of a shear-dependent mass-transfer mechanism for atherogenesis. *Proc. Roy. Soc. London (Biol.)* **177**, 109–159.

Curry, F-R. E. (1988) Mechanics and thermodynamics of transcapillary exchange. In *Handbook of Physiology—Cardiovascular System IV.* American Physiological Society, Bethesda, MD, Part I, pp. 309–374.

Davies, P. F., Remuzzi, A., Gordon, E. F., Dewey, C. F., Jr., and Gimbrone, M. A., Jr. (1986) Turbulent fluid shear stress iduces vascular endothelial cell turnover *in vitro*. *Proc. Natl. Acad. Sci.* **83**, 2114–2117.

Dewey, C. F., Bussolari, S. R., Gimbrone, M. A., and Davies, P. F. (1981) The dynamic response of vascular endothelial cells to fluid shear stress. *J. Biomech. Eng.* **103**, 177–185.

Dewey, C. F., Bussolari, S. R., Gimbrone, M. A., Jr., and Davies, P. F. (1981) The dynamic response of vascular endothelial cells to fluid shear stress. *J. Biomech. Eng.* **103**, 177–185.

Flaherty, J. T., Pierce, J. E., Ferrans, V. J., Patel, D. J., Tucker, W. K., and Fry, D. L. (1972) Endothelial nuclear patterns in the canine arterial tree with particular reference to hemodynamic events. *Circulation Res.* **30**, 23–33.

Fry D. L. (1968) Acute vascular endothelial changes associated with increased blood velocity gradients. *Circulation Res.* **22**, 165–197.

Fry D. L. (1969) Certain histological and chemical responses of the vascular interface to acutely induced mechanical stress in the aorta of the dog. *Circulation Res.* **24**, 93–108.

Fung, Y. C. (1965) *Foundations and Solid Mechanics.* Prentice-Hall, Englewood Cliffs, N.J.

Fung, Y. C., and Liu, S. Q. (1993) Elementary mechanics of the endothelium of blood vessels. *J. Biomech. Eng.* **115**, 1–12.

Gau, G. S., Ryder, T. A., and MacKenzie, M. L. (1980) The effect of blood flow on the surface morphology of the human endothelium. *J. Pathol.* **131**, 55–60.

Giddens, D. P., Zarins, C. K., and Glagov, S. (1990) Response of arteries to near-wall fluid dynamic behavior. *Appl. Mech. Rev.* **43**, S98–S102.

Hammersen, F. and Lewis, D. H. (eds.) (1985) *Endothelial Cell Vesicles.* Proc. Workshop, Karger, Basel, 1985.

Helmlinger, G., Geiger, R. V., Schreck, S., and Nerem, R. M. (1991) Effects of pulsatile flow on cultured vascular endothelial cell morphology. *J. Biomech. Eng.* **113**, 123–131.

Hsiung, C. C. and Skalak, R. (1984) Hydrodynamic and mechanical aspects of endothelial permeability. *Biorheology* **21**, 207–221.

Kamiya, A. and Togawa, T. (1980) Adaptive regulation of wall shear stress to flow change in the canine carotid artery. *Am. J. Physiol.* **239**, H14–H21.

Kamiya, A., Bukhari, R., and Togawa, T. (1984) Adaptive regulation of wall shear stress optimizing vascular tree function. *Bull. Math. Biol.* **46**, 127–137.

Kim, D. W., Gotlieb, A. L., and Langille, B. L. (1989) *In vivo* modulation of enthothelial *F*-actin microfilaments by experimental alterations in shear stress. *Arteriosclerosis* **9**, 439–445.

Kim, D. W., Langille, B. L., Wong, M. K. K., and Gotlieb, A. L. (1989) Patterns of endothelial microfilament distribution in the rabbit aorta in situ. *Circulation Res.* **64**, 21–31.

Koslow, A. R., Stromberg, R. R., Friedman, L. I., Lutz, R. J., Hilbert, S. L., and Schuster, P. (1986) A flow system for the study of shear forces upon cultured endothelial cells. *J. Biomech. Eng.* **108**, 338–341.

Levesque, M. J. and Nerem, R. M. (1985) The elongation and orientation of cultured endothelial cells in response to shear stress. *J. Biomech. Eng.* **107**, 341–347.

Markin, V. S. and Martinac, B. (1991) Mechano sensitive ion channels as reporters of bilayer expansion. A theoretical model. *Biophys. J.* **60**, 1–8.

Nerem, R. M. and Girard, P. R. (1990) Hemodynamic influence on vascular endothelial biology. *Toxicologic Pathol.* **18**, 572–582.

Nollert, M. U., Diamond, S. L., and McIntire, L. V. (1991) Hydrodynamic shear stress and mass transport modulation of endothelial cell metabolism. *Biotech. Bioeng.* **38**, 588–602.

Pappenheimer, J. R. (1953) Passage of molecules through capillary walls. *Physiol. Rev.* **33**, 387–423.

Repin, V. S., Dolgov, V. V., Zaikina, O. E., Novikov, I. A., Antonov, A. S., Nikolaeva, N. A., and Smirnov, V. N. (1984) Heterogeneity of endothelium in human aorta. *Atherosclerosis* **50**, 35–52.

Rhodin, J. A. G. (1980) Architecture of the vessel wall. In *Handbook of Physiology, Sec 2, Vascular Smooth Muscle*, D. F. Bohr, A. P. Samlyo, and H. V. Sparks, Jr. (eds.) American Physiological Society, Bethesda, MA Chap. 1, pp. 1–32.

Rodbard, S. (1970) Negative feedback mechanisms in the architecture and function of the connective and cardiovascular tissues. *Perspective Biol. Med.* **13**, 507–527.

Sakariassen, K. S., Aarts, P. A. M. M., Degroot, P. G., Houdijk, W. P. M., and Sixma, J. J. (1983) A perfusion chamber developed to investigate platelet interaction in flowing blood with human vessel wall cells. *J. Lab. Clin. Med.* **102**, 522–535.

Sato, M., Levesque, M. J., and Nerem, R. M. (1987) Application of the micropipette technique to the measurement of the mechanical properties of cultured bovine endothelial cells. *J. Biomech. Eng.* **109**, 27–34.

Sato, M., Theret, D. P., Wheeler, L. T., Ohshima, N., and Nerem, R. M. (1990) Application of the micropipette technique to the measurement of cultured porcine aortic endothelial cell viscoelastic properties. *J. Biomech. Eng.* **112**, 263–268.

Simionescu, M., Simionescu, N., and Palade, G. E. (1975) Segmental differentiations of cell junctions in the vascular endothelium. The microvasculature. *J. Cell Biol.* **67**, 863–885.

Simionescu, N., Simionescu, M., and Palade, G. E. (1976) Segmental differentiations of cell junctions in the vascular endothelium. Arteries and veins. *J. Cell Biol.* **68**, 705–723.

Simionescu, N. and Simionescu, M. (eds.) (1988) *Endothelial Cell Biology in Health and Disease.* Plenum Press, New York.

Singer, S. J. and Nicolson, G. L. (1972) The fluid mosaic model of the structure of cell membranes. *Science* **175**, 720–731.

Skalak, R., Tözeren, A., Zahalak, G., Elson, E., and Chien, D. (Scheduled 1993) *Biomechanics of Cells.* Springer-Verlag, New York.

Sprague, E. A., Steinbach, B. L., Nerem, R. M., and Schwartz, C. J. (1987) Influence of a laminar steady-state fluid-imposed wall shear stress on the binding, internalization, and degradation of low-density lipoproteins by cultured arterial endothelium. *Circulation* **76**, 648–656.

Strong, A. B., Absolom, D. R., Zingg, W., Hum, O., Ledain, C., and Thompson, B. E. (1982) A new cell for platelet adhesion studies. *Animals Biomed. Eng.* **10**, 71–82.

Theret, D., Levesque, M. J., Sato, M., Nerem, R. M., and Wheeler, L. T. (1988) The application of homogeneous half-space model in the analysis of endothelial cell micropipette measurements. *J. Biomech. Eng.* **110**, 190–199.

Thilo-Korner, D. G. S. and Freshney, R. I. (eds.) (1983) *International Endothelial Cell Symposium.* Karger, Basel.

Viggers, R. F., Wechezak, A. R., and Sauvage, L. R. (1986) An apparatus to study the response of cultured endothelium to shear stress. *J. Biomech. Eng.* **108**, 332–337.

Zarins, C. K., Zatina, M. A., Giddens, D. P., Ku, D. N., and Glagov, S. (1987) Shear stress regulation of artery lumen diameter in experimental atherogenesis. *J. Vascular Surg.* **5**, 413–420.

CHAPTER 6

Bioviscoelastic Fluids

6.1 Introduction

Most biofluids are viscoelastic. Our saliva, for example, behaves more like an elastic body than like water. Mucus, sputum, and synovial fluids are well known for their elastic behavior. Viscoelasticity is an important property of mucus. In the respiratory tract mucus is moved by cilia lining the walls of the trachea and bronchi. If the mucus were a Newtonian fluid, the ciliary motion will be less effective in moving it. Similar ciliary motion is responsible for the movement of the ovum from ovary to uterus through the fallopian tube.

Clinical observers have noted that in many obstructive pulmonary diseases such as chronic bronchitis and cystic fibrosis, there is an increase in the "viscosity" of the mucus secretion. Many treatment modalities have been adopted in an attempt to reduce the viscosity and hence increase the clearance rate of secretions. Thus measurement of viscoelasticity is a useful tool for clinical investigation, and artificial control of viscoelasticity of mucus secretion is an objective of drug research. The social importance of such studies becomes evident when one thinks of the working days lost in the nation due to broncho-pulmonary disease.

There are many ways viscoelasticity of a fluid can be revealed. For example, if one swirls a synovial fluid in a flask and stops the movement suddenly, the elastic recoil sets it rotating for a moment in the opposite direction. Another striking demonstration was given by Ogston and Stainier (1953). A drop of fluid was placed on an optically flat surface and a convex lens was set on top (see Fig. 6.1:1). The lens was pressed down and the interference patterns known as Newton's rings were used to measure the distance between the lens and the optical flat (cf. Problem 6.1 at the end of this chapter).

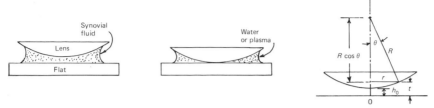

Figure 6.1:1 Ogston and Stainer's (1953) demonstration of the elasticity of synovial fluid by Newton's ring.

When the experiment was done with water, the lens was easily pressed into contact with the optical flat. The same happened with blood plasma diluted to have the same protein composition as synovial fluid. With glycerol, the lens moved very slowly down toward the flat. With synovial fluid, the lens stopped moving while it was still some distance above the optical flat. The distance depended on the load pressing it down. Moreover, when the load was removed, the lens moved slightly up again due to elastic recoil. This property of synovial fluid suggests that it may be impossible to squeeze it out entirely between the articulating surfaces in joints, just as it cannot be squeezed from between a lens and an optical flat. This would obviously be a valuable property for a lubricant.

Accurate quantitative measures of viscoelasticity can be obtained with a rheogoniometer (see Chapter 2, Sec. 2.14). But there are occasions in which it is difficult to collect a sufficient amount of test samples, or the samples may be quite inhomogeneous, or it may be suspected that testing in a rheometer would disturb the structure of the fluid too much from the *in vivo* condition for the results to be meaningful. In these situations special methods are used. A typical example is the study of protoplasm, which, like blood, tends to coagulate when removed from the cell, or when the cell is injured. Hence methods must be found to measure viscosity within the cell itself.

The properties of protoplasm, mucus, etc., tend to be very complex, and the literature is replete with various qualifying words which are supposed to describe the properties of these materials. Some knowledge of these terms is useful. In the first place, the term *rheology* is defined as the science of the deformation and flow of matter. It is a term coined by E. C. Bingham. The term *biorheology*, the rheology of biological material, was introduced by A. L. Copley. The meaning of *hemorheology*, the study of blood rheology, is obvious; but terms like *thixotropy*, *rheopexy*, *dilatancy*, *spinability*, etc., are sometimes confusing. The word *thixotropy* was introduced by H. Freundlich to describe sol-gel transformation of colloidal solutions. Many suspensions and emulsions (solid and liquid particles, respectively, dispersed in a liquid medium) and colloids (liquid mixture with very fine suspended particles that are too small to be visible under an ordinary light microscope) show a fall in viscosity as the shear stress is increased, and the fall in viscosity (or "consistency") is recovered slowly on standing. This recovery is called thixotropy. A typical example of a thixotropic material is kitchen jello, which in a gel form can be liquified if violently shaken, but returns to gel form when left standing for some time. Thixotropy is generally associated with a change from liquid state into a gel. Some rheologists, however, use the term *thixotropy* to describe all recoverable losses in consistency produced by shearing.

If the consistency or viscosity falls when shearing increases, the material is said to show *shear thinning*. The reverse effect, an increase in viscosity with increasing shear rate, is called *shear thickening*. If the material consistency is increased when sheared but "softens" when left to stand, it is said to show *reverse thixotropy* or *negative thixotropy*. Shear thinning is often explained by breaking up of the structure by shearing, as in the example of whole blood (see Chapter 3, Sec. 3.1). One of the possible reasons for shear thickening is the crowding of particles, as sand on a beach: the closely packed particles have to move apart into open packing before they can pass one another. This is called *dilatancy*.

The term *rheopexy* was used by some authors to describe certain thixotropic systems that reset more quickly if the vessel containing them is rotated slowly. But other authors use the term for negative thixotropy. Thus the meaning of the word is not unique.

Of course, the precise description of a rheological property is the constitutive equation. If a mathematical description is known, all looseness is gone.

6.2 Methods of Testing and Data Presentation

The most prevalent viscoelastic biofluid is protoplasm, the contents of all living cells. It is also the most complex and difficult to test. It is rheologically complex because in its various forms it shows almost all the modes of behavior that have been found in other fields. It is difficult to test because it coagulates when it is taken out of the cell, or when the cell membrane is injured. To test the mechanical properties of the protoplasm one should test it inside the living cell. The methods are therefore specialized and limited. We shall discuss some of them in Sec. 6.3.

Most other biofluids can be collected and tested in laboratory instruments. Generally speaking, there are two kinds of tests: (a) small perturbations from an equilibrium condition, and (b) measurements in a steady flow. In the first case we think of the material as a solid. We speak of strain. We measure the relationship between stress and strain history. Deviations from static equilibrium are kept small in the hope that the material behavior will be linear, that is, the stress will be linearly related to strain history. In the second case we think of the material as a fluid. We speak of flow and velocity gradient. We focus attention not only on viscosity, but also on normal stresses.

Many interesting features of rheological properties of biofluids are revealed in experiments not belonging to these two classes. Some call for large deformation from equilibrium. Others call for accelerated or decelerated flow, or transition from rest to steady motion. Pertinent experiments of this last category will be mentioned later when occasion arises. In the present section we shall consider the two main categories of experiments.

6.2.1 Small Deformation Experiments

The most common tests are creep, relaxation, and oscillations. The instruments mentioned in Chapter 2, Sec. 2.14, namely, the concentric cylinder viscometer, the cone-and-plate viscometer, and the rheogoniometer, can be used for this purpose.

For testing very small samples, say 0.1 ml or less, the oscillating magnetic microrheometer may be used. We shall discuss a model used by Lutz et al. (1973) in some detail to illustrate the principle of this instrument and the method of data analysis and presentation. A sketch of the instrument is shown in Fig. 6.2:1. In this instrument, the fluid is stressed by moving a microscopic sphere of iron (with a diameter of the order of 200 μm) through the fluid under the influence of a magnetic force. The motion of the sphere is transduced into an electric signal by an instrument called an "optron." The fluid is held in a cavity hollowed out in a copper block so that with circulating water good temperature control can be obtained.

An analysis of the instrument is as follows. The force that acts on the particle, F_m, in a magnetic field generated by an electromagnet, is proportional to the square of the current, I, passing through the coils; thus

$$F_m = cI^2, \tag{1}$$

where c is a constant. In the instrument shown in Fig. 6.2:1, two magnets were used, and the circuit is hooked up in such a way that the force is given by

$$F_m = c(I_1^2 - I_2^2), \tag{2}$$

where I_1, I_2 are the currents in the two magnets:

$$I_1 = I_0 + I_A \sin \omega t, \qquad I_2 = I_0 - I_A \sin \omega t, \tag{3}$$

so that

$$F_m = 4cI_0I_A \sin \omega t. \tag{4}$$

I_0 and I_A are, respectively, the d.c. and a.c. amplitudes of the currents from a signal generator; ω is the frequency. The constant c can be determined by calibration of the instrument with a particle in a Newtonian fluid.

Since the driving force $F_m(t)$ is sinusoidal, the movement of the sphere, $x(t)$, tracked by the optron, must be sinusoidal also, as long as the amplitude of oscillation is small, so that the system remains linear. Hence we can use

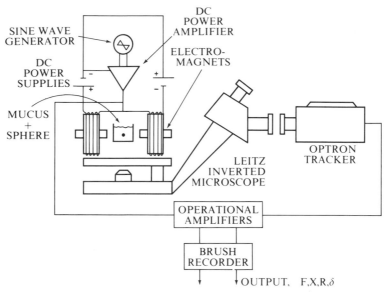

Figure 6.2:1 Schematic diagram of an oscillating magnetic microrheometer. From Lutz, Litt, and Chakrin (1973), by permission.

the complex representation of a harmonic oscillation (Chapter 2, Sec. 2.12) to write

$$F_m(t) = F_0 e^{i\omega t}, \qquad x(t) = x_0 e^{-i\delta} e^{i\omega t}, \qquad (5)$$

where F_0, x_0, and δ are real numbers. δ is the phase lag of the displacement x from the force $F_m(t)$. Recall that the physical interpretation of $F_m(t)$ is either the real part of $F_0 e^{i\omega t}$, namely, $F_0 \cos \omega t$, or the imaginary part, $F_0 \sin \omega t$. Similarly, $x_0 e^{-i\delta} e^{i\omega t}$ means

$$x(t) = x_0 \cos(\omega t - \delta) \qquad \text{or} \qquad x_0 \sin(\omega t - \delta). \qquad (5a)$$

The force $F_m(t)$ is balanced by the viscoelastic drag and inertia. To calculate the drag force acting on the sphere, we need the stress–strain relationship of the fluid. Let τ and γ be the shear stress and shear strain, both represented by complex numbers. We shall write the stress–strain relationship in the form

$$\tau = G(i\omega)\gamma, \qquad (6)$$

where $G(i\omega)$ is the *complex shear modulus of elasticity* (see Chapter 2, Sec. 2.12). If we write

$$G(i\omega) = G'(\omega) + iG''(\omega), \qquad (7)$$

then $G'(\omega)$ is called the *storage shear modulus of elasticity*, whereas $G''(\omega)$ is the *loss shear modulus of elasticity*. These names are associated with the fact that if the material obeys Hooke's law, G'' is zero, and the strain energy stored in the material is proportional to G'. On the other hand, if the material behaves like a Newtonian viscous fluid, $G' = 0$, and the energy dissipated is proportional to G''.

Equation (6) is also written frequently as

$$\tau = \mu\dot{\gamma} = i\omega\mu\gamma,$$ (8)

where

$$\mu(\omega) = \mu' + i\mu'' = \frac{1}{i\omega} G(i\omega).$$ (9)

Hence

$$G' = -\omega\mu'', \qquad G'' = \omega\mu'.$$ (10)

The advantage of the form given by Eq. (8) is that it is identical with the Newtonian viscosity law. Using it in the basic equations of motion and continuity, it can be shown (Problem 6.3) that if the amplitude of the motion is very small (so that the square of velocity is negligible), the fluid drag acting on the sphere is given by Stokes' formula [see *Biodynamics: Circulation* (Fung, 1984), pp. 250–256, for derivation]:

$$F_D = 6\pi\mu(\omega)rv,$$ (11)

where r is the radius of the sphere, and v is the velocity of the sphere relative to the fluid container which is much larger than the sphere. The difference of F_m and F_D is equal to the mass of the sphere times acceleration. Hence

$$\frac{4}{3}\pi r^3 \rho_s \frac{d^2x}{dt^2} = F_m - 6\pi\mu r \frac{dx}{dt}.$$ (12)

Here ρ_s is the density of the iron sphere. Substituting Eq. (5) into Eq. (12), we obtain

$$-\frac{4}{3}\pi r^3 \rho_s \omega^2 x_0 e^{-i\delta} = F_0 - 6\pi\mu r(i\omega)x_0 e^{-i\delta}.$$ (13)

Solving for μ, we have

$$\mu = \frac{-i}{6\pi r\omega}\left(\frac{F_0}{x_0} e^{i\delta} + \frac{4}{3}\pi r^3 \rho_s \omega^2\right)$$ (14)

and, by Eq. (9),

$$G(i\omega) = \frac{F_0}{6\pi r x_0} e^{i\delta} + \frac{2}{9}\rho_s r^2 \omega^2.$$ (15)

Note that if r is very small, so that the last term is negligible, then Eq. (15) yields

$$G' = \frac{F_0}{6\pi r x_0}\cos\delta, \qquad G'' = \frac{F_0}{6\pi r x_0}\sin\delta.$$ (16)

Experimental results for G' and G'' are often plotted against the frequency ω on a logarithmic scale. Typical examples will be shown in the following sections. The curves of G', G'' are often interpreted in terms of Maxwell and Voigt models and their combinations, as discussed in Chapter 2, Sec. 2.13.

Complementing the oscillation tests, there are relaxation and creep tests. These can be done with a rheogoniometer or similar instrument. Theoretically, information on creep, relaxation, and small oscillations are equivalent: from one the others can be computed.

6.2.2 Flow Experiments

Flow tests on viscoelastic fluids can be done by the same means as for ordinary viscous fluids considered in Chapter 2, Sec. 2.14. Attention is called, however, to the tensile stresses associated with shear flow. The coefficient of viscosity of a biofluid measured in a steady viscometric flow should not be expected to agree with the complex viscosity coefficient measured by oscillations of small amplitude. In a steady flow, the strain rate is constant; the strain therefore increases indefinitely. The molecular structure of the fluid will not be the same as that in the equilibrium condition. Hence the coefficient of viscosity will be different in these two cases.

6.3 Protoplasm

Protoplasm is the sum of all the contents of the living cell, not including the cell membrane that surrounds the cell. It consists of a continuous liquid phase, the cytoplasm, with various types of particles, granules, membraneous structures, etc., suspended in it. One hopes to gain a better understanding of the interaction of the various parts in the cell through a study of the rheological properties of the protoplasm.

Since protoplasm tends to coagulate when removed from the cell, it is desirable to measure its rheological properties within the cell *in vivo*. One way to do this is to measure the rate of movement of granules existing within the cell, or by introducing particles into the cell artificially and then to measure their rate of movement. In the latter case one has to show that the rheological behavior of the protoplasm is not unduly disturbed. A centrifuge can be used to drive the particles. However, if iron or nickel particles are inserted, a magnetic field can be used. The viscosity of the protoplasm can be computed from the force and the rate of movement of the particles. The Stokes formula (see Problem 6.2 at the end of this chapter) needs corrections for a particle of nonspherical shape, influence of the finite container (cell membrane), and the influence of neighboring particles. The last correction is especially important because small granules are present in large quantities in protoplasm. The specific gravity of the granules can be determined by floating them (if this can be done) in sugar solutions of different concentrations and finding the specific gravity of the solution in which they neither rise nor fall.

Another way for assessing the viscosity of protoplasm is based on Brownian motion of the granules. The usual technique is to centrifuge the granules to one side of the cell and measure their return. According to Albert Einstein (1905) the displacement D_x of a particle in Brownian motion is related to the time interval t, by the formula

$$D_x^2 = 14.7 \times 10^{-18} \frac{Tt}{\eta a}, \tag{1}$$

where T is the absolute temperature, a is the diameter of the particle, and η is the coefficient of viscosity. This formula was later improved by von Smoluchowski by multiplying the right-hand side by a factor of about 1.2. Many simplifying assumptions are involved, and the value for the viscosity obtained can be only approximate. For this and other formulas based on Brownian motion, see the review article by E. N. Harvey (1938).

By means of these methods, L. V. Heilbrunn (1958) and others have shown that protoplasm has a complex rheological behavior. Heilbrunn suggests that the granule-free cytoplasm of plant cells has a viscosity of about 5 cP (i.e., about five times the viscosity of water), and that the entire protoplasm, including granules, has a viscosity several times as great. For protozoa, amoeba and paramecium have been studied extensively. Paramecium protoplasm shows shear thickening. Slime moulds show even more complex features, shear thickening, shear thinning, thixotropy, self-excited oscillations, viscoelasticity, and complex inhomogeneity.

Marine eggs of sea urchins, starfish, worms, and clams are spherical in shape and have become standard objects of study. Heilbrunn (1926) determined the viscosity of the protoplasm of the egg of *Arbacia punculeta* (Lamarck) to be 1.8 cP, using the colorless granules that constitute a majority of the inclusions. These granules of various sizes interfere with one another, and the value for the viscosity of the entire protoplasm is given as 6.9 cP. The alternative Brownian movement method yielded a viscosity value of 5 cP.

6.4 Mucus from the Respiratory Tract

The viscoelasticity of mucus is strongly influenced by bacteria and bacterial DNA, and questions arise whether certain properties of sputum are basic to certain diseases or are due to the concomitant infection. Because of its very high molecular weight, DNA solution is elastic even at low concentration. It is very important to collect samples that can be called normal. For dogs this can be done by the tracheal pouch method (Wardell et al., 1970). In this preparation, a 5–6 cm segment of cervical trachea is separated, gently moved laterally, and formed *in situ* into a subcutaneously buried pouch. The respiratory tract is anastomosed to reestablish a patient airway. After the operation, the pouch retains its autonomic innervation and blood circulation, as it is histologically and histochemically normal reflection of the intact airway. Milliliter quantities of tracheal secretions that accumulate within the pouch may be collected periodically by aspiration through the skin.

With this method of collection, Lutz et al. (1973) showed that normal mucus of the respiratory tract is heterogeneous, and shows considerable variation in dry weight/volume and pH from sample to sample from the same animal from day to day, as well as from animal to animal. A typical example of the experimental results is shown in Fig. 6.4:1, which shows the storage modulus G' and loss modulus G'' as a function of frequency for whole secretions over a period of time. The variations of pH and weight concentration are also shown in the figure. At the lowest pH, the G' curve is almost flat, showing little indication of dropping off even at low frequency, and thus suggesting an entangled or cross-linked system. At higher pH the moduli are smaller. Samples from other dogs show that considerable variation exists also in concentration. At a weight concentration of 5%–6%, the values of G' at 1 rad/s are on the order of 100 dyn/cm^2.

To clarify the physical and chemical factors influencing mucus rheology, Lutz et al. collected mucus from dogs with similar blood type over a period of many weeks, and immediately freeze dried and stored the samples. After a sufficient amount of material had been collected, it was pooled, dialyzed against distilled water to remove electrolytes, and lyophilized, to obtain the dried polymer. The material was then reconstituted in 0.1 M tris buffer at the desired weight concentration, pH, and sodium chloride concentration. Testing of such reconstituted mucus makes it possible to study the individual chemical factors. Data show that lowering of the pH not only gives elevated values of G' and G'', but also that the curves do not fall off sharply with decreasing frequency. The effect of pH on G'' is smaller, implying that pH affects elastic components of the dynamic modulus, but does not as signiificantly change the viscous components. Ionic strength changes can also bring about significant changes of G' and G''. Interpretation of these changes in terms of the molecular configuration of the mucin glycoprotin is discussed by Lutz et al.

These examples show that the viscoelasticity of normal mucus varies over a quite wide range. Hence in clinical applications one must be careful on interpreting rheological changes.

Testing mucus by creep under a constant shear stress is one of the simplest methods for examining its viscoelastic behavior. An example of such a test result is shown in Fig. 6.4:2, from Davis (1973). A cone-and-plate rheogoniometer was used. A step load is applied at time zero. The load is removed suddenly at a time corresponding to the point D. The curve ABCD gives the creep function.

Mathematically, the creep function and the complex moduli $G'(\omega)$ and $G''(\omega)$ can be converted into each other through Fourier transformation. In practice, creep testing furnishes more accurate information on $G'(\omega)$, $G''(\omega)$ at low frequency (say, at $\omega < 10^{-2}$ rad/sec), than the direct oscillation test. On the other hand, the complex modulus computed from the creep function at high frequency (say, at $\omega > 10^{-1}$ rad/sec), is usually not so accurate. A combination of these two methods would yield better results over a wider

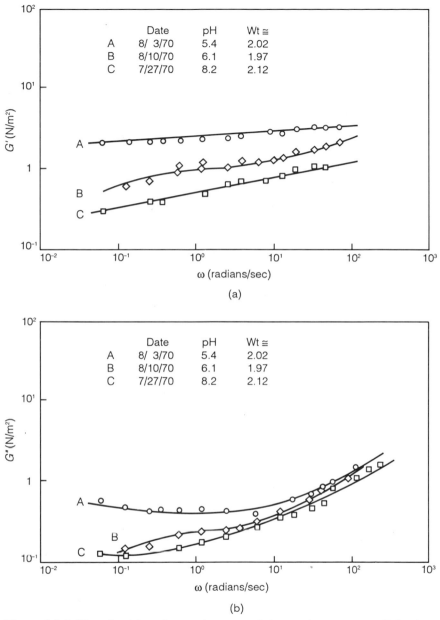

Figure 6.4:1 Viscoelasticity of normal mucus of the respiratory tract of the dog. (a) Storage modulus vs. frequency; (b) loss modulus. From Lutz, Litt, and Chakrin (1973), by permission.

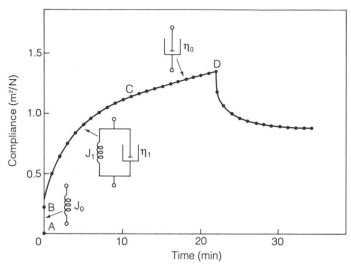

Figure 6.4:2 Creep characteristics for sputum at 25°C. The different segments of the curve are: A–B, elastic region; B–C, Voigt viscoelastic region; C–D, viscous region. From Davis (1973), by permission.

range of frequencies, with $G'(\omega)$, $G''(\omega)$ computed from the creep function at lower frequencies, and measured directly from oscillation tests at higher frequencies.

6.4.1 Flow Properties of Mucus

Mucus from the respiratory tract is so elastic that it is rarely tested in the steady flow condition, because one fears that the structure of the mucus would be very different in steady flow as compared with that *in vivo*. But the response of mucus to a varying shear strain that goes beyond the infinitesimal range is of interest, because it reveals certain features not obtainable from small perturbation tests. Figure 6.4:3 shows a record for sputum that was tested in a cone-and-plate viscometer, the control unit of which gave uniform acceleration of the cone to a preset maximum speed and then a uniform deceleration. The stress history shows a large hysteresis loop. There exists a *yield point* on the loading curve at which the mucus structure apparently breaks down. Beyond the yield point, the stress response moves up and down, very much like steels tested in the plastic range beyond their yield point. On the return stroke the curve is linear over a considerable part of its length, but does not pass through the origin. It cuts the stress axis at another point, D, which is a residual stress left in the specimen after the completion of a strain cycle.

It is not easy to explain the yield point in detail. Other investigators, using different instruments and different testing procedures, have obtained yield

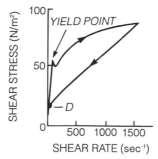

Figure 6.4:3 A typical rheogram for sputum subjected to varying shear strain at 25°C. Data from Davis (1973).

stresses that are much smaller than that of Davis. The usefulness of the information is not yet clear.

An excellent paper discussing the mechanics of clearance of mucus in cough is given by Basser et al. (1989).

6.5 Saliva

It is interesting to learn that saliva is elastic. A typical test result on a sample of saliva of man tested in a rheogoniometer is shown in Fig. 6.5:1, from

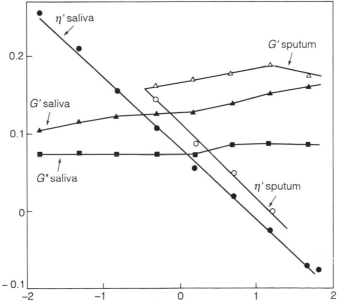

Figure 6.5:1 Viscoelasticity data for saliva and sputum (25°C) subjected to a small oscillation test in a Weissenberg rheogoniometer. Abscissa: log frequency (rad/sec). Ordinate: log dynamic viscosity μ' (P), log storage, and loss moduli (G', G'') (N/m²). Solid symbols: saliva. Open symbols: sputum. From Davis (1973), by permission.

Davis (1973). The value of the real part of the complex viscosity coefficient, μ' [$= G''/\omega$; see Eq. (10) of Sec. 6.2] is seen to be very dependent on frequency, having a value in excess of 100 P at low frequency, falling to less than 0.5 P at high frequency. The storage modulus G' gradually increases with frequency and has a value between 10 and 50 dyn/cm^2. Also shown in Fig. 6.5:1 are typical curves of the sputum obtained by the same author. The variations in G' and μ' with frequency are similar for both sputum and saliva, except that the latter is approximately half as elastic and half as viscous at any given frequency.

Since it is easy to collect saliva samples, it has been proposed to use it to test mucolytic drugs.

6.6 Cervical Mucus and Semen

Sex glands produce viscoelastic fluids. The mucus produced in glands in the uterine cervix is viscoelastic. Research on the viscoelasticity of cervical mucus has a long history, mainly because it was found that the consistency of the mucus changes cyclically in close relationship with various phases of ovarian activity. Gynecologists suggested that measurements of consistency may yield an efficient way to determine pregnancy. Veterinarians suggested that changes in cervical mucus viscoelasticity may indicate the presence of "heat" in cows and other animals. In extreme climatic conditions of low or high temperature, the physical symptoms of "in heat" may become obscure, and a mechanical test can be very helpful. For a farmer, being able to tell when a cow is in heat is to eliminate a waste of time if reproduction is desired.

A detailed account of the history of this topic is given by Scott Blair (1974). Figure 6.6:1, from Clift et al. (1950), shows the type of results to be expected in man. Samples were taken with great care from the os of the cervix. Clift et al. measured the varying pressure required to extrude the mucus from a capillary tube at a constant rate of extrusion. As an "index" of consistency, they took the logarithm of the slope of pressure vs. flow curve (i.e., dp/dQ, where Q is the flow rate). The time scale in Fig. 6.6:1 is in days (of menstrual cycle) for nonpregnant women, and in weeks for pregnant women. The data show a marked minimum of the index of consistency at the time of ovulation (as confirmed by endometrial biopsy and temperature), and a less marked minimum before menstruation. Some patients showing relatively low consistency in early pregnancy later aborted.

What is the meaning of the variation of viscoelasticity of the cervical mucus? Chemically, the changing viscoelasticity reflects the effect of hormones on the mucus composition. Physically, what purpose does the changing consistency serve? Many people have looked for an answer through a study of the swimming of spermatozoa in semen and in cervical mucus. Von Khreningen–Guggenberger (1933) showed that sperm cannot move vertically upward in a salt solution, but they do so in vaginal mucus. Lamar et

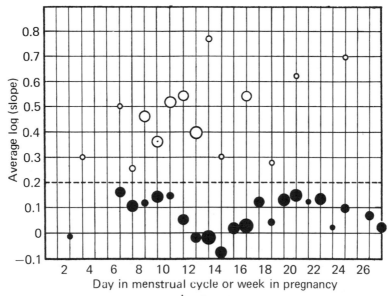

Figure 6.6:1 Average log (slope $dp/d\dot{Q}$) of cervical secretions at different stages of menstrual cycle or of pregnancy in 86 nonpregnant women (solid circles) and 35 pregnant women (open circles). Areas of circles are proportional to the number of cases included in the average, the smallest circle representing one case. The slope of the pressure (p) vs. flow ($\dot{Q}$) curve was obtained by extruding the mucus from a capillary tube. From Clift et al. (1950), by permission.

al. (1940) placed little "blobs" of cervical mucus and semen in a capillary tube, separated by air bubbles. The sperms thus had to pass along the very thin, moist layer on the wall of the tube to reach the cervical mucus. It was found that the penetration by the sperm was maximal for cervical mucus collected at about the 14th day of the menstrual cycle.

The swimming of spermatozoa is a problem of great interest to mechanics. G. I. Taylor (1951, 1952) first showed that sperm will save energy to reach a greater distance if they swim in shoals with their bodies nearly parallel to each other in executing wave motion. Following Taylor, much work has been done on the mechanics of swimming of microbes in general, and sperm in particular (see Wu et al., 1974).

6.7 Synovial Fluid

Our elbow and knee joints have a coefficient of friction which is much smaller than that of most man-made machines. It is much smaller than the friction between the piston and cylinder in an aircraft engine, or that between the shaft and journal bearing in a generator. What makes our joints so efficient? For one thing, the cartilage on the surfaces of the joints is an amazing material. Cartilage against cartilage has a coefficient of friction much lower than teflon against teflon. For another, there is the synovial fluid, whose rheological properties seem to be particularly suited for joint lubrication.

Synovial fluid fills the cavities of the synovial joints of mammals. It is similar in constitution to blood plasma, but it contains less protein. It contains some hyaluronic acid, which is a polysaccharide combined with protein, with a molecular weight in the order of 10 million. Synovial fluid is much more viscous than blood, and this seems to be due to the hyaluronic acid. A synovial fluid that contains less hyaluronic acid than normal has a lower viscosity. An enzyme that destroys the hyaluronic acid greatly reduces the viscosity. In a steady shear flow, which tends to straighten out the long chain molecules and line them up, the viscosity will decrease if the shear rate is increased.

The anticipated elasticity was confirmed by Ogston and Stanier's (1953) experiment on Newton's ring and the elastic recoil mentioned in Sec. 6.1. Balazs (1966) showed further that hyaluronic acid solutions above a certain minimum concentration and salt content exhibit extremely elastic properties at pH 2.5, and form a viscoelastic paste.

6.7.1 Small Deformation Experiments of Synovial Fluid

The viscoelasticity of hyaluronic acid solution can be measured by the two types of experiments discussed in Sec. 6.2: small perturbations from equilibrium and steady flow. A typical result of a test of the first kind is shown in Fig. 6.7:1, from Balazs and Gibbs (1970), who used an oscillating Couette rheometer. The real and imaginary part of the complex dynamic shear modulus $G^*(i\omega)$, i.e., the storage modulus G' and the loss modulus G'', are plotted against the angular frequency. It is seen that as the frequency of oscillation is increased, the modulus of rigidity of the fluid, the vector sum of G' and G'', increases. More importantly, the curves representing the loss and storage moduli cross each other. Thus hyaluronic acid solutions are predominantly viscous at low frequencies, and predominantly elastic at high frequencies. This means that at low frequencies configurational adjustments of the polysaccharide chain, through Brownian motion, are rapid enough to allow the chains to maintain the random configurations under the imposed strain and to slip by each other, resulting in viscous flow. At high frequencies the chains cannot adjust as rapidly as the oscillating strain and thus are deformed sinusoidally.

The magnitude of the moduli were found to increase with increasing concentration of hyaluronic acid. The frequency at which the crossover of G' and G'' occurs depends strongly on the concentration of hyaluronic acid, the pH, and the concentration of NaCl in the solution.

In a similar way, the viscoelasticity of synovial fluids from old and osteoarthrotic persons was tested by Balazs and his associates. Their results are

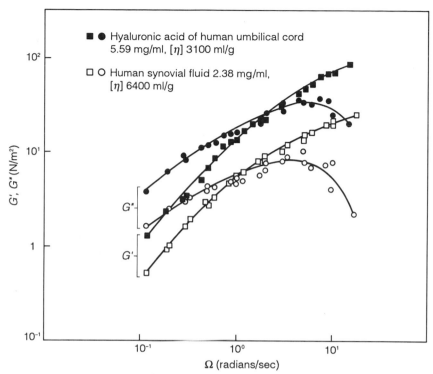

Figure 6.7:1 Dynamic storage moduli G' (open and solid squares) and dynamic loss of moduli G'' (open and solid circles) of a hyaluronic acid solution and synovial fluid from normal young male knee joints plotted against frequency, after reduction of data at various temperatures to 37°C. The concentration of hyaluronic acid and its limiting viscosity number, $[\eta]$, are given for each sample. From Balazs and Gibbs (1970), by permission.

shown in Fig. 6.7:2. The synovial fluids of the normal knee joints of young (27–39 years) and older (52–78 years) human donors did not show significant differences in concentrations of hyaluronic acid, proteins, and sialic acid. The moduli curves in both groups exhibited rapid transition from viscous to elastic behavior. However, the curves for young and older persons were significantly different at higher frequencies, at which the loss modulus G'' of the synovial fluid from young persons was found to be much smaller than that of fluid from older persons. Thus the ratio of energy stored to energy dissipated at high frequencies decreases with aging.

Balazs found that G' and G'' depend strongly on the concentration and size of the hyaluronic molecule. The logarithm of $|G^*(i\omega)|$ was shown to be linearly proportional to the logarithm of the product of the hyaluronic acid concentration and molecular weight. A 10-fold change of the latter product results in approximately a 100-fold change in $|G^*(i\omega)|$.

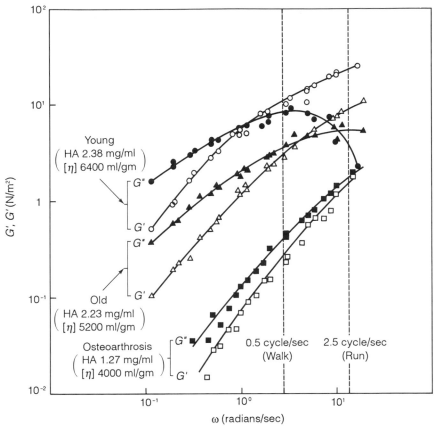

Figure 6.7:2 Dynamic storage moduli G' (open symbols) and dynamic loss moduli G'' (solid symbols) of three human synovial fluid samples plotted against frequency, after reduction of data from various temperatures to 37°C. Broken vertical lines indicate the frequencies that correspond approximately to the movement of the knee joint in walking and running. The concentration of hyaluronic acid and its limiting viscosity number, $[\eta]$, in each sample are given in parentheses. Reproduced from Balazs (1968), by permission.

Figure 6.7:2 also shows the moduli of synovial fluids from osteoarthritic patients. Chemically, these fluids have lower hyaluronic acid and higher protein concentrations than normal persons. Rheologically, the lower hyaluronic acid concentration resulted in a significantly lower shear modulus, and caused the crossover from viscous to elastic behavior to occur at a much higher frequency.

Tests of synovial fluids obtained from patients with gout, chondrocalcinosis, traumatic synovitis, and arthritis also exhibited a lower $G^*(i\omega)$ than normal fluids. In patients treated with corticosteroids, it was shown that the rheological properties are rapidly restored.

The viscoelastic properties of synovial fluid and hyaluronic acid solutions led Balazs to hypothesize that the biological function of synovial fluid is to serve as a shock absorber. It also led him to propose the use of hyaluronic acid as a viscoelastic paste to be injected into the joint when the normal function is impaired. Trials of such injections into race horses and human patients have claimed a certain measure of success.

6.7.2 Flow Properties of Synovial Fluid

Let us turn to the other type of test of synovial fluid as a liquid in steady flow. King (1966) tested synovial fluids of bullocks in a cone-and-plate rheogoniometer in the rotation mode. His results are illustrated in Fig. 6.7:3. It is seen that synovial fluid is strongly shear-thinning. Its viscosity is high at low shear rate, but decreases rapidly as the shear rate increases.

As suggested by the theory of "simple fluids," one expects the normal stress to develop in a viscometer flow. In a cone-and-plate experiment, the measurable normal stress is the difference between that normal to the plate, P_{11}, and the radial stress, P_{22}. King's results are shown in Fig. 6.7:4. It is seen that considerable normal stress develops in a pure shear flow. This very interesting phenomenon is seen in other polymer solutions. The theoretical formulation

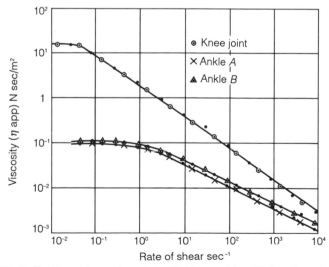

Figure 6.7:3 Bullock's ankle and knee joint fluids tested in a Weissenberg rheogoniometer in the rotation mode. At low shear rates the coefficients of apparent viscosity are constant and the fluids exhibit Newtonian characteristics. At higher shear rates they are shear-thinning. Note that the knee joint fluid has a much higher apparent viscosity. From King (1966), by permission.

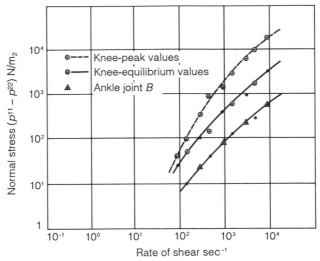

Figure 6.7:4 The normal stress associated with shear stress in the rotation mode. Fluids from bullock's ankle and knee joints. The peak values of the normal stress are the ones that developed within a few seconds after the clutch was suddenly turned on. The normal stress then fell immediately after the peak, and reached an equilibrium value some time later (around 50 sec). For the knee joint fluid the difference between the peak and equilibrium values is large. For the ankle joint, the peak value was only 30% above the equilibrium value on average, and only the equilibrium values are plotted. From King (1966), by permission.

of the constitutive equations of the synovial fluid is reviewed by Lai, Kuei, and Mow (1978).

Problems

6.1 *Newton's Rings.* If the convex surface of a lens is placed in contact with a plane glass plate with a drop of fluid between them, as in Fig. 6.1:1, a thin film of fluid is formed between the two surfaces. The thickness of the film is very small at the center. Such a film is found to exhibit interference colors, produced in the same way as colors in a thin soap film. The interference bands are circular and concentric with the point of contact. When viewed by reflected light, the center of the pattern is black, as it is in a thin soap film. When viewed by transmitted light, the center of the pattern is bright.

Show that, if light travels through the film vertically, dark fringe will occur whenever the thickness of the fluid is equal to integral multiples of half-wavelength, i.e., $\frac{1}{2}\lambda_0, \lambda_0$, etc. Hence by measuring the radius of a bright or dark ring, the thickness of the film can be found. With reference to Fig. 6.1:1, show that the thickness of the fluid film, t, is given by

$$t = h_0 + (R - R\cos\theta) \doteq h_0 + \frac{r^2}{2R}.$$

Hence, show that in monochromatic light in normal incidence, the radii of dark fringes are

$$r = \sqrt{mR\lambda - 2Rh_0}, \qquad m = 0, 1, 2, \ldots .$$

See M. Born and E. Wolf, *Principles of Optics*, 3rd Edition, Pergamon Press, New York, 1965, p. 289, for details.

6.2 Consider a sphere moving in an incompressible (Newtonian) viscous fluid at such a small velocity that the Reynolds number based on the radius of the sphere and velocity of the center is much smaller than 1. Write down the governing differential equations and boundary conditions. Derive Stokes' result (1850) that the total force of resistance, F, imparted on the sphere by the fluid is

$$F = 6\pi\mu aU,$$

where μ is the coefficient of viscosity, a is the radius of the sphere, and U is the velocity of the sphere.

This result applies only to a single sphere in an infinite expanse of homogeneous fluid. If the fluid container is finite in size or if there are other spheres in the neighborhood, or if the Reynolds number approaches 1 or larger, then the equation above needs "corrections." For Stokes' solution, see C. S. Yih (1977) *Fluid Mechanics*, pp. 362–365. For the correction for inertial forces in slow motion, see "Oseen's approximation" in the same book, pp. 367–372. Dynamic effects due to sphere oscillation, sudden release from rest, or variable speed are also discussed in Yih's book. For the correction for neighboring spheres, see E. Cunningham (1910) *Proc. Roy. Soc. London A* **83**, 357.

6.3 Prove that Stokes' formula, Eq. (11) of Sec. 6.2, is valid for a sphere oscillating at an infinitesimal amplitude in a linear viscoelastic fluid obeying the stress–strain relationship, Eq. (8) of Sec. 6.2.

   Hint. Write down the equations of motion, continuity, and the stress–strain relationship. Show that the basic equations are the same as those of Problem 6.2, except that μ is now a complex number.

6.4 Consider the magnetic microrheometer discussed in Sec. 6.2. The force F_0 in Eq. (16) of Sec. 6.2 cannot be measured directly, but can be calibrated by testing the sphere in a Newtonian viscous fluid. Work out details.

6.5 An important function of the mucus in the trachea is to help the cilia clear the dust particles carried in by air. Consider cilia motion as a mathematical problem, present all the equations needed to analyze the motion of the fluid around the moving cilia. Will your set of equations yield a unique solution? How can these equations solve the mucus clearing problem?

6.6. Design a viscometer that can measure the viscoelasticity of a body fluid of small quantity. Consistent with accuracy and reliability the smaller the amount of sample fluid needed, the better. The design should meet the following requirements:

(1) The storage and loss moduluses, i.e., the real and imaginary part of the dynamic modulus: $G(i\omega) = \mu + i\eta$, should be measured.
(2) The instrument should be reusable sample after sample. It should be able to be cleaned, disinfected, and calibrated.
(3) Any disposable parts should be inexpensive.

The report should include:

(1) A sketch of the instrument.
(2) A verbal description of its construction and operation.
(3) The formulas to be used to compute the viscosity or viscoelasticity from the quantities measured.
(4) Discussion of its possible applications.
(5) Comments on the manufacturing or economical aspects.

References

Balazs, E. A. (1966) Sediment volume and viscoelastic behavior of hyaluronic acid solutions. *Fed. Proc.* **25**, 1817–1822.

Balazs, E. A. (1968) *Univ. Michigan Med. Center J.* Special Issue, **34**, 225.

Balazs, E. A. and Gibbs, D. A. (1970) The rheological properties and biological function of hyaluronic acid. In *Chemistry and Molecular Biology of the Intercellular Matrix*, E. A. Balazs ed. Academic Press, New York, Vol. 3, pp. 1241–1253. For details see Gibbs et al. (1968) *Biopolymers* **6**, 777–791.

Basser, P. J., McMahon, T. A., and Griffith, P. (1989) The mechanics of mucus clearance in cough. *J. Biomech. Eng.* **111**, 289–297.

Bingham, E. C. and White, C. F. (1911) Viscosity and fluidity of emulsions, crystallin liquids, and colloidal solutions. *J. Am. Chem. Soc.* **33**, 1257–1268.

Burgers, J. M. (1935) *First Report on Viscosity and Plasticity*. Prepared by the Committee for the Study of Viscosity of the Academy of Sciences at Amsterdam. *Kon. Ned. Akad. Wet. Verhand* **15**, 1.

Burgers, J. M. (1938) *Second Report on Viscosity and Plasticity*. Prepared by the Committee for the Study of Viscosity of the Academy of Sciences at Amsterdam. *Kon Ned. Akad. Wet., Verhand* **16**, 1–287.

Clift, A. F., Glover, F. A., and Scott Blair, G. W. (1950) *Lancet* **258**, 1154–1155.

Davis, S. (1973) In *Rheology of Biological Systems*, H. L. Gabelnick and M. Litt eds. Charles C. Thomas, Springfield, IL, pp. 158–194.

Einstein, A. (1905) *Investigations on the Theory of Brownian Movement*, with notes by R. Fürth, translated into English from German by A. D. Cowper, Methuen, London (1926), Dover Publications (1956). Original paper in *Ann. Phys.* **17** (1905), p. 549.

Frey-Wyssling, A. (ed.) (1952) *Deformation and Flow in Biological Systems*. North-Holland, Amsterdam.

Fung, Y. C. (1984) *Biodynamics: Circulation*. Springer-Verlag, New York.

Gabelnick, H. L. and Litt, M. (eds.) (1973) *Rheology of Biological Systems*. Charles C. Thomas, Springfield, IL.

Gibbs, D. A., Merrill, E. W., and Smith, K. A. (1968) Rheology of hyaluronic acid. *Biopolymers* **6**, 777–791.

Harvey, E. N. (1938) Some physical properties of protoplasm *J. Appl. Phys.* **9**, 68–80.

Heilbrunn, L. V. (1926) The centrifuge method of determining protoplasmic viscosity. *J. Exp. Zool.* **43**, 313–320.

Heilbrunn, L. V. (1956) *The Dynamics of Living Protoplasm*. Academic Press, New York.

Heilbrunn, L. V. *The Viscosity of Protoplasm. Plasmatologia* Springer-Verlag, Wien, Vol. 2.

King, R. G. (1966) A rheological measurement of three synovial fluids. *Rheol. Acta* **5**, 41–44.

Kuethe, A. M. and Chow, C.-Y. *Foundations of Aerodynamics*, 4th edition. John Wiley, New York.

Lai, W. M., Kuei, S. C., and Mow, V. S. (1978) Rheological equations for synovial fluids. *J. Biomech. Eng.* **100**, 169–186.

Lamar, J. K., Shettles, L. B., and Delfs, E. (1940) Cyclic penetrability of human cervical mucus to spermatozoa in vitro. *Am. J. Physiol.* **129**, 234–241.

Lutz, R. J., Litt, M., and Chakrin, L. W. (1973) Physical–chemical factors in mucus rheology. In *Rheology of Biological Systems*, H. L. Gabelnick and M. Litt (eds.) Charles C. Thomas, Springfield, IL, pp. 119–157.

Ogston, A. G. (1970) The biological function of the glycosaminoglycans. In *Chemistry and Molecular Biology of the Intercellular Matrix*, E. A. Balazs (ed.) Academic Press, New York, pp. 1231–1240.

Ogston, A. G. and Stanier, J. E. (1953) The physiological function of hyaluronic acid in synovial fluid; viscous, elastic and lubricant properties. *J. Physiol. (London)* **119**, 244–252 and 253–258. See also, *Biochem. J.* (1952), **52**, 149–156.

Radin, E. L., Swann, D. A., and Weisser, P. A. (1970) Separation of a hyaluronate-free lubrication fraction from synovial fluid. *Nature* **228**, 377–378.

Scott Blair, G. W. (1974) *An Introduction to Biorheology.* Elsevier, New York.

Taylor, G. I. (1951) Analysis of the swimming of microscopic organisms. *Proc. Roy. Soc. London A* **209**, 447–461.

Taylor, G. I. (1952) The action of waving cylindrical tails in propelling microscopic organisms. *Proc. Roy. Soc. London A* **211**, 225–239.

von Khreningen-Guggenberger, J. (1933) Experimentelle Untersuchungen über die vertikale spermien wanderung. *Arch. Gynäk.* **153**, 64–66.

Wardell, J. R., Jr., Chakrin, L. W., and Payne, B. J. (1970) The canine tracheal pouch: A model for use in respiratory mucus research. *Am. Rev. Resp. Dis.* **101**, 741–754.

Wu, T. Y., Brokaw, C. J., and Brennen, C. (eds.) (1974) *Swimming and Flying in Nature*, 2 Vols. Plenum Press, New York.

Yih, C. S. (1977) *Fluid Mechanics, A Concise Introduction to the Theory*, corrected edition. West River Press, Ann Arbor, MI.

CHAPTER 7

Bioviscoelastic Solids

7.1 Introduction

This chapter is focused on soft tissues. We shall first consider some of the most elastic materials in the animal kingdom: abductin, resilin, elastin, and collagen. Collagen will be discussed in greater detail because of its extreme importance to human physiology. Then we shall consider the thermodynamics of elastic deformation, and make clear that there are two sources of elasticity: one associated wit change of internal energy, and another associated with change of entropy. Following this, we shall consider the constitutive equations of soft tissues. Results of uniaxial tension experiments will be considered first, leading to the concept of quasilinear viscoelasticity. Then we will discuss biaxial loading experiments on soft tissues, methods for describing three-dimensional stresses and strains in large deformation, and the meaning of the pseudo-strain energy function.

In Sec. 7.10 we give an example: the constitutive equation of the skin. In Sec. 7.11 we present equations describing generalized visoelastic relations. The chapter is concluded with a discussion of the method for computing strains from known stresses if the stress–strain relationship is given in the other way: stresses expressed as functions of strains.

In the chapters to follow, we consider in detail the blood vessels, skeletal muscle, heart muscle, and smooth muscles. Muscles are the materials that make biomechanics really different from any other mechanics. Finally, in Chapter 12, we discuss bone and cartilage.

7.2 Some Elastic Materials

7.2.1 Actin

Actin molecules are present in all muscles, leukocytes, red blood cells, endo-thelial cells, and many other cells. The strength of a single actin filament were measured by Kishino and Yanagida (1988). The measurement is based on the fact that a single actin filament ($\sim$ 7 nm in diameter) can be clearly seen by video-fluorescence microscopy. Actin monomers are globular. They polymerize into filaments. Actin filaments labelled with phalloidin–tetramethyl–rhodamine are stable. Both ends of a single actin filament were caught using two kinds of microneedles connected to micromanipulators under a fluorescence microscope. One of the needles, used for measuring force, was very flexible, and the other, used for pulling actin filaments, was stiff. Before the experiments, the needles were coated with monomeric myosin to increase their affinity with actin. The stiff needle was pulled until the filament broke. Force was calculated from the bending of the flexible needle. For filaments of length 4 to 32 μm, the tensile force of the actin filament was found to be 108 ± 5 (s.d., $n = 61$) pN without breaking, and almost independent of the filament length. This force is comparable with the force exerted on a single thin element in muscle cells during isometric contraction. The tensile strength of the actin filament is, on assuming a force of 108 pN sustained by a fiber of diameter 7 nm, at least 2.2×10^6 N/m^2, or 2.2 MPa.

7.2.2 Elastin

Elastin is the most "linearly" elastic biosolid materials known. If a cylindrical specimen of elastin is prepared and subjected to uniaxial load in a tensile testing machine, a tension-elongation curve as shown in Fig. 7.2:1 is obtained. The abscissa is the tensile strain defined as the change of length divided by the initial (unloaded) length of the specimen. The ordinate is the stress defined as the load divided by the initial cross-sectional area of the specimen at zero stress. Note that the loading curve is almost a straight line. Loading and unloading do lead to two different curves, showing the existence of an energy dissipation mechanism in the material; but the difference is small. Such elastic characteristics remain at least up to $\lambda = 1.6$.

Elastin is a protein found in vertebrates. It is present as thin strands in the skin and in areolar connective tissue. It forms quite a large proportion of the material in the walls of arteries and veins, especially near the heart. It is a prominent component of the lung tissue. The ligamentum nuchae, which runs along the top of the neck of horses and cattle, is almost pure elastin. Specimens for laboratory testing can be prepared from the ligamentum nuchae of ungu-lates (but cat, dog, and man have very small ligamentum nuchae). These ligaments also contain a small amount of collagen, which can be denatured by heating to 66°C or above. Heating to this degree and cooling again does not change the mechanical properties of elastin.

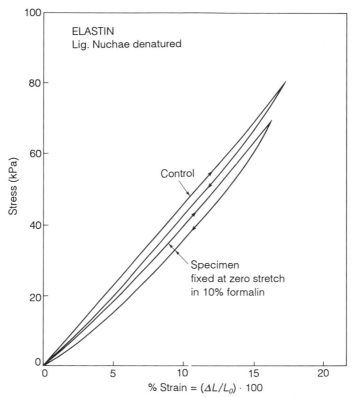

Figure 7.2:1 The stress–strain curve of elastin. The material is the ligamentum nuchae of cattle, which contains a small amount of collagen that was denatured by heating at 100°C for an hour. Such heating does not change the mechanical properties of elastin. The specimen is cylindrical with rectangular cross section. Loading is uniaxial. The curve labeled "control" refers to native elastin. The curve labeled "10% formalin" refers to a specimen fixed in formalin solution for a week without initial strain. From Fung and Sobin (1981). Reproduced by permission of ASME.

The function of the ligamentum nuchae in the horse is clear: it holds up the heavy head and permits its movement with little energy cost. If the horse depended entirely on muscles to hold its head up, energy for maintaining tension in the muscle would be needed.

Elastin in the arteries and lung parenchyma provides elasticity to these tissues. In skin it keeps the tissue smooth. In humans it is known that the gene responsible for synthesizing elastin is turned off at puberty.

7.2.3 Incomplete Fixation of Elastin in Aldehyde

One particular property of elastin has probably had a profound influence on our knowledge of anatomy and histology. In the microscopic examina-

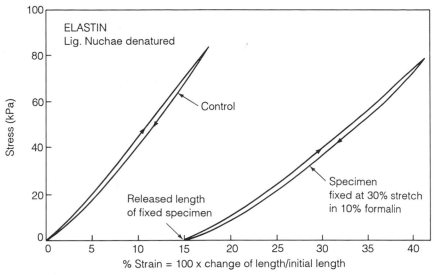

Figure 7.2:2 The stress–strain curve of a specimen of elastin that was first stretched 30% and then soaked in 10% formalin for two weeks. On releasing the stretch, the specimen shortened, but 15% of stretch remained. Subsequent loading produced the stress–strain curves shown on the right-hand side. These curves may be compared with the "control." The arrows on the curves show the direction of loading (increasing strain) or unloading (decreasing strain). From Fung and Sobin (1981). Reproduced by permission of ASME.

tion of a tissue, the tissue is usually fixed by formalin, formaldehyde, or glutaraldehyde; then embedded, sectioned, and stained. Elastin cannot be fixed: when elastin is soaked in these fixation agents for a long period of time, (hours, days, or weeks), it retains its elasticity. If an elastin specimen is stretched under tension and then soaked in these agents, upon release of the tension the specimen does not return to its unstretched length entirely, but it can recover 40%–70% of its stretch (depending on the degree of stretch), and then still behave elastically. An example is shown in Fig. 7.2:2, which refers to a specimen that was stretched to a length 1.3 times its unstressed length, soaked in formalin for two weeks, and released and tested for its stress–strain relationship. It is seen that the "fixed" specimen behaves elastically, although its Young's modulus is somewhat smaller.

If the strain (stretch ratio minus one) at which a specimen is stretched while soaked in the fixative agent is plotted against the retained strain after the "fixed" specimen is released (stress-free), we obtain Fig. 7.2:3. In this figure the abscissa shows the initial stretch during fixation, and the ordinate shows the retained stretch upon release. It is seen that elastic recovery occurs in elastin at all stretch ratios. In other words, what is commonly believed to be "fixed" is not fixed at all.

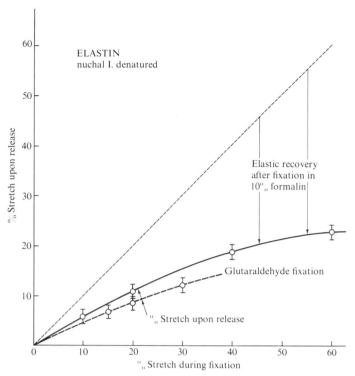

Figure 7.2:3 Elastic recovery of elastin after fixation in formalin and glutaraldehyde solutions. Specimens of elastin were stretched and then fixed in the solution. Upon release from the stretch, the retained elongation of the specimen was measured. The retained stretch is plotted against the initial stretch. The distance between the 45° line and the curves is the amount of strain recovered elastically. From Fung and Sobin (1981).

Now if a tissue is fixed in one of these fixing agents in a state of tension, e.g., an inflated lung, or a distended artery (as these organs are usually fixed by perfusion), and then sectioned under no load, the residual stress in the elastin fibers will be released, and the length of the elastic fibers will be shortened to its length at zero stress state. The fixed part of the tissue, which is inextensible, will be buckled (wrinkled) by the shortening of the elastin. As a consequence the tissue would appear buckled and uneven. This is illustrated in Figs. 7.2:4(a) and (b). In Fig. 7.2:4(a) is shown the lung parenchyma of a spider monkey, which was fixed in glutaraldehyde and embedded in wax. In Fig. 7.2:4(b) is shown parts of the same lung, which was embedded in celloidin, a hard plastic. The elastin fibers in the wax-embedded specimen was allowed to shrink when the wax melted at one time, whereas the elastin fibers in the celloidin-embedded specimen was never allowed to shrink. The difference in the appearance of the pulmonary alveoli in these two photographs is evident.

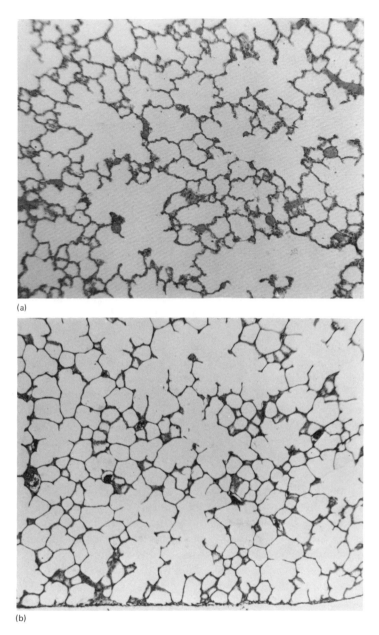

(a)

(b)

Figure 7.2:4 Photomicrographs of the lung tissue of a spider monkey. The lung was fixed in the chest by inspiration of glutaraldehyde solution into the airway. The two photomicrographs here were taken from histological slides prepared in two different ways. The photomicrograph shown in (a) was taken from a slide for which at one stage of its preparation the tissue was allowed to shrink in a stress-free state. The photomicrograph in (b) was taken from a tissue that was not allowed to shrink after it is "fixed" in situ. Courtesy of Dr Sidney Sobin.

Thus we may say that the wrinkled appearance of most published photomicrographs of the lung tissue is an artefact caused by the unsuspected elastic recovery of the elastin fibers in the tissue. The elastin fibers are stretched in the living condition. If the tension in the fibers is allowed to become zero during the preparation of the histological specimen, the fibers will contract and change the appearance of the tissue.

7.2.4 The Elastin Molecule

The molecular structure of the tropoelastin, a precursor molecule of elastin, has been sequenced (Bressan et al., 1987; Deak et al., 1988; Indik et al., 1987; Raju et al., 1987; Tokimitsu et al., 1987; Yeh et al., 1987). Mecham and Heuser (1991) have shown that tropoelastin is formed intracellularly and then cross-linked extracellularly. The mature, cross-linked elastin molecule is inert and so stable that in normal circumstances it lasts in the body for the entire life of the organism.

Repeating sequences in elastin molecule have been noted, and some of their analogs have been prepared chemically, and studied thermo-mechanically. Of these, poly (V PG VG), poly (V PG F GV G AG), and poly (VPGG) on γ-irradiation cross-linking have been shown to be elastic. Urry (1991, 1992) and his associates have shown that these polypeptides will self-assemble into more ordered molecular assemblies on raising temperature, i.e., they exhibit inverse temperature transitions. The molecular proocesses that correspond to the entropic elastomeric force in the self-assembling (nonrandom) systems have been studied in detail. Urry has invented some new bioelastic protein-based polymers on the basis of this research. He has also broadened the view that this inverse temperature transitions is a basic mechanism of biological free energy transduction.

The sources of elasticity of elastin, like those of other soft tissues, must be a decrease of entropy, or an increase of internal energy with increasing strain, (or both see Sec. 7.4). Hoeve and Flory (1958) explained elastin elasticity on the entropy theory. Urry (1985, 1986) identified a mechanism of libration or rocking of some peptide segments that contributes to the entropy. The self-assembling mechanism discussed by Urry (1991) has a critical temperature in the order of 25°C, above which more ordered aggregation forms. Hence Urry predicts a decrease of elasticity at temperature lower than about 25°C. He verified the phenomenon in the synthesized polypentapeptide named above. Debes and Fung (1992) examined the critical temperature problem very carefully in the lung tissue (parenchyma) of the rat, and did not find any critical temperature associated with a sudden change of mechanical properties. One may conclude that the inverse temperature transition phenomenon identified by Urry for a synthetic analog of a part of the elastin molecule may not be a major mechanism for the whole elastin.

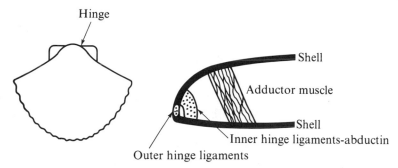

Figure 7.2:5 A plane view of a scallop (*Pecten*) and a sketch of the hinge and the adductor muscle.

Other models of elastin elasticity are proposed by Partridge (1969), Gray (1970), Weisfogh and Anderson (1970), Gosline (1978), and Fleming et al. (1980).

7.2.5 Resilin and Abductin

A biosolid similar to elastin in mechanical behavior but quite different in chemical composition is *resilin*. Resilin is a protein found in arthropods. It is hard when it is dry, but in the natural state it contains 50% to 60% water and is soft and rubbery. Dried resilin can be made rubbery again by soaking it in water. It can be stretched to three times its initial length. In the range of stretch ratio $\lambda = 1 - 2$, the Young's modulus is about $1.8 \times 10^7 \, \text{dyn/cm}^2$ or 1.8 MPa. The shear modulus G is 0.6 MPa. Insects use resilin as elastic joints for their wings, which vibrate as an elastic system. Fleas and locusts use resilin at the base of their hind legs as catapults in their jumping.

An elastic protein found in scallops' hinges is *abductin*. Scallops use it to open the valves (the adductor muscle is used to close the valves). See Fig. 7.2:5. The elastic moduli of abductin and elastin are about the same.

7.2.6 Elasticity Due to Entropy and Internal Energy Changes

Elasin, resilin, and abductin, like rubber, are constituted of long flexible molecules that are joined together here and there by cross-linking to form three-dimensional networks. The molecules are convoluted and thermal energy keeps them in constant thermal motion. The molecular configurations, hence the entropy, change with the strain. From entropy change elastic stress appears (see Sec. 7.4). With this interpretation, Treloar (1967) showed that the shear modulus, G, is related to the density of the material, ρ, the average value of the weight of the piece of molecule between one cross-link and the next, M, and the absolute temperature, T, according to the formula

$$G = \frac{\rho RT}{M}, \tag{1}$$

where R = gas constant = 8.3×10^7 erg/deg mol. The Young's modulus is related to the shear modulus G by the formula

$$E = 2(1 + v)G, \tag{2}$$

where v is the Poisson's ratio. If the material is volumetrically incompressible so that $v = \frac{1}{2}$, then $E = 3G$.

In using the formula above for rubbery protein, ρ should be the concentration of the protein in g/cm^3 of material. Water contributes to density, but not to shear modulus, hence its weight should be excluded from ρ. This formula is probably correct for those proteins which are already diluted with water at the time they were cross-linked. There is a different rule for materials that were not diluted until after they had been cross-linked. In the latter case, the dilution then stretches out the molecules so that they are no longer randomly convoluted. Rubber swollen with paraffin is such an example.

Crystalline materials derive their elastic stress from changes in internal energy. Their elastic moduli are related to the strain of their crystal lattices. Equation (1) does not apply to crystalline materials, neither does it apply to fibers whose elasticity comes partly from internal energy changes and partly from entropy changes.

Most biological materials that can sustain finite strain have rubbery elasticity. But not all. For example, hair can be stretched to 1.7 times its initial length, and will spring back, but this is because the protein keratin, of which it is made, can exist in two crystalline forms—one with tight α helices, and one with looser, β helices (Ciferri, 1963; Feughelman, 1963). When hair is stretched, some of the α helices are changed into β helices. Table 7.2:1 gives the average values of the Young's modulus and tensile strength for several common materials.

7.2.7 Fibers

Fibers form a distinctive class of polymeric materials. X-ray diffraction shows that fibers contain both crystalline regions, where the molecules are arranged in orderly patterns, and amorphous regions, where they are arranged randomly (see Hearle, 1963). A typical example is collagen bundle. Plant fibers and synthetic textile fibers belong to this class. Elastin "fibers" do not: they are thin strands of a rubbery material.

TABLE 7.2:1 Mechanical Properties of Some Common Materials

Material	Young's modulus (MPa)	Tensile strength (MPa)
Resilin	1.8	3
Abduction	1–4	
Elastin	0.6	
Collagen (along fiber)	1×10^3	50–100
Bone (along osteones)	1×10^4	100
Lightly vulcanized rubber	1.4	
Oak	1×10^4	100
Mild steel	2×10^5	500

7.2.8 Crystallization Due to Strain

Raw (unvulcanized) rubber can be stretched to several times its length and held extended. Stress relaxation is almost complete. On release it does not recoil to its original length. But this is not a viscous flow, because it can be made to recoil by heating. The stress relaxation of rubber is due to crystallization. Stretching extends the molecules so that they tend to run parallel to each other and crystallize. Heating disrupts the crystalline structure. Based on the same principle, crystallization of a polymer solution can be induced in other ways. A high polymer solution may be made to enter a tube in the liquid state, crystallize under a high shear strain rate, and emerge as a fiber. A biological example is silk, which contains two proteins: fibroin and sericin. Sericin is a gummy material which dissolves in warm water and is removed in the manufacturing of silk thread. Silkworm (*Bombyx*) has a pair of glands which spin two fibroin fibers enveloped in sericin. Fibroin taken directly from the silk glands is soluble in water and non-crystalline. It becomes a fiber when it passes through the fine nozzle of a spinneret. The Young's modulus of silk is about 10^4 MPa. It breaks at a stretch ratio of about 1.2. Spider's web is similar to silk.

7.3 Collagen

Collagen is a basic structural element for soft and hard tissues in animals. It gives mechanical integrity and strength to our bodies. It is present in a variety of structural forms in different tissues and organs. Its importance to man may be compared to the importance of steel to our civilization: steel is what most of our vehicles, utensils, buildings, bridges, and instruments are made of. Collagen is the main load carrying element in blood vessels, skin, tendons, cornea, sclera, bone, fascia, dura mater, the uterian cervix, etc.

The mechanical properties of collagen are therefore very important to biomechanics. But, again in analogy with steel, we must study not only the properties of all kinds of steels, but also the properties of steel *structures*; here we must study not only collagen molecules, but also how the molecules wind themselves together into fibrils, how the fibrils are organized into fibers, and fibers into various tissues. In each stage of structural organization, new features of mechanical properties are acquired. Since in physiology and biomechanics, our major attention is focused on the organ and tissue level, we must study the relationship between function and morphology of collagen in different organs.

7.3.1 The Collagen Molecules

A collagen is defined as a protein containing sizable domains of triple-helical conformation and functioning primarily as supporting elements in an extracellular matrix. The arrangement of amino acids in the collagen molecules is shown schematically in Fig. 7.3:1. Every third residue is glycine. Proline and OH-proline follow each other relatively frequently. The individual chains are left-handed helices with approximately three residues per turn. The chains are, in turn, coiled around each other following a right-handed twist with a pitch of approximately 8.6 nm. The three α chains are arranged with slight longitudinal displacements. The amino acids within each chain are displaced by a distance of 0.291 nm, with a relative twist of $-110°$, making the distance between each third glycine 0.873 nm.

To date 12 types of collagen have been identified. Figure 7.3:2 shows three types of collagen. The α chains of Type I are designated as $\alpha 1(I)$, $\alpha 2(I)$, etc. The amino acid composition of these chains are listed in Table 7.3:1. See Nimni (1988) for comprehensive data.

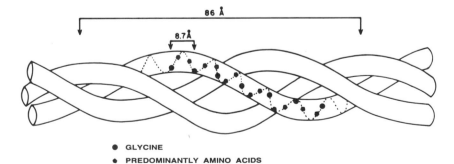

● GLYCINE

◆ PREDOMINANTLY AMINO ACIDS

Figure 7.3:1 Schematic drawing of collagen triple helix. The individual α chains are left-handed helices with approximately three residues per turn. The chains are, in turn, coiled around each other following a right-handed twist. Reproduced from Nimni (1988) by permission.

TYPE I

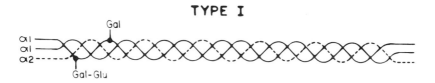

TYPE II

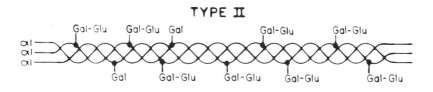

TYPE III

Figure 7.3:2 Diagram of three types of collagen, differing in chain composition and degrees of glycosylation. Disulfide cross-linked are only seen in Type III collagen. Reproduced from Nimni (1988), by permission.

TABLE 7.3:1 Amino Acid Composition of the Human Collagen Chains*

Amino acid	$\alpha1(I)$	$\alpha2(I)$	$\alpha1(II)$	$\alpha1(III)$
4-Hydroxyproline	108	93	97	125
Aspartic acid	42	44	43	42
Threonine	16	19	23	13
Serine	34	30	25	39
Glutamic acid	73	68	89	71
Proline	124	113	120	107
Glycine	333	338	333	350
Alanine	115	102	103	96
Valine	21	35	18	14
Leucine	19	30	26	22
Lysine	26	18	15	30
Arginine	50	50	50	46
Others[++]	38	63	72	18

* Residues per 1000 total residues.
[++] Others include 3-hydroxyproline, half-cystine, methionine, isoleucine, tyrosine, phenylalanine, hydroxylysine, histidine, gal-hydroxylysine, and glc-gal-hydroxylysine.
From Nimni, M. E. (1988), and *Semin Arthritis Rheum.* **8**, 1983. With permission.

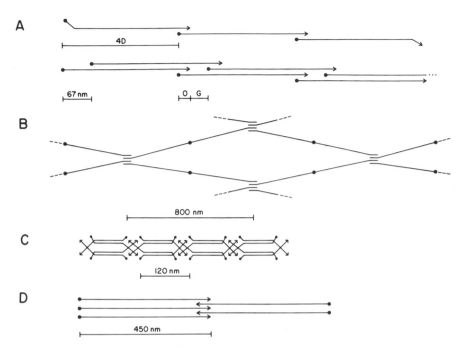

Figure 7.3:3 Molecular architecture of the aggregates formed by (A) the fiber forming collagens; (B) Type IV collagen; (C) Type VI collagen; and (D) Type VII collagen. In this illustration, filled circles an arrowheads are used to denote the directionality of individual molecules. From Miller (1988). Reproduced by permission from *Collagen*, © CRC Press, Inc., Boca Raton, FL.

7.3.2 Aggregate Structure

The close relation of function and structure of collagen aggregates, according to E. J. Miller (1988), is shown in Fig. 7.3:3. (A) shows the fiber-forming collagens of Types I, II, III, V, and K. (B) shows Type IV collagen, which is a major constituent of basement membranes. (C) shows type VI collagen, which is prevalent in placental villi. (D) shows Type VII collagen, whose distribution is unknown, but has been isolated from placental membranes.

Type I collagen is virtually ubiquitous in distribution. It can be isolated from virtually any tissue or organ, especially the bone, dermis, placental membranes, and tendon. Type II is located chiefly in hyaline cartilag, and cartilage-like tissues such as the nucleus pulposus of the vertebral body, and vitreous body of the eye. Type III collagen, along with Type I, is a major constituent of tissues such as the dermis and blood vessel walls, and other more distensible connective tissues. It is also ubiquitous. Type V is a relatively minor constituent in any tissue or organ, but has a distribution similar to that of Type I. Type K, which is XI, is distributed like Type II, chiefly in cartilage. Two of the chains of Type K collagen are highly homologous to those of Type II.

The collagens of Types IX and X are minor constituents of hyaline carti-lages. The are called *short-chain collagens* because their polypeptide chains are shorter than those of fibrillar procollagens. Type IX collagen molecules contain three relatively short triple-helical domains connected by nontriple-helical sequences, instead of a single, long triple-helical domain found in fibrillar collagens. Type IX collagen is also a proteoglycan in that one of its polypeptide subunits serve as the core protein for a chondroitin sulfate side chain. A collagen homologous to Type IX was identified by Gordon et al. (1987) and is named collagen Type XII. The structure, function, and distribu-tion of Type IX/XII collagens are reviewed by Olsen et al. (1988). It is suspected that the Type IX/XII class of molecules play a major role in the assembling of collagen fibrils.

7.3.3 Collagen Fibrils and Fibers

Consider first the fiber-forming collagen molecules. A collection of tropo-collagen molecules forms a collagen fibril. In an electron microscope, the col-lagen fibrils appear to be cross-striated, as is illustrated in the cases of tendon and skin in Fig. 7.3:4. The periodic length of the striation, D, is 64 nm in native fibrils and 68 nm in moistened fibrils. A model of the organization

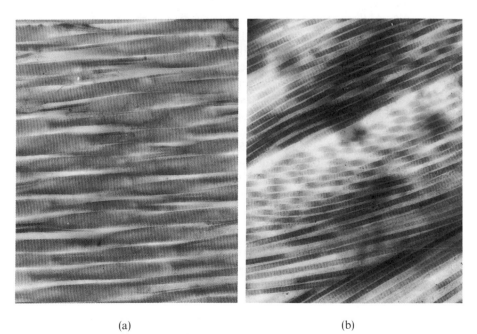

(a)   (b)

Figure 7.3:4 Electron micrographs of (a) parallel collagen fibrils in a tendon, and (b) mesh work of fibrils in skin ($\times$ 24 000). From Viidik (1973), by permission.

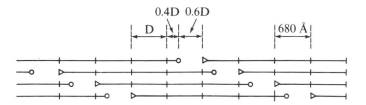

Figure 7.3:5 The concept of quarter-stagger of the molecules. The length of each molecule is 4.4 times that of a period (*D*). From Viidik (1973), and Viidik (1977), by permission.

of the fibrils is shown in Fig. 7.3:5. The length of each molecule is 4.4 times that of the period of the striation, *D*. Hence each molecule consists of five segments, four of which have a length *D*, whereas the fifth is shorter, of length 0.4*D*. In a parallel arrangement of these molecules, a gap of 0.6*D* is left between the ends of successive molecules. The gap appears as the lighter part of the striation. The alignment of the molecules is shown in Fig. 7.3:5 as perfectly straight and parallel; but another current view is that they are not so perfect, but are bent somewhat and have varying spacing between neighboring molecules, with the degree of bending varying with the attachment of water molecules.

The diameter of the fibrils varies within a range of 20 to 40 nm, depending on the animal species and the tissue.

Bundles of fibrils form fibers, which have diameters ranging from 0.2 to 12 μm. Two examples are shown in Fig. 7.3:4. In the light microscope they are colorless; they are birefringent in polarized light. In tendons, they are probably as long as the tendon itself. In connective tissues their length probably varies considerably; there is no definite information on this point.

Packaging of collagen fibers has many hierarchies, depending on the tissue. In parallel-fibered structures such as tendon, the fibers are assembled into primary bundles, or fascicles, and then several fascicles are enclosed in a sheath of reticular membrane to form a tendon. Figure 7.3:3 shows the hierarchy of the rat's tail tendon according to Kastelic et al. (1978). The fibers frequently anastomose with each other at acute angles, in contrast to the fibrils, which are considered not to branch at all in the native state.

7.3.4 The Wavy Course of the Fibers

Diamont, Baer, and their associates (1972) examined rat's tail tendon in the polarized light microscope and found a light and dark pattern with a periodicity of the order of 100 μm, which they interpreted as the waviness of the collagen fiber in the *fascicle* shown in Fig. 7.3:6. When the tendon is stretched, the amplitude of the waviness of the crimped fibers decreases. By rotating the

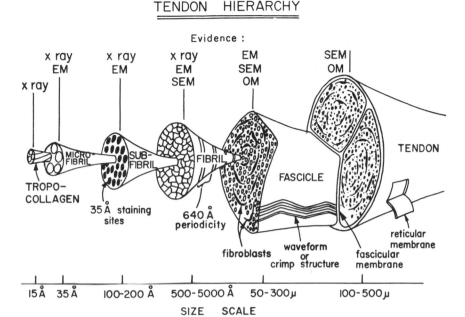

Figure 7.3:6 Hierarchy of structure of a tendon according to Kastelic et al. (1978). Reproduced by permission. Evidences are gathered from X ray, electron microscopy (EM), scanning electron microscopy (SEM), and optical microscopy (OM).

tendon specimen between crossed polaroids, Diamont et al. (1972) and Dale et al. (1972) showed that the wave shape of the crimps is planar. When the tendon was teased down to fine bundles, it was observed that the physical outlines of these subbundles followed the waveform that was deduced from the polarizing optics of the intact tendon bundle. The typical waveform parameters are given in Table 7.3:2. As the fiber is stretched, the "bending angle," θ_0, decreases and tends to zero when the fiber is straight. The fiber diameter is age dependent (see Torp et al., 1974). For example, the rat's tail tendon fiber diameter increases from 100 to 500 nm as the rate ages. The scatter of data for the wavelength l_0 and angle θ_0 can be considerable.

Thus the basic mechanical units of a tendon are seen to be bent collagen fibers. The question arises whether the fibers are intrinsically bent because of some fine structural features of the fibrils. Gathercole et al. (1974), using SEM to resolve the individual collagen fibrils about 100 nm in diameter as they follow the waveform in a rat's tail tendon, failed to find any specific changes in morphology and fine structure along the length of the waveform. It is then suggested that the curvature of the fibers might be caused by the shrinking of the noncollagen components or "ground substance" of the tendon, i.e., that the curvature is caused by the buckling of the fibers. This suggestion is consistent with the experience that the integrity of the ground substance is of great importance to the mechanical integrity of the tendon. Enzymatic di-

TABLE 7.3:2 Typical Wave Parameters from Various Tendons at Zero Strain (Unstretched Condition). From Dale et al. (1972)

Source and age	l_0 (μm)	θ_0 (deg)
Rat tail (14 months)	100	12
Human diaphragm (51 years)	60	12
Kangaroo tail (11.7 years)	75	8–9
Human achilles (46 years)	20–50	6–8

Definition of wave parameters:

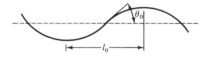

gestion directed at the noncollagen components can greatly change the mechanical properties of the tendon. The buckling model was investigated by Dale and Baer (1974), and it was suggested that hyaluronic acid, which is a major space filling material and which has a fairly high metabolic turnover rate, may be responsible for the buckling of the collagen fibers.

In some connective tissues, it has been suggested that elastin and collagen together form a unit of composite material. The straight elastic fibers are attached to the bent collagen fibers. In the pulmonary alveolar walls (interalveolar septa), however, this was found not to be the case (Sobin et al, 1988).

7.3.5 Ground Substance

Collagen fibers are integrated with cells and intercellular substance in a tissue. In a dense connective tissue, the cells are mostly fibrocytes; the intercellular substance consists of fibers of collagen, elastin, reticulin, and a hydrophilic gel called *ground substance*. Dense connective tissues contain a very small amount of ground substance; loose connective tissues contain a lot. The composition of the ground substance varies with the tissue, but it contains mucopolysaccharides (or glycosaminoglycans), and tissue fluid. The mobility of water in the ground substance is a problem of profound interest in biomechanics, but it is an extremely complex one. The hydration of collagen, i.e., the binding of water to the collagen molecules, fibrils, and fibers, is also an important problem in biomechanics with respect to the problem of movement of fluid in the tissues, as well as to the mechanical properties of the tissue.

7.3.6 Structure of Collagenous Tissues

Depending on how the fibers, cells, and ground substance are organized into a structure, the mechanical properties of the tissue vary. The simplest structure, from the point of view of collagen fibers, consists of parallel fibers, as in tendon and ligaments. The two- and three-dimensional networks of the skin are more complex, whereas the most complex are the structures of blood vessels, intestinal mucosa, and the female genital tracts. Let us consider these briefly.

The most rigorously parallel-fibered structure of collagen is found in each lamina of the cornea. In adjacent laminae of the cornea, the fiber orientation is varied. The transparency of the cornea depends on the strict parallelism of collagen fibers in each lamina.

Tissues whose function is mainly to transmit tension can be expected to adopt the parallel-fiber strucure. Tendon functions this way, and is quite regularly parallel fibered, as is shown in Figs. 7.3:4 and 7.3:6. The fiber bundles appear somewhat wavy in the relaxed condition, but become more straight under tension.

A joint ligament has a similar structure, but is less regular, with collagen fibers sometimes curved and often laid out at an angle oblique to the direction of motion. Different collagen fibers in the ligament are likely to be stressed differently in different modes of function of the ligament. Most ligaments are purely collagenous, the only elastin fibers being those that accompany the blood vessels. But the ligamenta flava of the spine and ligamentum nuchae of some mammals are mostly elastin.

A ligament has both ends inserted into bones, whereas a tendon has only one insertion. The transition from a ligament into bone is gradual; the rows of fibrocytes are transformed into groups of osteocytes, first arranged in rows and then gradually dispersed into the pattern of the bone, by way of an intermediate stage, in which the cells resemble chondrocytes. The collagen fibers are continuous and can be followed into the calcified tissue. The transition from a tendon into a bone is usually not so distinct; the tendon inserts broadly into the main fibrous layer of the periosteum.

The other end of a tendon is joined to muscle. Generally the tendon bundles are invaginated into the ends of the muscle fibers in the many terminal indentations of the outer sarcolemmal layer. Recent investigation suggests that collagenous fibrils, which are bound to the plasma membranes as well as to the collagen fibers, provide the junction.

Parallel fibers that are spread out in sheet form are found in those fasciae into which muscle inserts, or in those expanded tendons called *aponeuroses*, which are membraneous sheets serving as a means of attachment for flat muscles to the bone. The tendinous center of the diaphragm is similarly structured. These sheets appear white and shiny because of their tight structure. Other membranes that contain collagen but the fibers of which are not so regularly structured are opaque. To this group belong the periosteum, perichondrium, membrana fibrosa of joint capsules, dura mater, sclera, some fasciae, and some organ capsules. The cells in these membranes are irregular both in shape and in arrangement.

The structure of collagen fibers in the skin is more complex, and must be considered as a three-dimensional network of fibrils [see Fig. 7.3:4(b)], although the predominant fiber direction is parallel to the surface. These fibers are woven into a more or less rhombic parallelogram pattern, which allows considerable deformation without requiring elongation of the individual fibers. In the dermis, collagen constitutes 75% of skin dry weight, elastin about 4%.

The collagen fiber structure in blood vessels is three-dimensional. A more detailed picture is presented in Chapter 8, Sec. 8.2.

The female genital tract is a muscular organ, with smooth muscle cells arranged in circular and spiral patterns. Actually, in the human uterus only 30%–40% of its wall volume is muscle, and in the cervix only 10%; the rest is connective tissue. In the connective tissue of the genital tract the ground substance dominates, as the ratio of ground substance to fiber elements is 1.5:1 in the nonpregnant corpus and 5:1 in the cervix near full term. During pregnancy, the ground substance grows at the rate of the overall growth, while collagen increases more slowly, and elastin and reticulin almost not at all. Hence the composition of the connective tissue changes.

This brief sketch shows that collagen fibers are organized into many different kinds of structures, the mechanical properties of which are different.

7.3.7 The Stress–Strain Relationship

A typical load-elongation curve for a tendon tested in simple elongation at a constant strain rate is shown in Fig. 7.3:7. It is seen that the curve may be divided into three parts. In the first part, from O to A, the load increases exponentially with increasing elongation. In the second part, from A to B, the relationship is fairly linear. In the third part, from B to C, the relationship is nonlinear and ends with rupture. The "toe" region, from O to A, is usually the physiological range in which the tissue normally functions. The other regions, AB and BC, correspond to reserve strength of the tendon. The ultimate stress of human tendon at C lies in the range 50–100 MPa. The maximum elongation at rupture is usually about 10%–15%.

If the tissue is loaded at a finite strain rate and then its length is held constant, it exhibits the phenomenon of stress relaxation. An example is shown in Fig. 7.3:8, for an anterior cruciate ligament. The figure on the left-hand side (A) corresponds to the case in which the ligament was loaded to about one-third of its failure load and then unloaded immediately at a constant speed. The figure on the right-hand side (B) corresponds to the case in which the ligament was loaded to the same load F_0, and then the length was held constant. The load then relaxes asymptotically to a limiting value F_A.

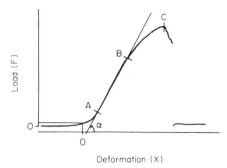

Figure 7.3:7 A typical load-elongation curve for a rabbit limb tendon brought to failure with a constant rate of elongation. The "toe" part is from O to A. The part A–B is almost linear. At point C the maximum load is reached. α is the angle between the linear part of the curve and the deformation axis. The slope, $\tan \alpha$, is taken as the "elastic stiffness," from which the Young's modulus listed in Table 7.2:1 is computed. From Viidik (1973), by permission.

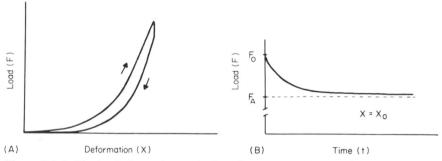

Figure 7.3:8 The load-elongation and relaxation curves of an anterior cruciate ligament specimen. In (A), the specimen was loaded to about one-third of its failure load and then unloaded at the same constant speed. In (B), the specimen was stretched at constant speed until the load reached F_0; then the stretching was stopped and the length was held constant. The load then relaxed. From Viidik (1973), by permission.

A third feature should be noted. If a tissue is taken from an animal, put in a testing machine, tested for a load-elongation curve by a cycle of loading and unloading at a constant rate of elongation, left alone at the unstressed condition for a resting period of 10 min or so until it has recovered its relaxed length, and then tested a second time following the same procedure, the load-elongation curve will be found to be shifted. Figure 7.3:9 shows an example. In the first three consecutive tests, the stress–strain curves are seen to shift to the right, with an increased region of the "toe." The first three relaxation curves (shown in the right-hand side panel), however, are seen to shift upward. If the test is repeated indefinitely, the difference between successive cycles is decreased, and eventually disappears. Then the specimen is said to have been *preconditioned*.

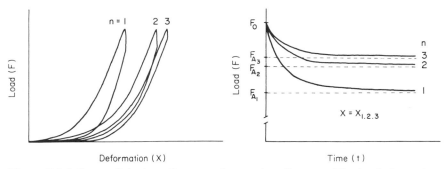

Figure 7.3:9 Preconditioning of an anterior cruciate ligament. The load-elongation and relaxation curves of the first three cycles are shown. From Viidik (1973), by permission.

The reason that preconditioning occurs in a specimen is that the internal structure of the tissue changes with the cycling. By repeated cycling, eventually a steady state is reached at which no further change will occur unless the cycling routine is changed. Changing the upper and lower limits of the cycling will change the internal structure again, and the specimen must be preconditioned anew.

These features: a nonlinear stress–strain relationship, a hysteresis loop in cyclic loading and unloading, stress relaxation at constant strain, and preconditioning in repeated cycles, are common to other connective tissues such as the skin and the mesentery. They are seen also in blood vessels and muscles; but the degrees are different for different tissues. The hysteresis loop is quite small for elastin and collagen, but is large for muscle. The relaxation is very small for elastin, larger for collagen, and very large for smooth muscle. Preconditioning can be achieved in blood vessels very quickly (in two or three cycles) if blood flow into the blood vessel wall (vasa vasorum) is maintained, but it may take many cycles if flow in the vasa vasorum is cut. We shall encounter these features again and again in the study of bioviscoelastic solids.

7.3.8 Change of Collagen Molecular Structure with Tension

In low-angle x-ray diffraction of collagen, a periodicity of about 67 nm can be seen. The length of this period (long period) increases when the specimen is stretched. The relationship between the long period and the mechanical properties of rat tail tendon with the age of the rat has been studied thoroughly by Riedl et al. (1980), and Nemetschek et al. (1980).

7.3.9 Change of Fiber Configuration with Strain

By the electron microscopy method, it can be shown that if a tendon is stretched 10%, the spacing of the characteristic light and dark pattern increases 9%. Thus 1% of the stretching is due to straightening of the fiber (Cowan et al., 1955). It is believed that in the basic alignment of the collagen molecules, the fifth segment (see Fig. 7.3:5) in which the amino acids are more or less randomly placed, contributes most of the stretch when a specimen is stressed.

7.3.10 Critical Temperatures

At 65°C, mammalian collagen shrinks to about one-third of its initial length. This is the basis of the technique of making shrunken human heads (Harkness, 1966). The shrinkage is due to breakdown of the crystalline structure. Shrunken collagen gives no X-ray diffraction pattern (Flory and Garrett, 1958). It is rubbery, with a Young's modulus of about 1 MPa.

7.3.11 Change with Life Cycle

As the function of an organ changes in life, the mechanical properties of its tissue change also. Perhaps no example is more impressive in this regard than the event of childbirth. M. L. R. and R. D. Harkness (1959a) investigated the uterine cervix of the rat during pregnancy and after birth. In nonpregnant rat the uterine cervix contains 5%–10% of collagen by weight. The rat cervix enters the vagina via two canals (horns of uterus). In a transversal section through the cervix of a rat uterus the canals appear as flattened ellipses with fibers arranged roughly concentric around each canal. The Harknesses used cervixes cut from rats at various stages of pregnancy. They slipped a rod through each of the canals, fixed one rod and applied a force to the other. They recorded the extension of the cervix under a constant load. Their results are shown in Fig. 7.3:10. Cervixes from rats that were not pregnant, or had been pregnant up to 12 days, were relatively inextensible and showed little creep. Later in pregnancy the cervix enlarges and becomes much more extensible, and creeps more under a constant load. At 21 days the cervix that is stretched by a sufficiently large constant load creeps at a constant rate. It seems that the stretching is not resisted by elastic restoring force, but instead by viscosity. Within a day after the young have been born the cervix reverts to its original properties, although it is still larger than it was before pregnancy.

The restoration of mechanical properties of the cervix was studied in greater detail by Viidik and Rundgren (see Viidik, 1973). Figure 7.3:11 shows both the load-length relationship and the "stress"–strain relationship. The

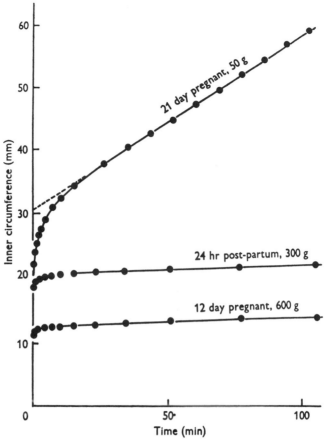

Figure 7.3:10 The creep curves of the cervixes of rats in various stages of pregnancy, stretched by the loads indicated. From Harkness and Harkness (1959), by permission.

change of length at zero load with days post partum shows the rapid change of the size of the cervix. The distensibility increases but the strength decreases with days post partum. If "stress" is calculated by excluding ground substance and including only the estimated collagen fibers' cross-sectional area, then Fig. 7.3:11(b) shows that the collagen framework *per se* is actually stronger at one day post partum than in the virgin state. Then the strength decreases during the resorptive and restorative phase to well below the values for the virginal animal.

It was suggested by the Harknesses (1959b) that the creep characteristics of the near-term uterine cervix are due to changes in the ground substance. They treated the cervix with the enzyme trypsin, which does not attack collagen, and found that the creep rate was greatly increased. They also found that the uterus of the nonpregnant rat undergoes cyclic changes in

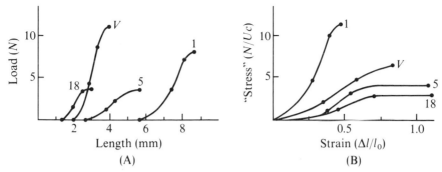

Figure 7.3:11 The mechanical behavior of the uterine cervix of the rat, virginal (V), and 1, 5, and 18 days post partum. Figure on the left (A) shows load vs. length, with specimens relaxed at zero load and then loaded until failure. Figure on the right (B) shows "stress–strain" curves for the same experiment. From Viidik (1973), who attributed the data to Dr. A. Rundgren. Reproduced by permission. Uc stands for the estimated total cross sectional area of the collagen fibers in the uterine cervix specimen.

water content but not in dry weight, whereas the maximum tensile strength decreases when the tissue is swollen. Thus a swollen cervix with increased water content and changed ground substance properties may be the reason for the dramatic distensibility of uterine cervix in the process of giving birth.

7.4 Thermodynamics of Elastic Deformation

In the preceding sections we considered a number of elastic solids. It would be interesting to consider the relationship between the elasticity and the internal constitution of the material: whether it is crystalline or amorphous. Without going into the details of molecular structure and material constitution, we can clarify the essence of the problem from thermodynamic considerations.

There are two sources of elastic response to deformation: change of internal energy, and change of entropy. To see this, let us consider the laws of thermodynamics connecting the specific internal energy $\mathscr{E}$ (internal energy per unit mass), specific entropy S (entropy per unit mass), absolute temperature T, pressure p, specific volume V, density ρ, stress σ_{ij}, strain e_{ij}, and stress and strain deviations σ'_{ij}, e'_{ij}. The first law of thermodynamics (law of conservation of energy) states that in a given body of material of unit volume an infinitesimal change of internal energy is equal to the sum of heat transfered to the body, dQ, and work done on the body, which is equal to the product of the stress σ_{ij} and the change of strain de_{ij}, i.e., σ_{ij} de_{ij}. Expressing all quantities in unit mass (mass in unit volume is ρ), we obtain

$$d\mathscr{E} = dQ + \frac{1}{\rho} \sigma_{ij} de_{ij}. \tag{1}$$

Note that the summation convention for indexes is used here: a repetition of the index i or j means summation over the range 1, 2, and 3.

The second law of thermodynamics states that the heat input dQ is equal to the product of the absolute temperature T and the change of entropy dS:

$$dQ = T \, dS. \tag{2}$$

Combining these expressions, we have

$$d\mathscr{E} = T \, dS + \frac{1}{\rho} \sigma_{ij} de_{ij}. \tag{3}$$

(See Y. C. Fung, *Foundations of Solid Mechanics*, p. 348 et seq. for further details.) In mechanics it is useful to separate out the pressure (negative of the mean stress) from the stress and write

$$\sigma_{ij} = -p\delta_{ij} + \sigma'_{ij}, \tag{4}$$

where σ'_{ij} represents the *stress deviations*. Correspondingly, the strain is separated into a volumetric change and a distortion part:

$$e_{ij} = \tfrac{1}{3} e_{\alpha\alpha} \delta_{ij} + e'_{ij}, \tag{5}$$

where $e_{\alpha\alpha}$ is the change of volume dV, and e'_{ij} represents the *strain deviations*. Using Eqs. (4) and (5), we can write Eq. (3) as

$$d\mathscr{E} = T \, dS - p \, dV + \frac{1}{\rho} \sigma'_{ij} de'_{ij}. \tag{6}$$

If $\mathscr{E}$ and S are considered as functions of the strain e_{ij} and temperature T, then we obtain from Eq. (3):

$$\sigma_{ij} = \rho \left(\frac{\partial \mathscr{E}}{\partial e_{ij}} - T \frac{\partial S}{\partial e_{ij}} \right)_T. \tag{7}$$

This equation shows that the stress σ_{ij} arises from two causes: the increase of specific internal energy with respect to strain, $\partial \mathscr{E}/\partial e_{ij}$, and the decrease of the specific entropy with respect to strain, $-\partial S/\partial e_{ij}$, both converted to values per unit volume of the material through multiplication by the density ρ. Alternatively, we may use the change of specific volume (vol. per unit mass), $dV = e_{\alpha\alpha}/\rho$ and the strain deviations to describe the deformation; then the use of Eq. (6) yields

$$\sigma'_{ij} = \rho \left(\frac{\partial \mathscr{E}}{\partial e'_{ij}} - T \frac{\partial S}{\partial e'_{ij}} \right)_T, \tag{8}$$

$$-p = \left(\frac{\partial \mathscr{E}}{\partial V} - T \frac{\partial S}{\partial V} \right)_T. \tag{9}$$

From Eq. (7) we see that if the entropy does not change (isentropic process), then

$$\sigma_{ij} = \rho \frac{\partial \mathscr{E}}{\partial e_{ij}} \bigg|_{S,T}. \tag{10}$$

But if the internal energy does not change, then the stresses arise from the entropy,

$$\sigma_{ij} = -\rho T \frac{\partial S}{\partial e_{ij}} \bigg|_{\mathscr{E},T}. \tag{11}$$

According to statistical mechanics, entropy is proportional to the logarithm of the number of possible configurations that can be assumed by the atoms in a body. The more the atoms are randomly dispersed or in random motion, the higher the entropy. If order is imposed, the entropy decreases. The entropy source of stress arises from the increased order or decreased disorder of the atoms in a material when strain is increased.

There are two ways with which the entropy of a body can be changed: by conduction of heat through the boundary, and by internal entropy production through internal irreversible processes such as viscous friction, thermal currents between crystals, polymer chain changes, and structural configuration changes. In laboratory experiments isentropic condition is not easy to achieve. If the internal entropy production does not vanish, then to maintain a constant entropy in the specimen a certain exact amount of heat must be conducted away from the surface. Hence in practice, the use of Eq. (10) is limited.

On the other hand, an isothermal (constant temperature) condition is easier to maintain. If temperature and strain are considered as the independent variables, we can transform Eq. (3) by introducing a new dependent variable

$$F = \mathscr{E} - TS, \tag{12}$$

which is called the *specific free energy* (i.e., free energy per unit mass). Differentiating Eq. (12) and using Eq. (3), we obtain

$$dF = d\mathscr{E} - T\,dS - S\,dT$$

$$= \frac{1}{\rho} \sigma_{ij} de_{ij} - S\,dT. \tag{13}$$

In analogy with Eq. (7), we obtain

$$\sigma_{ij} = \rho \left(\frac{\partial F}{\partial e_{ij}} + S \frac{\partial T}{\partial e_{ij}} \right)_S, \tag{14}$$

which is another way of stating the sources of stress: through change of free energy and temperature. In the particular case of an isothermal process, $T = $ const., we obtain

$$\sigma_{ij} = \rho \frac{\partial F}{\partial e_{ij}}\bigg|_T . \tag{15}$$

Equations (10), (11), and (15) can also be derived by observing that for any function F of T and e_{ij}, we must have

$$dF = \frac{\partial F}{\partial T}\bigg|_e dT + \frac{\partial F}{\partial e_{ij}}\bigg|_T de_{ij}, \tag{16}$$

which, when compared with Eq. (13), yields

$$\frac{1}{\rho} \sigma_{ij} = \frac{\partial F}{\partial e_{ij}}\bigg|_T , \tag{17}$$

$$-S = \frac{\partial F}{\partial T}\bigg|_e . \tag{18}$$

Differentiating Eq. (15) with respect to T and Eq. (18) with respect to e_{ij}, we see that they are both equal to $\partial^2 F / \partial T \partial e_{ij}$. Hence

$$\frac{\partial}{\partial T}\left(\frac{\sigma_{ij}}{\rho}\right) = -\frac{\partial S}{\partial e_{ij}} . \tag{19}$$

Substituting back into Eq. (7), we have

$$\sigma_{ij} = \rho \frac{\partial \mathscr{E}}{\partial e_{ij}}\bigg|_T + T \frac{\partial}{\partial T}(\sigma_{ij})\bigg|_{e_{ij}} . \tag{20}$$

The factor ρ in the last term can be canceled because ρ does not change when e_{ij} is held constant.

Equation (20) is a transformation of Eq. (7) into a form more convenient for laboratory experimentation. We see from Eqs. (20) and (19) that the contribution to stress due to entropy change can be measured by measuring the change of stress σ_{ij} with respect to temperature under the condition of constant strain. Let a piece of material be held stretched at constant strain while the temperature T is altered. The equilibrium stress σ_{ij} is measured, which gives σ_{ij} as a function of T. We can then compute $T(\partial \sigma_{ij}/\partial T)$, which gives us the contribution of entropy change to stress response to changing strain. Note that $T(\partial \sigma_{ij}/\partial T)$ is equal to $(\partial \sigma_{ij}/\partial \ln T)$; hence if σ_{ij} at constant strain is plotted against $\log_e T$, then the slope of the curve of σ_{ij} vs. $\log_e T$ is equal to the stress due to entropy, the second term of Eq. (20).

Experiments of this sort have been done on rubber, and it has led to the conclusion that rubber elasticity is derived mainly from entropy change. Similar experiments can be done on biological materials, except that the range of temperature at which a living tissue can remain viable is usually very limited, and the range of log T could be too small to be useful. If the material changes with changing temperature, then again the method cannot be applied. For example, it does not work for table jelly, whose crosslinks break as the temperature rises. It does not work for elastin, which takes up water from the surroundings as the temperature changes.

Engineers are familiar with the concept of *strain energy function*. If a material is elastic and has a strain energy function W, which is a function of the strain components $e_{11}, e_{22}, e_{33}, e_{12}, e_{23}, e_{31}$, then the stress can be obtained from the strain energy function by differentiation:

$$\sigma_{ij} = \frac{\partial W}{\partial e_{ij}}. \tag{21}$$

This equation appears very similar to Eqs. (10) and (17). Hence the strain energy function can be identified with the internal energy per unit volume in an isentropic process, or the free energy per unit volume in an isothermal process. In a more general situation, entropy and internal energy both change, or free energy and temperature both change; then Eq. (21) would have to be identified with Eqs. (7) or (14). Thus the strain energy function depends on the thermodynamic process. The identification of Eq. (21) with the thermodynamic equations (7), (10), (11), (14), and (15) is considered to be a justification of the assumption of the existence of the strain energy function.

Finally, note that all the discussion above presupposes that the state variables are T, S, and e_{ij}, that is, the internal energy and the free energy are functions of the strain e_{ij}, and not of the history of strain, or strain rate, or any other factors such as pH, electric charges, chemical reaction, etc. If these other variables are also significant, then the stress will depend not only on strain, but also on the other variables. In other words, the discussion above is applicable only to elastic bodies and elastic stress. The concept of strain energy function embodied in Eq. (21) is applicable only to elastic bodies.

7.5 Behavior of Soft Tissues Under Uniaxial Loading*

So far we have discussed some more or less "pure" biological materials. From here on we shall consider tissues that are composed of several of these materials and ground substances.

From the point of view of biomechanics, the properties of a tissue are known if its constitutive equation is known. The constitutive equation of a material can only be determined by experiments.

---

* The following material up to p. 285 is taken from the author's paper "Stress–Strain-History Relations of Soft Tissues in Simple Elongation," in *Biomechanics: Its Foundations and Objectives*, Y. C. Fung, N. Perrone, and M. Anliker (eds.) Prentice-Hall. 1972.

The simplest experiment that can be done on a biosolid is the uniaxial tension test. For this purpose a specimen of cylindrical shape is prepared and stretched in a testing machine. The load and elongation are recorded for prescribed loading or stretching histories. From these records we can deduce the stress–strain relationship of the material under uniaxial loading.

That the stress–strain relationship of animal tissues deviates from Hooke's law was known to Wertheim (1847), who showed that the stress increases much faster with increasing strain than Hooke's law predicts. It is also known that tissues in the physiological state are usually not unstressed. If an artery is cut, it will shrink away from the cut. A broken tendon retracts away; the lung tissue is in tension at all times.

If a segment of an artery is excised and tested in a tensile testing machine by imposing a cyclically varying strain, the stress response will show a hysteresis loop with each cycle, but the loop decreases with succeeding cycles, rapidly at first, then tending to a steady state after a number of cycles (see Fig. 7.5:1). The existence of such an initial period of adjustment after a large disturbance seems common to all tissues. From the point of view of mechanical testing, the process is called *preconditioning*. Generally, only mechanical data of preconditioned specimens are presented.

The hysteresis curves of a rabbit papillary muscle (unstimulated) are shown in Fig. 7.5:2. Simple tension was imposed by loading and unloading at the rates indicated in the figure. It is seen that the hysteresis loops did not depend very much on the rate of strain. In general, this insensitivity with respect to the strain rate holds within at least a 10^3- fold change in strain rate.

Figure 7.5:3 shows a stress-relaxation curve for the mesentery of the rabbit. The specimen was strained at a constant rate until a tension T_1 was

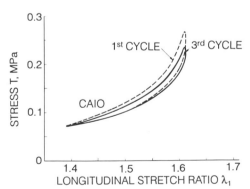

Figure 7.5:1 Preconditioning. Cyclic stress response of a dog's carotid artery, which was maintained in cylindrical configuration (by appropriate inflation or deflation) when stretched longitudinally. λ_1 is the stretch ratio referred to the zero-stress length of the segment; 37°, 0.21 cycles/min. Physiological length $L_p = 4.22$ cm. $L_p/L_0 = 1.67$. Diameter at physiological condition = 0.32 cm. $A_0 = 0.056$ cm^2. Dog wt., 18 kg. From Lee, Frasher, and Fung (1967).

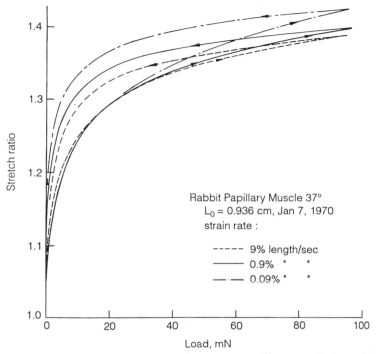

Figure 7.5:2 The length-tension curve of a resting papillary muscle from the right ventricle of the rabbit. Strain rates 0.09% length/sec; 0.9% length/sec; and 9% length/sec. Length at 9 mg force = 0.936 cm. 37°C. $A_0 = 1.287$ mm^2. From Fung (1972), by courtesy of Dr. John Pinto.

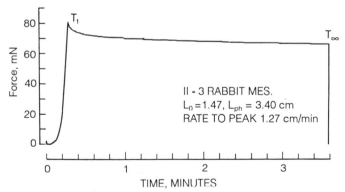

Figure 7.5:3 Relaxation curve of a rabbit mesentery. The specimen was stressed at a strain rate of 1.27 cm/min to the peak. Then the moving head of the testing machine was suddenly stopped so that the strain remained constant. The subsequent relaxation of stress is shown. From Fung (1967).

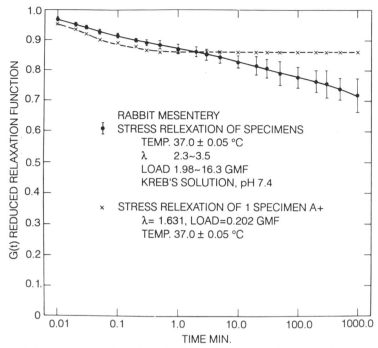

Figure 7.5:4 Long-term relaxation of rabbit mesentery. Solid curve shows the mean reduced relaxation function for 16 different initial stress values. The dashed curve refers to one test at a much lower stress level. By H. Chen; reproduced from Fung (1972).

obtained. The length of the specimen was then held fixed and the change of tension with time was plotted. In a linear scale of time only the initial portion of the relaxation curve is seen. Relaxation in a long period of time is shown in Fig. 7.5:4, in which the abscissa is log t. It is seen that in 17 hours a large portion of the initial stress was relaxed. If the initial stress is sufficiently high, the relaxation curve does not level off even at $t = 10^5$ sec (see the solid curve in Fig. 7.5:4). If, however, the initial stress is lower than a certain value, the stress levels off at the elapse of a long time, as shown by the dashed curve in Fig. 7.5:4. Apparently, at the higher stress levels, the relaxation has not stopped even at 10^5 sec. The reduced relaxation function $G(t)$ shown here is defined by Eq. (1) *infra*.

Figure 7.5:5 illustrates the creep characteristics of the papillary muscles of the rabbit loaded in the resting state under a constant weight. The change of length was recorded and replotted on a logarithmic scale of time. The creep characteristics depend very much on the temperature and the load.

These features of hysteresis, relaxation, and creep at lower stress ranges ("physiological") are seen on all materials tested in our laboratory: the mesentery of the rabbit and dog, the femoral and carotid arteries and the veins of the dog, cat, and rabbit, the ureter of several species of animals, in-

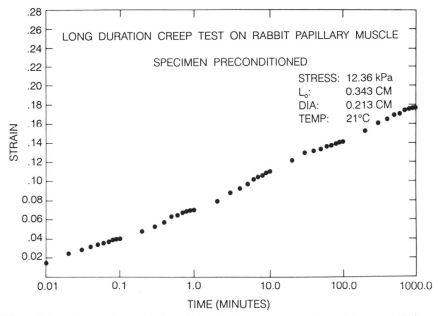

Figure 7.5:5 Creep characteristics of the papillary muscle of the rabbit. By J. Pinto; reproduced from Fung (1972).

cluding human, and the papillary muscles at the resting state. The major differences among these tissues are the degree of distensibility. In the physiological range, the mesentery can be extended 100%–200% from the relaxed (unstressed) length, the ureter can be stretched about 60%, the resting heart muscle about 15%, the arteries and veins about 60%, the skin about 40%, and the tendons 2%–5%. Beyond these ranges the tissues usually have a large reserve strength before they rupture and fail.

7.5.1 Stress Response in Loading and Unloading

In the following we speak of stress and strain in the Lagrangian sense. For a one-dimensional specimen loaded in tension, the tensile stress T is the load P divided by the cross-sectional area of the specimen at the zero stress state, A_0; whereas the "stretch ratio" λ is the ratio of the length of the specimen stretched under the load, L, divided by the initial length at the zero stress state, L_0. Thus

$$T = \frac{P}{A_0}, \qquad \lambda = \frac{L}{L_0}. \tag{1}$$

The zero stress state of the specimen must be identified. The identification may not be easy because in the neighborhood of the zero stress state a soft tissue can be very soft and difficult to handle. But this step cannot be omitted.

For an incompressible material, the cross-sectional area of a cylindrical specimen is reduced by a factor $1/\lambda$ when the length of the specimen is increased by a factor λ. Hence the Eulerian stress σ (which is referred to the cross section of the deformed specimen) is λ times T:

$$\sigma = \frac{P}{A} = \frac{P}{A_0}\lambda = T\lambda. \tag{2}$$

Consider first the relationship between load and deflection in the *loading* process. If the slope of the T vs. λ curve as shown in Fig. 7.5:2 is plotted against T, the result as shown in Fig. 7.5:6 is obtained. As a first approximation, we may fit the experimental curve by a straight line in the range of T exhibited and by the equation

$$\frac{dT}{d\lambda} = \alpha(T + \beta). \tag{3}$$

Then, an integration gives

$$T + \beta = ce^{\alpha\lambda}. \tag{4}$$

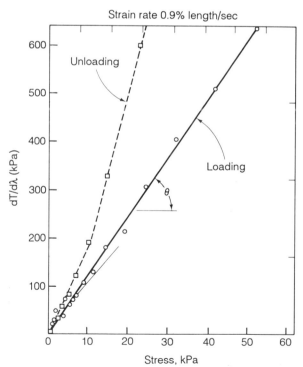

Figure 7.5:6 The variation of the Young's modulus with stress at a strain rate of 0.9% length/sec, illustrating the method of determining the constants $\hat{\alpha}_1, \hat{\beta}_1$. Near the origin, a different straight line segment is required to fit the experimental data. From Fung (1972).

The integration constant can be determined by finding one point on the curve, say $T = T^*$ when $\lambda = \lambda^*$. Then

$$T = (T^* + \beta)e^{\alpha(\lambda - \lambda^*)} - \beta. \tag{5}$$

Note that if λ is referred to the natural state, we must have, by definition, $T = 0$ when $\lambda = 1$; this is possible only if

$$\beta = \frac{T^* e^{-\alpha(\lambda^* - 1)}}{1 - e^{-\alpha(\lambda^* - 1)}}. \tag{6}$$

Equation (3) includes Hookean materials, for which

$$\frac{dT}{d\lambda} = \text{const.} \tag{7}$$

A more refined representation of the experimental data shown in Fig. 7.5:6 can be made by several straightline segments, e.g.,

$$\frac{dT}{d\lambda} = \alpha_1(T + \beta_1) \qquad \text{for } 0 \le T \le T_1, \quad 1 < \lambda \le \lambda_1, \tag{8a}$$

$$\frac{dT}{d\lambda} = \alpha_2(T + \beta_2) \qquad \text{for } T_1 \le T \le T_2, \quad \lambda \ge \lambda_1. \tag{8b}$$

This is often a good practical choice because in the analysis of organ function the quantity $dT/d\lambda$ appears frequently, and it is desirable to have it represented by as simple a form as possible. In this case, the integrated curve must be represented by two expressions in the form of Eq. (5), matched at the juncture $T = T_1$.

Unloading at the same strain rate results in similar straight lines with different slopes, as shown by the dotted curve in Fig. 7.5:6.

7.5.2 Other Expressions

One of the best known approaches to the elasticity of bodies capable of finite deformation is to postulate the form of an elastic potential, or strain energy function, W. For example, if a body is elastically isotropic, the strain energy function must be a function of the strain invariants. Well-known examples of strain-energy functions are those of Mooney (1940), Rivlin (1947), and Rivlin and Saunders (1951). See the treatise of Green and Adkins (1960), where references are given.

Valanis and Landel (1967) presented a strain-energy function

$$W = \sum_{i=1}^{3} f(\ln \lambda_i), \tag{9}$$

where $\lambda_1, \lambda_2, \lambda_3$ are the principal stretch ratios and f is a certain function. Specific applicaton of Valanis's form to soft tissues was made by Blatz et al.

(1969), who proposed the following:

$$f(\ln \lambda_i) = C(\lambda_i^\alpha - 1). \tag{10}$$

When applied to the rabbit's mesentery, Blatz et al. (1969) found $\alpha = 18$. For skeletal muscle in the resting state, they found $\alpha = 8$; for latex rubber, $\alpha = 1.5$.
    Blatz et al. (1969) proposed also the following:

$$f(\ln \lambda_i) = C[e^{\alpha(\lambda_i^2-1)} - 1], \tag{11}$$

for which the uniaxial tension case of an incompressible material gives

$$\sigma = \left(\frac{G}{\alpha + 1}\right)\left\{\lambda e^{\alpha(\lambda^2-1)} - \frac{1}{\lambda^2} e^{\alpha[(1/\lambda)-1]}\right\}. \tag{12}$$

    Veronda and Westmann (1970) proposed the following form for the strain potential of an isotropic material expressed in terms of the strain invariants I_1, I_2, I_3:

$$I_1 = \lambda_1^2 + \lambda_2^2 + \lambda_3^2, \quad I_2 = \lambda_1^2\lambda_2^2 + \lambda_2^2\lambda_3^2 + \lambda_3^2\lambda_1^2, \quad I_3 = \lambda_1\lambda_2\lambda_3,$$
$$W = C_1[e^{\beta(I_1-3)} - 1] + C_2(I_2 - 3) + g(I_3), \tag{13}$$

where C_1, C_2, β, are constants and g is a function which becomes zero if the material is incompressible, $g(1) = 0$. For cat's skin, they suggested the following constants:

$$C_1 = 0.00394, \quad \beta = 5.03, \quad C_2 = -0.01985.$$

Equation (13) presupposes isotropy. However, most biological tissues are not isotropic.
    Other expressions proposed, but not reduced to the form of strain energy, and not pretending to be generally valid for three-dimensional stress states, are the following:

Wertheim (1847)	$\varepsilon^2 = a\sigma^2 + b\sigma,$
Morgan (1960)	$\varepsilon = a\sigma^n,$
Kenedi et al. (1964)	$\sigma = k\varepsilon^d,$
	$\sigma = B[e^{m\varepsilon} - 1],$ (14)
Ridge and Wright (1964)	$\varepsilon = C + k\sigma^b,$
	$\varepsilon = x + y\log\sigma.$

    For example, cornea is a clear window comprising the most anterior portion of the eye. It is comprised of five layers: the epithelium, Bowman's membrane, the stroma, Descemet's membrane, and the endothelium. The fibers in each layer are parallel, and in successive layers run in alternate orthogonal directions. Hoeltzel et al. (1992) measured the mechanical properties of bovine, rabbit, and human corneas under uniaxial tension. Cyclic tensile tests were performed over the physiological range, up to a maximum

TABLE 7.5:1 Hoeltzel et al.'s (1992) Results on the Uniaxial Tensile Stress–Strain Relationship of Cornea in the Third Cycle of Loading

	Human	Rabbit	Bovine
Parameter α (MPa)	99	134	121
Exponent β	1.98	1.99	2.10
Correlation R^2	0.997	0.996	0.995
Av. thickness (mm)	0.82	0.50	1.53
Time postmortem when tested (hrs)	36	< 12	36–48

of 10% strain beyond slack strain. The stress–strain relationships were found to be nonlinear. An empirical formula is used to fit the experimental data:

$$\ln \sigma = \ln \alpha + \beta \ln(\varepsilon - \varepsilon_s),$$

where σ is stress, ε is strain, ε_s represents the slack strain (the difference between zero strain and the smallest strain to initiate load bearing in the specimen), and α and β are constants. This is

$$\sigma = \alpha(\varepsilon - \varepsilon_s)^\beta.$$

Table 7.5:1 shows the numerical results. R^2 is the correlation coefficient. Edema, environmental factors, and change of dimensions during test are discussed in the original paper.

7.6 Quasi-Linear Viscoelasticity of Soft Tissues

The experimental results illustrated in Figs. 7.5:1 through 7.5:6 show that biological tissues are not elastic. The history of strain affects the stress. In particular, there is a considerable difference in stress response to loading and unloading.

Most authors discuss soft tissue experiments in the framework of the linear theory of viscoelasticity relating stress and strain on the basis of the Voigt, Maxwell, and Kelvin models, Sec. 2.11. Buchthal and Kaiser (1951) formulated a continuous relaxation spectrum that corresponds to a combination of an infinite number of Voigt and Maxwell elements. A nonlinear theory of the Kelvin type was proposed by Viidik (1966) on the basis of a sequence of springs of different natural length, with the number of participating springs increasing with increasing strain.

It is reasonable to expect that for oscillations of small amplitude about an equilibrium state, the theory of linear viscoelasticity should apply. For finite deformations, however, the nonlinear stress–strain characteristics of the living tissues must be accounted for.

Instead of developing a constitutive equation by gradual specialization of a general formulation, I shall go at once to a special hypothesis. Let us consider a cylindrical specimen subjected to tensile load. If a step increase in elongation (from $\lambda = 1$ to λ) is imposed on the specimen, the stress developed will be a function of time as well as of the stretch λ. The history of the stress response, called the *relaxation function*, and denoted by $K(\lambda, t)$, is assumed to be of the form

$$K(\lambda, t) = G(t)T^{(e)}(\lambda), \qquad G(0) = 1, \tag{1}$$

in which $G(t)$, a normalized function of time, is called the *reduced relaxation function*, and $T^{(e)}(\lambda)$, a function of λ alone, is called the *elastic response*. We then assume that the stress response to an infinitesimal change in stretch, $\delta\lambda(\tau)$, superposed on a specimen in a state of stretch λ at an instant of time τ, is, for $t > \tau$:

$$G(t - \tau) \frac{\partial T^{(e)}[\lambda(\tau)]}{\partial \lambda} \delta\lambda(\tau). \tag{2}$$

Finally, we assume that the superposition principle applies, so that

$$T(t) = \int_{-\infty}^{t} G(t - \tau) \frac{\partial T^{(e)}[\lambda(\tau)]}{\partial \lambda} \frac{\partial \lambda(\tau)}{\partial \tau} d\tau, \tag{3}$$

that is, the tensile stress at time t is the sum of contributions of all the past changes, each governed by the same reduced relaxation function.

Rewriting Eq. (3) in the form

$$T(t) = \int_{-\infty}^{t} G(t - \tau)\dot{T}^{(e)}(\tau) d\tau, \tag{4}$$

where a dot denotes the rate of change with time, we see that the stress response is described by a linear law relating the stress T with the elastic response $T^{(e)}$. The function $T^{(e)}(\lambda)$ plays the role assumed by the strain ε in the conventional theory of viscoelasticity. Therefore, the machinery of the well-known theory of linear viscoelasticity (see Chapter 2, Sec. 2.11, and the author's *Foundations of Solid Mechanics*, Chapter 15) can be applied to this hypothetical material.

The inverse of Eq. (4) may be written as

$$T^{(e)}[\lambda(t)] = \int_{-\infty}^{t} J(t - \tau)\dot{T}(\tau) d\tau, \tag{5}$$

which defines the *reduced creep function* $J(t)$. Let $T^{(e)}(\lambda) = F(\lambda)$ and $\lambda = F^{-1}(T^{(e)})$ be the inverse of $F(\lambda)$, i.e., the stretch ratio corresponding to the tensile stress $T^{(e)}$. Then for a unit step change of the tensile stress T at $t = 0$, the time history of the stretch ratio is

$$\lambda(t) = F^{-1}[J(t)]. \tag{6}$$

The lower limits of integration in Eqs. (3), (4), and (5) are written as $-\infty$ to mean that the integration is to be taken before the very beginning of the motion. If the motion starts at time $t = 0$, and $\sigma_{ij} = e_{ij} = 0$ for $t < 0$, Eq. (3) reduces to

$$T(t) = T^{(e)}(0+)G(t) + \int_0^t G(t-\tau)\frac{\partial T^{(e)}[\lambda(\tau)]}{\partial \tau}\, d\tau. \tag{7}$$

If $\partial T^{(e)}/\partial t$, $\partial G/\partial t$ are continuous in $0 \le t < \infty$, the equation above is equivalent to

$$T(t) = G(0)T^{(e)}(t) + \int_0^t T^{(e)}(t-\tau)\frac{\partial G(\tau)}{\partial \tau}\, d\tau \tag{8a}$$

$$= \frac{\partial}{\partial t}\int_0^t T^{(e)}(t-\tau)G(\tau)\, d\tau. \tag{8b}$$

Equation (8a) is suitable for a simple interpretation. Since, by definition, $G(0) = 1$, we have

$$T(t) = T^{(e)}[\lambda(t)] + \int_0^t T^{(e)}[\lambda(t-\tau)]\frac{\partial G(\tau)}{\partial \tau}\, d\tau. \tag{9}$$

Thus the tensile stress at any time t is equal to the instantaneous stress response $T^{(e)}[\lambda(t)]$ decreased by an amount depending on the past history, because $\partial G(\tau)/\partial \tau$ is generally of negative value. The question of experimental determination of $T^{(e)}(\lambda)$ and $G(t)$ will be discussed in the following sections.

7.6.1 The Elastic Response $T^{(e)}(\lambda)$

By definition, $T^{(e)}(\lambda)$ is the tensile stress instantaneously generated in the tissue when a step function of stretching λ is imposed on the specimen. Strict laboratory measurement of $T^{(e)}(\lambda)$ according to this definition is difficult, because at a sudden application of loading, transient stress waves will be induced in the specimen and a recording of the stress response will be confused by these elastic waves. However, if we assume that the relaxation function $G(t)$ is a continuous function, then $T^{(e)}(\lambda)$ may be approximated by the tensile stress response in a loading experiment with a sufficiently high rate of loading. In other words, we may take the $T(\lambda)$ obtained in Sec. 7.5 as $T^{(e)}$.

A justification of this procedure is the following. The relaxation function $G(t)$ is a continuously varying decreasing function as shown in Fig. 7.5:4 (normalized to 1 at $t = 0$). Now, if by some monotonic process λ is increased from 0 to λ in a time interval ε, then at the time $t = \varepsilon$ we have, according to Eq. (9),

$$T(\varepsilon) = T^{(e)}(\lambda) + \int_0^\varepsilon T^{(e)}[\lambda(\varepsilon-\tau)]\frac{\partial G(\tau)}{\partial \tau}\, d\tau. \tag{10}$$

But, as τ increases from 0 to ε, the integrand never changes sign, hence

$$T(\varepsilon) = T^{(e)}(\lambda)\left[1 + \varepsilon\frac{\partial G}{\partial \tau}(c)\right], \tag{11}$$

where $0 \leq c \leq \varepsilon$. Since $\partial G/\partial \tau$ is finite, the second term tends to 0 with ε. Therefore, if ε is so small that $\varepsilon|\partial G/\partial \tau| \ll 1$, then

$$T^{(e)}(\lambda) \doteq T(\varepsilon). \tag{12}$$

7.6.2 The Reduced Relaxation Function $G(t)$

It is customary to analyze the relaxation function into the sum of exponential functions and identify each exponent with the rate constant of a relaxation mechanism; thus

$$G(t) = \frac{\sum C_i e^{-v_i t}}{\sum C_i}. \tag{13}$$

Two important points are often missed:

(1) If an experiment is cut off prematurely, one may mistakenly arrive at an erroneous limiting value $G(\infty)$, which corresponds to $v_0 = 0$ in Eq. (13).
(2) The exponents v_i should not be interpreted literally without realizing that the representation of empirical data by a sum of exponentials is a non-unique process in practice.

As an example of the first point, the relaxation curves of Buchthal and Kaiser (1951) terminate at 100 ms. However, the data in Fig. 7.5:4 show that relaxation goes on beyond 1000 min! Examples of the second kind lead to the observation that a measured characteristic time of a relaxation experiment often turns out to be the length of the experiment. Lanczos (1956, p. 276) gives an example in which a certain set of 24 decay observations was analyzed and found that it could be fit equally well by three different expressions for x between 0 and 1:

$$f(x) = 2.202e^{-4.45x} + 0.305e^{-1.58x},$$
$$f(x) = 0.0951e^{-x} + 0.8607e^{-3x} + 1.5576e^{-5x}, \tag{14}$$
$$f(x) = 0.041e^{-0.5x} + 0.79e^{-2.73x} + 1.68e^{-4.96x}.$$

Lanczos comments, "It would be idle to hope that some other modified mathematical procedure could give better results, since the difficulty lies not with the manner of evaluation but with the extraordinary sensitivity of the exponents and amplitudes to very small changes of the data, which no amount of least-square or other form of statistics could remedy."

Realizing these difficulties, we conclude first that for a living tissue, a viscoelasticity law based on the fully relaxed elastic response, $G(t)$ as $t \to \infty$, is unreliable. In fact, a formulation based on $G(\infty)$ may run into a true difficulty because often it seems that $G(t) \to 0$ when $t \to \infty$. Secondly, one should look into other experiments, such as creep, hysteresis, and oscillation, in order to determine the relaxation function.

7.6.3 A Special Characteristic of Hysteresis of Living Tissues

The hysteresis curves of most biological soft tissues have a salient feature: the hysteresis loop is almost independent of the strain rate within several decades of the rate variation. This insensitivity is incompatible with any viscoelastic model that consists of a finite number of springs and dashpots. Such a model will have discrete relaxation rate constants. If the specimen is strained at a variable rate, the hysteresis loop will reach a maximum at a rate corresponding to a relaxation constant. A discrete model, therefore, corresponds to a discrete hysteresis spectrum, in opposition to the feature of living tissues that we have just described. This suggests at once that one should consider a continuous distribution of the exponents v_i: thus passing from a discrete spectrum α_i associated with v_i to a continuous spectrum $\alpha(v)$ associated with a continuous variable v between 0 and ∞.

7.6.4 $G(t)$ Related to Hysteresis

Let us consider a *standard linear solid* (Kelvin model), for which the governing differential equation is [see Chapter 2, Sec. 2.11, Eq. (2.11:7)]

$$T + \tau_\varepsilon \dot{T} = E_R[T^{(e)} + \tau_\sigma \dot{T}^{(e)}], \tag{15}$$

where τ_ε, τ_σ, E_R are constants. Equation (15) is subjected to the initial condition

$$\tau_\varepsilon T(0) = E_R \tau_\sigma T^{(e)}(0). \tag{16}$$

However, our special definition of the reduced relaxation function requires that

$$T^{(e)}(0) = T(0), \qquad G(0) = 1, \tag{17}$$

hence

$$E_R = \frac{\tau_\varepsilon}{\tau_\sigma}. \tag{18}$$

By integrating Eq. (15) with the initial condition specified above, we obtain the response $T^{(e)}$ to a unit step increase in stress $T(t) = \mathbf{1}(t)$: the creep function

$$J(t) = \frac{1}{E_R}\left[1 - \left(1 - \frac{\tau_\varepsilon}{\tau_\sigma}\right)e^{-t/\tau_\sigma}\right]\mathbf{1}(t). \tag{19}$$

Conversely, the relaxation function is obtained by integrating the differential equation for stress T for a unit step increase in the elastic response:

$$G(t) = E_R\left[1 + \left(\frac{\tau_\sigma}{\tau_\varepsilon} - 1\right)e^{-t/\tau_\varepsilon}\right]\mathbf{1}(t). \tag{20}$$

With these expressions the meaning of the constants τ_σ, τ_ε, E_R are clear. τ_σ

is the time constant for creep at constant stress, τ_ε is the time constant for relaxation at constant strain, and E_R is the "residual," the fraction of the elastic response that is left in the specimen after a long relaxation, $t \to \infty$. These physical constants are related through Eq. (18).

If a sinusoidal oscillation is considered, so that in Eq. (3) we substitute

$$T(t) = T_0 e^{i\omega t}, \qquad T^{(e)}(t) = T_0^{(e)} e^{i\omega t}, \tag{21}$$

then, on changing $t - \tau$ to ξ and ξ to t, we obtain

$$T_0/T_0^{(e)} = i\omega \int_0^\infty G(t) e^{-i\omega t}\, dt = \mathcal{M} = |\mathcal{M}| e^{i\delta}, \tag{22}$$

where $\mathcal{M}$ is the complex modulus. By using Eq. (20), or, more simply, Eq. (15), we obtain

$$\mathcal{M} = \frac{1 + i\omega\tau_\sigma}{1 + i\omega\tau_\varepsilon} E_R, \qquad |\mathcal{M}| = \left(\frac{1 + \omega^2\tau_\sigma^2}{1 + \omega^2\tau_\varepsilon^2}\right)^{1/2} E_R. \tag{23}$$

δ is the phase shift, and $\tan \delta$ is a measure of "internal damping":

$$\tan \delta = \frac{\omega(\tau_\sigma - \tau_\varepsilon)}{1 + \omega^2(\tau_\sigma \tau_\varepsilon)}. \tag{24}$$

When the modulus $|\mathcal{M}|$ and the internal damping $\tan \delta$ are plotted against the logarithm of ω, curves as shown in Fig. 7.6:1 are obtained. The damping reaches a peak when the frequency ω is equal to $1/\sqrt{\tau_\sigma \tau_\varepsilon}$. Correspondingly, the elastic modulus $|\mathcal{M}|$ has the fastest rise for frequencies in the neighborhood of $1/\sqrt{\tau_\sigma \tau_\varepsilon}$. If the internal friction (hysteresis loop) is insensitive to

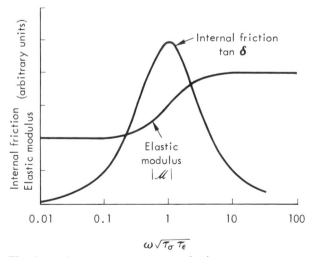

Figure 7.6:1 The dynamic modulus of elasticity $|\mathcal{M}|$ and the internal damping $\tan \delta$ plotted as a function of logarithm of frequency ω, for a standard linear solid. From Fung (1972).

frequency ω, we must spread out the peak. This can be done by superposing a larger number of Kelvin models. This is the basic reason for introducing a continuous spectrum of relaxation time into our problem.

7.6.5 Continuous Spectrum of Relaxation

To implement the idea of continuous spectrum, let us first rewrite the relaxation function in a different notation. Let

$$S = \frac{\tau_\sigma}{\tau_\varepsilon} - 1, \qquad E_R = \frac{1}{1 + S}. \tag{25}$$

Then substituting into Eqs. (20) and (23), we obtain

$$G(t) = \frac{1}{1 + S}[1 + Se^{-t/\tau_\varepsilon}], \tag{26}$$

$$\mathcal{M} = \frac{1}{1 + S}\left(1 + S\frac{\omega\tau_\varepsilon}{\omega\tau_\varepsilon + \frac{1}{\omega\tau_\varepsilon}} + iS\frac{1}{\omega\tau_\varepsilon + \frac{1}{\omega\tau_\varepsilon}}\right). \tag{27}$$

Now, let τ_ε be replaced by a continuous variable τ, and let $S(\tau)$ be a function of τ. For a system with a continuous spectrum, we replace S by $S(\tau)\,d\tau$ in Eqs. (26), (27), and (19) and integrate with respect to τ to obtain the following generalized reduced relaxation function and complex modulus:

$$G(t) = \left[1 + \int_0^\infty S(\tau)e^{-t/\tau}\,d\tau\right]\left[1 + \int_0^\infty S(\tau)\,d\tau\right]^{-1}, \tag{28}$$

$$\mathcal{M}(\omega) = \left[1 + \int_0^\infty S(\tau)\frac{\omega\tau\,d\tau}{\omega\tau + \frac{1}{\omega\tau}} + \int_0^\infty iS(\tau)\frac{d\tau}{\omega\tau + \frac{1}{\omega\tau}}\right]\left[1 + \int_0^\infty S(\tau)\,d\tau\right]^{-1}. \tag{29}$$

A normalization factor is added to each of these formulas to meet the definition that $G(t) \to 1$, $J(t) \to 1$ when $t \to 0$.

Our task is to find the function $S(\tau)$ that will make $G(t)$, $J(t)$, and $\mathcal{M}(\omega)$ match the experimental results. In particular, we want $\mathcal{M}(\omega)$ to be nearly constant for a wide range of frequency.

A specific proposal is to consider

$$S(\tau) = \frac{c}{\tau} \qquad \text{for} \quad \tau_1 \leq \tau \leq \tau_2, \tag{30}$$

$$= 0 \qquad \text{for} \quad \tau < \tau_1, \tau > \tau_2, \tag{31}$$

where c is a dimensionless constant. Then

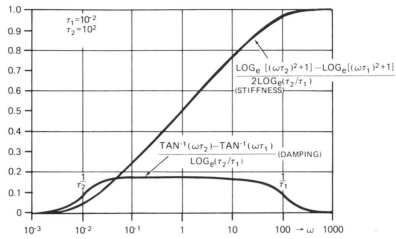

Figure 7.6:2 The stiffness (real part of the complex modulus $\mathcal{M}$) and the damping plotted as functions of the logarithm of the frequency ω; corresponding to a continuous relaxation spectrum $S(\tau) = c/\tau$ for $\tau_1 \leq \tau \leq \tau_2$ and zero elsewhere. $\tau_1 = 10^{-2}$, $\tau_2 = 10^2$. From Neubert (1963).

$$\mathcal{M}(\omega) = \left\{ 1 + \int_{\tau_1}^{\tau_2} \left[c \frac{\omega\tau}{1 + (\omega\tau)^2} + \frac{ic}{1 + (\omega\tau)^2} \right] d(\omega\tau) \right\} \left\{ 1 + \int_{\tau_1}^{\tau_2} \frac{c}{\tau} d\tau \right\}^{-1}$$

$$= \left\{ 1 + \frac{c}{2} \left[\ln(1 + \omega^2\tau_2^2) - \ln(1 + \omega^2\tau_1^2) \right] \right.$$

$$\left. + ic[\tan^{-1}(\omega\tau_2) - \tan^{-1}(\omega\tau_1)] \right\} \left\{ 1 + c \ln \frac{\tau_2}{\tau_1} \right\}^{-1}. \qquad (32)$$

This results in a rather constant damping for $\tau_1 \leq 1/\omega \leq \tau_2$, as can be seen from Fig. 7.6:2 for an example in which $\tau_1 = 10^{-2}$, $\tau_2 = 10^2$. The variable part of the stiffness [the real part of $\mathcal{M}(\omega)$] is also plotted in Fig. 7.6:2. The maximum damping occurs when the frequency is $\omega_n = 1/\sqrt{(\tau_1\tau_2)}$. The stiffness rises the fastest in the neighborhood of ω_n. At ω_n, the maximum damping is proportional to

$$\tan^{-1}\left[\frac{1}{2} \left(\sqrt{\frac{\tau_2}{\tau_1}} - \sqrt{\frac{\tau_1}{\tau_2}} \right) \right], \qquad (33)$$

which varies rather slowly with τ_2/τ_1 ratio, as can be seen from Fig. 7.6:3, in which the above quantity divided by $\ln(\tau_2/\tau_1)$ is plotted.

The corresponding reduced relaxation function can be evaluated in terms of the exponential integral function which is tabulated. Figure 7.6:4 shows an example in which $G(t)$ is plotted as a function of time. The portion in the middle, $\tau_1 \ll t \ll \tau_2$, is nearly a straight line; therefore the stresses decreases

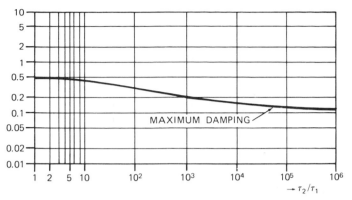

Figure 7.6:3 The maximum damping as a function of the ratio τ_2/τ_1 for a solid corresponding to a continuous relaxation spectrum. $S(\tau) = c/\tau$ for $\tau_1 \leq \tau \leq \tau_2$, and zero elsewhere. From Neubert (1963).

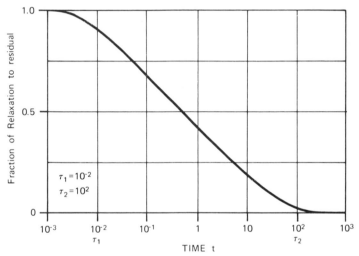

Figure 7.6:4 The reduced relaxation function $G(t)$ of a solid with a continuous relaxation spectrum $S(\tau) = c/\tau$ for $\tau_1 \leq \tau \leq \tau_2$, and zero elsewhere. From Neubert (1963).

with $\ln t$ in that segment. This specific spectrum, therefore, gives us the features desired. We are left with three parameters, c, τ_1, τ_2, to adjust with respect to the experimental data.

On substituting Eq. (30) into Eq. (28), and evaluating the integrals, we obtain the reduced relaxation function:

$$G(t) = \left\{1 + c\left[E_1\left(\frac{t}{\tau_2}\right) - E_1\left(\frac{t}{\tau_1}\right)\right]\right\}\left[1 + c\ln\left(\frac{\tau_2}{\tau_1}\right)\right]^{-1}, \qquad (34)$$

where $E_1(z)$ is the exponential integral function defined by the equation

$$E_1(z) = \int_z^\infty \frac{e^{-t}}{t}\, dt \qquad (|\arg z| < \pi) \tag{35}$$

and tabulated in Abramowitz and Stegun (1964). For $t \to \infty$, $E_1(t/\tau_2)$ and $E_1(t/\tau_1) \to 0$, and

$$G(\infty) = \left[1 + c\ln\left(\frac{\tau_2}{\tau_1}\right)\right]^{-1}. \tag{36}$$

For values of t of the order of 1 sec,

$$G(t) = [1 - c\gamma - c\ln(t/\tau_2)][1 + c\ln(\tau_2/\tau_1)]^{-1}, \tag{37}$$

where γ is the Euler's constant. In an interval within (τ_1, τ_2), the slopes of the relaxation and creep curves vs. the logarithm of time are

$$\frac{dG}{d(\ln t)} = \frac{c}{1 + c\ln(\tau_2/\tau_1)}, \tag{38}$$

To obtain the reduced creep function $J(t)$ corresponding to the relaxation function given by Eq. (34), Dortmans et al. (1987) used the method of Laplace transformation. The Laplace transform of a function $f(t)$ is given by

$$\bar{f}(s) = \int_0^\infty f(t)e^{-st}\, dt \tag{39}$$

Multiplying Eqs. (4) and (5) by $e^{-st}\, dt$ and integrating from 0 to ∞, one obtains

$$\bar{T}^{(e)}(s) = \frac{1}{s\bar{G}(s)}\, T(s) = s\bar{J}(s)\, T(s) \tag{40}$$

Hence

$$\bar{J}(s)\bar{G}(s) = \frac{1}{s^2} \tag{41}$$

which shows that the reduced creep function is related to the reduced relaxation function as the inverse of Eq. (41):

$$\int_0^t J(t - \tau)G(\tau)\, d\tau = t \tag{42}$$

From Eq. (34), we have

$$\bar{G}(s) = G(\infty)s^{-1}\{1 + c\ln[(1 + s\tau_2)/(1 + s\tau_1)]\} \tag{43}$$

Hence, from Eq. (41),

$$\bar{J}(s) = J(\infty)s^{-1}\{1 + c\ln[(1 + s\tau_2)/(1 + s\tau_1)]\}^{-1} \tag{44}$$

where

$$J(\infty) = 1/G(\infty) = 1 + c\ln(\tau_2/\tau_1) \tag{45}$$

The creep function $J(t)$ can be obtained from Eq. (44) by taking the inverse Laplace transform. In this way Dortmans et al (1987) found

$$J(t) = J(\infty)\left\{1 + \frac{(1 + s_0\tau_2)(1 + s_0\tau_1)}{cs_0(\tau_2 - \tau_1)}e^{s_0t}\right.$$

$$\left. + c\int_{1/\tau_1}^{1/\tau_2} \frac{1}{(c\pi)^2 + [1 + c\ln(x\tau_2 - 1)/(1 - x\tau_1)]^2}\frac{e^{-xt}}{x}dx\right\} \quad (46)$$

In practical applications of these formulas to living tissues. I have the experience that the relaxation spectrum given by Eqs. (30), (31) and the associated the relaxation function given by Eq. (34), the complex modulus given by Eq. (32), and the damping given by Eq. (33) work very well in virtually all cases we know, but the creep function given by Eqs. (45) and (46) does not work so well (they are quite good for papillary muscle, but not for blood vessels and lung tissue). I have a feeling that creep is fundamentally more nonlinear, and perhaps does not obey the quasi-linear hypothesis. The microstructural process taking place in a material undergoing creep could be quite different from that undergoing relaxation or oscillation. Analogous situation is known for metals at higher temperature.

7.6.6 A Graphic Summary

The models presented above are summarized in Fig. 7.6:5. In the top row are shown the well-known linear spring-damper models of Maxwell, Voigt, and Kelvin (Sec. 2.11). In the second row are shown the hysteresis-log frequency relationships of the three models directly above. The significant variation of hysteresis with frequency is seen in each of these models. In the third row is our model which is a generalization of Kelvin's model in three ways: First, it is a combination of a large number of Kelvin's units in series. Second, the elastic springs in every component of this model are nonlinear. The tension in each spring is a nonlinear function of the stretch ratio, with a constitutive equation of the type described in Sec. 7.5. All springs have the same type of nonlinearity; but are of different sizes. The dampers are unusual: they are linear with respect to the tension in the springs. Third, the sizes of the springs and dampers are varied in such a way that the characteristic frequencies of the successive Kelvin units form an almost continuous spectrum, and the characteristic peak hysteresis of all the units are approximately the same. The soft-tissue has a curve of hysteresis vs. log frequency as shown in the bottom row, which is pretty flat over a wide range of frequency. The flat curve is the sum of an infinite number of bell-shaped curves. Such a model of soft tissue has a continuous relaxation spectrum.

7.6.7 Historical Remarks

Although we found Eq. (30) for biological tissues according to the reasoning presented above (Fung, 1972), we found later that this spectrum has had a

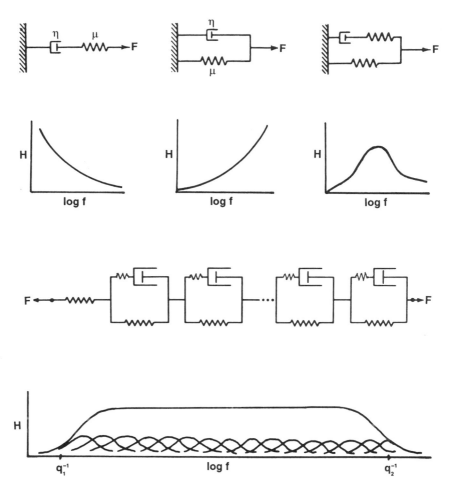

Figure 7.6:5 A summary of the principal features of viscoelastic models. Three standard viscoelastic models, namely, the Maxwell, Volgt and Kelvin models are shown
in the top row, and a mathematical model of the viscoelasticity of biological soft tissues
is shown in the third row. Figures in the second row show the relationships between
the hysteresis (H) and the logarithm of frequency (ln f) of the three models immediately
above. The figure in the bottom row shows the general hystersis-log frequency relationship of most living soft tissues, corresponding to the model shown in the third row.
For the soft tissue model the springs are nonlinear, and each Kelvin unit contributes
a small bell-shaped curve, the sum of which is flat over a wide range of frequencies.

long history. Hysteresis insensitive to frequency was described by Becker and Föppl (1928) in their study of electromagnetism in metals [see Becker and Döring (1939)]. Wagner (1913) described it in dielectricity. Theodorsen and Garrick (1940) introduced it into the theory of airplane flutter; it is now called "structural" damping by engineers, Knopoff (1965) showed that the earth's crust has an internal friction which is independent of the frequency. Routbart and Sack (1966) showed that the internal friction for nonmagnetic materials is either constant or decreases slightly with frequency in the range 1–40 kHz. Mason (1969) attributed the reason for this to the existence of a kind of "kink" in the dislocation line and the associated kink energy barrier. Bodner (1968) formulated a special plasticity theory to account for the phenomena.

The concept of a continuous relaxation spectrum was considered by Wagner and Becker. Wagner (1913) investigated the function

$$S(\tau)\,d\tau = \frac{kb}{\sqrt{\pi}}e^{-b^2z^2}\,dz \qquad \text{where} \qquad z = \log\left(\frac{\tau}{t_0}\right), \tag{47}$$

and k, b, t_0 are constants. Becker and Foppl (1928) introduced the spectrum given in Eq. (30). Neubert (1963) developed the Becker theory thoroughly. Guth et al. (1946) have shown that the viscoelasticity of rubber follows Eq. (30) also.

7.6.8 Oscillatory Stretch

If the elastic response $T^{(e)}$ is assumed to oscillate harmonically, the corresponding oscillation of the stretch ratio is anharmonic. Figure 7.6:6(a) shows the time course of $T^{(e)}$ and λ when $T^{(e)}$ oscillates harmonically. If λ oscillates sinusoidally, then $T^{(e)}$ oscillates anharmonically as shown in Fig. 7.6:6(b). If the amplitude of oscillation is small, an approximate linear relationship holds, and both stress and strain will oscillate harmonically. Patel et al. (1970) show that linear viscoelasticity applies to arteries as long as the superimposed sinusoidal strain remains below 4% (on top of $\lambda - 1.6$), and that their data were reproducible up to 16 h following removal of tissue. Many other workers reached similar conclusions.

7.6.9 An Example of Another Approach: Collagen Fibers in Uniaxial Extension

The experimental results shown in Figs. 7.3:7 and 7.3:9 in Sec. 7.3 can be described by several regimes: (1) the small strain "toe" region in which the Lagrangian stress is a nonlinear function of the stretch ratio, λ; (2) the almost linear regime in which the Lagrangian stress increases linearly with increasing

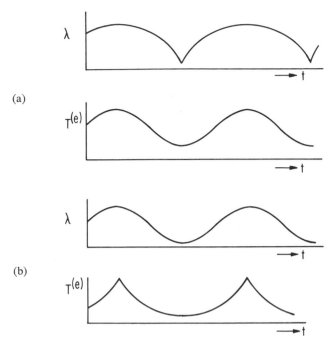

(a)

(b)

Figure 7.6:6 The quasi-linear stress–strain-history relationship proposed in Eq. (9).
(a) The time course of the extension λ when the elastic response $T^{(e)}$ oscillates
sinusoidally. (b) The time course of the elastic response $T^{(e)}$ when the extension λ varies
sinusoidally. From Fung (1972).

λ; and (3) the nonphysiological, overly extended, and failing regime at larger
stretch.

The stress–strain relationship of collagen in the "toe" region can be described by an exponential expression like Eq. (4) of Sec. 7.5:

$$T = C(e^{\alpha\lambda} - e^{\alpha}) \qquad \text{for } 1 < \lambda \le \lambda_0. \tag{49}$$

A Hooke's law, or a neo-Hookean law of finite deformation may be used for
the "linear" region. Johnson et al. (1992), however, used the following formula
according to Wineman's (1972) representation of a Mooney–Rivlin material
for both the toe and linear regions:

$$T = C_0\left(1 + \mu\frac{1}{\lambda}\right)\left(\lambda^2 - \frac{1}{\lambda}\right)\frac{1}{\lambda}. \tag{50}$$

Here T is the Lagrange stress, λ is the stretch ratio, C_0 and μ are constants.
The constant μ is finite in the toe regime where λ lies between 1 and λ_0; whereas
$\mu = 0$ in the *linear* regime, corresponding to a neo-Hookean material.

Mooney–Rivlin material is *isotropic*. Collagen fibers are intrinsically *transversely orthotropic*. Let the x_3 axis be the axis of the fiber, then Green and
Adkins (1960) have shown that the strain energy function of a transversely

orthotropic material must be a polynomial of the strain invariants I_1, I_2, I_3, and the Green's strains E_{ij} in the form

$$W = W(I_1, I_2, I_3, E_{33}, E_{33}^2 + E_{31}^2 + E_{32}^3). \tag{51}$$

This fact can be important in generalizing the uniaxial formula to a three-dimensional tensorial equation which will be needed in treating composite materials of which collagen is a part.

Collagen is viscoelastic. If the stress T given by Eq. (50) is considered to be the *elastic response* $T^{(e)}$ of Sec. 7.6, then, according to the *quasi-linear theory* of viscoelasticity presented in Sec. 7.6, the stress at time t due to a strain history $\lambda(s)$ is

$$T(t) = \int_{-\infty}^{t} G(t - s) \frac{\partial T^{(e)}[\lambda(s)]}{\partial \lambda(s)} \frac{\partial \lambda(s)}{\partial s} ds, \tag{52}$$

where $G(t - s)$ is the *reduced relaxation function* defined in Sec. 7.6, with $G(0) = 1$. Johnson et al. (1992), starting with a theory of Pipkin and Rogers (1968), taking the first term of a series of n terms, and using a special choice of the form of the relaxation function made by Wineman (1972), obtained the same result as given by Eq. (52).

7.6.10 Limitations and Extensions

The quasi-linear constitutive equation presented above is, of course, only an abstraction. It has been found to work reasonably well for the skin, arteries, veins, tendons, ligaments, lung parenchyma, pericardium, and muscle and ureter in the relaxed state. Even for these soft tissues there are cetain ranges of stresses, strains, rates of strains, and frequencies of oscillations in which the formula does not represent a specific tissue accurately. The strain rate effect, especially, can be a problem. This effect is represented by a relaxation spectrum which is smooth and flat over a very wide range of frequencies in our theory. In reality, any specific tissue may have a spectrum with a number of localized peaks and valleys which is not taken into account in the quasi-linear formulaton. Finally, although it has been acknowledged that the relaxation function should depend on the invariants of the tensors of the stress, strain, and strain rate, no experimental identification of such a constitutive equation is known.

A general theory of the constitutive equation of nonlinear viscoelastic materials has been given by Green and Rivlin (1957) and Green, Rivlin and Spencer (1959) from the point of view of tensorial power series expansion. The nth term of the series is an n-tuple integral of the history of the strain tensor. Another theory is given by Pipkin and Rogers (1968) from the point of view of successive step changes of strain. The nth term of the series is the response to n previous steps of change. Wineman (1972) and others have studied the leading terms of the Pipkin and Rogers series from the point of view of successive approximation. So far, only Young, Vaishnav, and Patel (1977) have made an attempt to experimentally determine all the constants in a double integral approach to the canine aorta. All other attempts have used a single integral which is the first term in Pipkin and Rogers' formulation. Johnson et al. (1992) have shown that the single inegral approach of Wineman is equivalent to our method presented in Sec. 7.6.

7.7 Incremental Laws

We have shown that the mechanical properties of soft tissues such as arteries, muscle, skin, lung, ureter, mesentery, etc., are qualitatively similar. They are inelastic. They do not meet the definition of an elastic body, which requires that there be a single-valued relationship between stress and strain. These tissues show hysteresis when they are subjected to cyclic loading and unloading. When held at a constant strain, they show stress relaxation. When held at a constant stress, they show creep. They are anisotropic. Their stress–strain-history relationships are nonlinear. Arterial properties vary with the sites along the arterial tree, aging, short- or long-term effects of drugs, hypertension, and innervation or denervation. When all these factors are coupled, the problem of how to describe the mechanical properties of tissues in a simple and accurate mathematical form becomes quite acute.

A popular approach to nonlinear elasticity uses the incremental law: a linearized relationship between the incremental stresses and strains obtained by subjecting a material to a small perturbation about a condition of equilibrium. This approach was applied to the arteries in the 1950s. But the elastic constants so determined are meaningful only if the initial state from which the perturbations are applied is known, and are applicable only to that state. It turns out that these incremental moduli are strongly dependent on the initial state of stress (and, for some tissues, on strain history). Usually the incremental law is derived for a selected meaningful equilibrium state, such as the state of uniform expansion in all directions, or a well defined physiological state.

An example is shown in Fig. 7.7:1, in which the little loops inside the large loops are stress–strain relations obtained when a dog's mesentery was subjected to small cyclic variations of strain.

An important feature of the incremental stress–strain relationship is revealed in Fig. 7.7:1. The small loops are not parallel to each other. Neither are they tangent to the loading or unloading curves of the large loop. Thus the incremental law does vary with the level of stress, and is *not equal* to the tangent of the loading or unloading curve of finite strain. This important observation must be borne in mind when one considers the relationship between the incremental law and the stress–strain relation in finite deformation. There are many authors who derive a relationship between stress and strain in finite strain (usually in a loading process) and carelessly identify its

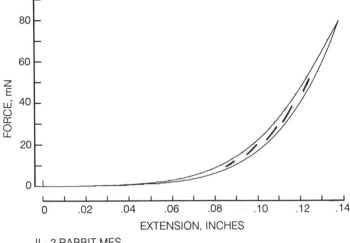

II - 2 RABBIT MES.
L_0 0.58" L_{ph} 1.34"
STRAIN RATE ± 0.1 in./min.

Figure 7.7:1 Typical loading–unloading curves of the rabbit mesentery tested in uniaxial tension at 25°. The large loops were obtained when the specimen was stretched and unloaded between $\lambda = 1$ and $\lambda = 1.24$. The small loops were obtained by first stretching the specimen to the desired strain, then performing loading and unloading loops of small amplitude. Preconditioning was done in every case. From Fung (1967).

derivative with the modulus of the incremental law. One should expect that in general they would not agree. The incremental moduli should be determined by incremental experiments.

Successful incremental laws have been developed by Wilson (1972) and Lai-Fook et al. (1976, 1977) for lung tissue in small perturbation from a state of uniform expansion, and by Bergel (1961), Patel and Vaishnav (1972), and others for large blood vessels.

Biot (1965) has shown how the incremental law can be used step by step to deal with a type of finite deformation in which the strains are small but the rotations are large. For biological applications in which the strains are finite the use of incremental laws are not really convenient because one has to change the constants as the deformation progresses. The pseudo-elasticity laws discussed in the next section is often simpler for the full range of deformation.

7.8 The Concept of Pseudo-Elasticity

We now introduce a drastic simplication to reduce the quasi-linear viscoelastic constitutive equation of a living tissue to a *pseudo-elastic* constitutive equation. This simplification is possible only if a tissue can be *preconditioned*. To test a tissue in any specific procedure of loading and unloading, it is necessary to perform the loading cycle a number of times before the stress–strain relationship is repeatable. This process is called *preconditioning*. Confining our attention to preconditioned tissues subjected to cyclic loading and unloading at constant strain rates, we see that by the definition of preconditioning, the stress–strain relationship is well defined, repeatable, and predictable. For the *loading branch* and the *unloading branch* separately, the stress–strain relationship is unique.

Since stress and strain are uniquely related in each branch of a specific cylic process, *we can treat the material as one elastic material in loading, and another elastic material in unloading.* Thus we can borrow the method of the theory of elasticity to handle an inelastic material. To remind us that we are really dealing with an inelastic material, we call it *pseudo-elasticity.*

Pseudo-elasticity is, therefore, not an intrinsic property of the material. It is a convenient description of the stress–strain relationship in specific cyclic loading. By its use the very complex property of the tissue is more simply described. Fortunately, the usefulness of the concept of pseudo-elasticity to biomechanics is greatly enhanced by the fact that the hysteresis of living tissues is rather insensitive to the strain rate. By virtue of rate insensitivity, pseudo-elasticity acquires a certain measure of independence.

The strain-rate insensitivity has been demonstrated in Sec. 7.5; e.g., in Fig. 7.5:2. A large amount of data exists on various soft tissues which shows that for a 1000-fold change in strain rate that can be easily achieved in most laboratories, the change in stress for a given strain usually does not exceed a factor of 1 or 2. The extreme situation is encountered in ultrasound experiments. It is known that with few exceptions the energy dissipation does not change more than a factor of 2 or 3 when the frequency changes from 1000 Hz to 10^7 Hz. Dunn et al. (1969) reviewed the subject critically, and showed that for most tissues the attenuation per cycle is virtually independent of frequency (see their curves of α/f vs. f, where f is the frequency, α is the attenuation per unit distance, and α/f is the attenuation per wave). Although stress fluctuation is very small in the ultrasound experiments, as compared with the stress variation in the arterial wall due to pulsatile blood flow, yet it is interesting to observe this uniform behavior in a frequency range from nearly 0 to millions of Hz. Translated into the language of the relaxation spectrum presented in Eq. (30) of Sec. 7.6, the ultrasound experiments suggest that the lower limit of relaxation time, τ_1, is very small for most living tissues; perhaps in the range of 10^{-8} sec.

Joint ligaments' insensitivity to strain rate in the 0.1 to 1.0 m/sec range is shown by Mabuchi et al. (1991) for canine stifle joint. At lower ranges of strain rate, Haut and Little (1969) and Noyes et al. (1974) reported increasing stiffness with increasing strain rate.

The use of pseudo–elasticity for soft tissues is very convenient, but, of course, it is only an approximation. A detailed account of the frequency effect is not simple. All who have looked for the effect of strain rate on the stress–strain relationship of living tissues found it. McElhaney (1966), in his study of the dynamic response of muscles, shows that there is a maximum of about 2.5-fold increase in stress at any given strain when the strain rate is increased from 0.001 to 1000 per second, an increase of 10^6-fold. Van Brocklin and Ellis (1965) state that in tendons there is no strain-rate effect when the rate is small, but the effect becomes significant when the rate is high. Collins and Hu (1972), using explosive methods (as in a shock tube) to impose a high strain rate ($\dot{\varepsilon}$) onto human aortic tissue (a study relevant to the safety against automobile crash problem), obtained the result

$$\sigma = (0.28 + 0.18\dot{\varepsilon})(e^{12\varepsilon} - 1) \quad \text{for} \quad \dot{\varepsilon} < 3.5 \text{ sec}^{-1}. \tag{1}$$

Bauer and Pasch (1971), working with rat tail artery, found that the dynamic Young's modulus is independent of frequency from 0.01 to 10 Hz, but the loss coefficient does vary with frequency. All these do not give a uniform picture, but our suggestion of pseudo-elasticity representation does offer a great simplification.

The mathematical formulation presented in Sec. 7.6 is consistent with the concept of pseudo-elasticity. We introduced the concept of "elastic response" $T^{(e)}$ and formulated a quasi-linear viscoelasticity law. By the introduction of a broad continuous spectrum of relaxation time into the viscoelasticity law we made the hysteresis nearly constant over a wide range of frequencies. In this way we unified the phenomena of the nonlinear stress-strain relationship, relaxation, creep, hysteresis, and pseudo-elasticity into a consistent theory. The utility of the pseudo-elasticity concept, however, lies in the possibility to describe the stress–strain relationship in loading or unloading by a law of elasticity, instead of viscoelasticity. Further simplification is obtained if one assume that a strain energy function exists. We shall discuss this in Sec. 7.11, for which the following sections are preparatory.

7.9 Biaxial Loading Experiments on Soft Tissues

The uniaxial loading experiments discussed so far cannot provide the full relationship between all stress and strain components. To obtain the tensorial relationship, it is necessary to perform biaxial and triaxial loading tests.

Figure 7.9:1 shows a possible way of testing a rectangular specimen of uniform thickness in biaxial loading. Figure 7.8:2 shows the schematic view of the equipment developed in the author's laboratory. The specimen floats in physiological saline solution contained in an open-to-atmosphere upper compartment of a double compartment tray. The lower compartment is a part of a thermoregulation system. Water of specified temperature is supplied by a temperature regulator. The solution is slowly circulated, compensated for evaporation from time to time, and maintained at pH 7.4 by bicarbonate buffer.

The specimen is hooked along its four edges by means of small staples. Each hook is connected by means of a silk thread to a screw on one of the four force-distributing platforms (see Fig. 7.8:1). This setup allows individual

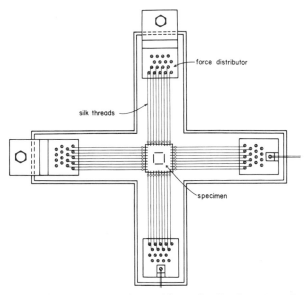

Figure 7.9:1 Setup of specimen hooking and force distribution. From Lanir and Fung (1974a).

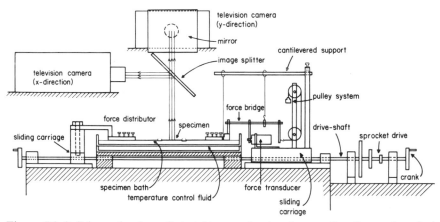

Figure 7.9:2 Schematic view of stretching mechanism in one direction and optical system. From Lanir and Fung (1974a).

adjustment of the tension of each thread. The force-distributing platforms are bridged over the edges of the saline tray in an identical manner for both directions (Fig. 7.9:2). One platform is rigidly mounted to the carriage of a sliding mechanism, the opposite platform is horizontally attached to a force transducer. Both the support and transducer are rigidly connected to the carriage of another sliding unit. A pulley system on this carriage allows the force distributor to be pulled by a constant weight on top of the force exerted by the transducer. The carriages of the opposite sliding units are displaced

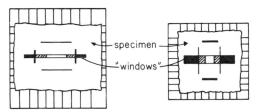

Figure 7.9:3 Typical display of the specimen on the two VDA monitors. From Lanir and Fung (1974a).

by means of an interconnected threaded drive-shaft. The threads of the left and right sliding units are pulled by the drive shaft at an equal rate in opposite directions, so that the specimen can be stretched or contracted without changing its location.

The loading strings are approximately parallel when the equipment is in operation. For a specimen varying in size from 3×3 to 6×6 cm the maximum deviation of the loading strings from the centerline is less than 0.05 rad. Thus at the very corners of the rectangular specimen a shear stress of the order of $\frac{1}{20}$th of the normal stress may exist at the maximum stretch. However, the distance between the "bench marks" (the black rectangular marks at the center of the specimen in Fig. 7.9:1) whose dimensional changes are measured is approximately half of the overall dimensions; therefore the maximum shear stress acting at the corners of the bench marks is expected to be no more than 2% or 3% of the normal stress. These strings are "tuned" (in the manner of piano tuning by turning a set of screws) at the beginning (during the preconditioning process) in such a way that the rectangular benchmarks made on a relaxed specimen remain rectangular upon loading, without visible distortion.*

The dimensions of the specimen between the bench marks are continuously measured and monitored by a video dimension analyzer (VDA). This system consists of a television camera, a video processor, and a television monitor. The video processor provides an analog signal proportional to the horizontal distance between independently selectable levels of optical density, in two independent areas (windows) in the televized scene (Fig. 7.9:3). The video processor utilizes the vertical and horizontal synchronization pulses to define the X-Y coordinates of the windows as well as their height and width, and measures the distance between the markers.

The strain in the specimen can be measured in two perpendicular directions.

The stretching mechanism can stretch the tissue in two directions either independently or coordinated according to a prescribed program, see Lanir

---

* Initially, the bench marks are painted with an indelible ink. After tuning, we often make another rectangular mark inside the painted one by means of four very slender L-shaped wires of stainless steel. One leg of the L is impaled into the tissue; the other leg rests on top of the specimen. The four legs resting on top makes a rectangle which is excellent for video monitoring. See Debes and Fung (1992).

and Fung (1974a). Applications to the testing of skin are presented in Lanir and Fung (1974b). Applications to the testing of lung tissue, with an additional method for thickness measurement, are presented in Vawter, Fung, and West (1978), Zeng et al. (1987), Yager et al. (1992), Debes and Fung (1992), and Debes (1992). A more recent model uses a digital computer to control the stretching in the two directions, and record and analyze the data. Similar equipment used to perform biaxial loading tests on arteries is described by Fronek et al. (1976), a sketch of which is given in Fig. 2.14:5.

The biaxial-loading testing equipment has to be much more elaborate than the uniaxial one because of the need to control boundary conditions. The edges must be allowed to expand freely. In the target region, the stress and strain states should be uniform so that data analysis can be done simply. The method described above is aimed at this objective. The target region is small and away from the outer edges in order to avoid the disturbing influence of the loading device. Strain is measured optically to avoid mechanical disturbance.

Triaxial loading experiments on cuboidal specimens of lung tissue using hooks and strings for loading have been done by Hoppin et al. (1975). But there is no way to observe a smaller region inside the cuboidal specimen in order to avoid the large edge effects.

7.9.1 Whole Organ Experiments

An alternative to testing excised tissues of simple geometry is to test whole organs. For the lung, one observes a whole lobe *in vivo* or *in vitro*. For the artery, one measures its deformation when internal or external pressures are changed, or when longitudinal tension is imposed. In whole organ experiments, the tissues are not subjected to traumatic excision, and one can measure their mechanical properties in conditions closer to *in vivo* conditions. But usually there are difficulties in analyzing whole organ data, either because of difficulties in knowing the stress and strain distributions, or because some needed pieces of data cannot be measured. On the other hand, no experiment on an excised tissue can be considered complete without an evaluation of the effect of the trauma of dissection. Therefore, whole organ experiments and excised tissue experiments complement each other.

7.10 Description of Three-Dimensional Stress and Strain States

To avoid getting into overly complicated situations, let us consider a rectangular plate of uniform thickness and of orthotropic material as shown in Fig. 7.10:1. Two pairs of forces F_{11}, F_{22}, act on the edges of the plate. No shear stress acts on these edges; hence we say that the coordinate axes x, y are the *principal axes*. Let the original size of the rectangle at the zero stress state

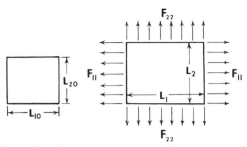

Figure 7.10:1 Deformation of a rectangular membrane. The membrane is subjected to tensile forces in the x and y directions. There is no shear force acting on the edges. The directions x, y are, therefore, the principal directions. From the forces and the deformation, various stresses and strains are defined in Eqs. (1) through (5).

be L_{10}, L_{20}. After the imposition of the forces the plate becomes bigger, and the edge lengths become L_1, L_2. Let the thickness of the original and the deformed plate be h_0 and h, respectively. Then we define the stresses

$$\sigma_{11} = \frac{F_{11}}{L_2 h}, \qquad \sigma_{22} = \frac{F_{22}}{L_1 h}, \qquad T_{11} = \frac{F_{11}}{L_{20} h_0}, \qquad T_{22} = \frac{F_{22}}{L_{10} h_0},$$

$$S_{11} = \frac{1}{\lambda_1} T_{11} = \frac{\rho_0}{\rho} \frac{1}{\lambda_1^2} \sigma_{11}, \qquad S_{22} = \frac{1}{\lambda_2} T_{22} = \frac{\rho_0}{\rho} \frac{1}{\lambda_2^2} \sigma_{22}, \tag{1}$$

where σ_{11}, σ_{22} are stresses defined in the sense of Cauchy and Euler. T_{11}, T_{22} are stresses defined in the sense of Lagrange and Piola, and S_{11}, S_{22} are stresses defined in the sense of Kirchhoff. ρ_0 and ρ are the densities of the material in the zero stress state and the deformed state, respectively. Clearly these three stresses are convertible to each other. All three are used frequently. Lagrangian stresses are the most convenient for the reduction of laboratory experimental data. Kirchhoff stresses are directly related to the strain energy function. Eulerian or Cauchy stresses are used in equations of equilibrium or motion. Thus in practice the little effort of remembering all three definitions can be well repaid.

For the description of deformation, the ratios

$$\lambda_1 = \frac{L_1}{L_{10}}, \qquad \lambda_2 = \frac{L_2}{L_{20}} \tag{2}$$

are defined as the principal stretch ratios. The strains

$$E_1 = \tfrac{1}{2}(\lambda_1^2 - 1), \qquad E_2 = \tfrac{1}{2}(\lambda_2^2 - 1) \tag{3}$$

are defined and used according to the method of Green and St. Venant, whereas

$$e_1 = \frac{1}{2}\left(1 - \frac{1}{\lambda_1^2}\right), \qquad e_2 = \frac{1}{2}\left(1 - \frac{1}{\lambda_2^2}\right) \tag{4}$$

are defined and used according to Almansi and Hamel. See Chapter 2, Sec. 2.3. The strain measures

$$\varepsilon_1 = \frac{L_1 - L_{10}}{L_{10}} = \lambda_1 - 1, \qquad \varepsilon_2 = \frac{L_2 - L_{20}}{L_{20}} = \lambda_2 - 1 \qquad (5)$$

are called infinitesimal strains. Again, use of any one of these strain measures is sufficient. But the literature is strewn with all of them, and it pays to know their differences. They are different numerically. For example, if $L_1 = 2$ and $L_{10} = 1$, then $E_1 = \frac{3}{2}$, $e_1 = \frac{3}{8}$, and $\varepsilon_1 = 1$. But if the deformation is small, then E, e, and ε are approximately the same. For example, if $L_1 = 1.01$, $L_{10} = 1.00$, then $E_1 \doteq e_1 \doteq \varepsilon \doteq 0.01$. Green's strain and Almansi's strain both reduce to the infinitesimal strain when the deformation is infinitesimal.

In the description of a large deformation, it turns out that the most convenient quantity to consider is the square of the distance between any two points, because to the square of distances Pythagoras' rule applies. This is why the Green and Almansi strains are introduced. Some authors call $\log \lambda$ the "true" strain, which is also a useful strain measure.

7.11 Strain-Energy Function

Theoretical analysis of bodies subjected to finite deformation is quite complex. A brief description is given in the author's *Foundations of Solid Mechanics*, Chapter 16. One of the most important results proved there is the use of the *strain potential*, or synonymously, the *strain-energy function*. Let W be the strain energy per unit mass of the tissue, and ρ_0 be the density (mass per unit volume) in the zero-stress state, then $\rho_0 W$ is the strain energy per unit volume of the tissue in the zero stress state. Let W be expressed in terms of the nine strain components $E_{11}, E_{22}, E_{33}, E_{12}, E_{21}, E_{23}, E_{32}, E_{31}, E_{13}$, and be written in a form that is symmetric in the symmetric components E_{12} and E_{21}, E_{23} and E_{32}, and E_{31} and E_{13}. The nine strain components are treated as independent variables when partial derivatives of $\rho_0 W$ are formed. Then when such a strain-energy function exists, the stress components S_{ij} can be obtained as derivatives of $\rho_0 W$:

$$S_{ij} = \frac{\partial(\rho_0 W)}{\partial E_{ij}}. \qquad (1)$$

More explicitly, for the rectangular element shown in Fig. 7.10:1, we have

$$S_{11} = \frac{\partial(\rho_0 W)}{\partial E_{11}}, \qquad S_{22} = \frac{\partial(\rho_0 W)}{\partial E_{22}}, \qquad S_{33} = \frac{\partial(\rho_0 W)}{\partial E_{33}}. \qquad (2)$$

Not all elastic materials have a strain-energy function. Those that do are called *hyperelastic* materials. (For definitions of elasticity, hyperelasticity, and hypoelasticity, see Fung, *Foundations of Solid Mechanics*, Sec. 16.6.)

It can be shown that these formulas are equivalent to the following. Let W be expressed not in terms of E_{ij} but in the nine components of the *deformation gradient tensor* $\partial x_i/\partial a_j$, where (a_1, a_2, a_3) denote the coordinates of a material particle in the zero-stress state of the body, and (x_1, x_2, x_3) are the coordinates of the same particle in the deformed state of the body, both referred to a rectangular cartesian frame of reference. Then the Lagrangian stresses are

$$T_{ij} = \frac{\partial(\rho_0 W)}{\partial(\partial x_i/\partial a_j)}. \tag{3}$$

Note that since $\partial x_i/\partial a_j$ is, in general, not equal to $\partial x_j/\partial a_i$, the tensor component T_{ij} is in general not equal to T_{ji}, i.e., T_{ij} is not symmetric. Referring to the rectangular element shown in Fig. 7.10:1, we have

$$T_{11} = \frac{\partial(\rho_0 W)}{\partial \lambda_1}, \qquad T_{22} = \frac{\partial(\rho_0 W)}{\partial \lambda_2}, \qquad T_{33} = \frac{\partial(\rho_0 W)}{\partial \lambda_3}, \tag{4}$$

where

$$\lambda_1 = \frac{\partial x_1}{\partial a_1}, \qquad \lambda_2 = \frac{\partial x_2}{\partial a_2}, \qquad \lambda_3 = \frac{\partial x_3}{\partial a_3} \tag{5}$$

are the stretch ratios. It is useful also to record the general relationship between the Cauchy, Lagrange, and Kirchhoff stresses (see Fung, *Foundations of Solid Mechanics*, Sec. 16.2):

$$T_{ij} = S_{ip} \frac{\partial x_j}{\partial a_p}, \qquad S_{ij} = \frac{\partial a_i}{\partial x_\alpha} T_{j\alpha}, \tag{6}$$

$$\sigma_{ji} = \frac{\rho}{\rho_0} \frac{\partial x_i}{\partial a_p} T_{pi} = \frac{\rho}{\rho_0} \frac{\partial x_i}{\partial a_\alpha} \frac{\partial x_j}{\partial a_\beta} S_{\beta\alpha}, \tag{7}$$

$$T_{ji} = \frac{\rho_0}{\rho} \frac{\partial a_j}{\partial x_m} \sigma_{mi}, \qquad S_{ji} = \frac{\rho_0}{\rho} \frac{\partial a_i}{\partial x_\alpha} \frac{\partial a_j}{\partial x_m} \sigma_{m\alpha}. \tag{8}$$

The question arises whether a strain-energy function exists for a given material. If the material is perfectly elastic, then the existence of a strain-energy function can often be justified on the basis of thermodynamics (see Sec. 7.4). But living tissues are not perfectly elastic. Therefore, they cannot have a strain-energy function in the thermodynamic sense. We take advantage, however, of the fact mentioned in Sec. 7.7, that in cyclic loading and unloading the stress–strain relationship after preconditioning does not vary very much with the strain rate. If the strain-rate effect is ignored altogether; then the loading curve and the unloading curve (they are unequal) can be separately treated as a uniquely defined stress–strain relationship, which is associated with a strain-energy function. We shall call each of these curves a *pseudoelasticity* curve, and the corresponding strain-energy function the *pseudo strain-energy function*. The existence of a pseudo-strain energy is an assumption that must be justified by experiments consistent with the accepted degree of approximation.

The formulas (6) and (8) need modification if the material is incompressible, i.e., if its volume does not change with stresses. For an incompressible material the *pressure* in the material is not related to the strain in the material. It is an indeterminate quantity as far as the stress–strain relationship is concerned. The pressure in a body of incompressible material can be determined only by solving the equations of motion or equilibrium and the boundary conditions. The general relationship in this case is

$$S_{ij} = \frac{\partial(\rho_0 W)}{\partial E_{ij}} - p\frac{\partial a_i}{\partial x_r}\frac{\partial a_j}{\partial x_r}, \tag{9}$$

where p is the pressure (see Fung, *Foundations of Solid Mechanics*, Sec. 16.7).

7.12 An Example: The Constitutive Equation of Skin

Although biaxial experiments alone are not sufficient to derive a three-dimensional stress–strain relationship, for membranous material they are sufficient to yield a two-dimensional constitutive equation for plane states of stress. Such an equation connects the three components of stress in the plane of the membrane with the three components of strains in the membrane.

Biaxial experiments on rabbit skin have been reported by Lanir and Fung (1974a,b) from whose data Tong and Fung (1976) gleaned the pseudo-strain-energy function for a loading process (increasing strain). The skin specimens were taken from the rabbit abdomen. It was first verified that the mechanical property was orthotropic, and that in cyclic loading and unloading at constant strain rates the stress–strain relationship is essentially independent of the strain rate. Hence the argument discussed in Sec. 7.7 applies and a pseudo-strain-energy function can be defined for either loading or unloading. Let the x_1 axis refer to direction along the length of the body, from head to tail, and the x_2 axis the direction of the width. Let $E_1 (= E_{11})$, $E_2 (= E_{22})$ be the strains defined in the sense of Green [see Eq. (3) of Sec. 7.10]. Tong and Fung (1976) chose the following form for the pseudo-strain-energy function for skin (the reason for this choice will be explained later (following Eq. (3) in this section):

$$\begin{aligned}\rho_0 W^{(2)} = \tfrac{1}{2}(&\alpha_1 E_1^2 + \alpha_2 E_2^2 + \alpha_3 E_{12}^2 + \alpha_3 E_{21}^2 + 2\alpha_4 E_1 E_2) \\ + \tfrac{1}{2}c\,&\exp(a_1 E_1^2 + a_2 E_2^2 + a_3 E_{12}^2 + a_3 E_{21}^2 + 2a_4 E_1 E_2 \\ + \,&\gamma_1 E_1^3 + \gamma_2 E_2^3 + \gamma_4 E_1^2 E_2 + \gamma_5 E_1 E_2^2),\end{aligned} \tag{1}$$

where the α's, a's, γ's, and c are constants, and E_{12} is the shear strain [see Eq. (15) of Sec. 2.3] which is included in Eq. (1), but was kept at zero in the experiments. This is a two-dimensional analog of the strain-energy function discussed in Sec. 7.10. If, for a membrane, we are concerned only with the plane state of stress caused by stretching in the plane of the membrane, then we may use such a two-dimensional strain-energy function in the manner of Eq. (1) of Sec. 7.11 (with the index i, j limited to 1 and 2). Thus, from Eq. (1) above and Eqs. (1) or (2) of Sec. 7.11 we obtain the Kirchhoff stresses:

$$S_{11} = \frac{\partial(\rho_0 W^{(2)})}{\partial E_1} = \alpha_1 E_1 + \alpha_4 E_2 + cA_1 X,$$

$$S_{22} = \frac{\partial(\rho_0 W^{(2)})}{\partial E_2} = \alpha_4 E_1 + \alpha_2 E_2 + cA_2 X, \tag{2}$$

$$S_{12} = \frac{\partial(\rho_0 W^{(2)})}{\partial E_{12}} = \alpha_3 E_{12} + ca_3 E_{12} X,$$

where

$$A_1 = a_1 E_1 + a_4 E_2 + \tfrac{3}{2}\gamma_1 E_1^2 + \gamma_4 E_1 E_1 + \tfrac{1}{2}\gamma_5 E_2^2,$$
$$A_2 = a_4 E_1 + a_2 E_2 + \tfrac{3}{2}\gamma_2 E_2^2 + \tfrac{1}{2}\gamma_4 E_1^2 + \gamma_5 E_1 E_2, \tag{3}$$
$$X = \exp(a_1 E_1^2 + a_2 E_2^2 + a_3 E_{12}^2 + a_3 E_{21}^2 + 2a_4 E_1 E_2$$
$$+ \gamma_1 E_1^3 + \gamma_2 E_2^3 + \gamma_4 E_1^2 E_2 + \gamma_5 E_1 E_2^2).$$

The first term in Eq. (1) was introduced because the data appear to be "biphasic." We use the second term to express the behavior of the material

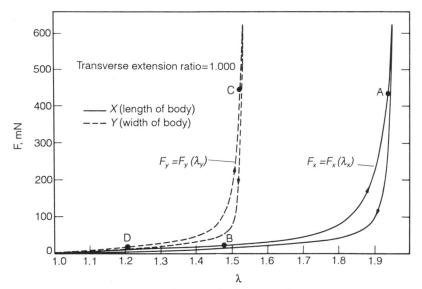

Figure 7.12:1 Force vs. stretch ratio curves in the x experiments (λ_y was fixed while λ_x varied; solid curves) and in the y experiments (λ_x fixed; dotted). The choice of the points A, B, C, D is illustrated. In this example, $\lambda_y = 1$ in the x experiment, and $\lambda_x = 1$ in the y experiment. The difference of the curves is due to anisotropy. Rabbit abdominal skin: The x or x_1 axis points in the direction from the head to the tail. From Tong and Fung (1976), by permission. Experimental results are from Lanir and Fung (1974b).

at a high stress level, and use the first term to remedy the situation at a lower stress level.

From Eqs. (2) we can compute the derivatives $\partial S_1/\partial E_1$, $\partial S_2/\partial E_2$, and $\partial S_1/\partial E_2 = \partial S_2/\partial E_1$. We then determine the constants denoted by the α's, a's, and c by requiring that S_{11}, S_{22} and their derivatives measured experimentally are fit exactly by the equations named above at some selected points. The choice of these points is illustrated in Fig. 7.12:1, in which the stress–strain relationships with E_2 fixed are shown in the solid curves, and those with E_1 fixed are shown by the dashed curves. A pair of points A and C are located on the curves in a region where stresses change rapidly, whereas the points B and D are located in a region in which the strains are relatively small. The constants α_1, α_2, α_4 are determined essentially by experimental data at B and D, whereas a_1, a_2, a_4, and c are determined essentially by the data at A and C. The exact locations of the points A, B, C, and D are not very important. The values of the constants change very little by choosing different (reasonable) points on any two sets of data curves.

A typical comparison between the experimental data and the mathematical expression is shown in Fig. 7.12:2. In Fig. 7.12:2(a) the longitudinal force F_x is plotted against the longitudinal stretch ratio λ_x, while the transverse direction is kept at the natural length so that $\lambda_y = 1$, whereas in the other direction, F_y is plotted against λ_y, while λ_x is kept at 1. The experimental data points are shown as squares. Data computed from our theoretical formulas (1)–(7) are plotted as circles for case A (all γ's $= 0$), and as crosses for case B ($\gamma_4 = \gamma_5$). The fit is good for the entire curve. In Fig. 7.12:2(b) the stress in the transverse direction is plotted against the longitudinal stretch ratio, while the transverse strain is kept at zero, and vice versa. Again the fit between the mathematical formula and the experimental data is good. Although these good fittings are not surprising because the constants α_1, α_2, etc., are determined from these two sets of experimental data, the ability of the mathematical formulas to fit the entire set of data when data at only four points on the curves are used is nontrivial. Considering the tremendous nonlinearity of these curves, one might say that the fitting is remarkably good.

One may conclude that the pseudo-strain-energy function, Eq. (1) is suitable for rabbit skin for stresses and strains in the physiological range. For all practical purposes, the third-order terms in the exponential function [the last line of Eq. (1)] involving the constants γ are unimportant; and there is no significant loss of accuracy if we set all γ's equal to zero. Hence we may simplify Eq. (1) to the form

$$\rho_0 W^{(2)} = f(\alpha, E) + c \exp[F(a, E)], \tag{4}$$

where

$$f(\alpha, E) = \alpha_1 E_{11}^2 + \alpha_2 E_{22}^2 + \alpha_3 E_{12}^2 + \alpha_3 E_{21}^2 + 2\alpha_4 E_{11} E_{22}, \tag{5}$$

$$F(a, E) = a_1 E_{11}^2 + a_2 E_{22}^2 + a_3 E_{12}^2 + a_3 E_{21}^2 + 2a_4 E_{11} E_{22}. \tag{6}$$

Finally, if one is concerned mainly with higher stresses and strains in the

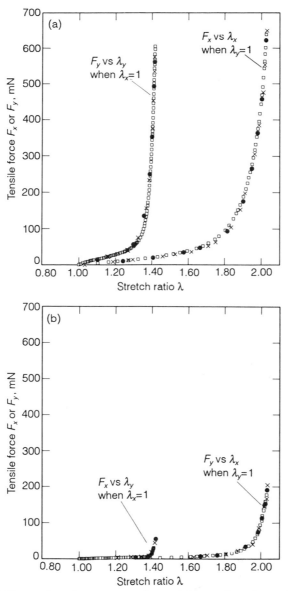

Figure 7.12:2 Comparison between experimental data and mathematical expression. The tensile forces F_x, F_y are given in milli Newton. Lagrange stress T_x is equal to F_x divided by A_x, the cross-sectional area perpendicular to the x axis. Squares: experimental data. Circles: from Eqs. (1) and (2) with $\alpha_1 = \alpha_2$, all γ's = 0. $a_1 = 3.79$, $a_2 = 12.7$, $a_4 = 0.587$, $c = 0.779$ N/m^2, $\alpha_1 = \alpha_2 = 1{,}020$ N/m^2, and $\alpha_4 = 254$ N/m^2. Crosses: From Eq. (2) with $\alpha_1 = \alpha_2$, $\gamma_1 = \gamma_2 = 0$, $\gamma_4 = \gamma_5 \neq 0$. $a_1 = 3.79$, $a_2 = 18.4$, $a_4 = 0.587$, $c = 0.779$ N/m^2, $\alpha_1 = \alpha_2 = 1{,}020$ N/m^2, and $a_4 = 254$ N/m^2. $\gamma_4 = \gamma_5 = 15.6$. From Tong and Fung (1976), by permission. Experimental data are from Lanir and Fung (1974b).

physiological range, and does not care for great accuracy at very small stress levels, then the first term in Eq. (4) can be omitted, and we have simply

$$\rho_0 W^{(2)} = c \exp[F(a, E)]. \tag{7}$$

7.12.1 Other Examples

The pseudo-strain-energy function for arterial walls presented in Chapter 8, Sec. 8.6, is also of the form of Eq. (7). That for lung tissue is given by Fung (1975), Fung et al. (1978), and Vawter et al. (1979) and is also of the form of Eq. (7).

The genesis of Eqs. (1) and (7) lies in a general feature of soft tissues in simple elongation, described in Sec. 7.6. As shown in Eq. (2) of Sec. 7.6, the slope of the stress–strain curve is proportional to the tensile stress. As a consequence [Eq. (3) of Sec. 7.6], the stress is an exponential function of the strain. Generalizing this, Fung (1973) proposed a pseudo-strain-energy function in the form

$$
\begin{aligned}
W = \tfrac{1}{2}\alpha_{ijkl}E_{ij}E_{kl} \\
+ (\beta_0 + \beta_{mnpq}E_{mn}E_{pq})\exp(\gamma_{ij}E_{ij} + \kappa_{ijkl}E_{ij}E_{kl} + \cdots),
\end{aligned} \tag{8}
$$

where α_{ijkl}, β_0, β_{mnpq}, γ_{ij}, and κ_{ijkl} are constants to be determined empirically, and the indices $i, j, k \ldots$, range over 1, 2, 3. The summation convention is used (Chapter 2, Sec. 2.2). Equations (1) and (7) are simplified versions of Eq. (8). It is amusing to note that it took several years before we realized that a better fit with experimental data can be obtained by dropping the first-order terms ($\gamma_{ij}E_{ij}$) in the exponential function in Eq. (8).

The pseudo-elasticity of excised visceral pleura has been measured by Humphrey, Vawter, and Vito (1987). For the analysis of the lung, the visceral pleura contributes a significant part of the pressure-volume curve at larger lung volume. Humphrey and Yin (1991) (see Sec. 9.10) also studied the mechanical properties of the visceral pericardium. The visceral pericardium has a significant effect on heart mechanics at larger ventricular volume.

7.13 Generalized Viscoelastic Relations

Since we regard biosolids as viscoelastic bodies, it seems appropriate to conclude our discussion with a proposal for their constitutive equations. We have discussed the quasi-linear viscoelasticity of tissues in uniaxial state of stress in Sec. 7.6. It is logical to generalize the results to two or three dimensions by changing Eq. (7) of Sec. 7.6 into a tensor equation:

$$S_{ij}(t) = S_{kl}^{(e)}(0+)G_{ijkl}(t) + \int_0^t G_{ijkl}(t-\tau)\frac{\partial S_{kl}^{(e)}[\mathbf{E}(\tau)]}{\partial\tau}\,d\tau.$$

This generalization is analogous to the classical relation, Eq. (24) of Sec. 2.11. Here S_{ij} is the Kirchhoff stress tensor; $\mathbf{E}$, with components E_{ij}, is the Green's strain tensor; and G_{ijkl} is the reduced relaxation function tensor, the word "reduced" referring to the condition $G_{ijkl} = 1$ when $t = 0$. $S_{ij}^{(e)}$ is the "elastic" stress tensor corresponding to the strain tensor $\mathbf{E}$. It is a function of the strain components E_1, E_{22}, E_{12}, etc. $S_{ij}^{(e)}$ is the stress that is reached instantaneously when the strains are suddenly increased from 0 to E_{11}, E_{22}, etc. Following the arguments of Sec. 7.6, we assume that the "elastic" responses $S_{ij}^{(e)}$ can be approximated by the pseudoelastic stresses. The range of the indexes is 1, 2 if the material is a membrane subjected to a plane state of stress, and 1, 2, 3 if it is a three-dimensional body.

Although G_{ijkl} is a tensor or rank 4, it will have only two independent components if the material is isotropic. If the material is anisotropic, a careful consideration of the material symmetry as presented in Green and Adkins (1960) *Large Elastic Deformation* will be helpful. It is likely that G_{ijkl} will have a relaxation spectrum of the same kind as discussed in Sec. 7.6.

7.14 The Complementary Energy Function: Inversion of the Stress–Strain Relationship

If stresses are known analytic functions of strains, one should be able to express strains as analytic functions of stresses. In linear elasticity Hooke's law can be expressed both ways. In nonlinear elasticity, however, it is not always simple to invert a stress–strain law. For example, if the stress s is a cubic function of the strain ε:

$$s = a\varepsilon + b\varepsilon^2 + c\varepsilon^3.$$

We know that ε cannot be expressed as a rational function of s, a, b, and c. If we think of ε, s, a, b, c as tensors, then an inversion is more difficult; in fact, unknown.

However, the exponential strain energy function given in Eq. (7) of Sec. 7.12 can be easily inverted. Such an inversion is useful whenever one wishes to calculate strains from known stresses; or when one wishes to use the complementary energy theorem in numerical analysis. Inversion is needed also if one wishes to use the method of stress functions, which is still applicable in finite deformation, because the equations of equilibrium remain linear when expressed in terms of Cauchy stresses.

The general theory is presented by Fung (1979). We express stresses in terms of strains in a constitutive equation

$$S_{ij} = \frac{\partial \rho_0 W}{\partial E_{ij}}, \tag{1}$$

where S_{ij} and E_{ij} are the Kirchhoff stress tensor and Green's strain tensor, respectively, W is the strain energy function (expressed in terms of strain) per unit mass, and ρ_0 is the density of the material in the initial, unstressed condition, so that $\rho_0 W$ is the strain energy per unit volume. Then according to the theory of continuum mechanics (see, for example, the author's *Foundations of Solid Mechanics*, Sec. 16.7), it is known that if the constitutive equation (1) can be inverted, then there exists a *complementary energy function* $W_c(S_{11}, S_{12}, \ldots)$ *of the stress components*, such that

$$\rho_0 W_c = S_{ij} E_{ij} - \rho_0 W, \tag{2}$$

in which the summation convention over repeated indexes is used, and

$$E_{ij} = \frac{\partial \rho_0 W_c}{\partial S_{ij}}. \tag{3}$$

On the other hand, if a complementary energy function can be found, then Eq. (3) at once gives the inversion of the stress–strain relation. The problem of inversion consists of finding $\rho_0 W_c$ from $\rho_0 W$ through the contact transformation given in Eqs. (2) and (3).

Let us relabel the six independent components of E_{ij} as $E_1, E_2, \ldots, E_6$, and the six independent components of stresses S_{ij} as $S_1, S_2, \ldots, S_6$. Let a square matrix $[a_{ij}]$ be symmetric and nonsingular. We define a quadratic form

$$Q = \sum_{i=1}^{6} \sum_{j=1}^{6} a_{ij} E_i E_j. \tag{4}$$

We shall show that if *the strain energy function* $\rho_0 W$ *is an analytic function of* Q,

$$\rho_0 W = f(Q), \tag{5}$$

then it can be inverted. We proceed as follows. Since

$$S_i = \frac{\partial \rho_0 W}{\partial E_i} = \frac{df}{dQ} \frac{\partial Q}{\partial E_i} = 2 \frac{df}{dQ} \sum_{j=1}^{6} a_{ij} E_j \tag{6}$$

is a linear equation for E_i, we can solve for E_i. Let us use matrix notations, with $\{\ \}$ denoting a column matrix, $[\ ]$ denoting a square matrix, T denoting transpose, and -1 power denoting inverse. Then we have

$$\{E_i\} = \left(2 \frac{df}{dQ}\right)^{-1} [a_{ij}]^{-1} \{S_j\}. \tag{7}$$

A substitution into Eq. (4) gives

$$Q = \left(2 \frac{df}{dQ}\right)^{-2} \{S_i\}^T [a_{ij}]^{-1} \{S_j\}. \tag{8}$$

If we define a polynomial P of stresses by

$$P = \tfrac{1}{4}\{S_i\}^T[a_{ij}]^{-1}\{S_j\}, \tag{9}$$

then Eq. (8) becomes

$$P = Q\left(\frac{df}{dQ}\right)^2. \tag{10}$$

When $f(Q)$ is given, Eq. (10) can be solved for Q as a function of P. Now according to Eqs. (2), (1), and (5),

$$\rho_0 W_c = \sum_{i=1}^{6} \frac{df}{dQ}\frac{\partial Q}{\partial E_i} E_i - \rho_0 W. \tag{11}$$

Since Q is a homogeneous function of E_i of degree 2, we know by Euler's theorem for a homogeneous function that $E_i(\partial Q/\partial E_i) = 2Q$. Hence

$$\rho_0 W_c = 2Q\frac{df(Q)}{dQ} - f(Q). \tag{12}$$

When the solution of Eq. (10) is substituted into the right-hand side of Eq. (12), it becomes a function of P, i.e., of stresses. The inversion is thus accomplished.

7.14.1 Example

Although the derivation given above refers to three-dimensional cases, it applies equally well to two dimensions if appropriate changes are made in the range of the indexes. As an example, consider the strain-energy function given in Eq. (3) of Sec. 8.5, for arteries, for which

$$Q = a_1 E_1^2 + a_2 E_2^2 + 2a_4 E_1 E_2, \tag{13}$$

$$\rho_0 W = f(Q) = \frac{c}{2}\, e^Q. \tag{14}$$

Then Eq. (8) is reduced to

$$Q = c^{-2}(a_1 a_2 - a_4^2)^{-1} e^{-2Q}(a_2 S_1^2 + a_1 S_2^2 - 2a_4 S_1 S_2). \tag{15}$$

If we define in this case

$$P = c^{-2}(a_1 a_2 - a_4^2)^{-1}(a_2 S_1^2 + a_1 S_2^2 - 2a_4 S_1 S_2), \tag{16}$$

then P and Q are related by the equation

$$Qe^{2Q} = P, \tag{17}$$

which is independent of the material constants. Reading Q as a function P through Eq. (17) renders Eq. (12) a function of P. Q.E.D.

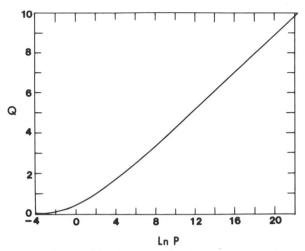

Figure 7.14:1 The universal function $Q(P)$ for inverting a nonlinear stress–strain relationship of the exponential type with the pseudo-strain-energy function given in Eqs. (13) and (14). From Fung (1979).

7.14.2 The Universal Function $Q(P)$

Note that Eq. (17) does not contain material constants, and it defines a universal relationship between P and Q for an exponential strain-energy function. Taking the logarithm of both sides, we obtain

$$\ln Q + 2Q = \ln P. \tag{18}$$

The function $Q(P)$ is best exhibited in terms of $\ln P$, as is given in Fig. 7.14:1.

7.15 Constitutive Equation Derived According to Microstructure

One of the ambitious programs in mechanics is to build up the constitutive equations of the continua defined according to the successive levels of sizes considered in Sec. 2.1. One way is to build from the ground up: from elementary particles to atoms, from atoms to molecules, from molecules to macromolecules, to proteins, cells, tissues, and organs. Another way is to analyze from the top down, in the reversed direction. At any given level, one feels that a greater understanding is obtained if the relationship between the constitutive equations of materials in successive levels of sizes is known.

In biomechanics, for example, one would like to "explain" the constitutive equation of the skin in terms of its microstructure or ultrastructure, or to "derive" it from the microstructure up. Viidik, Lanir, Shoemaker, Fung, Mow, Lai, Tözeren, Woo, and many others have made substantial contributions to this effort. Viidik (1968) first introduced the concept that the number of collagen fibers in a soft tissue that are pulled straight and brought into action increases as the tensile stress in the tissue increases. His latest review is Viidik (1990). Lanir (1979a,b) assumes that the tissue is composed of a network of collagen and another network of elastin. At zero stress state, the elastin fibers are straight and the collagen fibers are coiled, bent, or buckled. Lanir (1983) introduced thermodynamic considerations. Shoemaker (1984) followed with an exhaustive irreversible thermodynamics analysis. Shoemaker and Fung (1986) presented a formula derived from a collagen–elastin model, and used it to fit the experimental results on human skin obtained by Schneider (1982), and on dog pericardium obtained by Lee et al. (1985). The authors obtained good fitting, but the fiber structure of the tissues was hypothetical, unknown in quantitative details. Future work would have to be based on a more solid foundation of morphometric data on the collagen and elastin fibers, fibroblasts, blood vessels, and ground substances in the tissues of interest.

Kwan and Woo (1989) derived the nonlinear stress–strain relationship of the tendons and ligaments not strictly on the basis of microstructural data, but hinting at the waviness of the collagen fibers to suggest that the tension-elongation curve of each single fiber consists of two straight line segments: the first segment with a smaller slope representing the wavy regime, the second segment with a layer slope representing the taut regime. Assembling many parallel fibers with a certain statistics of length of fibers and transition points between waviness and tautness yields good fitting with experimental results.

Lai, Lanir, Mow, and others derived properties of cartilage on the basis of multiphasic theory. See Sec. 12.10.

The attempts by Lanir and his associates, and Humphrey and Yin on the heart are discussed in Sec. 10.8.

Problems

7.1 Take a tube of gas or a piece of steel and give it a quick stretch while there is no heat added. The material will cool down. Take a rubber band or a strip of muscle and give it a quick stretch. It will heat up. (A very easy demonstration is to take a rubber band, touch it with your forehead to feel its temperature. Then quickly stretch the rubber band and touch the skin on the forehead again. It will be found that it has heated up.)

Explain these phenomena on the basis that gas and steel mainly derive their stress from internal energy whereas rubber and muscle mainly derive their stress from entropy changes.

7.2 A well-known result of Hardung (1952), Bergel (1961), and others is illustrated in Fig. P7.2. It shows a rather small change in phase angle over a range of frequencies from 1 to 10 Hz, and a gentle stiffening of the dynamic modulus. Could this be explained by a Maxwell model of viscoelasticity? If not, then what kind of model would be consistent with the experimental findings?

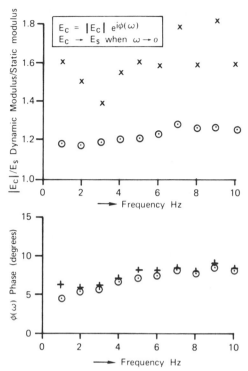

Figure P7.2 The dynamic modulus and damping phase angle ϕ [angle δ in Eq. (22)] of arteries from the data of Hardung (1952) and Bergel (1961). Circles: aorta. Crosses: other arteries. From Westerhof and Noordergraf (1970).

7.3 The exponential integral $E(z)$ is defined by Eq. (38) of Sec. 7.6. It enters naturally into the relaxation function of a material with a continuous relaxation spectrum specified in Eq. (30) of Sec. 7.6. How does $E(z)$ behave when z is either very large or very small? Derive asymptotic formulas for this function when $z \to 0$ and $z \to \infty$.

7.4 Is there a two-dimensional strain-energy function that corresponds to the stress–strain relation for the red blood cell membrane proposed in Eq. (7) of Sec. 4.7 in Chapter 4?

7.5 The following question is concerned with the practical process of data reduction. Experimental data on stress are usually tabulated at specified values of stretch ratio. Let the stretch ratios be uniformly spaced at $\lambda_0 + \Delta\lambda,\ \lambda_0 + 2\Delta\lambda,\ \ldots$ and the corresponding stresses be $T_1,\ T_2,\ \ldots$. From this we can compute the finite differences $\Delta T_n = T_n - T_{n-1}$, and evaluate the ratio $\Delta T/\Delta\lambda$ at $\lambda = \lambda_n$ as $\Delta T_n/\Delta\lambda$. Assume that a differential equation as given by Eq. (2) of Sec. 7.5 is valid. What is the corresponding difference equation (i.e., one expressed in terms of $\Delta T_n/\Delta\lambda$ and T_n)? In practice we examine the difference equation and then infer whether the differential equation is valid. The material constants α and β can be determined directly from the finite differences without going through the process of numerical differentiation.

7.6 Let the relaxation function be

$$G(t) = e^{-vt}$$

and the elastic response be given by Eq. (4) of Sec. 7.5:

$$T^{(e)}(\lambda) = (T^* + \beta)e^{\alpha(\lambda - \lambda^*)} - \beta.$$

Find the tension history $T(t)$ for the following cases according to Eq. (4) of Sec. 7.6:
(a) λ is a step function of amplitude c, i.e., $\lambda = 1 + c\,\mathbf{1}(t)$;
(b) λ is a ramp function, $\lambda = 1 + ct$;
(c) λ is oscillating, $\lambda = \lambda_0 + c\cos\omega t\,\mathbf{1}(t)$.

7.7 If the relaxation function $G(t)$ is given by Eqs. (28) and (30) of Sec. 7.6, and $T^{(e)}(\lambda)$ is given by Eq. (4) of Sec. 7.5, determine the tension history $T(t)$ according to Eq. (4) of Sec. 7.6 for the three cases named in Problem 7.6. For the case (c), work out the answer only for an oscillation of small amplitude, $c \ll 1$, so that the dependence on c can be linearized.

7.8 The stress–strain relationship of a viscoelastic material is

$$\sigma_{ij}(x_1, x_2, x_3, t) = \int_{-\infty}^{t} G_{ijke}(t - \tau)\frac{\partial e_{kl}(x_1, x_2, x_3, \tau)}{\partial \tau}\,d\tau,$$

where σ_{ij} are the stresses, e_{kl} are the strains, G_{ijke} are the relaxation functions; x_1, x_2, x_3 are Euclidean coordinates, and t is time. Derive an equation of motion for this material analogous to the Navier equation for Hookean solids.

7.9 A piece of connective tissue is subjected to a uniaxial stretch. Let the entropy per unit volume of the tissue be S. Let the stretch ratio be λ. Assume that the entropy decreases with the increasing stretch according to the relation

$$S = -Ce^{\alpha(\lambda^2 - 1)},$$

where C and α are constants. What would be the relationship between the tension in the specimen and the stretch ratio λ as far as the contribution of entropy is concerned? What other source of elasticity is there?

7.10 A surgeon has to excise a certain dollar-sized patch of skin on a patient with a skin tumor. How would you recommend to excise the diseased skin and suture the wound? What are the stress and strain distribution in the skin after surgery? What is the tension in the suture? Obviously, stress concentration is to be avoided. Design a good way to operate. In skin surgery there is a technique called Z plasty, in which a Z-shaped incision is made, the middle segment of the Z being over the tumor, and the triangular flaps are then rotated so the apices cross the line of the tumor. Explain why it is a good technique.

7.11 Skin is anisotropic. In the skin there is a natural direction along which most of the collagen fibers are aligned. Healing will be faster if a cut is made in such a direction as compared with one cut perpendicular to this direction. Explain this qualitatively and formulate a mathematical theory of Z plasty cognizant of this feature.

7.12 Consider the models of tissues shown in Figs. P7.12(a), (b), and (c), respectively. In (a) and (b), fibers are embedded in a homogeneous isotropic material. In (c), spherical bubbles are packed and glued together.

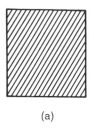

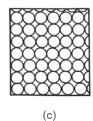

(a)   (b)   (c)

Figure P7.12

Propose a stress–strain relationship for each of these composite materials. In solving this problem, answer first whether there are axes of symmetry for the material, whether there is a special choice of coordinate axes that will simplify the stress–strain relationship. Use these special coordinate systems to obtain specific results. Then rotate coordinates to an arbitrary orientation to obtain general relationships.

References

Abramowitz, M. and Stegun, A. (1964) *Handbook of Mathematical Functions*, Ser. 55. U.S. Government Printing Office, Washington, D.C.

Alexander, R. M. (1968) *Animal Mechanics*. University of Washington Press, Seattle.

Banga, G. (1966) Structure and function of elastin and collagen. Asso. Pry. 1st Inst. of Pathological Anatomy and Exp. Cancer Res., Med. University, Budapest.

Bauer, R. D. and Pasch, T. H. (1971) The quasistatic and dynamic circumferential elastic modulus of the rat tail artery. *Pflügers Arch.* **330**, 335–345.

Becker, E. and Föppl. O. (1928) Dauerversuche zur Bestimmung der Festigkeitseigenschaften, Beziehungen zwischen Baustoffdämpfung und Verformungs geschuwindigkeit. *Forschung Gebiete Ingenieurwesens.* V.D.I., No. 304.

Becker, R. and Döring, W. (1939) *Ferronmagnetismus*. Springer, Berlin, Chapter 19.

Bergel, D. H. (1961) The dynamic elastic properties of the arterial wall. *J. Physiol.* **156**, 458–469.

Bergel, D. H. (1961) The static elastic properties of the arterial wall. *J. Physiol.* **156**, 445–457.

Biot, M. A. (1965) *Mechanics of Incremental Deformations*. Wiley, New York.

Blatz, P. J., Chu, B. M., and Wayland, H. (1969) On the mechanical behavior of elastic animal tissue. *Trans. Soc. Rheol.* **13**, 83–102.

Bodner, S. R. (1968) In *Mechanical Behavior of Materials under Dynamic Loads*, U. S. Lindhom (ed.) Springer, New York, pp. 176–190.

Bressan, G. M., Argos, P., and Stanley, K. K. (1987) Repeating structures of Chick tropoelastin revealed by complementary DNA cloning. *Biochem.* **26**, 1497–1503.

Buchtal, F. and Kaiser, E. (1951) *The Rheology of the Cross Striated Muscle Fibre with Particular Reference to Isotonic Conditions*. Det Kongelige Danske Videnskabernes Selskab, Copenhagen, *Dan. Biol. Medd.* **21**, No. 7, p. 328.

Chen, H. Y. L. and Fung, Y. C. (1973) In *Biomechanics Symposium*, ASME Publ. No. AMD-2, American Society of Mechanical Engineers, New York, pp. 9–10.

Chu, B. M. and Blatz, R. J. (1972) Cumulative microdamage model to describe the hysteresis of living tissues. *Annu. Biomed. Eng.* **1**, 204–211.

Ciferri, A. (1963) The $\alpha \leftrightharpoons \beta$ transformation in keratin. *Trans. Faraday Soc.* **59**, 562–569.

Collins, R. and Hu, W. C. (1972) Dynamic constitutive relations for fresh aortic tissue. *J. Biomech.* **5**, 333–337.

Cowan, P. M., North, A. C. T., and Randall J. T. (1955) X-ray diffraction studies of collagen fibers. *Symp. Soc. Exp. Biol.* **9**, 115–126.

Dai, F., Rajagopal, K. R., and Wineman, A. S. (1992) Nonuniform extension of a nonlinear viscoelastic slab. *Int. J. Solids Struct.* **29**, 911–930.

Dale, W. C., Baer, E., Keller, A., and Kohn R. R. (1972) On the ultrastructure of mammalian tendon. *Experientia* **28**, 1293–1295.

Dale, W. C. and Baer, E. (1974) Fiber-buckling in composite systems: a model for the ultrastructure of uncalcified collagen tissues. *J. Mater. Sci.* **9**, 369–382.

Deak, S. B., Pierce, R. A., Belsky, S. A., Riley, D. J., and Boyd, C. D. (1988) Rat tropoelastin is synthesized from a 3.5 kilobase mRNA. *J. Biol. Chem.* **263**, 13504–13507.

Debes, J. (1992) The mechanical properties of pulmonary parenchyma and arteries. Ph. D. thesis, University of California, San Diego.

Debes, J. and Fung, Y. C. (1992) The effect of temperture on the biaxial mechanics of excised lung parenchyma of the dog. *J. Appl. Physiol.* **73**, 1171–1180.

Diamant, J., Keller, A., Baer, E., Litt, M., and Arridge, R. G. C. (1972) Collagen; ultrastructure and its relation to mechanical properties as a function of ageing *Proc. Roy. Soc. London B* **180**, 293–315.

Dortmans, L. J. M. G., Ven, A. A. F. van de, and Sauren, A. A. H. J. (1987) A note on the reduced creep function corresponding to the quasi–linear visco-elastic model proposed by Fung. Private Communication.

Dunn, F., Edmonds, P. D., and Fry, W. J. (1969) Ultrasound. In *Biological Engineering*, H. P. Schwan (ed.) McGraw-Hill, New York, p. 205.

Emery, A. H. and White, M. L. (1969) A single-integral constitutive equation. *Trans. Soc. Rheolo.* **13**, 103–110.

Feughelman, M. (1963) Free-energy difference between the alpha and beta states in keratin. *Nature* **200**, 127–129.

Flory, P. J. and Garrett, R. R. (1958) Phase transitions in collagen and gelatin systems. *J. Am. Chem. Soc.* **80**, 4836–4845.

Fronek, K., Schmid-Schönbein, G., and Fung, Y. C. (1975) A noncontact method for three-dimensional analysis of vascular elasticity *in vivo* and *in vitro*. *J. Appl. Physiol.* **40**, 634–637.

Fung, Y. C. (1965) *Foundations of Solid Mechanics*. Prentice-Hall, Englewood Cliffs, NJ.

Fung, Y. C. (1967) Elasticity of soft tissues in simple elongation. *Am. J. Physiol.* **213**, 1532–1544.

Fung, Y. C. (1968) Biomechanics: Its scope, history, and some problems of continuum mechanics is physiology. *Appl. Mech. Rev.* **21**, 1–20.

Fung, Y. C. (1972) Stress–strain-history relations of soft tissues in simple elongation. In *Biomechanics: Its Foundations and Objectives*, Y. C. Fung, N. Perrone, and M. Anliker (eds.) Prentice-Hall, Englewood Cliffs, NJ.

Fung, Y. C. (1973) "Biorheology of soft tissues." Presented on September 5, 1972 to the International Congress of Biorheology, Lyon, France. (*Biorheology* **10**, 139–155.)

Fung, Y. C. (1975) Stress, deformation, and atelectasis of the lung. *Circulation Res.* **37**, 481–496.

Fung, Y. C. (1977) *A First Course in Continuum Mechanics*. Prentice-Hall, Englewood Cliffs, NJ.

Fung, Y. C., Tong, P., and Patitucci, P. (1978) Stress and strain in the lung. *J. Eng. Mech. Div. Am. Soc. Civil Eng.* **104(EMI)**, 201–224.

Fung, Y. C. (1979) Inversion of a class of nonlinear stress–strain relationships of biological soft tissues. *J. Biomech. Eng.* **101**, 23–27.

Fung, Y. C. and Sobin, S. S. (1981) The retained elasticity of elastin under fixation agents. *J. Biomech. Eng.* **103**, 121–122.

Fung, Y. C. (1990) *Biomechanics: Motion, Flow, Stress, and Growth*. Springer-Verlag, New York.

Gathercole, L. J., Keller, A., and Shah, J. S. (1974) The periodic wave pattern in native tendon collagen: Correlation of polarizing with scanning electron microscopy. *J. Microscopy* **102**, 95–105.

Gizdulich, P. and Wesseling K. H. (1988) Forearm arterial pressure-volume relationships in man. *Clin. Phys. Physiol. Meas.* **9**, 123–132.

Gordon, M. K., Gerecke, D. R., and Olsen, B. R. (1987) Type XII collagen: Distinct extracellular matrix component discovered by cDNA cloning. *Proc. Nat. Acad. Sci. U. S. A.* **84**, 6040–6044.

Gosline, J. M. (1978) Hydrophobic interaction and a model for the elasticity of elastin. *Biopolymers*, **17**, 677–695.

Gray, W. R. (1970) Some kinetic aspects of crosslink biosynthesis. *Adv. Exp. Med. Biol.* **79**, 285–290.

Green, A. E. and Adkins, J. E. (1960) *Large Elastic Deformations*. Oxford University Press, New York.

Guth, E., Wack, P. E., and Anthony, R. L. (1946) Significance of the equation of state for rubber. *J. Appl. Physiol.* **17**, 347–351.

Hardung, V. (1952) Ueber eine methode zur messung der dynamischen elastizität und viskosität kautschukähnlicher körper, insbesondere von Blutgefäszen und anderen elastischen geweheteilen. *Helv. Physiol. Pharm. Acta* **10**, 482–498.

Harkness, M. C. R., Harkness, R. D., and McDonald, D. A. (1957) The collagen and elastin content of the arterial wall in the dog. *Proc. Roy. Soc. London B* **146**, 541–551.

Harkness, M. L. R. and Harkness, R. D. (1959a) Changes in the physical properties of the uterine cervix of the rat during pregnancy. *J. Physiol.* **148**, 524–547.

Harkness, M. L. R. and Harkness, R. D. (1959b) Effect of enzymes on mechanical properties of tissues. *Nature* **183**, 821–822.

Harkness, R. D. (1966) Collagen, *Sci. Progr.* **54**, 257–274.

Hart-Smith, L. J. and Crisp, J. D. C. (1967) Large elastic deformations of thin rubber membranes. *Int. J. Eng. Sci.* **5**, 1–24.

Hearle, J. W. S. (1963) Fiber structure. *J. Appl. Polym. Sci.* **7**, 172–192, 207–223.

Hoeltzel, D. A., Altman, P., Buzard, K., and Choe, K.-I. (1992) Strip extensiometry for comparison of the mechanical response of bovine, rabbit, and human corneas. *J. Biomech. Eng.* **114**, 202–215.

Hoeve, C. A. J. and Flory, P. J. (1958). The elastic properties of elastin. *J. Am. Chem. Soc.* **80**, 6523–6526.

Hoppin, F. G., Lee, J. C., and Dawson, S. V. (1975) Properties of lung parenchyma in distortion. *J. Appl. Physiol.* **39**, 742–751.

Humphrey, J. D., Vawter, D. L., and Vito, R. P. (1987) Pseudoelasticity of excised visceral pleura. *J. Biomech. Eng.* **109**, 115–120.

Johnson, G. A., Rajagopal, K. R., and Woo, S. L.-Y. (1992) A single integral finite strain viscoelastic model of ligaments and tendons. To be published.

Kastelic, J., Galeski, A., and Baer, E. (1978) The multicomposite structure of tendon. *J. Connective Tissue Res.* **6**, 11–23.

Kenedi, R. M., Gibson, T., and Daly, C. H. (1964) Bioengineering studies of the human skin; the effects of unidirectional tension. In *Structure and Function of Connective and Skeletal Tissue*, S. F. Jackson, S. M. Harkness, and G. R. Tristram (eds.) Scientific Committee, St. Andrews, Scotland, pp. 388–395.

Kenedi, R. M., Gibson, T., Evans, J. H., and Barbenel, J.G. (1975) Tissue mechanics. *Phys. Med. Biol.* **20**, 699–717.

Kishino, A. and Yanagida, T. (1988) Force measurements by micromanipulation of a single actin filament by glass needles. *Nature*, **334**, 74–76.

Knopoff, L. (1965) Attenuation of elastic waves in the Earth. In *Physical Acoustics*, W. P. Mason (ed.) Academic Press, New York, Vol. IIIB, Chapter 7, pp. 287–324.

Kwan, M. K. and Woo, S. L.-Y. (1989) A structural model to describe the nonlinear stress–strain behavior for parallel-fibered collagenous tissues. *J. Biomech. Eng.* **111**, 361–363.

Lai-Fook, S. J. (1977) Lung parenchyma described as a prestressed compressible material. *J. Biomech.* **10**, 357–365.

Lai-Fook, S. J., Wilson, T. A., Hyatt, R. E., and Rodarte, J. R. (1976) Elastic constants of inflated lobes of dog lungs. *J. Appl. Physiol.* **40**, 508–513.

Lanczos, C. (1956) *Applied Analysis*. Prentice-Hall, Englewood Cliffs, NJ.

Langewouters, G. J., Wesseling, K. H., and Goedhard, W. J. A. (1984) The static elastic properties of 45 human thoracic and 20 abdominal aortas *in vitro* and the parameters of a new model. *J. Biomech.* **17**, 425–435.

Langewouters, G. J., Wesseling, K. H., and Goedhard, W. J. A. (1985) The pressure dependent dynamic elasticity for 35 thoracic and 16 abdominal human aortas *in vitro* described by a five component model. *J. Biomech.* **18**, 613–620.

Langewouters, G. J., Zwart, A., Busse, R., and Wesseling, K. H. (1986) Pressure-diameter relationships of segments of human finger arteries. *Clin. Phys. Physiol. Meas.* **7**, 43–55.

Lanir, Y. (1979a) A structural theory for the homogeneous biaxial stress–strain relationship in flat collagenous tissues. *J. Biomech.* **12**, 423–436.

Lanir, Y. (1979b) The rheological behavior of the skin: Experimental results and a structural model. *Biorheology* **16**, 191–202.

Lanir, Y. and Fung, Y. C. (1974) Two-dimensional mechanical properties of rabbit skin. I. Experimental system. *J. Biomech.* **7**, 29–34. II. Experimental results. ibid., **7**, 171–182.

Lee, J. S., Frasher, W. G., and Fung, Y. C. (1967) Two-Dimensional Finite-Deformation on Experiments on Dog's Arteries and Veins. Tech. Rept. No. AFOSR 67-1980, University of California, San Diego, California.

Lee, J. S., Frasher, W. G., and Fung, Y. C. (1968) Comparison of the elasticity of an artery *in vivo* and in excision. *J. Appl. Physiol.* **25**, 799–801.

Lee, M. C., LeWinter, M. M., Freeman, G., Shabetai, R., and Fung, Y. C. (1985) Biaxial mechanical properties of the pericardium in normal and volume overloaded dogs. *Am. J. Physiol.* **249**, H222–H230.

Majack, R. H. and Bornstein, P. (1985) Heparin regulates the collagen phenotype of vascular smooth muscle cells: Induced synthesis of an MPVrPV 60,000 collagen. *J. Cell Biol.* **100**, 613–619.

McElhaney, J. H. (1966) Dynamic response of bone and muscle tissue. *J. Appl. Physiol.* **21**, 1231–1236.

Mecham, R. P. and Heuser, J. E. (1991). In *Cell biology and extracellular matrix.* 2nd ed. (ed. by E. D. Hay), Chapter 3. Plenum Press, New York.

Miller, E. J. (1988) Collagen types: Structure, distribution, and functions. *Collagen* **1**, 139–154.

Mooney, M. (1940) A theory of large elastic deformation. *J. Appl. Phys.* **11**, 582–592.

Morgan, F. R. (1960) The mechanical properties of collagen fibres: Stress–strain curves. *J. Soc. Leather Trades Chem.* **44**, 171–182.

Nemetschek, T., Riedl, H., Jonak, R., Nemetschek-Gansler, H., Bordas, J., Koch, M. H. J., and Schilling, V. (1980) Die viskoelastizität parallelsträngigen bindegewebes und ihre bedeutung für die function. *Virchows Arch. A Path. Anat. Histol.* **386**, 125–151.

Neubert, H. K. P. (1963) A simple model representing internal damping in solid materials. *Aeronaut. Q.* **14**, 187–197.

Nimni, M. E. (1988) *Collagen.* 4 Vols: 1. *Biochemistry*; 2. *Biochemistry and Biomechanics*; 3. *Biotechnology*; 4. *Molecular Biology*, B. R. Olsen (co-ed.) CRC Press, Boca Raton, FL.

Olsen, B. R., Gerecke, D., Gordon, M., Green, G., Kimura, T., Konomi, H., Muragaki, Y., Ninomiya, Y., Nishimura, I., and Sugrue, S. (1988). A new dimension in the extracellular matrix. In *Collagen*, M. Nimni (ed.) CRC Press, Boca Raton, FL, Vol. 4.

Patel, D. J., Carew, T. E., and Vaishnav, R. N. (1968) Compressibility of the arterial wall. *Circulation Res.* **23**, 61–68.

Patel, D. J., Tucker, W. K., and Janicki, J. S. (1970) Dynamic elastic properties of the aorta in radial direction. *J. Appl. Physiol.* **28**, 578–582.

Patel, D. J. and Vaishnav, R. N. (1972) The rheology of large blood vessels. In *Cardiovascular Fluid Dynamics*, D. H. Bergel (ed.) Academic, New York, Vol. 2, pp. 1–64.

Pereira, J. M., Mansur, J. M., and Davis, B. R. (1991) The effects of layer properties on shear disturbance propagation in skin. *J Biomech. Eng.* **113**, 30–35.

Pinto, J. and Fung, Y. C. (1973) Mechanical properties of the heart muscle in the passive state, and stimulated papillary muscle in quick-release experiments. *J. Biomech.* **6**, 597–616, 617–630.

Pipkin, A. C. and Rogers, T. G. (1968) A nonlinear integral representation for viscoelastic behavior. *J. Mech. Phys. Solids* **16**, 59–74.

Raju, K. and Anwar, R. A. (1987) Primary structures of the bovine elastin *a*, *b*, and *c* deduced from the sequences of cDNA clones. *J. Biol. Chem.* **262**, 5755–5762.

Ramachandran, G. N. (ed.) (1967) Treatise on collagen. Vol. 1. Chemistry of Collagen. Vol. 2. Biology of Collagen. Vol. 3. Chemical Pathology of Collagen. Univ. Madra, India. Academic Press, New York.

Ridge, M. D. and Wright, V. (1964) The description of skin stiffness. *Biorheology* **2**, 67–74.

Ridge, M. D. and Wright, V. (1966) Mechanical properties of skin: A bioengineering study of skin texture. *J. Appl. Physiol.* **21**, 1602–1606.

Riedl, H., Nemetschek, T., and Jonak, R. (1980) A mathematical model for the changes of the long-period structure of collagen. In *Biology of Collagen*, A. Viidik and J. Vuust (eds.) Academic Press, San Diego, pp. 289–296.

Rivlin, R. S. (1947) Torsion of a rubber cylinder. *J. Appl. Phys.* **18**, 444–449.

Rivlin, R. S. and Saunders, D. W. (1951) Large elastic deformations of isotropic materials VII. Experiments on the deformation of rubber. *Philos. Trans. Soc. London A* **243**, 251–288.

Routbart, J. L. and Sack, H. S. (1966) Background internal friction of some pure metals at low frequencies. *J. Appl. Phys.* **37**, 4803–4805.

Schneider, D. (1982) Viscoelasticity and tearing strength of the human skin. Ph. D. Dissertation. Department of AMES/Bioengineering. University of California, San Diego.

Shoemaker, P. A. (1984) Irreversible thermodynamics, the constitutive law, and a constitutive model for two-dimensional soft tissues. Ph. D. Dissertation, University of California, San Diego.

Shoemaker, P. A., Schneider, D., Lee, M. C., and Fung, Y. C. (1986) A constitutive model for two-dimensional soft tissues an its application to experimental data. *J. Biomech.* **19**, 695–702.

Sidrick, N. (1976) Constitutive equation of rabbit skin subjected to shear stress. The torsion test. M. S. Thesis, AMES Bioengineering Department. University of California, San Diego.

Snyder, R. W. (1972) Large deformation of isotropic biological tissue. *J. Biomech.* **5**, 601–606.

Sobin, S. S., Fung, Y. C., and Tremer, H. M. (1988) Collagen and elastin fibers in human pulmonary alveolar walls. *J. Appl. Physiol.* **64**: 1659–1675.

Theodorsen, T. and Garrick, E. (1940) *Mechanism of Flutter.* Rept. 685, U. S. Nat. Adv. Comm. Aeronaut.

Tong, P. and Fung, Y. C. (1976) The stress–strain relationship for the skin. *J. Biomech.* **9**, 649–657.

Tong, P. and Fung, Y. C. (1976) The stress–strain relationship for the skin. *J. Biomech.* **9**, 649–657.

Torp, S., Arridge, R. G. C., Armeniades, C. D., and Baer, E. (1974) Structure-property relationships in tendon as a function of age. In *Proc. 1974 Colston Conference*, Dept. of Physics, University of Bristol, U. K., pp. 197–222. See also, pp. 223–250.

Treloar, L. R. G. (1967) *The Physics of Rubber Elasticity*, 2nd edition. Oxford University Press, New York.

Urry, D. W. (1985) Protein elasticity based on conformations of sequential polypeptides: the biological elastic fiber. *J. Protein Chem.* **3**, 403–436.

Urry, D. W., Haynes, B., and Harris, R. D. (1986) Temperature dependence of length of elastin and its polypentapeptide. *Biochem. and Biophys. Res. Commun.* **141**, 749–755.

Urry, D. W. (1991) Thermally driven self-assembly, molecular structuring, and entropic mechanisms in elastomeric polypeptides. In *Molecular Conformation and Biological Interaction* (G. N. Ramachandran Festschrift), P. Balaram and S. Ramaseshan (eds.) Indian Academy of Science, Bangalore, India, pp. 555–583.

Urry, D. W. (1992) Free energy transduction in polypeptides and proteins based on inverse temperature transitions. *Progr. Biophys. Mol. Biol.* **57**, 23–57.

Valanis, K. C. and Landel, R. I. (1967) The strain-energy function of hyperelastic material in terms of the extension ratios. *J. Appl. Phys.* **38**, 2997–3002.

Van Brocklin, J. D. and Ellis, D. (1965) A study of the mechanical behavior of toe extensor tendons under applied stress. *Arch. Phys. Med. Rehab.* **46**, 369–375.

Vawter, D., Fung, Y. C., and West, J. B. (1978) Elasticity of excised dog lung parenchyma. *J. Appl. Physiol.* **45**, 261–269.

Vawter, D., Fung, Y. C., and West, J. B. (1979) Constitutive equaton of lung tissue elasticity. *J. Biomech. Eng. Trans. ASME* **101**, 38–45.

Veronda, D. R. and Westmann, R. A. (1970) Mechanical characterizations of ski-finite deformations. *J. Biomech.* **3**, 111–124.

Viidik, A. (1966) Biomechanics and functional adaptation of tendons and joint ligaments. In *Studies on the Anatomy and Function of Bone and Joints*, F. G. Evans (ed.) Springer-Verlag, New York, pp. 17–39.

Viidik, A. (1968) A rheological model for uncalcified parallel-fibered collagenous tissue. *J. Biomech.* **1**, 3–11.

Viidik, A. (1978) On the correlation between structure and mechanical function of soft connective tissues. *Verh. Anat. Ges.* **72**, 75–89.

Viidik, A. and Vuust, J. (eds.) (1980) *Biology of Collagen* Academic Press, New York. Chapter 17, by Viidik, Mechanical Properties of Parallel-fibered Collagenous Tissues, pp. 237–255; Chapter 18, by Viidik, Interdependence between Structure and Function in Collagenous Tissues, pp. 257–280.

Viidik, A. (1990) Structure and function of normal and healing tendons and ligaments. In *Biomechanics of Diarthrodial Joints*, Mow, Ratcliffe, and Woo (eds.) Springer-Verlag, New York, pp. 3–38.

Wagner, K. W. (1913) Zur theorie der unvoll Kommener dielektrika. *Ann. Phys.* **40**, 817–855.

Wertheim, M. G. (1847) Memoire sur l'elasticite et la coheison des principaux tissus du corps humain. *Ann. Chimie Phys. Paris* (Ser. 3), **21**, 385–414.

Westerhof, N. and Noodergraaf, A. (1970) Arterial viscoelasticity: A generalized model. *J. Biomech.* **3**, 357–379.

Wilson, T. A. (1972) A continuum analysis of a two-dimensional mechanical model for the lung parenchyma. *J. Appl. Physiol.* **33**, 472–478.

Wineman, A. S. (1972) Large axially symmetric stretching of a nonlinear viscoelastic membrane. *Int. J. Solids Struct.* **8**, 775–790.

Wineman, A., Wilson, D., and Melvin, J. W. (1979) Material identification of soft tissue using membrane inflation. *J. Biomech.* **12**, 841–850.

Wineman, A. S. and Rajagopal, K. R. (1990) On a constitutive theory for materials undergoing microstructural changes. *Arch. Mech.* **42**, 53–75, Warszawa.

Woodhead-Galloway, J. (1980) Collagen: The anatomy of a protein. Arnold, London.

Yager, D., Feldman, H., and Fung, Y. C. (1992) Microscopic vs. macroscopic deformation of the pulmonary alveolar duct. *J. Appl. Physiol.* **72**, 1348–1354.

Zeng, Y. J., Yager, D., and Fung, Y. C. (1987) Measurement of the mechanical properties of the human lung tissue. *J. Biomech. Eng.* **109**, 169–174.

Young, J. T., Vaishnav, R. N., and Patel, D. J. (1977) Nonlinear anisotropic viscoelastic properties of canine arterial segments. *J. Biomech.* **10**, 549–559.

CHAPTER 8

Mechanical Properties and Active Remodeling of Blood Vessels

8.1 Introduction

Blood vessels belong to the class of soft tissues discussed in Chapter 7. They do not obey Hooke's law. Figure 7.5:1 in Chapter 7, Sec. 7.5 demonstrates the nonlinearity of the stress–strain relationship and the existence of hysteresis. They also creep under constant stress and relax under constant strain. These mechanical properties have a structural basis. In Sec. 8.2 we consider the structure of the blood vessel wall. From Sec. 8.3 on, however, our attention will be concentrated on the mathematical description of the mechanical properties. In seeking simplification whenever it is justifiable, we take advantage of the fact that most blood vessels are thin-walled tubes deforming axisymmetrically (including inflation, longitudinal stretching, and torsion), and that as far as hemodynamics is concerned, we need to know only the relationship between the blood pressure and the inner diameter of the tube. In this situation we may treat the vessel wall as a membrane. The stresses of concern are circumferential and longitudinal, the principal strains are also circumferential and longitudinal. The vessel wall may be treated as a two-dimensional body, the constitutive equation is biaxial. In Secs. 8.3–8.5, we formulate a two-dimensional quasi-linear viscoelastic constitutive equation for blood vessels, using the pseudo-elasticity concept introduced in Chapter 7. In Secs. 8.7 and 8.8, we treat the blood vessel wall as a three-dimensional body, and study the differences between the mechanical properties of the intima-media layer and those of the adventitia. The results can then be applied to general three-dimensional problems such as bifurcation, aneurysm, surgery, etc. In Secs. 8.9–8.11, we consider the mechanical properties of arterioles, capillary blood vessels, and veins.

Finally, in Secs. 8.12–8.16, we turn to another feature of living tissue: its ability to remodel itself when the stress and strain acting on the tissue are changed from homeostatic values. Remodeling of tissues due to stress changes in any organ is usually nonhomogeneous because stress distribution is in general nonhomogeneous. Hence remodeling is a three-dimensional phenomenon, to which the discussions of Secs. 8.7–8.9 are relevant.

8.2 Structure and Composition of Blood Vessels

The architecture of blood vessel wall is sketched in Fig. 8.2:1. The blood vessel wall consists of three layers: the intima, media, and adventitia. The intima is the innermost layer and contains the endothelial cells. The media is the middle layer and contains the smooth muscle cells. The adventitia is the outermost layer and is mainly collagen fibers and ground substances. Figure 8.2:1 shows that the proportions of these three layers vary according to the size of the vessel. The exact definition of the intima has some uncertainty. According to Rhodin

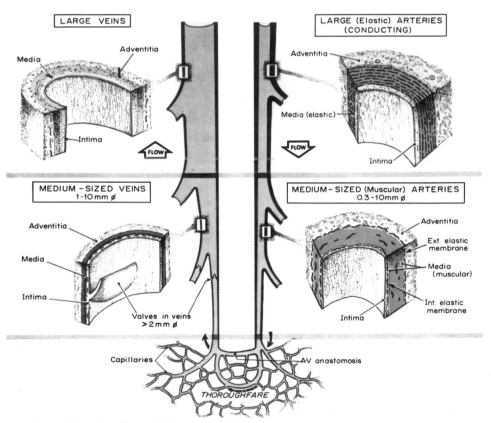

Figure 8.2:1 Rhodin's (1980) sketch of mammalian blood vessels showing the various components of the vascular wall. ϕ, diameter. From *Handbook of Physiology, Sec. 2. The Cardiovascular System.* Vol. II, p. 2. Reproduced by permission of the author and the American Physiological Society.

(1980), biochemists and physiologists often consider intima synonymous with endothelium, whereas pathologists use the word for the subendothelial layer alone. Most anatomists and cell biologists define the intima as composed of the endothelial cells, the basal lamina (~ 80 nm thick), and the subendothelial layer composed of collagenous bundles, elastic fibrils, smooth muscle cells, and perhaps some fibroblasts. The subendothelial layer is usually present only in the large elastic arteries such as human aorta, whereas in the majority of other blood vessels the intima consists of only endothelial cells and basal lamina.

The tunica media is made up of smooth muscle cells, a varied number of elastic sheets, bundles of collagenous fibrils, and a network of elastic fibrils. Its dividing line with the adventitia is a layer of elastin.

The adventitial layer contains collagen fibers, ground substances, and some fibroblasts, macrophages, blood vessels (vasa vasorum), myelinated nerves, and nonmyelinated nerves.

Figure 8.2:2 is a photomicrograph of the intima-media layer of the thoracic aorta of the rat at the zero stress state. Note the elastic layer between muscle cells.

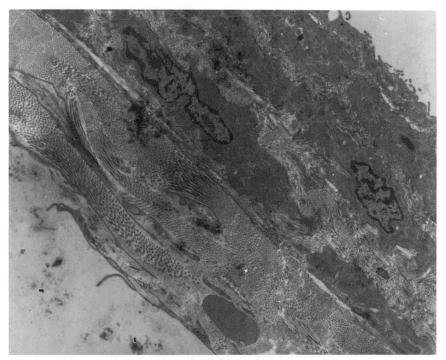

Figure 8.2:2 A photograph of a region of the intima media of the thoracic artery wall of the rat, showing the major components. Vascular smooth muscle cells, elastin layers, collagen fibrils, and ground substance.

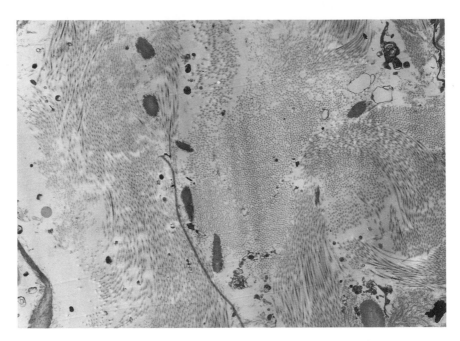

Figure 8.2:3 A photograph of a region of the adventitia of the thoracic artery wall of the rat, showing collagen fibrils, ground substance, and some fibroblasts.

TABLE 8.2:1 Percentage Composition of the Media and Adventitia of Several Arteries at In Vivo Blood Pressure (Mean ± S.D.)

	Pulmonary artery	Thoracic aorta	Plantar artery
Media			
Smooth muscle	46.4 ± 7.7	33.5 ± 10.4	60.5 ± 6.5
Ground substance	17.2 ± 8.6	5.6 ± 6.7	26.4 ± 6.4
Elastin	9.0 ± 3.2	24.3 ± 7.7	1.3 ± 1.1
Collagen	27.4 ± 13.2	36.8 ± 10.2	11.9 ± 8.4
Adventitia			
Collagen	63.0 ± 8.5	77.7 ± 14.1	63.9 ± 9.7
Ground substance	25.1 ± 8.3	10.6 ± 10.4	24.7 ± 9.3
Fibroblasts	10.4 ± 6.1	9.4 ± 11.0	11.4 ± 2.6
Elastin	1.5 ± 1.5	2.4 ± 3.2	0

Figure 8.2:3 is a photomicrograph of the adventitia of the same vessel at zero stress state at a higher magnification. Figure 8.2:4 shows the histogram of the diameter of the collagen fibrils in the adventitia. It is seen that the diameters of the collagen fibrils spread over quite a wide range. Table 8.2:1

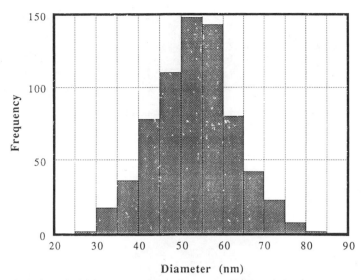

Figure 8.2:4 A typical histogram of the diameter of collagen fibrils in the adventitia of the thoracic artery of the rat.

shows the composition of the media and adventitia layers, i.e., the percentage of volume of the smooth muscle cells, collagen bundles, elastin aggregates, fibroblast cells, and ground substances in media and adventitia. Morphometric data of this kind are essential when one attempts to correlate the mechanical properties of the vessel with the structure of the materials. The collagen in the adventitia and media of blood vessel is mostly Type III, some Type I, and a trace of Type V. The collagen of the basal lamina is Type IV.

The blood vessel wall itself is supplied with blood flow (except the smallest blood vessels, whose cells are within a short distance, say 25 μm, from blood). The blood vessels perfusing larger blood vessel walls are called *vasa vasorum*. An illustration of the vasa vasorum in a pulmonary artery is given in Fig. 8.2:5. Arteries seem to be somewhat less dependent on the vasal supply for their integrity. Veins, however, rapidly degenerate when the vasal supply is interrupted.

The structure of blood vessels varies along the arterial tree. In large arteries the number of lamellar layers increases with wall thickness. In smaller arteries, the relative wall thickness is increased, the elastin is less prominent in the media, and with increasing distance from the aorta, eventually only the inner and outer elastic laminae can be clearly seen. The muscle fibers increase in amount, arranged in quasi-concentric layers with prominent muscle–muscle attachment, and in a helical pattern that has finer pitch in the more peripheral vessels. In the capillaries only the endothelium remains.

The basic structure of the veins is similar to that of the arteries. The relative wall thickness is generally lower than in the arteries, and the media contains very little elastic tissue. The intimal surface of most larger veins has

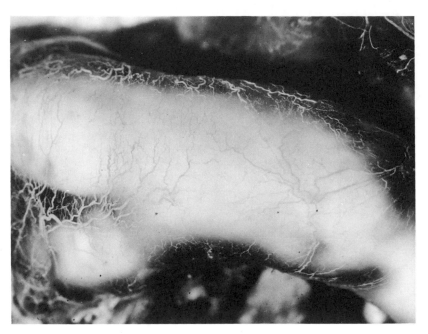

Figure 8.2:5 Photograph of the pulmonary artery of a rabbit, showing the vasa vasorum in the wall. The wall substance has been rendered transparent by glycerin immersion. The microvessels have been injected with silicone elastomer. The white background is an injected mass filling the cavity of the artery. From W. G. Frasher (1966), by permission.

crescentic semilunar valves, generally arranged in pairs and associated with a distinct sinus or local widening of the vessel. The adventitia of veins is relatively thick and contains much collagen. In the abdominal vena cava and its main tributaries, and in the mesenteric veins, there are prominent longitudinal muscle fibers. Veins contain a relatively high amount of collagen; the elastin/collagen ratio is about 1:3.

Roach and Burton (1957) studied vessel elasticity before and after digestion with formic acid or trypsin to remove elastin and collagen, respectively. They showed that the slopes of the distensibility curves of the trypsin treated specimens were similar to those of the untreated specimens in the low stress region, whereas in the high stress region, whether the elastin was depleted or not did not seem to matter. By this method of attack they correlated the overall mechanical properties of a vessel with the content and structure of the material components.

8.3 Arterial Wall as a Membrane: Behavior Under Uniaxial Loading

The arterial wall is, of course, a three-dimensional body. Stresses vary across the thickness of the wall. But blood vessel is a tube subjected to blood pressure.

As it can be seen from the theory of thin shells, the stresses and strains in the vessel wall can be separated into two parts: the mean values and the deviations from the mean. The mean circumferential, longitudinal, and shear stresses are uniform across the wall. If transverse shear can be neglected, then the differential equations governing these mean stresses are exactly the same as those equations governing a shell whose wall thickness is very small; i.e., a curved membrane coinciding with the midsurface of the shell. The constitutive equation needed is the one relating the membrane stress with the membrane strain. Thus, if we are interested only in determining the mean stresses in the shell under the hypothesis that the transverse shear vanishes, then the mathematical problem is a two-dimensional one. This reduction of a three-dimensional problem into a two-dimensional one is a great simplification. Hence we shall devote this and the following four sections to the study of the arterial wall as a membrane.

There are, of course, other occasions in which the determination of the deviation of stresses from the mean is important. Then it is necessary to consider the arterial wall as a three-dimensional body. All problems involving bending of the wall must consider the shell as a three-dimensional body. Problems in which the external load is nonaxisymmetric, localized, or concentrated are three-dimensional. Problems concerning curved or branching arteries or vessels with stenosis or aneurysin belong to this category. We devote Secs. 8.7 and 8.8 to the study of arterial wall as a three-dimensional body.

Considering the arterial wall as a membrane, the simplest experiment that can be done is the uniaxial test. Take a longitudinal or circumferential strip of a vessel wall with the shape of a rectangular parallelopiped. Pull it lengthwise, and record the force-elongation relationship. The lateral sides are left free. At the no-load condition, let the length of the specimen be L_0, the width W_0, and the thickness H_0. Under a load, the length becomes L. The ratio of L to the initial length L_0 is the *stretch ratio*, λ. The load divided by the initial cross-sectional area A_0 yields the tensile stress T:

$$T = \frac{\text{load}}{A_0}. \tag{1}$$

An example of the stress–strain relationship, $T(\lambda)$, for a specimen of a dog's aorta which was loaded at a constant rate until a maximum tension was reached and then immediately unloaded at the same constant rate (triangular strain history), is shown in Fig. 7.5:1.

After a sufficient number of repeated cycling the stress–strain loop is stabilized and does not change any further. Then the stress can be related to the strain in the form

$$T = f_1(\lambda) \qquad \text{when} \quad \frac{d\lambda}{dt} > 0, \tag{2}$$

$$T = f_2(\lambda) \qquad \text{when} \quad \frac{d\lambda}{dt} < 0, \tag{3}$$

where f_1, f_2 are two functions. If $f_1(\lambda) = f_2(\lambda)$, then the material is *elastic*. If $f_1(\lambda) \neq f_2(\lambda)$, then the material is *inelastic*. Figure 7.5:1 shows that arteries are inelastic.

As is discussed in Chapter 7, Sec. 7.6, the hysteresis loop of arteries is rather insensitive to strain rate. Mathematically this is equivalent to saying that $f_1(\lambda)$, $f_2(\lambda)$ depend on the instantaneous values of λ and that their dependence on the rate of change of λ may be ignored in many cases.

A convenient way to examine the functions $f_1(\lambda)$ and $f_2(\lambda)$ is to plot the slope $dT/d\lambda$ against T. An example is shown in Fig. 8.3:1 [from Tanaka and Fung (1974)]. If a material obeys Hooke's law, such a plot will show a horizontal straight line. The example shows that the artery does not obey Hooke's law. For tensile stress T greater than 20 kPa and less than about 60 kPa (a physiological range, but much below the breaking strength of the aorta) the curves in Fig. 8.3:1 can be represented by straight lines, and the following approximation is valid:

$$dT/d\lambda = \alpha(T + \beta) = \alpha T + E_0, \tag{4}$$

whre α, β, and $E_0 = \alpha\beta$ are constants. This implies an exponential stress–strain relationship (Fung, 1967)

$$T = (T^* + \beta)e^{\alpha(\lambda - \lambda^*)} - \beta \tag{5}$$

for $20 < T < 60$ kPa, where (λ^*, T^*) represents one point on the stress–strain curve in the region of validity of Eq. (4).

The mean values of the physiological constants α and E_0, defined for tensile stress in the range 20–60 kPa, are presented in Fig. 8.3:2 for aortic specimens obtained by cutting strips in the longitudinal and circumferential directions along the aortic tree of the dog. The parameter α represents the rate of increase of the Young's modulus with respect to increasing tension (corresponding to the slope of the curves in Fig. 8.3:1 at stress T greater than 20 kPa). Figure 8.3:2 shows that for circumferential segments of arteries, α increases toward the periphery, i.e., arteries become more highly nonlinear further away from the heart. The corresponding data on α for the longitudinal segments show a less regular variation along the aortic tree. The parameter E_0 is the intercept of the straight-line segment extended to $T = 0$. Figure 8.3:2 shows that E_0 varies greatly along the aortic tree; and its value for longitudinal segments can be very different from that of the circumferential segments; the more so toward the periphery, further away from the heart. These differences undoubtely reflect the changes in material composition and configuration of collagen, elastin, and smooth muscle fibers in the arterial wall along the length of the aorta.

Frank (1920) was the first to study the relationship between Young's modulus $(dT/d\lambda)$ and stress (T). The linear relationship given in Eq. (4) is not inconsistent with the curves published by Frank [see Kenner (1967) and Wetterer and Kenner (1968)]. Laszt (1968), Hardung (1953), and Bauer

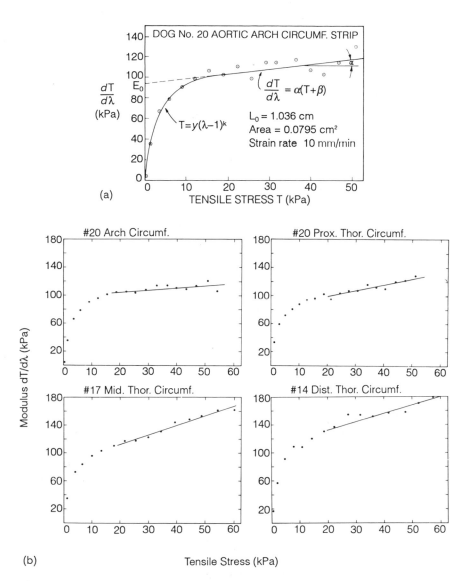

Figure 8.3:1 Plot of the Young's modulus (tangent modulus, $dT/d\lambda$) vs. the tensile stress (T) in a specimen of thoracic aorta of the dog in a loading process. Part (a) shows a power law for small T and an approximate straight line for T greater than 20 kPa. Part (b) shows the straight-line representation for various segments of the aorta. The straight line represents Eq. (5). From Tanaka and Fung (1974), by permission.

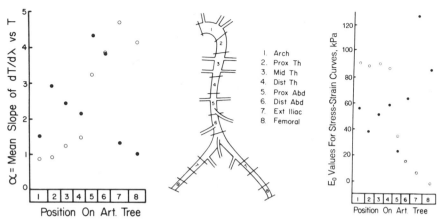

Figure 8.3:2 Middle panel: Sketch of arterial segments. Left panel: variation of α along the aortic tree. Open circles, circumferential segments. Filled circles, longitudinal segments. Right panel: variation of the constant E_0 of Eq. (4) describing the stress–strain curve in an exponential form. Open circles: circumferential segments. Filled circles: longitudinal segments. Numbering of segments: see middle panel. From Tanaka and Fung (1974), by permission.

and Pasch (1971) presented similar data. Tanaka and Fung (1974) pointed out that such a relationship is valid only if the stress is sufficiently large; for low stress levels they showed that a power law fits the experimental data. A power law was used by Wylie (1966) in a theoretical analysis of arterial waves. The variation of the mechanical properties of blood vessels along the arterial tree has been studied by McDonald (1968), Anliker (1972), and others by means of arterial pulse waves, and by Azuma, Hasegawa, and Matsuda (1970) and Azuma and Hasegawa (1971) by *in vitro* experiments.

8.3.1 Stress Relaxation at a Constant Strain

As is the case with other tissues, the relaxation function of the aorta depends on the stress level. In the physiological range with stress $T > 20$ kPa, the relaxation function is normalizable; i.e., it is possible to express the stress history in response to a *step change in strain* in the form

$$T(t, \lambda) = G(t) * T^{(e)}(\lambda) \qquad G(0) = 1, \tag{6}$$

where $G(t)$ is a function of time, and $T^{(e)}(\lambda)$ is a function of strain, the $*$ symbol being a multiplication sign. If the strain history were continuously variable, it may be regarded as consisting of a superposition of many step functions, and the resulting stress history would still be given by Eq. (6), but the symbol $*$ has to be interpreted as a convolution integration. The function $G(t)$

is the *normalized relaxation function*. The function $T^{(e)}(\lambda)$ is the *pseudo-elastic stress* corresponding to the instantaneous stress ratio λ.

The mean normalized relaxation function $G(t)$ of the dog's aorta is shown in Fig. 8.3:3 (from Tanaka and Fung, 1974), in which the horizontal axis indicates the time elapsed after the step increase of strain, plotted on a logarithmic scale, whereas the vertical axis is the ratio of the tensile stress at time t to that at the end of step strain. Figure 8.3:3 refers to circumferential segments of the arteries. Longitudinal segments are similar. The relaxation function varies from specimen to specimen. Figure 8.3:3 shows the mean and standard deviation at each instant of time following the step load. The variability, indicated by the standard deviation, is the least for the aortic arch, and increases toward the periphery. Since the standard deviation is caused by differences in age, sex, etc., the results shown in Fig. 8.3:3 suggest that the effect of these factors on the mechanical properties of arteries is bigger in the smaller arteries.

The value of the normalized relaxation function of $G(t)$ at a given value of t is smaller for smaller arteries. This means that the smaller arteries relax faster and more fully. Stress relaxation is very fast at the beginning; then it slows down logarithmically. When $t = 1$ sec, 9% of the tension in an aortic arch is relaxed, whereas in a femoral artery 20% of the tension is relaxed. At $t = 300$, most of the relaxation has taken place. In the literature it is often postulated that the normalized relaxation function $G(t)$ decreases with time like a logarithm of t. Thus

$$G(t) = \mathcal{R} \log t + d. \tag{7}$$

From Fig. 8.3:3, it is seen that this representation is not very good for the arteries. Nevertheless, it is a useful approximation for relatively small value of t, say for $1 < t < 100$ sec.

The variation of the viscoelastic property of the artery along the arterial tree can be examined by comparing certain characteristic constants of the curves of relaxation. One of the characteristic numbers is the slope of the relaxation curve vs. $\log t$ at a specific value of t. This is the constant $\mathcal{R}$ of Eq. (7). With $\mathcal{R}$ evaluated at $t = 1$ sec, the results are shown in the right-hand panel of Fig. 8.3:4. It is seen that $\mathcal{R}$ is negative, and for circumferential segments, the absolute value of $\mathcal{R}$ increases toward the periphery away from the heart. For the longitudinal segments the absolute value of $\mathcal{R}$ decreases at first until the level of abdominal aorta, then increases toward smaller arteries. Another characteristic parameter is the value of $G(t)$ at a larger value of t. For $G(t)$ at 300 sec, the results are shown in the left-hand panel of Fig. 8.3:4. Here we see that smaller arteries relax more.

The variation of arterial elasticity and viscoelasticity along the arterial tree can be explained qualitatively by the composition and structure of the arterial wall. Elastin is soft, elastic, and relaxes very little, while collagen is also elastic, but has a much larger Young's modulus than elastin, and relaxes more than elastin. Smooth muscle has a low modulus of elasticity,

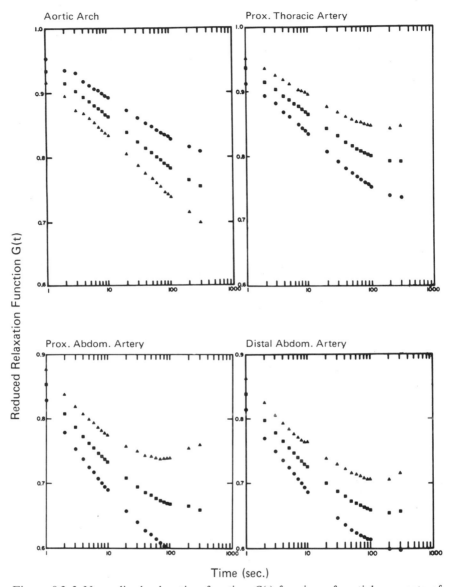

Figure 8.3:3 Normalized relaxation function $G(t)$ for circumferential segments of arteries. Mean $\pm$ s.d. ($n = 10$). Note that the vertical coordinates range over 0.6 to 1.0 in the upper panels; but they are from 0.6 to 0.9 in the lower panels. $G(t)$ is dimensionless. From Tanaka and Fung (1974), by permission.

Note: By definition, $G(0)$ is 1. At each value of the time t, the middle data point (square) represents the mean value of $G(t)$ in 10 experiments, whereas the circle and triangle symbols represent the mean $\pm$ one standard deviation. Note the very large standard deviations in abdominal arteries at large t. The mean value of $G(t)$ (as well as each individual relaxation curve) never increases with increasing time. Do not read $G(t) \pm$ s.d. as $G(t)$.

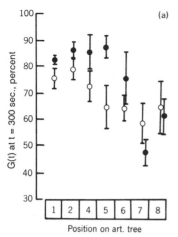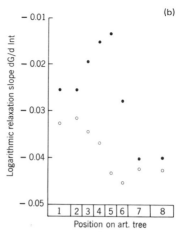

Figure 8.3:4 (a) Variation of the percentage value of the normalized relaxation function $G(t)$ at $t = 300$ sec along the aortic tree. Open circles: circumferential segments. Filled circles: longitudinal segments. Numbering of segments: see Fig. 8.3:2. (b) Variation of the slope of the curve of normalized relaxation function $G(t)$ vs. log t evaluated at $t = 1$ sec. Open circles: circumferential segments. Filled circles: longitudinal segments. Numbering of segments: see Fig. 8.3:2. From Tanaka and Fung (1974), by permission.

but a tremendous relaxation. In terms of the parameters discussed above, smooth muscle has a large negative $\mathscr{R}$, and a very small $G(t)$ at $t = 300$ sec. The results shown in Fig. 8.3:4 reflect the change in composition of the arterial wall along the arterial tree. Azuma and his associates (1970, 1971, 1973) have explained qualitative correlation between mechanical properties and arterial wall composition. Moritake et al. (1973) and Hayashi et al. (1971, 1974) have examined the mechanical properties and degenerative process of brain arteries in post-mortem specimens of brain aneurysm. They showed that aging and aneurysm are associated with an increase of collagen and a decrease of smooth muscle in these vessels, while the percentage of elastin remained almost constant.

We have based the discussion above on Eqs. (6) and (7), which are useful approximations. For greater accuracy the relaxation function $G(t)$ depends somewhat on the stretch ratio λ. See Sharma (1974), and Wetterer, Bauer, and Busse (1978). Such a dependence exists because the structure of the arterial wall changes with the stress level.

8.3.2 Creep Under a Constant Load

A typical curve showing creep under a constant load is shown in Fig. 8.3:5, in which the abscissa indicates time on a logarithmic scale, and the vertical axis is the change of strain after the imposition of a step loading. It is seen

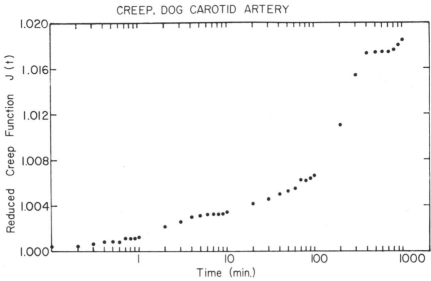

Figure 8.3:5 A typical creep curve, plotted as a reduced creep function $J(t) = [\lambda(t) - 1]/[\lambda(\theta) - 1]$. Dog carotid artery. From Tanaka and Fung (1974), by permission.

that the creep of the dog's artery is remarkably small. In the *in vitro* tests, the creep process was irreversible; after a long period of creep, the specimen could not return to its original length after the load was removed.

8.3.3 Unification of Hysteresis and Relaxation

Since relaxation, hysteresis, and the difference between static and dynamic elasticity are revelations of the same phenomenon, there should exist a unified relation among them. As is discussed in Chapter 7, Sec. 7.6, a method proposed by Fung (1972) consists of interpreting the multiplication sign * in Eq. (6) as a convolution operator, so that the right-hand side of Eq. (6) represents a convolution integral between $G(t)$ and $T^{(e)}[\lambda(t)]$. The factor $T^{(e)}[\lambda]$ is nonlinear. The convolution operator is linear. The equation is therefore quasi-linear.

It is intuitively clear that we have no right to expect the quasi-linear relationship to be valid for the full range of the inelastic phenomenon in a large deformation. In the analogous case of metals, rubber, and other high polymers, it is known that the viscoelastic law [Eq. (6)] works for small strains, but not for creep to failure. We expect the same to be true with biological tissues.

Fung (1972) reasoned that since the stress–strain relationship of arteries is found to be insensitive to strain rate over a wide range of the rates, the relaxation spectrum must be broad. If the relaxation function is written in the form

$$G(t) = \frac{1}{A}\left[1 + \int_0^\infty S(\tau)e^{-t/\tau}\,d\tau\right], \tag{8}$$

where

$$A = \left[1 + \int_0^\infty S(\tau)\,d\tau\right]$$

is a normalization factor, then the spectrum $S(\tau)$ is expected to be a continuous function of the relaxation time τ. A special form of $S(\tau)$ is proposed:

$$S(\tau) = c/\tau \qquad \text{for} \quad \tau_1 \le \tau \le \tau_2$$
$$= 0 \qquad \text{for} \quad \tau < \tau_1, \tau > \tau_2. \tag{9}$$

where c is a dimensionless constant. The theoretical relaxation and creep functions corresponding to Eq. (9) are given in Sec. 7.6. The theoretical slope of the relaxation curve vs. $\log t$ [Eq. (7)] is

$$\mathscr{R} = -c\left[1 + c\ln\left(\frac{\tau_2}{\tau_1}\right)\right]^{-1} \qquad \text{for} \quad \tau_1 < \tau < \tau_2. \tag{10}$$

When this procedure is applied to the aorta, the results listed in Table 8.3:1 are obtained. The comparison between the theoretical relaxation function and the experimental one is good. We must say, however, that the virtue of Eq. (9) does not lie only in the fact that it fits a specific relaxation

TABLE 8.3:1 The Constants c, τ_1, τ_2, Defining the Relaxation Spectrum of the Aorta in Eq. (9). From Tanaka and Fung (1974)

	c	τ_2(sec)	τ_1(sec)
Circumferential			
Arch	0.0424	434	0.367
Prox Th	0.0399	192	0.260
Mid Th	0.0459	286	0.211
Dist Th	0.0512	230	0.118
Prox Abd	0.0655	162	0.059
Dist Abd	0.0687	98.6	0.051
Ext Iliac	0.0726	282	0.015
Femoral	0.0646	119	0.040
Longitudinal			
Arch	0.0311	451	0.431
Prox Th	0.0297	93.9	0.137
Mid Th	0.0230	245	0.101
Dist Th	0.0178	757	0.051
Prox Abd	0.0153	428	0.064
Dist Abd	0.0373	452	0.0599
Ext Iliac	0.0832	2480	0.0065
Femoral	0.0638	107.5	0.0096

curve, but that, in addition, it makes the stress–strain relationship insensitive to the strain rate over a very broad range (between a strain rate of 0.001 to 1.0 length/sec in these examples).

8.4 Arterial Wall as a Membrane: Biaxial Loading and Torsion Experiments

A normal artery is a circular cylindrical tube subjected to a biaxial loading of an internal pressure, a longitudinal stretch, and a shear. Hence biaxial experiments on arteries are very popular. One instrument used by the author is shown in Fig. 1.14:4 in Chapter 1. Another instrument designed by Deng et al. is shown in Fig. 8.4:1. Similar instruments are used by other authors. In these experiments a segment of artery is soaked in physiological saline, tied at both ends to hollow cannulas, inflated by internal pressure, and stretched longitudinally. The diameter and the length between two gauge marks are measured by optical and electronic instruments without touching the tissue. End effect is minimized by selecting cannulas of suitable diameters and allow the gauge marks to be sufficiently far away from the ties.

The instrument designed by Deng et al. (1992) can impose a torsion on the specimen in addition to stretching and inflation, Fig. 8.4:1. This allows a shear stress-shear strain relationship to be measured in addition to the longitudinal and circumferential stresses and strains, as will be discussed at the end of this section.

Results of biaxial experiments are interpreted by regarding the arterial wall as a membrane. We use r, θ, and z for the radial, circumferential, and axial coordinates, respectively, in a cylindrical polar frame of reference with the z axis located at the center of the tube. We consider the mean stresses $\sigma_{\theta\theta}$ and σ_{zz} in the circumferential and axial directions as uniformly distributed, and assume that the transverse shear stresses $\sigma_{r\theta}$, σ_{rz} are zero. The radial stress σ_{rr} is considered either as zero or as a constant.

A membrane is a *two-dimensional* continuum. We can apply the strain-energy method discussed in Secs. 7.11 and 7.4 to the membrane. We assume that there exists a two-dimensional strain-energy function $W^{(2)}$ per unit mass, or $\rho_o W^{(2)}$ per unit volume at the zero stress state, ρ_o being the mass density, at zero stress. Then the circumferential, longitudinal, and shear stresses, $S_{\theta\theta}$, S_{zz}, and $S_{\theta z}$, respectively, can be derived from the strain-energy function as follows:

$$S_{\theta\theta} = \frac{\sigma_{\theta\theta}}{\lambda_\theta^2} = \frac{\partial(\rho_o W^{(2)})}{\partial E_{\theta\theta}},$$

$$S_{zz} = \frac{\sigma_{zz}}{\lambda_z^2} = \frac{\partial(\rho_o W^{(2)})}{\partial E_{zz}}, \tag{1}$$

$$S_{\theta z} = S_{z\theta} = \frac{\partial(\rho_o W^{(2)})}{\partial E_{\theta z}} = \frac{\partial(\rho_o W^{(2)})}{\partial E_{z\theta}}.$$

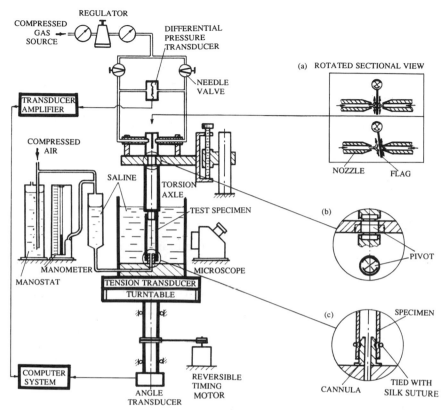

Figure 8.4:1 A test equipment for torsion, longitudinal stretching, and circumferential inflation of blood vessels in the author's laboratory.The torque is measured by the differential air pressure in the jets of an impinging on a little flag, and holding the flag in a null position. The specimen is twisted at the lower end. The torque generated is transmitted up through a hinge. Longitudinal force is measured by a force transducer. Internal and external pressures are controlled. The specimen is soaked in a saline bath. Equipment designed by Dr. Deng Shanxi in the author's lab.

Here λ_θ and λ_z are the stretch ratios of the middle surface of the blood vessel wall in the circumferential and axial directions, respectively, and

$$E_{\theta\theta} = \tfrac{1}{2}(\lambda_\theta^2 - 1), \qquad E_{zz} = \tfrac{1}{2}(\lambda_z^2 - 1) \qquad (2)$$

are the circumferential and axial strains defined in the sense of Green. $S_{\theta\theta}$, S_{zz}, and $S_{\theta z}$ are Kirchhoff's stresses, $\sigma_{\theta\theta}$, σ_{zz}, $\sigma_{\theta z}$ are Cauchy's stresses, $\rho_o W^{(2)}$ is a function of $E_{\theta\theta}$, E_{zz} and $E_{\theta z}$, $E_{z\theta}$. The experimental results are used to determine the two-dimensional strain energy function $\rho_o W^{(2)}$. Note that the blood vessel wall, regarded as a two-dimensional body, is not "incompressible" (i.e., its area can change). Hence there is no indeterminacy in the mean stress.

Let subscripts i and o denote the inner and outer surfaces, respectively. Then, for a tube of an inner radius r_i and an outer radius r_o, subjected to an inner pressure p_i and an outer pressure p_o, the equilibrium condition yields

$$\text{average value of } \sigma_{\theta\theta} = \frac{p_i r_i - p_o r_o}{h},$$

$$\text{average value of } \sigma_{zz} = \frac{F + p_i \pi r_i^2 - p_o \pi r_o^2}{\pi(r_o^2 - r_i^2)}, \tag{3}$$

where F is the force applied at the ends of the vessel, and h is the thickness of the wall. See Problem 1.4 in Chapter 1. The mathematical analysis is presented below.

8.4.1 Shear Modulus of Elasticity

Torsion experiment with the equipment sketched in Fig. 8.4:1 is important because it provides the shear modulus of elasticity of the blood vessel wall. When the vessel wall is regarded as a membrane in biaxial state of stress, inclusion of shear in the constitutive equation completes the equation in such a way that it can then be used for general biaxial problems.

For an isotropic elastic material obeying a linear stress–strain relationship, there is a well-known relationship between the shear modulus of elasticity G, the Young's modulus E, and the Poisson's ratio v (Eq. (2.8:5)):

$$G = \frac{E}{2(1 + v)}.$$

But this formula does not work for anisotropic materials. Further, for a soft tissue, the shear modulus increases with increasing level of both the shear and normal stresses.

8.4.2 Principle of Torsion Experiment

Consider a blood vessel segment as a circular cylindrical tube subjected to a transmural pressure p, a longitudinal force F, and a torque T (Fig. 8.4:2). Under loading, the vessel wall has outer and inner radii r_o and r_i, respectively, a cross-sectional area A, a length L, a total angle of twist θ, and an angle of twist per unit length, θ/L. The vessel wall material is considered incompressible, so that the volume of the wall under loading is constant:

$$AL = A_o L_o, \tag{4a}$$

where A_o and L_o are, respectively, the cross-sectional area and length of the vessel at the no-load condition ($p = F = T = 0$). Now,

$$A = \pi(r_o^2 - r_i^2). \tag{4b}$$

Hence we obtain

$$r_i = [r_o^2 - L_o A_o/(\pi L)]^{1/2}. \tag{5}$$

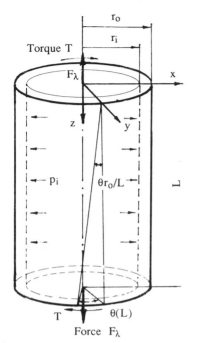
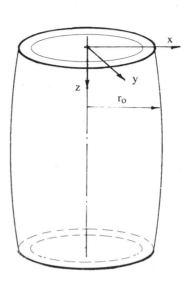

Figure 8.4:2 Torsion of a segment of blood vessel. Notations and symbols for analysis.

The ratio L/L_o is the *longitudinal stretch ratio* and is designated by the symbol λ_z. Let F_λ be the longitudinal force due to λ_z; F_θ be the longitudinal force due to torsion; F_p the longitudinal force due to blood pressure p. (F_p exists only in test specimens whose ends are plugged off.) Then the resultant longitudinal force acting on the vessel test specimen is

$$F = F_\lambda + F_\theta + F_p. \tag{6}$$

The force F_p in a specimen with plugged ends is

$$F_p = \pi p r_i^2. \tag{7}$$

The forces F_λ, F_θ have to be measured. The average longitudinal stress, σ_{zz}, is

$$\bar{\sigma}_{zz} = F/A. \tag{8}$$

The average circumferential stress, $\bar{\sigma}_{\theta\theta}$, is given by the Laplace formula

$$\bar{\sigma}_{\theta\theta} = p r_i/(r_o - r_i). \tag{9}$$

The torque in the cylinder, T, is equal to the product of the average shear stress in the wall, $\bar{\sigma}_{z\theta}$, the wall cross-sectional area, A, and the mean radius $(r_o + r_i)/2$. Hence,

$$\bar{\sigma}_{z\theta} = \bar{\sigma}_{\theta z} = T[A(r_o + r_i)/2]^{-1}. \tag{10}$$

The objective of our experiment is to determine the relationship between T and θ/L with fixed values of longitudinal and circumferential stresses, $\bar{\sigma}_{zz}$

and $\bar{\sigma}_{\theta\theta}$. One of the significant facts found in our experiment is that T is linearly proportional to θ/L in a range of interest of the variable θ/L. In this range, then, the shear stress $\bar{\sigma}_{z\theta}$ is linearly proportional to θ/L. Now, the shear strain at a radius r is, by definition,

$$e_{z\theta} = r\theta/2L. \tag{11}$$

Thus, the linearity named above implies that

$$\bar{\sigma}_{z\theta} = 2Ge_{z\theta} = Gr\theta/L. \tag{12}$$

Here r is the radius and G is the *shear modulus of elasticity*. G is independent of the shear strain. Although Eq. (12) looks like one, it is not a Hooke's law because G is a function of $\bar{\sigma}_{zz}$ and $\bar{\sigma}_{\theta\theta}$, and the total stress–strain relationship is nonlinear. The torque is given by an integral of the product of $\bar{\sigma}_{z\theta}$ and the moment arm r and the area $2\pi r\, dr$:

$$T = 2\pi \int_{r_i}^{r_o} G\frac{r\theta}{L}r^2\, dr = \frac{\pi}{2}\frac{G\theta}{L}(r_o^4 - r_i^4). \tag{13}$$

Writing

$$J = \frac{\pi}{2}(r_o^4 - r_i^4) \tag{14}$$

which is the polar moment of inertia of the vessel cross section, we obtain from Eq. (13) the familiar formula

$$T = GJ\frac{\theta}{L} \quad \text{or} \quad G = \frac{LT}{J\theta} \quad \text{or} \quad \frac{\theta}{L} = \frac{T}{GJ}. \tag{15}$$

8.4.3 A Necessary Refinement for the Real Case

The analysis presented above is made for a circular cylindrical tube of constant radius. The test specimens, however, have their ends tied to cannulas of fixed diameter which cannot vary with the inflation pressure. Hence, in reality, the shape of the specimen is not cylindrical, but bulged or necked in as sketched in the right-hand panel of Fig. 8.4:2. The outer and inner radii, r_o and r_i, are functions of z. The analysis must be refined to reflect this condition.

Even in the case of variable diameter, we assume the deformed tube to be axisymmetric in shape. Then, by the equations of static equilibrium, we know that the torque, T, remains independent of z, because there is no external torque acting on the cylinder. But since r_i and r_o are variable, the cross-sectional area, A, the polar moment of intertia, J, and the longitudinal stretch, λ_z, will vary with z. The rate of twist per unit length of the tube, which was written as θ/L before, now must be replaced by $d\theta/dz$, which is variable with z. The analysis presented above is valid for a segment of infinitesimal length. Equations (11) to (13) remain valid if we replace θ/L by $d\theta/dz$ and regard r_i,

r_o, λ_z, A, and J as functions of z. It follows that $\bar{\sigma}_{zz}$, $\bar{\sigma}_{\theta\theta}$, $\bar{\tau}_{z\theta}$, $G(\bar{\sigma}_{zz}, \bar{\sigma}_{\theta\theta})$ are now functions of z, whereas Eq. (15) yields

$$\frac{d\theta}{dz} = \frac{T}{G(z)J(z)}. \tag{16}$$

An integration yields

$$\theta(L) - \theta(0) = T \int_0^L [G(z)J(z)]^{-1} \, dz, \tag{17}$$

where $\theta(L)$ is the angle of twist at $z = L$, $\theta(0)$ is that at $z = 0$. $\theta(L)$ was denoted by θ before; $\theta(0)$ is defined as zero.

There are two ways to use these equations. If $d\theta/dz$ can be measured accurately, then $G(z)$ can be calculated from Eq. (16) to yield

$$G(z) = T[J(z)(d\theta/dz)]^{-1}. \tag{18}$$

Alternatively, if L can be varied and $\theta(L)$ can be measured accurately as a function of L, then Eq. (17) can be used to determine $G(z)$ as the solution of an integral equation.

TABLE 8.4:1 Shear Modulus G of Normal Rat Thoracic Aorta at a Physiological Pressure of 16 kPa (120 mm Hg)

Rat No.	1	2	3	4	5	6	Mean s.d.
G (kPa) at $\lambda_z = 1.2$	69	102	100	141	97	68	96 ± 29
G (kPa) at $\lambda_z = 1.3$	107	230	159	223	169	130	122 ± 48
Rat weight (gm)	558	525	532	474	476		513 ± 37
Rat age (days)	149	150	151	154	155	155	152 ± 3

TABLE 8.4:2 The Shear Modulus of the Thoracic Aorta of Normal Rats and Rats Subjected to Specified Days of Hypertension Caused by Aortic Constriction at the Celiac Artery Level. Internal Pressure, 16 kPa (120 mm Hg), External Pressure, 0 (Atmospheric)

| | | Shear modulus G | | Cross-sectional area |
| | | $\lambda_z = 1.2$ | $\lambda_z = 1.3$ | A_0 |
Days	No. of rats (n)	Mean s.d.	Mean s.d.	Mean s.d.
0	(6)	96 ± 29	169 ± 49	1.67 ± 0.08
2	(6)	122 ± 48	202 ± 102	1.68 ± 0.59
6	(3)	107 ± 5	170 ± 20	1.77 ± 0.05
10	(3)	139 ± 71	160 ± 0	1.94 ± 0.28
20	(3)	163 ± 66	97 ± 56	1.84 ± 0.04
30	(3)	128 ± 40	205 ± 123	1.94 ± 0.30

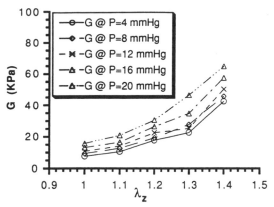

Figure 8.4:3 The shear modulus of the thoracic aorta of the pig as a function of the longitudinal stretch and transmural pressure.

The results obtained by Drs. S. X. Deng, J. Tomioka, and J. Debes in the author's lab are given in Table 8.4:1 for normal rat thoracic aorta, and Table 8.4:2 for hypertensive rat thoracic aorta. Using the same instrument, Debes (1992) obtained the shear modulus of dog pulmonary artery with results shown in Fig. 8.4:3.

8.4.4 The Strain Energy Function Including Shear

With non-vanishing shear stress, the strain-energy function may be assumed to be of the form

$$\rho_0 W^{(2)} = P + \frac{C}{2} e^Q, \tag{19}$$

where

$$Q = a_1 E_{\theta\theta}^2 + a_2 E_{zz}^2 + a_3(E_{\theta z}^2 + E_{z\theta}^2) + 2a_4 E_{\theta\theta} E_{zz}, \tag{20}$$

$$P = \tfrac{1}{2}[b_1 E_{\theta\theta}^2 + b_2 E_{zz}^2 + b_3(E_{\theta z}^2 + E_{z\theta}^2) + 2b_4 E_{\theta\theta} E_{zz}]. \tag{21}$$

E_{ij} are the Green's strains. a_i, b_i are material constants. Debes (1992) has shown that for the pulmonary arteries of the dog, the stress–strain relationship is linear so that it is sufficient to use

$$\rho_0 W^{(2)} = P. \tag{22}$$

He found G in the order of 10 to 60 kPa for the dog pulmonary artery. See Fig. 8.4:3.

For the aorta of the pig, Yu et al. (1993) and Xie et al. (1993) have shown that in the neighborhood of the zero stress state, the stress–strain relationship is linear. Regarding the strains E_{ij} as small quantities of the first order then the linearized form of Eq. (19) is

$$\rho_0 W^{(2)} = P + \frac{C}{2}(1 + Q) = \frac{C}{2} + \left(P + \frac{C}{2}Q\right).$$
(23)

In the physiological strain range, aorta needs the full equation (Eq. (19)).

8.5 Arterial Wall as a Membrane: Dynamic Modulus of Elasticity from Flexural Wave Propagation Measurements

Seismologists deduce the structure of the earth and the elastic constants of various parts inside the earth by examining the propagation of seismic waves around the globe. By analogy, a physiologist should be able to deduce the properties of a blood vessel by watching the propagation of waves in the blood vessels.

If we consider progressive waves of radial motion of long wavelength and infinitesimal amplitude propagating in an infinitely long circular cylindrical vessel filled with an inviscid fluid and having a linearly elastic wall, then the speed c of the wave is governed by the Moens–Korteweg equation

$$c^2 = \frac{Eh}{2\rho_f a}.$$
(1)

Here ρ_f is the density of the fluid, E is Young's modulus of the vessel wall in the circumferential direction, h is the wall thickness of the vessel, and a is its radius. The long string of qualifying words preceding Eq. (1) is necessary for the validity of this equation. Thus blood viscosity and initial stresses are ignored. The wavelength must be very long, and the wave amplitude must be very small compared with the vessel radius; otherwise the bending rigidity of the vessel wall and the nonlinear convective fluid inertia would have to be accounted for. See *Biodynamics* (Fung, 1984), Chapter 3 for the derivation of Eq. (1).

The many restrictions imposed on the Moens–Korteweg formula suggest that the formula is of limited use. The most serious limitation is that the waves must be progressive. For vessels of finite length, waves will be reflected and refracted at the ends and what one can observe are not harmonic progressive waves. In the human body, with the wave motion generated by the heart, the distance between the heart and the capillary blood vessels at the periphery is approximately a quater wavelength. In this distance the vessel diameter is reduced rapidly toward the periphery, and many generations of branching exist. These factors make it necessary to abandon this beautifully simple formula, and to resort to a much more complex mathematical analysis of the arterial waves.

Many other types of wave motion are possible in a blood vessel. For example, axial waves and torsion waves in the vessel wall have been predicted and confirmed. Nonaxisymmetric waves are found to be strongly dispersive.

Anliker et al. (1968, 1969) and Anliker (1972) have studied the wave motion in arteries by superimposing, on the naturally occurring pulse wave, artificially induced transient signals in the form of finite trains of high frequency sine waves. In particular, this method is used to study the attenuation of the wave amplitude along the path of propagation, thus deducing the viscoelastic properties of the blood vessel. Using pressure waves in the 20–200 Hz range in a dog's aorta, the wave train may be considered as progressive waves. They find that the amplitude attenuates exponentially:

$$A = A_0 e^{-k(\Delta x/\lambda)}, \tag{2}$$

where A_0 is the amplitude at the proximal transducer, A is the amplitude at the distal transducer, k is the logarithmic decrement, Δx is the distance between the transducers, and λ is the wavelength. For pressure waves propagating down stream (away from the heart), k is of the order of 0.7 to 1.0. The logarithmic decrement k for the torsion and axial waves are higher, in the range 3.5–4.5. In all cases the logarithmic decrements are essentially independent of frequency between 40 and 200 Hz. Wave speed increases with increasing blood pressure.

Langewouters et al. (1984, 1985, 1986) have measured the static elastic properties of human thoracic and abdominal aortas *in vitro*, and proposed the following empirical relation between the cross-sectional area of the lumen (A) and the pressure in the vessel (p):

$$A(p) = A_m \left\{ \frac{1}{2} + \frac{1}{\pi} \tan^{-1} \left(\frac{p - p_0}{p_1} \right) \right\}, \tag{3}$$

in which A_m, p_0, and p_1 are parameters that vary with location. This is based on the observation that the pressure-area relation may be written as

$$\frac{dp}{dA} = a + bp + cp^2, \tag{4}$$

which becomes Eq. (3) upon integration. They tabulated the empirical parameters and the incremental Young's modulus computed from these formulas, as well as the characteristic impedance Z_0 and the pulse wave velocity, V_p:

$$Z_0 = \sqrt{\frac{\rho \Delta p}{A \Delta A}}, \tag{5}$$

$$V_p = \sqrt{\frac{A \Delta p}{\rho \Delta A}}. \tag{6}$$

They modeled creep with a generalized Kelvin model, with two Maxwell elements in parallel with a spring, and published the parameters. Similar data for the finger arteries and forearm arteries are published by Gizdulich and Wesseling (1988) and Imholz et al. (1991).

8.6 Mathematical Representation of the Pseudo-Elastic Stress–Strain Relationship

Although many authors have proposed a large variety of mathematical expressions to describe stress–strain relationship in uniaxial tests (see Chapter 7, Sec. 7.5), for biaxial and triaxial tests there are essentially only two schools: one uses polynomials (Patel and Vaishnav, 1972; Vaishnav et al., 1972; Wesley et al., 1975), while the other uses exponential functions (Ayorinde et al., 1975; Brankov et al., 1974, 1976; Demiray, 1972; Fung, 1973, 1975; Fung et al., 1979; Gou, 1970; Hayashi et al., 1974; Kasyanov, 1974; Snyder, 1972; Tanaka and Fung, 1974). Both schools utilize strain-energy functions to simplify the mathematical analysis. Strain-energy functions are discussed in Chapter 7, Sec. 7.9. In the two-dimensional case the strain energy function $\rho_0 W^{(2)}$ is a function of the strains $E_{\theta\theta}$ and E_{zz} if shear is not involved. We shall consider the case without shear first, and add shear later in this section. Then, the form of $\rho_0 W^{(2)}$ advocated by the polynomial school is, according to Patel and Vaishnav (1972),

$$\rho_0 W^{(2)} = Aa^2 + Bab + Cb^2 + Da^3 + Ea^2 b + Fab^2 + Gb^3, \tag{1}$$

where $a = E_{\theta\theta}$, $b = E_{zz}$, and $A, B, \ldots, G$ are material constants. If all coefficients except A, B, C vanish, then we obtain Hooke's law of the linear theory. The seven-constants form shown in Eq. (1) is the simplest polynomial to be devised for the nonlinear theory. If the fourth degree terms are included the number of material constants is increased to 12. Patel and Vaishnav (1972) have shown that the accuracy of the function is not much improved by the inclusion of the fourth degree terms.

Most other authors use exponential functions. The form we prefer is the following (Fung, 1973; Fung, Fronek, Patitucci, 1979):

$$\rho_0 W^{(2)} = \frac{C}{2} \exp[a_1(E_{\theta\theta}^2 - E_{\theta\theta}^{*2}) + a_2(E_{zz}^2 - E_{zz}^{*2}) + 2a_4(E_{\theta\theta}E_{zz} - E_{\theta\theta}^* E_{zz}^*)], \tag{2}$$

where C (with the unit of stress, N/m^2) and a_1, a_2, a_4 (dimensionless) are material constants, and $E_{\theta\theta}^*$, E_{zz}^* are strains corresponding to an arbitrarily selected pair of stresses $S_{\theta\theta}^*$, S_{zz}^* (usually chosen in the physiological range). In principle it is unnecessary to specify the asterisk quantities, because they can be absorbed into the constant C to yield a form

$$\rho_0 W^{(2)} = \frac{C'}{2} \exp[a_1 E_{\theta\theta}^2 + a_2 E_{zz}^2 + 2a_4 E_{\theta\theta}E_{zz}]. \tag{3}$$

But in practice it is very helpful to introduce $E_{\theta\theta}^*$, E_{zz}^*. Not only are the values of $E_{\theta\theta}^*$, E_{zz}^* corresponding to $S_{\theta\theta}^*$, S_{zz}^* very important information, but also their use makes the identification of the constants C, a_1, a_2, a_4 from experimental data easier for each set of test specimens for computational reasons.

TABLE 8.6:1 Material Constants for Four Arteries of the Rabbit. The Protocol "p" Refers to Inflation with Increasing Internal Pressure. From Fung et al. (1979)

(a) Mean values for all tests with protocol "p": exponential strain-energy function

| | C (10 kPa) | a_1 | a_2 | a_4 | $E^*_{\theta\theta}$ | E^*_{zz} | $S^*_{\theta\theta}$ (10 kPa) | S^*_{zz} (10 kPa) |
		(nondimensional)			(nondimensional)			
Carotid	2.9307	2.5084	0.4615	0.1764	0.5191	0.9939	3.4000	1.3000
Left iliac	2.1575	8.1674	1.2173	1.0546	0.2418	0.8834	4.5000	1.9000
Lower aorta	2.1744	9.5660	3.0913	0.8805	0.2743	0.6495	4.9000	3.0000
Upper aorta	3.3856	2.8173	0.5239	0.5790	0.4061	0.9566	5.2000	1.9000

(b) Mean values for all tests with protocol "p": polynomial strain-energy function

	A	B	C	D (10 kPa)	E	F	G
Carotid	−7.1889	3.1255	0.1911	1.3711	10.2775	−3.3677	0.0787
Left iliac	−16.3871	−0.3854	2.9122	16.0463	29.6790	1.1872	−2.2552
Lower aorta	−14.7220	−4.1606	4.4821	16.5753	29.9390	6.2093	−1.8999
Upper aorta	−12.0062	5.1405	−1.5936	−2.1292	23.5706	−7.3431	2.2069

The efficacy of the mathematical representation can be tested on two grounds: (a) its ability to fit the experimental data over the full range of strains of interest, and (b) the usefulness of the parameters in distinguishing the members of a family of stress–strain curves. To obtain the needed data for evaluation, Fung et al. (1979) tested selected rabbit arteries. The results of two series of tests are analyzed. In one series, each artery was first stretched longitudinally to an approximate *in vivo* length (λ_z equal to 1.6 for the carotid artery, 1.5 for the iliac and lower abdominal aorta, 1.4 for upper abdominal aorta); then it was inflated with internal pressure, and the pressure, diameter, gauge length, and axial force were recorded. In another series, the internal pressure was kept at zero (i.e., same as the external pressure) and the vessel was stretched while the diameter, gauge length, and axial force were recorded. These experimental data are then matched with the theoretical stress–strain relationship derived from Eqs. (1) and (2). See Sec. 8.4.

For the exponential strain-energy function, the constants C, a_1, a_2, a_4 are determined by Fung et al. (1979) according to a modified Marquart's (1963) nonlinear least squares algorithm—by minimizing the sum of the squares of the differences between the experimental and theoretical data. For the polynomial, the ordinary least squares procedure suffices. Some results for the exponential strain-energy function are given in Table 8.6:1. $S_{\theta\theta}^*$, S_{zz}^* correspond to about 13.3 kPa (100 mm Hg) internal pressure and $\lambda_z = 1.6$. The overall fit is quite good; the average correlation coefficients are 0.988 for $S_{\theta\theta}$ and 0.861 for S_{zz} for the carotid artery.

The results of matching experimental data with polynomial Eq. (1) are given in Table 8.6:1. The standard deviations of the material constants $A, B, \ldots$ are sometimes very large compared with the mean. Some of the coefficients change sign in different runs of the same specimen. The average correlation coefficients are 0.971 for $S_{\theta\theta}$ and 0.878 for S_{zz} for the same carotid artery.

In Table 8.6:1 we present the material constants for four arteries of the rabbit to illustrate the systematic variation of these constants along the arterial tree. The entries are mean values for all tests with protocol "p". The change of sign of the polynomial coefficients from one artery to another is an unpleasant feature of the polynomial approach. These changes are caused by the sensitivity of the coefficients to relatively small changes in the shape of the stress–strain curves. The overall smaller coefficients of variation (ratio of standard deviation to the mean) indicate that the exponential form of the strain-energy function is preferred.

8.6.1 The Question of Parameter Identification

The question discussed above is a question of parameter identification—determination of physical parameters from a given set of experimental data. It should be noted that although the experimental data can be fitted accurately, the parameters themselves may not have much meaning. For example, in one experiment [rabbit carotid artery, Exp. 71, protocol $P + L$ in Fung et al. (1979)], we obtained the following when Eq. (1) is used:

$$\rho_0 W^{(2)} = -2.4385a^2 - 0.3589ab - 0.1982b^2$$
$$+ 4.6334a^3 + 3.2321a^2b + 0.3743ab^2 + 0.3266b^3,$$

which correlates with a set of experimental data with correlation coefficients 0.971 for $S_{\theta\theta}$, and 0.878 for S_{zz}. Here $a = E_{\theta\theta}$, $b = E_{zz}$, and $\rho_0 W^{(2)}$ is in units of 10 kPa. For the same arterial test specimen but in another test (Exp 71.1, protocol p), the pseudo-strain-energy function obtained was

$$\rho_0 W^{(2)} = -8.1954a^2 + 2.5373ab + 0.2633b^2$$
$$+ 2.7949a^3 + 11.1749a^2b - 3.0092ab^2 - 0.1166b^3,$$

which has the correlation coefficients 0.958 for $S_{\theta\theta}$ and 0.794 for S_{zz}. It is amazing that two polynomials so disparate can represent the same artery. Although the coefficients A, B, ... are very different, the stress–strain curves are quite similar in the experimental range. In other words, the coefficients are very sensitive to small changes in the data on stress–strain relationship. Workers in viscoelasticity and other branches of physics are familiar with this situation. (See another example in Chapter 7, Sec. 7.6.2.)

8.6.2 The Meaning of the Material Constants in the Exponential Pseudo-Strain-Energy Function

The question is often asked: What is the "physical" meaning of the material constants C, a_1, a_2, and a_4 in Eq. (2) or (4)? If we interpret the question as how do the stresses change when these constants are changed? Then it can be answered quite easily. We have chosen to require the stress–strain curves to pass through two states: the zero stress state ($S_{ij} = E_{ij} = 0$) and the state characterized by S_{ij}^* and E_{ij}^*. Between these two states the stress–strain curve is more curved if a_1, a_2 is larger. If E_{zz} is zero, then for a larger a_1, the curve of $S_{\theta\theta}$ vs. $E_{\theta\theta}$ will leave the origin closer to the strain axis, and then rise more rapidly to the point ($S_{\theta\theta}^*, E_{\theta\theta}^*$). This is illustrated in Fig. 8.6:1. Similarly, a_2 would affect the curve for S_{zz} vs. E_{zz}. The constant a_4 specifies the crosstalk between the circumferential and longitudinal directions. The constant C fixes the scale on the stress axis. The larger the values of a_1, a_2, and a_4, the smaller would be C.

The exponential form of the strain energy can be derived from the following reasoning. Strain is a tensor. Strain energy is a scalar. Therefore, the strain-energy function must be a function of a scalar measure of the strain tensor. Let a scalar measure be

$$Q = a_1 E_{\theta\theta}^2 + a_2 E_{zz}^2 + 2a_4 E_{\theta\theta} E_{zz}. \tag{4}$$

Since the structure of the arterial wall is not isotropic, we do not expect Q to

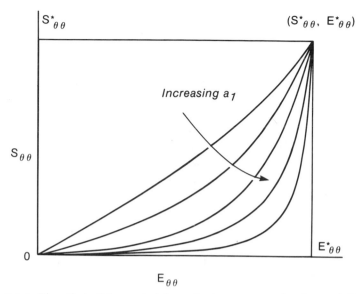

Figure 8.6:1 The effect of the material constants a_1, a_2, on the shape of the stress–strain curves. The scales of the coordinates are so chosen that all curves pass through the point S^*, E^*. The curves pass through the origin if $E_{zz} = 0$; otherwise they do not.

be a symmetric function of $E_{\theta\theta}$, E_{zz}; i.e., we expect $a_1 \neq a_2$. Now, if Q changes and the rate of change of the strain energy with respect to the change of Q is proportional to the current value of the strain energy, then we can write

$$\frac{d\rho_0 W^{(2)}}{dQ} = \rho_0 W^{(2)} \tag{5}$$

by absorbing the constant of proportionality in Q. Then, an integration of Eq. (5) gives

$$\rho_0 W^{(2)} = \frac{c'}{2} e^Q, \tag{6}$$

which is exactly Eq. (2), c' being a contant of integration. Thus, our proposed strain-energy function can be interpreted as using Q as a measure of the deformation and assuming that the rate of change of $\rho_0 W$ with respect to Q is proportional to $\rho_0 W$.

8.7 Blood Vessel Wall as a Three-Dimensional Body: The Zero Stress State

Blood vessels are, of course, three-dimensional bodies. Whenever the biaxial approach cannot be justified, one has to deal with the three-dimensional continuum. For the blood vessel, the first question we have to answer is: What is its zero stress state?

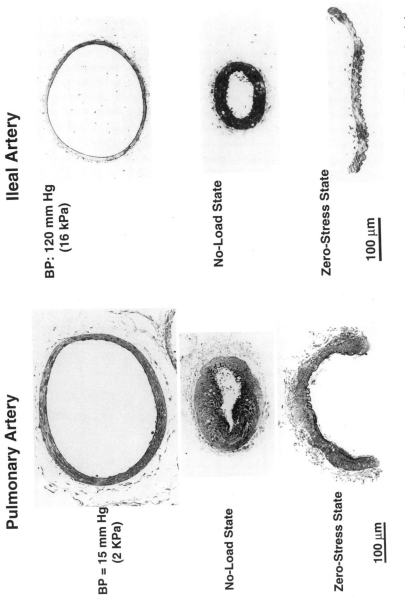

Figure 8.7:1 The zero stress, no-load, and *in vivo* states of two arteries: thoracic on the left, and ileal on the right. Photographs on the top row are the cross section of arteries fixed *in vivo* at normal physiological pressure. Those in the midrow are the cross sections fixed at zero transmural pressure and zero longitudinal load. The configurations at zero stress state are shown at the bottom. From Fung and Liu (1992), by permission.

A body in which there is no stress is at the zero stress state. If strain is calculated with respect to the zero stress state, then the strain is zero when the stress is zero, and vice versa. This is an important feature of the constitutive equation. Hence the analysis of stress and strain begins with the identification of the zero stress state.

At the top of Fig. 8.7:1, there are shown the cross sections of two arteries (a pulmonary artery and an ileal artery) that were obtained from anesthetized rats by perfusing the animal with glutaraldehyde *at normal blood pressure*, and fixing its tissue in situ. In the middle row of Fig. 8.7:1 are shown the cross sections of these arteries at the *no-load condition*. The vessels were reduced to no-load by excising them from the anesthetized animal, reducing the blood pressure and the longitudinal tension to zero, then putting the specimen into a glutaraldehyde solution and fixing it at no-load. In the bottom row of Fig. 8.7:1 are cross sections of these vessels at *zero stress*. These sections were obtained by first excising the arteries at no-load condition, cutting short segments transversely to obtain rings illustrated in the middle row, then cutting the rings open by a radial section. The rings sprang open into sectors, which were then fixed with glutaraldehyde.

From the photographs shown in Fig. 8.7:1 we see that the arterial walls were smooth in situ, and also at zero stress, but the intima were very much disturbed, wrinkled, and distorted at the no-load state. This was so because the vessel wall tissue was subjected to compressive strain in the inner wall region at the no-load state, and buckled. Compressive residual stress acted in the intima-media region in the circumferential direction, tensile residual stress acted in the adventitial region. Thus the no-load condition is the least suitable condition for morphometry of the geometric structure of tissues.

To verify that the stress in the opened sector is zero everywhere, one should make further arbitrary cuts and show that no further change of strain results. This was done by Han and Fung (1991) who showed that indeed no further measurable change of strain was found.

For the convenience of characterizing an open sector we define an *opening angle* as the angle subtended by two radii drawn from the midpoint of the inner wall (endothelium) to the tips of the inner wall of the open section, see angle α, Fig. 8.7:2. If the sector were circular, then the opening angle is independent of the location of the radial cut in the ring of the no-load condition. If the sector were not circular, then the opening angle does depend on the locaton where the cut is made, and thus is not a unique characterization. Zhou has shown, in Yu et al (1993), however, that the angle between the tangents at the tips of the section, the angle ψ in Fig. 8.7:2, is an *intrinsic* measure of the curved sector, because

$$\Psi = 2\pi - \int K \, ds,$$

where K denotes the curvature and s denotes a curvilinear coordinate representing the distance measured along the endothelial surface of the arterial

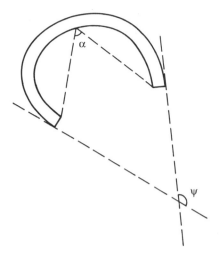

Figure 8.7:2 The opening angle α is defined as the angle between two radii joining the midpoint of the inner wall to the tips of the inner wall. If the vessel is not axisymmetric in vivo, then the opening angle α depends on the point at which the section is cut. But the angle between the tangents at the tips of the open sector, ψ, does not depend on the point of cutting.

sector. The integral is extended from one tip to the other. In practice, use of ψ requires experimental determination of the tangents, and is far less convenient than the use of the opening angle α.

The opening angles of blood vessels vary with the organ in which they are located, with the vessel size, thickness-radius ratio, curvature of the vessel centerline, and tissue remodeling. The opening angle is larger where the vessel is more curved, or thicker. For example, the opening angle of normal rat artery is about 160° in the ascending aorta, 90° in the arch, 60° in the thoracic region, 5° at the diaphragm, 80° in the abdomen, 100° in the iliac, dropping down to 50° in the popliteal artery, then rising again to about 100° in the tibial and plantar arteries (Fung and Liu, 1992).

There are similar spatial variations of opening angles in the aorta of the pig and dog (Han and Fung, 1991; Vasoughi et al., 1985), pulmonary arteries (Fung and Liu, 1991), systemic and pulmonary veins (Xie et al., 1991), and trachea (Han and Fung, 1991).

8.8 Blood Vessel Wall as a Three-Dimensional Body: Stress and Strain, and Mechanical Properties of the Intima, Media, and Adventitia Layers

An important task for the mechanical property determination of blood vessels is to obtain the constitutive equations for the intima, media, and adventitia as three separate layers. Since these layers are parallel and contiguous, it is best to use bending experiments which impose different strains in different layers while the stress distribution is revealed by bending moment and resultant force. Uniform stretching or shear experiments described in Secs. 8.3–8.5 are unsuitable for this purpose because the strains are the same in all the layers while the stress resultants cannot be separated into forces in different layers.

Since the blood vessel segments at the zero stress state (see Fig. 8.7:1 bottom) appear as curved beams, it is possible to test vessel specimens as beams. Two experimental arrangements are sketched below. In Fig. 8.8:1, the specimen is tested as a simply supported beam. In Fig. 8.8:2, the specimen is tested as a cantilever beam. In either case a large deformation can be imposed on the specimen, but the strains remain quite small. The strains in the specimen shown in Fig. 8.8:1 were measured by spraying the surfaces of the specimen with micro-dots (2 to 20 μm in size) of black indelible ink with a tooth brush. See Fig. 8.8:3. The coordinates of a set of smaller dots were measured and digitized from photographs of these dots taken through a stereomicroscope, first at the zero stress state, then at deformed states under increasing loads. From every set of three neighboring dots we can use the formulas given in Sec. 2.3, Eqs. (2.3:17), to compute the Lagrangian strains E_{ij} on the surface in a small area containing these three points. These strains can be correlated with the loading on the specimen. The specimen is very soft in the neighborhood of the zero stress state, hence a very small load must be applied. In the setup shown in Fig. 8.8:1 the load is applied by a fine wire as a cantilever beam whose deflection is calibrated for loading.

In the setup shown in Fig. 8.8:2 the specimen is clamped at one end and loaded at the other end, again by a fine wire used as a cantilever beam and the deflection at the tip is calibrated to read the loading. The deflection of the tissue is photographed and digitized.

The analyses of the results of these experiments are given in Yu, Zhou, and Fung (1993) and Xie, Zhou, and Fung (1993). It is shown that in both cases the stress–strain relationship is linear for strains up to the size that allow the rounding up of the specimen into a closed curve. If we write the relationship between the uniaxial Lagrangian stress T and the stretch ratio λ of the tissue as

$$T = E_i(\lambda_2 - 1) \text{ for the intima-media layer,}$$
$$T = E_o(\lambda_2 - 1) \text{ for the adventitial layer,} \tag{1}$$

where the subscript "i" means inner, "o" means outer, "2" means the circumferential direction, then it was found theoretically that E_i, E_o, and the location of the neutral axis as a fraction of the wall thickness can be determined by the load-deflection relationship, the measured strain distribution, and the thickness of the intima-media (lumped into one) layer as a fraction of the total wall thickness. When the thicknesses of the intima-media and adventitia are equal, it was found that for the thoracic aorta of the pig, the neutral axis is located at about 30% of the wall thickness from the inner wall, and the mean values of the Young's moduli are

$$E_i = 43.25 \text{ kPa for intima-media layer,}$$
$$E_o = 4.70 \text{ kPa for the adventitia layer.} \tag{2}$$

Thus the Young's moduluses of these two layers are almost an order of magnitude apart. In the neighborhood of the zero stress state the media of the arterial wall is much stiffer than the adventitia.

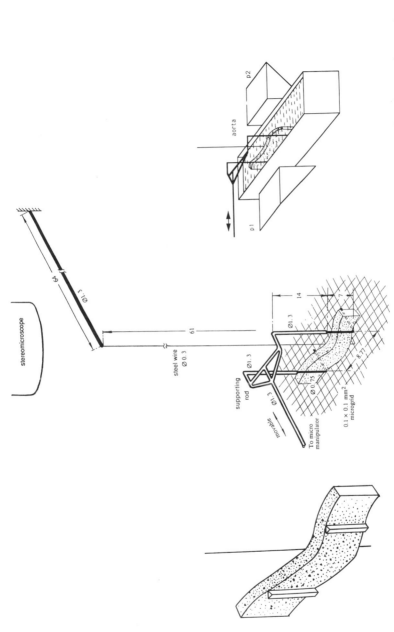

Figure 8.8:1 A sketch of our apparatus for measuring the strains in an arterial specimen subjected to bending. Microdots on three orthogonal surfaces are observed and photographed through a stereomicroscope with the help of a trough and two right optical prisms. The central sketch gives the dimensions. The sketch at the left shows our method of applying a force on a simple-supported specimen. The force is measured by the displacement of the tip of a thin wire, the top end of which is clamped and the clamp is moved by a micromanipulator. The lower end of the wire is observed by a microscope against the grid of an optical grating. The sketch on the right-hand side shows the use of two prisms to obtain photographs of the dots on the endothelial and adventitial sides of the specimen. Designed by Qilian Yu.

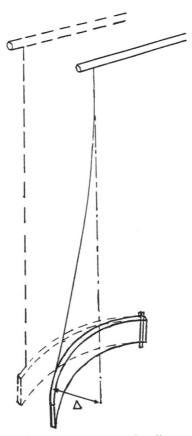

Figure 8.8:2 The test equipment to measure bending strains in a cantilevered condition. Designed by Jai Pin Xie.

The analysis of the results of the setup shown in Fig. 8.8:2 is more complex. The deflection curve of the test specimen has to be digitized and fitted with a smooth mathematical expression before it can be differentiated to obtain the curvature of the deflection curve which correlates with the bending moment. For the ascending aorta, using Eq. (1), and on assuming the thickness of the intima-media layer to be 50% of the total wall thickness and the neutral axis is located at 35% of wall thickness from the endothelial surface, we obtained

$$E_i = 447.5 \text{ kPa for intima-media layer,}$$
$$E_o = 111.9 \text{ kPa for the adventitial layer.}$$
(3)

For the descending aorta, under the same assumed thickness distribution and neutral axis location, the Young's moduli are

$$E_i = 247.8 \text{ kPa for intima-medial layer,}$$
$$E_o = 68.7 \text{ kPa for adventitia layer.}$$
(4)

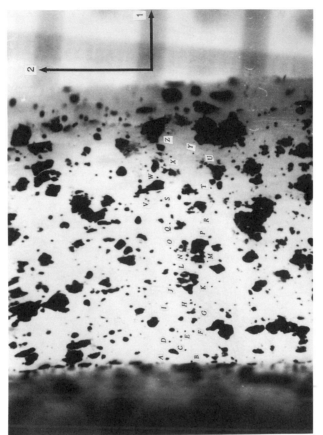

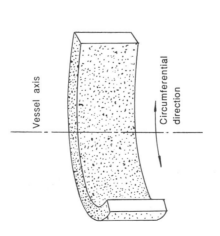

Figure 8.8:3 Left: a sketch of a strip of an arterial wall at the zero stress state which is pointed with microdots of indelible ink from a ball-point pen with a tooth brush. Right: a photograph of microdots on a test specimen. Dots are selected and numbered. Every group of three dots forming a triad provides three equations to compute the three strains in the triangle. The computed strains are assumed uniform in the triangle, and the location of the triangle is defined by its centroid. Photography by Qilian Yu in the author's lab.

Note that these moduli for the ascending and descending aorta are much larger than the corresponding moduli of the thoracic aorta. These differences reflect the differences of structure and composition of these blood vessels.

The good fit of the linear uniaxial stress–strain relationship given in Eq. (1) to the experimental results in the neighborhood of the zero stress level suggest that a similar good fit can be obtained for biaxial experiments which can be done on the biaxial testing machine described in Sec. 8.4. From these a three-dimensional stress–strain relationship can be derived if the material is incompressible by the method of Chuong and Fung (1983).

The next step is to match the linear constitutive equation valid for small strain to the nonlinear constitutive equation which is known to be valid in the physiological range of finite strain. A theoretical form suggested by the author is Eq. (8) of Sec. 7.10. This strain-energy function has been applied to the skin, see Sec. 7.10. It is applicable to the intima-media and adventitial layers of the blood vessel.

8.9 Arterioles. Mean Stress–Mean Diameter Relationship

Arterioles are the last generations of branches on the arterial tree. They are muscular vessels with very little adventitia. In many they are vessels with inner diameters in the order of 80 μm or less. By their smooth muscle action they regulate blood flow and control the blood pressure.

By measuring the inner and outer radii of the blood vessels simultaneously with pressure in the isolated, autoperfused mesentery of 14 cats during changes in systemic blood pressure, Gore (1972, 1974, 1984) was able to obtain the relationship between the mean diameter and the wall stress in several groups of vessels. Figure 8.9:1 shows Gore's results. Here the wall stress was computed from the formula

$$\langle \sigma_\theta \rangle = p \frac{r_i}{h}, \tag{1}$$

where p is the functional mean pressure, r_i is the inner radius of the vessel, and h is the wall thickness. $\langle \sigma_\theta \rangle$ is the average circumferential stress in the wall only under the assumption that the pressure acting on the outside of the vessel is zero, because the full formula is

$$\langle \sigma_\theta \rangle = \frac{p_i r_i - p_o r_o}{h}, \tag{2}$$

where the subscripts i and o refer to inside and outside of the vessel, respectively. The mean diameter was computed as the sum of inner and outer diameters divided by 2. Figure 8.9:1 shows that the stress-diameter relationship of arterioles (shown by the broken lines) is similar to that of the larger arteries.

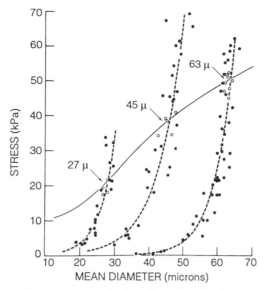

Figure 8.9:1 Stress-diameter curves (broken lines) identify three types of arterioles in the cat mesentery: 63 μ large arterioles, 45 μ arterioles, and 27 μ terminal arterioles. Open circles show stress states existing in these vessel segments when the mean systemic blood pressure was 13.3 $\pm$ 1.3 (s.d.) kPa (100 $\pm$ 10 (s.d.) mm Hg). Solid line is the stress distribution curve for vessels between a mean diameter of 12 to 70 μ when the blood pressure was 13.3 $\pm$ 1.3 kPa. From Gore (1974), by permission.

The solid line in Fig. 8.9:1 shows the mean circumferential stress in precapillary microvessels when the mean systemic blood pressure was 13.3 $\pm$ 1.3 (s.d.) kPa (100 $\pm$ 10 (s.d.) mm Hg). It is seen that the stress level decreases with decreasing vessel diameter. In an earlier paper, Gore (1972) showed that the response of frog microvessels to norepinephrine depends on the mean wall stress $\langle \sigma_\theta \rangle$ in the vessel. During iontophoretic application of constant, supramaximal doses of norepinephrine, maximum constriction of the frog mesenteric vessels occured when $\langle \sigma_\theta \rangle$ was in an optimum range S_t^* of 10–15 kPa. The magnitude of constriction was less in arterioles and arteries with $\langle \sigma_\theta \rangle$ greater or less than S_t^*. Gore suggested that $\langle \sigma_\theta \rangle$ in arterioles is normally equal to S_t^*, so they develop maximum force to overcome distending pressure; whereas $\langle \sigma_\theta \rangle$ in terminal arteries and arteries is normally greater than S_t^*, so these larger vessels cannot develop their full constrictive potential.

The mechanical response of the smooth muscles in the arterioles to changes in stretch, stress (or blood pressure), oxygen, norepinephrine, or other factors is the key to the regulation of blood flow. But this important subject remains hazy. A well-known result was given by Baez et al. (1960) who isolated and perfused the rat mesoappendix and measured the diameters of the small vessels

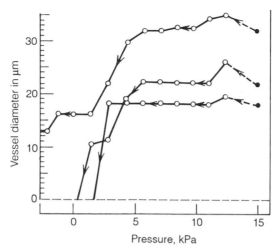

Figure 8.9:2 The change of vessel diameter with blood pressure measured by Baez et al. in the meso-appendix of the rat. Nerve-intact preparation. The phenomenon of unchanging diameter in certain ranges of pressure is exhibited. From Baez et al. (1960), by permission.

as the static perfusion pressure was varied. Figure 8.9:2 shows their results on a nerve-intact preparation. Note that there is a region of pressure in which the metarteriole (middle curve) does not change its diameter with pressure. For a small precapillary vessel (lower curve) this flat region is even larger. These flat regions are interpreted as due to the active contraction of the vascular smooth muscle. As the internal pressure is reduced below a certain limit, the muscle contraction closes the vessel. Since automatic closing is not possible without an active contraction, the closing phenomenon reflects smooth muscle action.

The results shown in Fig. 8.9:2 were obtained in a static condition. The vessels were perfused, but the arterial and venous pressures were equalized so that there was no flow. Related *in vivo* data with normal blood flow were given by Johnson (1968), who studied the change of arteriole diameter with respect to large-artery pressure in the cat mesentery. In experiments on 34 arterioles, 10 showed a simple decrease in diameter as pressure was reduced; 24 showed a biphasic response. In the latter group, the vessel narrowed for the first 5–15 sec, but then the vessel dilated. In 12 of these the diameter became larger in hypotension than it has been at normal arterial pressure. When pressure was suddenly restored, the vessel dilated at first but returned to the control size later. These results show how complex arteriolar behavior can be. See Johnson (1968, 1980).

The mechanics of smooth muscles in general, and vascular smooth muscle in particular, is a major frontier waiting for development.

8.10 Capillary Blood Vessels

Capillary blood vessels are the smallest blood vessels of the mammalian circulation system. Their diameters are about the same as that of the red blood cells. Some are smaller, e.g., in retina. Some are larger, e.g., in mesentery. Most are long thin tubes embedded in tissue. Some form very dense sheet-like networks, e.g., in the lung. Some form sinuses. There is no smooth muscle in their wall, no definable adventitia. The endothelium and basement membrane is the vessel wall. Through the wall, O_2, CO_2, water, ions, nutritional and waste molecules move.

In vivo observation of the elasticity of capillary blood vessels in the mesentery and some skeletal muscles in isolated preparation can be made under varying perfusion pressure without flow. Burton (1966), in summarizing Jerrad's unpublished data, stated that the frog mesenteric capillaries were less distensible than 0.2% per mm Hg change of blood pressure. Zweifach and Intaglietta (1968) stated that there was no measurable change in the diameters of the mesenteric capillaries of the rat and the rabbit when blood pressure was changed from the arterial to venous values. In other words, the mesenteric capillaries behave like rigid tubes. Fung (1966a,b) suggested that capillaries in the mesentery behave like a tunnel in a gel and that capillary behavior cannot be tested independently of this surrounding gel. Then Fung, Zweifach, and Intagletta (1966) measured the stress–strain relation of the mesentery, and Fung (1966b) used the result to compute the contribution of the surrounding tissue to the rigidity of the capillary blood vessels contained therein. He showed that when the mesentery was stretched to the extent used in most physiological experiments, the surrounding media contributed over 99% of the rigidity to the capillary blood vessels in the mesentery, with less than 1% contributed by the endothelium and basement membrane if the elastic modulus of the basement membrane were similar to that of the small artery or arteriole. It follows that the compliance of the capillaries depends on the amount of surrounding tissue that is integrated with the blood vessel, and to the degree the surrounding tissue is stressed. If the surrounding tissue is large compared with the capillary, and is stressed to the degree used in Zweifach and Intaglietta's (1968) observations, then the rigidity of the capillary would be derived mostly from the surrounding tissue. If the surrounding tissue is small compared with the capillary, and the tissue is much more relaxed, then the capillary will be more distensible.

An alternative hypothesis that can explain the observed rigidity of the capillary is that the basement membrane is thicker than previously assumed and has an elastic modulus as large as that of a taut tendon. This hypothesis remains to be verified. The basement membrane is made of collagen Type IV, (see Sec. 7.3.2), whose mechanical property is unknown. Unfortunately, Type IV collagen is not known to occur in larger quantity in any other tissues that are available for mechanical experimentation.

The tunnel-in-gel concept was further examined from the point of view of water movement between the capillary blood vessel and the surrounding media when the pressure in the capillary changes during micromanipulations. Intaglietta and Plomb (1973) showed that the concept is consistent with the observed fluid movement data.

One important corollary to the tunnel-in-gel concept of the capillary is that the capillary blood vessel will not collapse under a uniform external compression. A capillary in a compressed tissue can remain open for the same reason that an underground tunnel can remain open under tremendous earth pressure.

On the other hand, there are small blood vessels that do not receive much support from the surrounding tissue. A typical example is the capillary blood vessels of the pulmonary alveoli, which are separated from the air by a wall less than 1 μm thick. These vessels are found to be distensible.

Networks of pulmonary capillaries are organized in sheet form in the interalveolar septa. These are shown in Figs. 5.2:1 and 5.2:2 and again in Fig. 8.10:1. In the plane view [Fig. 8.10:1(a)], the vessels are crowded against each other; the "post" space between the vessels is filled with connective tissue. Hence in the plane view, there is no space to expand if the blood pressure changes; indeed, the blood vessels appear rigid with respect to blood pressure changes. However, in cross section [Fig. 8.10:1(b)], the interalveolar septa (capillary sheets) are thin-walled structures. Under increased blood pressure, the membranes will bulge out and the average thickness of the sheet will increase. This is illustrated in Fig. 8.10:2 [Sobin et al. (1972)]. Experi-

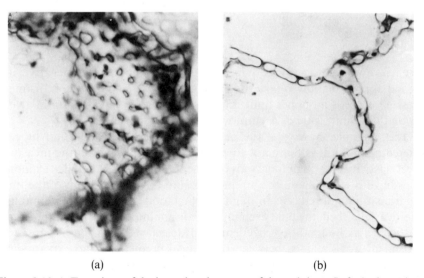

(a)   (b)

Figure 8.10:1 Two views of the interalveolar septa of the cat's lung. Left: A plane view. Right: A cross-sectional view. The vascular space was perfused with a catalyzed silicone elastomer, so that no red cells can be seen. From Fung and Sobin (1972), by permission.

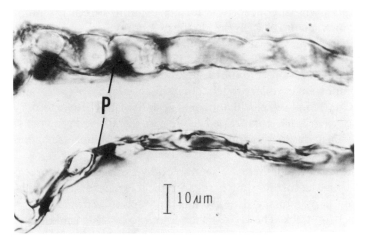

Figure 8.10:2 Photomicrograph of cat lung interalveolar septa cut in cross section and aligned parallel to optical axis. Stained basement membrane can be clearly seen. P = post; 20 m thick section. Top: sheet thickness = 10.5 μm at Δp = 2.69 kPa (27.4 cm H_2O). Bottom: 6.0 μm at Δp = 0.62 kPa (6.3 cm H_2O). From Sobin et al. (1972), by permission.

mental results on the variation of the thickness of the alveolar septa of the cat's lung with respect to the transmural pressure (Δp = local pressure of the blood minus the alveolar gas pressure) are shown in Fig. 8.10:3. These results may be summarized as follows: when the transmural pressure is positive, the thickness, h, increases linearly with increasing pressure according to the formula

$$h = h_0 + \alpha \, \Delta p, \tag{1}$$

where h_0 and α are constants. When the transmural pressure is negative, it is a good approximation to take the thickness h as zero. When the transmural pressure exceeds an upper limit, Eq. (1) ceases to be valid. As Δp increases beyond that limit, h tends asymptotically to be a constant.

This example shows that the pulmonary capillaries should not be considered as tubes. It is more apt to call them *sheets*. Based on the measured sheet elasticity of the pulmonary alveolar septa (Fig. 8.10:3), detailed examinations have been made on the unique pressure-flow relationship of the lung, the transit time of red cells in the lung, the blood volume in the pulmonary microvascular bed, the fluid exchange, the input impedance, the parenchyma elasticity, and the conditions of edema and atelectasis of the lung. A consistent agreement was obtained between theoretical predictions and experimental results. References to this literature can be found in Fung and Sobin (1977).

Thus the distensibility of capillary blood vessels with respect to blood pressure depends on the relationship of these vessels to the surrounding media. The mesentery and the lung provide two extreme examples. Larger blood vessels, such as the arterioles, venules, arteries, and veins, have the same

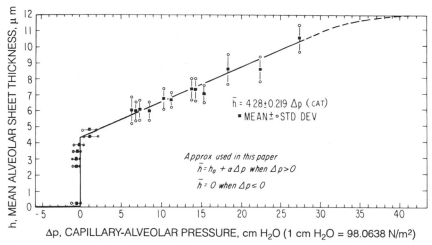

Figure 8.10:3 The sheet thickness (the average vascular space thickness in the inter-alveolar septa) of the cat's lung, plotted as a function of the transmural pressure p (the blood pressure minus the alveolar air pressure). The alveolar air pressure was maintained at 10 cm H_2O (980 N/m^2) above the intrapleural pressure, which was atmospheric (taken as 0). A composite plot with data from Sobin et al. (1972) and Fung and Sobin (1972).

problem; their compliances are also influenced by the surrounding tissue to varying degrees. Capillary blood vessels in different organs have different topologies and special structures and as a consequence have special mechanical properties.

In 1979, Bouskela and Wiederhielm found that the capillaries in the bat's wing are very distensible. The bat's wing is thin and unstretched so that the surrounding tissue is relaxed. The capillaries apparently do not receive much support from the surrounding tissue. (Remember that the Young's modulus of a soft tissue is nearly proportional to the stress in the tissue.) This points out the importance of controlling the tension in the tissue in physiological experiments involving capillaries.

8.11 Veins

The structure of veins is not very different from that of arteries. The constitutive equation is similar. The stress–strain curves of veins, as shown in Fig. 8.11:1, from Azuma and Hasegawa (1973), are similar to those of arteries. Figure 8.11:2 shows a comparison of the stress–strain curves of veins with other tissues. The zero–stress state of veins is presented by Xie et al (1991).

To determine the constitutive equation, the same kind of test procedure and precautions as discussed in Secs. 8.1 and 8.3 must be followed. To obtain a steady-state response in cyclic loading and unloading, preconditioning is

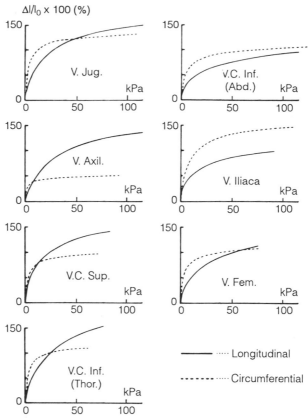

Figure 8.11:1 The stress–strain curves for the veins of the dog. Abscissas: stress. Ordinates: percent strain. From Azuma and Hasegawa (1973), by permission.

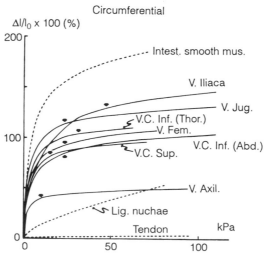

Figure 8.11:2 A comparison of the stress–strain relationships of various veins of the dog with that of the ligament nuchae, tendon, and intestinal smooth muscle. Abscissas: tensile stress. Ordinates: strain. From Azuma and Hasegawa (1973), by permission.

necessary. The stress–strain curve corresponding to an *increasing stretch* is called a *loading* curve. That corresponding to a *decreasing stretch* is called an *unloading* curve. For preconditioned specimens, the loading and unloading curves are stabilized, and are essentially independent of the *strain rate* (i.e., the speed) at which the cycling is done. Hence the concept of *pseudo-elasticity*, as discussed in Chapter 7, Sec. 7.7, is applicable to veins.

Data presented in Figs. 8.11:1 and 8.11:2 refer to uniaxial tests on strips of tissue in which only one component of stress does not vanish. If $\rho_0 W$ represents a three-dimensional strain-energy function, it can be reduced to the unaxial condition through the conditions $S_{11} = S$, $S_{22} = S_{33} = S_{12} = S_{23} = S_{31} = 0$. As an alternative, one may derive a one-dimensional strain-energy function $\rho_0 W^{(1)}$, as a function of the uniaxial strain E, whose derivative with respect to E yields the uniaxial stress (Kirchhoff stress S, Lagrangian stress T, and Cauchy stress σ):

$$S = \frac{\partial(\rho_0 W^{(1)})}{\partial E}, \qquad T = \lambda S, \qquad \sigma = \lambda^2 S, \tag{1}$$

where

$$E = \tfrac{1}{2}(\lambda^2 - 1),$$

and λ is the uniaxial stretch ratio, i.e., the ratio of the changed length of the specimen divided by the reference length of the specimen. Following the same reasoning as in Sec. 8.5, we prefer the following form for $\rho_0 W^{(1)}$:

$$\rho_0 W^{(1)} = \tfrac{1}{2}C \exp[\alpha(E^2 - E^{*2})], \tag{2}$$

where C and α are material constants, and E^* is the strain that corresponds to a selected value of stress S^*. The Lagrangian stress T corresponding to Eq. (2) is

$$T = \alpha E C \lambda \exp[\alpha(E^2 - E^{*2})], \tag{3}$$

$$T^* = \alpha C E^* \lambda^*. \tag{4}$$

It is easy to see that α determines how curved the stress–strain curve is. The more nonlinear the stress–strain relationship is, the larger the constant α. The product of the constants αC is similar to the familiar Young's modulus, provided that the modulus is defined as the ratio of the Kirchhoff stress to Lagrangian strain, S^*/E^*, where $S^* = T^*/\lambda^*$.

A typical stress–strain curve for cyclic loading and unloading at a constant strain rate is shown in Fig. 8.11:3. This figure refers to data obtained by P. Sobin (1977) in the author's laboratory on human vena cava from autopsy material. The dots represent the experimental values; the solid curves are those of Eq. (3). The reference length is defined as that length at which the stress is returned to zero in the cyclic loading process. Table 8.11:1 presents the numerical values of the constants for human inferior vena cava at a reference stress of $T^* = 250$ g/cm^2 (24.52 kPa), for the loading process only. It is seen that the venous mechanical property (as reflected by these parame-

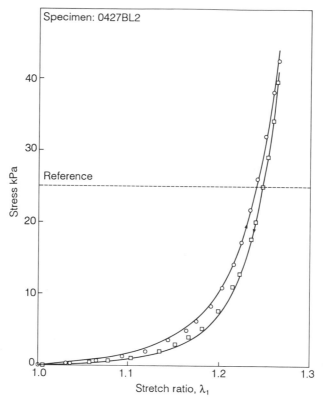

Figure 8.11:3 The stress–stretch ratio relationship of a post mortem human vena cava specimen tested in a uniaxial tensile loading condition. λ_1 is the axial stretch ratio. Circles: experimental data on loading (increasing strain). Squares: unloading. Solid 11.3 g/cm² (1.108 kPa), $\alpha = 180.4$, $\lambda^* = 1.24$, $T^* = 250$ g/cm² (24.5 kPa). From P. Sobin (1977).

ters) varies a great deal from one individual to another, and for the same individual from one location to another.

Test results on veins subjected to a biaxial stress field are reported by Wesley et al. (1975). A comparison of the incremental elastic moduli between a segment of human saphenous vein and a canine carotid artery is presented in Table 8.11:2. This comparison is of interest because in certain surgical operations a vein is used to substitute for an artery. Normally, of course, the internal pressure of the vein is smaller than that in the artery.

Wave propagation in veins has been studied by Anliker et al. (1969). General features were the same as those in arteries, except that local changes of wave speeds were found to be significant, indicating the existence of local variation in geometry, mechanical properties of the vessel wall, and tethering. Wave speed increases with increasing transmural pressure (e.g., in the vena cava during inspiration). The logarithmic decrement lies in the range 0.6–3.3

TABLE 8.11:1 Material Constants Based on Fitting Eq. (3) to Experimental Data Obtained from Post Mortem Human Vena Cava Specimens Subjected to Uniaxial Tension. From P. Sobin (1977)

			A. Circumferential			
Specimen	Sex	Age (yrs)	C (kPa)	α	λ^*	Corr. coeff.
0427BC1	M	65	0.990	153.034	1.16	0.999
0427BC2	M	65	1.113	180.401	1.12	0.998
0511LC1	F	11	0.620	365.308	1.11	0.999
0511LC2	F	11	1.437	45.127	1.37	0.999
0511BC1	F	11	1.190	43.715	1.425	0.997
0523BC1	M	20	1.544	157.891	1.115	0.998
0523BC2	M	20	1.727	130.712	1.13	0.999
0523CL1	M	20	1.166	85.197	1.25	0.998
0523CL2	M	20	1.261	74.177	1.26	0.999
0523BC1	M	20	1.667	43.593		0.998
0525LC1	M	49	1.588	44.542		1.000
0525LC2	M	49	1.168	87.114		1.000

			B. Longitidunal			
0427BL1	M	65	1.707	61.561	1.225	1.000
0427BL2	M	65	3.132	30.939	1.24	1.000
0511LL2	F	11	1.753	134.399	1.104	0.997
0523BL1	M	20	2.861	37.964	1.21	0.998
0523BL2	M	20	4.317	14.202	1.36	0.998
0523LL1	M	20	7.152	8.310	1.345	0.999
0523LL2	M	20	13.528	3.136	1.450	0.998
0525BL1	M	49	5.813	7.287		0.991
0525BL2	M	49	4.930	9.419		0.990
0525LL2	M	49	7.464	3.124		0.994

per wavelength, again essentially independent of frequency between 40 and 200 Hz.

Compared with arteries, veins work in a much lower pressure regime, in which the Young's modulus is highly dependent on the tensile stress and has low values. The low elastic modulus (i.e., high compliance) is the main reason why the capacity of the veins is so sensitive to postural changes and neural and pharmacological controls.

Veins are rich in smooth muscle, which responds to neural, humoral, pharmacological, and mechanical stimuli. The functional response of the veins to these stimuli is very important in physiology, because the veins contain 75% or more of the total blood volume. Any change in pressure or muscle tension or contraction will change the blood volume in the veins and

TABLE 8.11:2 The Properties of Veins and Carotid Artery of the Dog. From Wesley et al. (1975)

Pressure (cm H$_2$O)	Extension ratio		Incremental venous elastic modulus		Carotid artery increment modulus[a]	
	λ_θ	λ_z	E_θ (N/m^2 × 10^5)	E_z (N/m^2 × 10^5)	E_θ (N/m^2 × 10^5)	E_z (N/m^2 × 10^5)
			Canine jugular vein			
10	1.457	1.481	15 ± 3[b]	1.2 ± 0.18[c]	7.62	5.16
25	1.463	1.530	47 ± 6[c]	4.4 ± 0.3[c]	8.39	7.15
50	1.472	1.597	88 ± 7[c]	11.8 ± 2.1	9.51	10.69
75	1.478	1.646	98 ± 7[c]	46 ± 13[b]	10.37	13.97
100	1.482	1.675	117 ± 10[c]	67 ± 25	10.92	16.24
125	1.484	1.686	134 ± 24[c]	39 ± 32	11.16	17.17
150	1.484	1.686	171 ± 9[c]	113 ± 1[c]	11.16	17.17
			Human saphenous vein			
10	1.357	1.169	0.27 ± 0.12[c]	1.61 ± 0.32[b]	5.30	0.017
25	1.417	1.206	0.65 ± 0.13[c]	2.03 ± 0.39[b]	5.82	0.328
50	1.500	1.266	1.89 ± 0.41[c]	2.75 ± 0.78	6.00	0.735
75	1.561	1.325	9.85 ± 1.6	3.18 ± 0.76	9.66	1.80
100	1.602	1.381	15.0 ± 2.6	3.56 ± 0.58	12.77	3.15
125	1.621	1.430	20.4 ± 1.6[b]	3.98 ± 0.96	14.79	4.59
150	1.621	1.470	25.1 ± 7.5	4.75 ± 1.2	15.51	5.93

[a] The incremental modulus of the carotid artery is computed at the same extension ratios ($\lambda_\theta, \lambda_z$) as experienced by the veins at the pressure listed. Carotid moduli are quoted as a function of vessel strain, not as a function of pressure.

[b] $P < 0.05$ for the comparison between the venous and carotid moduli.

[c] $P < 0.01$ for the comparison between the venous and carotid moduli.

hence the cardiac output. The panorama of neural, humoral, and pharmacological control of veins is summarized in the magnificient book of Shepherd and Vanhoutte (1975).

Because veins are thin-walled, they are easily collapsible when they are subjected to external compressions. Many interesting phenomena occur in venous blood flow because of this.

A notable paper by Gaehtgens and Uekermann (1971) reports the distensibility of mesenteric venous microvessels, which contain some 25% of the total blood volume of the body. Working with the dog's mesentery, they measured the diameters of venous vessels of 22–148 μm internal diameter in response to arterial and venous pressure changes. Over a range of arterial pressure between 0 and 170 mm Hg (22.7 kPa) (while venous outflow pressure was 0) venular diameter changed by 31.8 ± 8.8% and venular length by 6.3 ± 4.4%. With a venous pressure elevation from 0 to 30 mm Hg (4.0 kPa) (while arterial pressure = 0) an increase of the volume of the venous micro-vessels of about 360% was measured. Thus the venous microvessel appear to be one of the most distensible elements of the vascular system.

8.12 Effect of Stress on Tissue Growth

In the following, we shall discuss the changes in geometry, structure, and mechanical properties that occur in the cardiovascular system as the stress in the system changes. These changes are the results of tissue remodeling. They occur when an organ is subjected to environmental changes such as hypoxia, zero gravity, hyperbaric conditions, disease, drugs, injury, surgery, healing, or rehabilitation. Historically, orthopedic surgeons were the first to pay attention to the role played by stress in the healing of bone fracture. In 1866, G. H. Meyer presented a paper on the structure of cancellous bone and demonstrated that "the spongiosa showed a well-motivated architecture which is closely connected with the statics of bone." A mathematician, C. Culmann, was in the audience. In 1867, Culmann presented to Meyer a drawing of the principal stress trajectories on a curved beam similar to a human femur. The similarity between the principal stress trajectories and the trabecular lines of the cancellous bone was remarkable. In 1869, J. Wolff claimed that there is a perfect mathematical correspondence between the structure of cancellous bone in the proximal end of the femur and the trajectories in Culmann's crane. In 1880, W. Roux introduced the idea of "functional adaptation." A strong line of research followed Roux. Pauwels, beginning in his paper in 1935 and culminating in his book of 1965, which was translated into English in 1980, turned these ideas to practical arts of surgery. Recently, Carter (1987, 1988) and his associates (1987, 1988) published a hypothesis about the relationship between stress and calcification of the cartilage into bone. Cowin (1984, 1986) and his associates (1979, 1981, 1985) have developed a mathematical theory of Wolff's law. Fukada (1974), Yasuda (1974), Bassett (1978), Saltzstein and Pollack (1987), and others have studied piezoelectricity of bone and developed the use of electromagnetic waves to assist healing of bone fracture. Lund (1921, 1978), Becker (1961), Smith (1974) and others have studied the effect of electric field on the growth of cells and on the growth of an amputated limb of a frog. See Fung (1990), *Biomechanics: Motion, Flow, Stress, and Growth*, Chapter 13, for references and a more detailed account.

The best known example of soft tissue remodeling due to change of stress is the hypertrophy of the heart caused by a rise in blood pressure. Another famous example was given by Cowan and Crystal (1975) who showed that when one lung of a rabbit was excised, the remaining lung expanded to fill the thoracic cavity, and it grew until it weighed approximately the initial weight of both lungs.

On the other hand, animals exposed to the weightless condition of space flight have demonstrated skeletal muscle atrophy. Leg volumes of astronauts are diminished in flight. In space flight vigorous daily exercise is necessary to keep astronauts in good physical fitness over a longer period of time, see references in Fung (1990).

8.13 Morphological and Structural Remodeling of Blood Vessels Due to Change of Blood Pressure

The systemic blood pressure can be changed in a number of ways: by drugs, high salt diet, constricting the flow of blood to the kidney, etc. If the aorta is constricted severely by a stenosis above the renal arteries, the aorta above the stenosis will become hypertensive, the whole upper body supplied by the upper aorta will become hypertensive, whereas the aorta below the stenosis will become hypotensive at first, but the reduced blood flow to the kidney will cause the kidney to secrete more of the enzyme renin into the blood stream and raise the blood pressure. If the stenosis was below the kidney arteries and is sufficiently severe, then the lower body will become hypotensive. Such a constriction can be imposed with a metal clamp, which is used in experiments.

The pulmonary blood pressure can also be changed by a number of ways. A most convenient way in the laboratory is to change the oxygen concentration of the gas breathed by the animal. If the oxygen concentration of the gas is reduced from normal (i.e., hypoxic), the pulmonary blood pressure will go up. This is the reaction human encounters when a person living at sea level goes to a high altitude.

An example of blood vessel remodeling when the blood pressure changes is given by Fung and Liu (1991). They created high pulmonary blood pressure in a rat by putting the animal into a low oxygen chamber. The chamber's oxygen concentration was 10% (about the same as that at the Continental Divide of the Rocky Mountains in Colorado). Nitrogen was added so that the total pressure was the same as the atmospheric pressure at sea level. When a rat entered such a chamber, its systolic blood pressure in the lung went up from the normal 15 mm Hg (2.0 kPa) to 22 mm Hg (2.93 kPa) within minutes, and maintains in the elevated pressure of 22 mm Hg of a week, then gradually rises to 30 mm Hg (4.0 kPa) in a month. See Fig. 8.13:1. The systemic blood pressure remains essentially unchanged in the meantime. Under such a step rise in blood pressure in the lung, its pulmonary blood vessel remodels. To examine the change, a rat is taken out of the chamber at a scheduled time. It was anesthetized immediately by an intraperitoneal injection of pentobarbital sodium according to a procedure and dosage approved by the University, NIH, and Department of Agriculture, then dissected according to an approved protocol. The specimens were fixed first in glutaraldehyde, then in osmium tetraoxide, embedded in Medcast resin, stained with toluidine blue O, and examined by light microscopy.

Figure 8.13:2 shows how fast the remodeling proceeds. In this figure, the photographs in each row refers to a segment of the pulmonary artery as

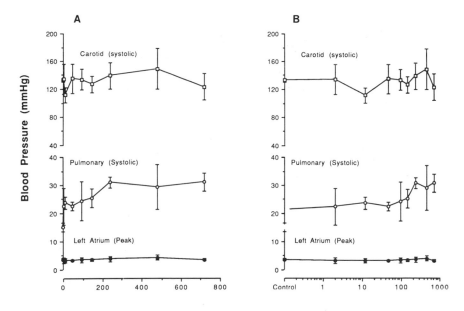

Hours of Exposure to Hypoxia

Figure 8.13:1 The course of change of the pulmonary arterial pressure in response to a step decrease of oxygen tension in the breathing gas. Note that the systemic blood pressure in the aorta changed very little. From Fung and Liu (1991), by permission. 1 mm Hg at 0°C = 133.32 N/m².

indicated by the leader line. The first photograph of the top row shows the cross secton of the arterial wall of the normal three month old rats. The specimen was fixed at the no-load condition. In the figure, the endothelium is facing upward. The vessel lumen is on top. The endothelium is very thin, of the order of a few micrometers. The scale of 100 μm is shown at the bottom of the figure. The dark lines are elastin layers. The upper, darker half of the vessel wall is the media. The lower, lighter half of the vessel wall is the adventitia. The second photo in the first row shows the cross section of the main pulmonary artery 2 hrs after exposure to lower oxygen pressure. There is evidence of small fluid vesicles and some accumulation of fluid in the endothelium and media. There is a biochemical change of elastin staining in vessel wall at this time.The third photograph shows the wall structure 12 hrs later. It is seen that the media is greatly thickened, while the adventitia has not changed very much. At 96 hrs of exposure to hypoxia, the photograph in the fourth column shows that the adventitia has thickened to about the same thickness as the media. The next two photos show the pulmonary arterial wall structure when the rat lung is subjected to 10 and 30 days of lowered oxygen concentration. The major change in these later periods is the continued thickening of the adventitia.

The photographs of the second row of Fig. 8.13:2 show the progressive changes in the wall of a smaller pulmonary artery. The third and fourth rows

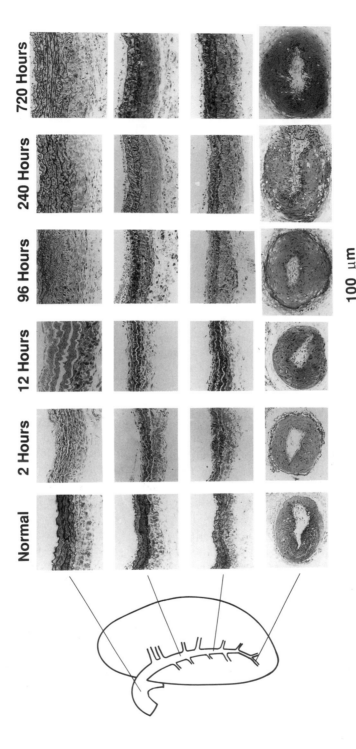

Figure 8.13:2 Indicial response of the pulmonary arterial structure to a step increase of pulmonary blood pressure. From Fung and Liu (1991), by permission.

are photographs of arteries of even smaller diameter. The inner diameter of the arteries in the fourth row is of the order of 100 μm, approaching the range of sizes of the arterioles. The remodeling of the vessel wall is evident in pulmonary arteries of all sizes.

Thus we see that the active remodeling of a blood vessel wall is nonuniform in space and proceeds quite fast. Histological changes can be identified within hours. The maximum rate of change occurs within a day or two.

8.14 Remodeling the Zero Stress State of a Blood Vessel

In association with the material and structural remodeling discussed in the preceding section, the zero-stress state of the tissue changes. Take the pulmonary artery as an example. Figure 8.14:1 shows the zero-stress state of various sections of the pulmonary arteries. The photograph in the middle of Fig. 8.14:1 shows an exposed view of the left lung of the rat. Different regions of the artery are labeled as Region 1, Region 2, ..., Region 8 as indicated in the figure. If a segment of the pulmonary artery in the ascending portion of Region 1 were cut, the cross section will open up as shown in the top photograph on the left. The endothelium of the inner wall faces downward in the photograph, so the opening angle α defined in Fig. 8.7:2 is about 270° here. At the top of

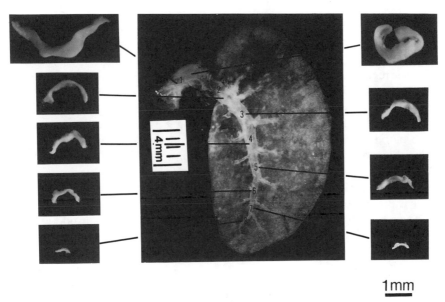

1mm

Figure 8.14:1 Variation of the opening angle of the pulmonary artery of the rat with location along the artery. Photographs are arterial sections at the zero stress state at the indicated locations, with the endothelial surface (original inner wall) facing downard. Note that the opening angle of pulmonary artery at the most curved region is greater than 360°. The vessel in this region turned itself inside out on relieving of its residual stresses. From Fung and Liu (1991), by permission.

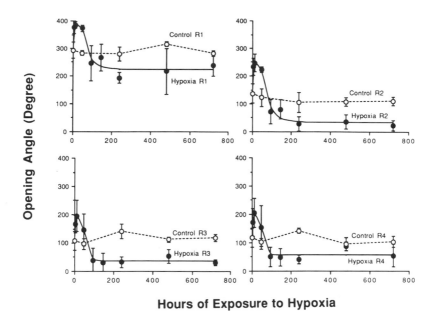

Figure 8.14:2 Indicial response of the pulmonary arterial opening angle to a step increase of blood pressure. From Fung and Liu (1991), by permission.

the arch in Region 1 where the artery is most curved, the zero-stress state is shown in the photograph at the top of the right hand column of Fig. 8.14:1. The opening angle of that section is about 360°! Similarly, the opening angles of other regions are shown by photographs.

The states shown in Fig. 8.14:1 are normal. When hypertension is induced in the pulmonary artery the opening angles will change with time. The courses of change in various regions are shown in Fig. 8.14:2. Since the input—the high blood pressure—is essentially a step function, the responses are the indicial responses in various regions. It is seen that the indicial responses of the opening angles are larger than the controls at first, then in due time become smaller than the controls. These changes are significant because the zero-stress state is the fundamental state for any analysis of stress and strain.

The remodeling of the zero-stress state of the aorta due to a sudden onset of hypertension has been reported by Fung and Liu (1989) and Liu and Fung (1989), see Fung (1990), Chapter 11, Secs. 11.2 and 11.3.

8.15 Remodeling of Mechanical Properties

Together with the remodeling of the material, structure, and zero-stress state, the mechanical properties of the tissue are expected to change. The changes can be described by a change of the constitutive equaton, or, usually, by changes of the material constants in the constitutive equation if its form does not need to be altered.

As an example, let us assume that a *pseudo-elastic strain-energy function* exists, denoted by the symbol $\rho_0 W$, and expressed as a function of the nine components of strain, E_{ij} ($i = 1, 2, 3, j = 1, 2, 3$), which is symmetric with respect to E_{ij} and E_{ji}, so that the stress components can be derived by a differentiation (Sec. 8.5):

$$S_{ij} = \frac{\partial \rho_0 W}{\partial E_{ij}}. \tag{1}$$

Here ρ_0 is the density of the material at the zero stress state, W is the strain energy per unit mass, $\rho_0 W$ is the strain energy per unit volume, E_{ij} are strains measured with respect to the material configuration at zero stress state. We assume the following form for $\rho_0 W$ (see Sec. 8.5):

$$\rho_0 W = C \exp[a_1 E_{11}^2 + a_2 E_{22}^2 + 2a_4 E_{11} E_{22}], \tag{2}$$

where C, A_1, a_2, a_4 are material constants, E_{11} is the circumferential strain, E_{22} is the longitudinal strain, both referred to the zero stress state.

Experiments have been done on rat arteries during the course of development of diabetes after a single injection of streptozocin. When the vessel wall

TABLE 8.15:1 Coefficients C, a_1, a_2 and a_4 of the Stress-Strain Relationship of the Thoracic Aorta of 20-Day Diabetic and Normal Rats. a_4 Was Fixed as the Mean Value from the Normal Rats*

Group	C (n/cm²)	a_1	a_2	a_4
Normal Rats				
Mean ± SD	12.21 ± 3.32	1.04 ± 0.35	2.69 ± 0.95	0.0036
20-day Diabetic Rats				
Mean ± SD	15.32 ± 9.22	1.53 ± 0.92	3.44 ± 1.07	0.0036

* From Liu, S. Q. and Fung, Y. C. (1992).

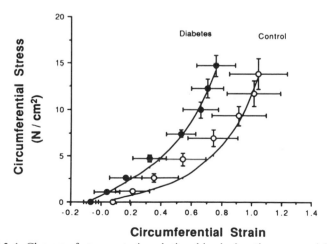

Figure 8.15:1 Change of stress–strain relationship during tissue remodeling. From Liu and Fung (1992), by permission.

is treated as one homogeneous material, the results are presented in Table 8.15:1 and Fig. 8.15:1, from Liu and Fung (1992). It is seen that the material constants change with the development of diabetes. In Liu and Fung (1992), the corresponding remodeling of the zero stress state is shown. These are examples of tissure remodeling in response to a disease.

8.16 A Unified Interpretation of the Morphological, Structural, Zero Stress State, and Mechanical Properties Remodeling

The remodeling of living tissues broadens the scope of constitutive equations. An overall perspective is as follows: The cells in the tissues live under stress. They respond to change of stress by changing their mass, metabolism, internal structure, production or resorption of proteins, and building or reabsorb extracellular structures. In such a response to change of stress, the cells may change their mechanical properties, sizes, structures, and their interaction with each other. In normal condition, a living organism has an equilibrium configuration which is called a *homeostatic state*. When the state of stress deviates from that of the homeostatic state, the rate of change of the mass of the tissue, or some components of it, may become positive or negative, depending on the magnitude of the deviation from the homeostatic condition.

Let the tissue be an aggregate of N materials. Let σ_{ij} and e_{ij}, with $i, j = 1, 2, 3$, represent the three-dimensional stress and strain components. Let ρ_n, L_n and μ_n represent, respectively, the mass density per unit volume, the characteristic length, and the chemical potential per unit mass of the nth material of the tissue, $n = 1, 2, \ldots, N$. Let τ be time: $\tau = 0$ initially, $\tau = t$ at present. Let

$$\mathop{F}_{\tau=0}^{t} [\;]$$

represent a functional of the entire history of the variables in the brackets, from $\tau = 0$ to $\tau = t$. Then

$$\sigma_{ij}(t) = \mathop{F}_{\tau=0}^{t} [e_{ij}(\tau), \rho_1(\tau), \ldots, \rho_N(\tau), L_1(\tau), \ldots, L_N(\tau), \mu_1, \ldots, \mu_N], \tag{1}$$

$$\rho_n(t) = \rho_n(0) + \int_0^t \dot{\rho}_n(\tau)\, d\tau, \tag{2}$$

$$\dot{\rho}_n(t) = G_n[\rho_1, \ldots, \rho_N, L_1, \ldots, L_N, e_{ij}, \sigma_{ij}, \text{and biochemical factors}]. \tag{3}$$

$$L_n(t) = L_n(0) + \int_0^t \dot{L}_n(\tau)\, d\tau \tag{4}$$

$$\dot{L}_n(t) = H_n[\rho_1, \ldots, \rho_N, L_1, \ldots, L_N, e_{ij}, \sigma_{ij}, \text{etc}] \tag{5}$$

Equation (1) states that stress is a functional of the strain and material composition (with specific molecular and ultrastructures implied, because molecules of the same molecular weight but different stereostructure may be counted as different materials). For a composite material, Eq. (1) must be written for each aggregate, as well as for the system as a whole. Equation (2) says that the density of a material is the result of its past history of growth. Equation (3) is the stress-growth law of the nth material. G_n is a function of the material composition, stress and strain, and biochemical and environmental factors. Equations (4) and (5) say the same for the geometry and dimensions.

Equations (1)–(5) are the constitutive equations. The examples given in Chapters 2–7 do not consider tissue growth and remodeling. When tissue growth and remodeling are involved, new boundary-value problems must be formulated on the basis of physical laws, boundary conditions, and Eqs. (1)–(5).

Problems

8.1 Consider the equilibrium of a vertical column of liquid of density ρ, height h, in a gravitational field with gravitational acceleration g. Let the pressures at the top and bottom of the column be p_0 and p_1, respectively. Consider the balance of forces and show that

$$p_1 - p_0 = \rho g h.$$

8.2 Apply the results of preceding problems to calculate the hydrostatic pressure in the arteries and veins of man. In physiology, pressure is *actually measured* most frequently with a mercury manometer, and is stated in terms of mm Hg. Show that the mean pressure distribution in the blood vessels of a standing man as shown in Fig. P8.2 is roughly correct. The atmospheric pressure is taken as zero.

Note. The density of blood is approximately 1 g cm^{-3}. The density of mercury is $\rho = 13.6$ g cm^{-3}. $g = 980$ cm sec^{-2}. 1 mm Hg pressure $= 133.32$ N/m$^2 = 1333$ dyn/cm^2.

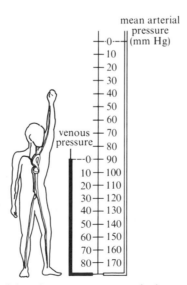

Figure P8.2 Mean arterial and venous pressures in human circulation, relative to atmospheric pressure. After Rushmer (1970), by permission.

8.3 Using the distensibility data of veins shown in Fig. 8.11:3, estimate the ratio of the volume of segments of veins in the leg of man when he/she is standing to that when he/she is lying down. The pressure external to the veins may be assumed to be atmospheric (taken as zero). The venous blood pressure at the level of the heart may be assumed zero in both postures.

Note. This may explain *fainting* that occurs sometimes when a person stands up from a reclining position or suddenly straightens up from stooping. The abnormal increase of the blood volume in the leg veins will decrease the venous filling pressure of the heart. This leads to a fall in cardiac output and thus of blood supply to the brain.

One of the reasons why everyone does not faint on standing is that reflex constriction of the smooth muscles in veins in the legs normally occurs so that their ability to act as a reservoir is greatly reduced. Another reason is that the smooth muscles in the arterioles also contract upon sudden increase of tension in the arteriole wall, so that the resistance to blood flow is increased and the arterial blood flow to the legs does fall. If smooth muscle action is impaired (e.g., by certain drugs or diseases), fainting or dizziness on standing will become common.

If the veins are maximally dilated as in hot climates or after a hot bath, the smooth muscles are relaxed and fainting is more common. On the other hand, a pilot executing a "high-g" turn will accentuate the problem by increasing the values of g, and black-out sometimes occurs. In this connection, how would you design an "anti-g" suit for pilots?

8.4 The average tensile stress in a blood vessel wall is given by the "law of Laplace" (see Chapter 1, Sec. 1.9):

$$\langle \sigma_\theta \rangle = \frac{p_i r_i - p_o r_o}{h},$$

where r_i and r_o are the inner and outer vessel radii, p_i and p_o are the internal and external pressures, respectively, and h is the vessel wall thickness. With the elasticity of the artery presented in Secs. 8.3 and 8.4, derive a formula for the arterial radius as a function of the pressures p_i and p_o.

Discuss the variation of p_i and p_o for the blood vessels in the body. In addition to the hydrostatic pressure distribution considered in Problems 8.1–8.3, consider the pulsatile pressure due to the heart, and the variation of pleural pressure in the thorax due to breathing.

8.5 When a vein is excised so that the pressure is reduced to zero, its radius is reduced by 40% and its length is reduced by 50%. Assume that the no-load state is the zero stress state of the vein, Fig. P8.5.
(a) What are the Green's strains $E_{\theta\theta}$, E_{zz} at the *in vivo* condition?
(b) If the strain-energy function of the vessel wall is given by

$$\rho_0 W = \frac{c}{2} \exp\{a_1 E_{\theta\theta}^2 + a_2 E_{zz}^2 + 2a_4 E_{\theta\theta} E_{zz}\}$$

what are the stresses $S_{\theta\theta}$ and S_{zz} *in vivo*?

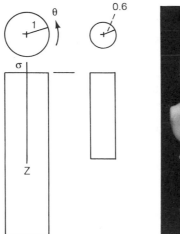

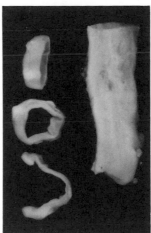

Figure P8.5   Figure P8.6

8.6 The zero stress states of some veins and vena cava of the rat are shown in Fig. P8.6. Now look back at Problem 8.5, what are the values of $E_{\theta\theta}$, E_{zz} at the no-load state, and at the *in vivo* condition?

8.7 A thin-walled cylindrical vessel of radius R is subjected to an internal pressure p, which produces a hoop tension of T, see Fig. P1.4 of Problem 1.4 in Chapter 1. Let T obey the law

$$T(t) = \int_{-\infty}^{t} G(t - \tau)\frac{\partial T^{(e)}[\lambda(\tau)]}{\partial \lambda}\frac{\partial \lambda(\tau)}{\partial \tau}\,d\tau,$$

$$T^{(e)}(\lambda) = (T^* + \beta)e^{\alpha(\lambda - \lambda^*)} - \beta,$$

$$G(t) = \frac{1}{A}\left[1 + c\int_{s_1}^{s_2}\frac{1}{s}e^{-t/s}\,ds\right],$$

$$A = 1 + c\int_{s_1}^{s_2}\frac{1}{s}\,ds.$$

If the vessel is suddenly inflated by an injection of fluid so that the radius is increased from the initial value $R = \lambda^* a$ to a new value $R = \lambda a$, with $\lambda > \lambda^*$, and a being the radius of the vessel when $p = 0$. What would be the history of the tension T? and of the pressure p? See I. Nigul and U. Nigul (1987), "On algorithms of evaluation of Fung's relaxation function parameters." *J. Biomech.* **20**, 343–352.

8.8 Let the experiment named in Problem 8.7 be changed by inflating the tube in such a way that the radius is increased uniformly in time at a constant rate k,

$$r(t) = a + kt.$$

What is the corresponding pressure history?

8.9 A vessel whose viscoelastic characteristics is described by Eq. (20) of Sec. 7.6.4, p. 281 (as a standard linear solid), and an elastic stress $T^{(e)}$ as shown in Problem 8.7 is subjected to a harmonic oscillation in radius. What is the tension oscillation in the vessel wall? What is the pressure oscillation?

Note. The following mathematical details may help. Using Eq. (7) of Sec. 7.6, and $T^{(e)}(\lambda)$ from Prob. 8.7, we have

$$T(t) = T^{(e)}(0^+)G(t) + \int_0^t E_R\left[1 - \left(1 - \frac{\tau_\sigma}{\tau_\varepsilon}\right)e^{-(t-\tau)/\tau_\varepsilon}\right]\alpha(T^* + \beta)e^{(\alpha\lambda - \alpha\lambda^*)}\frac{d\lambda}{d\tau}\,d\tau$$

Writing v for $1/\tau_\varepsilon$, we see that the right hand side involves the integral

$$F(t) = \int_0^t e^{v\tau}e^{\alpha\lambda}\frac{d\lambda}{d\tau}\,d\tau.$$

The difficult part of the problem is the evaluation of this integral, $F(t)$, when the stretch ratio λ oscillates harmonically:

$$\lambda = \lambda_0 + c\cos\omega t \cdot \mathbf{1}(t).$$

Since the viscoelasticity property is nonlinear, the real and imaginary parts of a complex representation of λ induces different tension, (not only a shift of phase angle), and has to be worked out separately. It suffices to consider the real part. Then

$$F(t) = \int_0^t e^{v\tau}e^{\alpha(\lambda_0 + c\cos\omega\tau)}(-c\omega)\sin\omega\tau\,d\tau$$

$$= -c\omega\,e^{\alpha\lambda_0}\int_0^t e^{v\tau}\sin\omega\tau\,e^{\alpha c\cos\omega\tau}\,d\tau.$$

In the theory of Bessel function, there are the formulas

$$e^{iz\sin\phi} = \sum_{n=-\infty}^{\infty} e^{in\phi}J_n(z)$$

$$= J_0(z) + 2\sum_{n=1}^{\infty} J_{2n}(z)\cos(2n\phi) + 2i\sum_{n=1}^{\infty} J_{2n-1}(z)\sin[(2n-1)\phi],$$

$$e^{iz\cos\phi} = \sum_{n=-\infty}^{\infty} i^n e^{in\phi}J_n(z)$$

$$= J_0(z) + 2\sum_{n=1}^{\infty} i^n J_n(z)\cos(n\phi).$$

See Erdelyi, Magnus, Oberhettinger, and Tricomi (1953), *Higher Transcendental Functions.* McGraw-Hill, New York, Vol. 2, p. 7, Eqs. (26) and (27). Hence

$$e^{\alpha c\cos\omega\tau} = J_0\left(\frac{\alpha c}{i}\right) + 2\sum_{n=1}^{\infty} i^n J_n\left(\frac{\alpha c}{i}\right)\cos(n\omega\tau).$$

But by definition of the Bessel function I_n:

$$I_n(x) = i^{-n}J_n(ix),$$

$$i^n J_n\left(\frac{\alpha c}{i}\right) = i^{-2n}(-1)^n i^n J_n\left(\frac{\alpha c}{i}\right)$$

$$= (-1)^n I_n(-\alpha c),$$

$$e^{\alpha c\cos\omega\tau} = I_0(-\alpha c) + (-1)^n 2\sum_{n=1}^{\infty} I_n(-\alpha c)\cos(n\omega\tau).$$

Hence

$$F(t) = -c\omega\, e^{\alpha\lambda_0} \int_0^t e^{vt} \sin \omega\tau \left[I_0(-\alpha c) + (-1)^n 2 \sum_{n=1}^{\infty} I_n(-\alpha c) \cdot \cos(n\omega\tau) \right] dt.$$

We need to evaluate the following integrals:

$$A_0 = \int_0^t e^{vt} \sin \omega\tau \, d\tau$$

$$= \frac{1}{\omega} \frac{1}{(v/\omega)^2 + 1} \left[e^{vt}\left(\frac{v}{\omega}\sin \omega t - \cos \omega t \right) + 1 \right],$$

$$A_n = 2 \int_0^t e^{vt} \sin \omega\tau \cos(n\omega\tau) \, d\tau \qquad (n = 1, 2, \ldots)$$

$$= \int_0^t e^{vt} [\sin(1 + n)\omega\tau + \sin(1 - n)\omega\tau] \, dt.$$

For $n \neq 1$,

$$A_n = \frac{1}{(1+n)\omega} \frac{1}{(v^2/(1+n)^2\omega^2)+1} \left[e^{vt}\left(\frac{v}{(1+n)\omega}\sin(1+n)\omega t - \cos(1+n)\omega t \right) + 1 \right]$$

$$+ \frac{1}{(1-n)\omega} \frac{1}{(v^2/(1-n)^2\omega^2)+1}$$

$$\times \left[e^{vt}\left(\frac{v}{(1-n)\omega}\sin(1-n)\omega t - \cos(1-n)\omega t \right) + 1 \right].$$

For $n = 1$,

$$A_1 = \frac{1}{2\omega} \frac{1}{(v^2/4\omega^2)+1} \left[e^{vt}\left(\frac{v}{2\omega}\sin 2\omega t - \cos 2\omega t \right) + 1 \right].$$

Hence, by substitution,

$$F(t) = -c\omega\, e^{\alpha\lambda_0} \left[I_0(-\alpha c) A_0 + (-1)^n \sum_{n=1}^{\infty} I_n(-\alpha c) A_n \right].$$

Having done this, it is simple to generalize the result to consider the continuous spectrum of Eq. (8.3–8).

8.10 A simplified analysis of residual stress in a thick-walled tube of Hookean material. At the zero stress state the body is a circular sector with a polar angle of θ, an inner radius of a_0, and an outer radius of b_0, Fig. P8.10. Assuming the material to be incompressible, compute the residual stress in the vessel wall when the vessel is rounded up into a circular tube at no transmural pressure, and the radii become a_1, b_1.

Now let an internal pressure p be imposed on the vessel, so that the vessel diameter is increased. Ignoring the residual stress, compute the new strain (or stretch ratio) distribution, and the new stress distribution in the vessel.

Compute a value of p at which the stress distribution in the vessel wall, with the residual stress condidered, will be approximately uniform, so that the circumferential stresses at the inner and outer walls are equal.

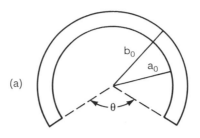

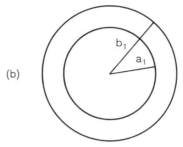

Figure P8.10

8.11 Modify the solution of Problem 8.10 by returning to the more accurate stress–strain relation

$$\sigma = (T^* + \beta)e^{\alpha(\lambda - \lambda^*)} - \beta$$

in which λ is the stretch ratio relative to the zero stress state. Are there any significant differences caused by the nonlinear stress–strain relationship?

8.12 If the material of a blood vessel wall is incompressible and has a strain-energy function,

$$\rho_0 W = P + \frac{C}{2}e^Q,$$

$$P = b_{ij}E_iE_j,$$

$$Q = a_{ij}E_iE_j,$$

where C, a_{ij}, b_{ij} are constants, E_1 = circumferential strain, E_2 = longitudinal strain, E_3 = radial strain, all in Green's sense. Derive the pressure-diameter relationship of the vessel if the zero stress state is a homogeneous circular cylindrical shell of inner radius R_i, outer radius R_o, and length L. In a deformed state, the radii and length are r_i, r_o, L, respectively.

8.13 Do the same as in the preceding problem if the zero stress state of the vessel is a circular arc of opening angle α.

8.14 Consider a theoretical problem: It is observed that anatomical structure and the materials of construction of the pulmonary arteries are quite similar to those of the systemic arteries of similar diameters. The major differences are that the ratio of the wall thickness to the lumen diameter of the pulmonary artery is usually smaller than that of the systemic vessels, and that the pulmonary vessels are embedded in the lung tissue, (tethered by the honeycomb-like interalveolar septa), whereas the aorta and some other systemic arteries are free standing (only lightly tethered by loose connective tissues to neighboring organs). Hence we are motivated to assume that the constitutive equations of the pulmonary and systemic arteries are similar, except that the material constants in the equations may be different. With this hypothesis, discuss the change of the diameters of the pulmonary arteries with respect to longitudinal stretch and the external stresses acting on the inner and outer walls of the vessel.

The external load acting on a pulmonary artery in the direction perpendicular to the vessel wall consists of the pulmonary blood pressure (p) acting on the inner wall, a tensile load imposed on the outer wall due to the tension in the interalveolar septa that are attached to the external wall of the pulmonary artery, and the pressure of the alveolar gas (p_A) acting on that part of the outer wall which is not covered by the interalveolar septa. The tensile load in the interalveolar septa, if averaged over an area which is much larger than the average cross sectional area of the individual alveoli, is called the *tissue stress of the lung parenchyma*, (symbol, σ_T). For a large pulmonary artery whose diameter is much larger than the diameter of the alveoli, σ_T may be considered as a uniform tensile stress. In this case, the external normal loads acting on the pulmonary artery are:

On the inner wall: p_i = a compressive stress p

$$\tag{1}$$

On the outer wall: p_o = a compressive stress $\alpha p_A - \sigma_T$

where α is the fraction of outer wall that is not covered by interalveolar septa. These are the p_i and p_o used in Eqs (6) and (7) of Sec. 1.9.

If the difference in the surface areas of the inner and outer walls of a blood vessel can be ignored, then the pressure difference $p_i - p_o$ is often called the *transmural pressure* and denoted by p_{TM}:

$$p_{TM} = p_i - p_o = p_i - \alpha p_A + \sigma_T \tag{2}$$

For example, the normal alveolar diameter of the lungs of the dog, cat, and man are in the ranges of 50–60 μm, 80–120 μm, 100–300 μm respectively, hence for these animals a vessel of diameter greater than 1 or 2 mm can be analyzed with the concept of tissue stress. On this basis, discuss the relationship between the arterial diameter and the stresses p, p_A, and σ_T for larger pulmonary vessels.

Pulmonary arteries with diameters smaller than the alveolar diameter are usually supported (tethered) by three or four interalveolar septa more or less equally spaced around the circumference. The outer surface of the vessel is then loaded by these septa. Part of the outer surface not covered by the septa is subjected to alveolar gas pressure p_A. This geometry suggests that the blood vessel cross section is non-circular under load.

Another equation involving the tissue stress σ_T can be derived by considering the equilibrium of a small element of the pulmonary pleura—the outer surface of the lung. (See Fung, *Biomechanics, Motion, Flow, Stress and Growth* 1990, Fig. 11.6:13.) The pleura is subjected to the pleural pressure (p_{PL}) on the outside, and alveolar gas pressure (p_A) and tissue stress (σ_T) in the inside. The pleural surface is curved, and has membrane tension in the wall. If the principal curvatures of the pleural membranes are κ_1 and κ_2 and the corresponding principal membrane tensions are N_1 and N_2, then the equivalent radial force per unit area is $N_1\kappa_1 + N_2\kappa_2$. The equation of equilibrium is (see Fung, *Biomechanics*, 1990, Sec. 11.6, Eq. 11.6; 11, p. 417):

$$\sigma_T + N_1\kappa_1 + N_2\kappa_2 + P_{PL} - \alpha'p_A = 0 \tag{3}$$

which connects σ_T to the pleural pressure and pleural tension, α', is the fraction of the inner surface of the pleura that is acted on by the alveolar gas pressure. Combining Eqs (1) and (3), we can express the forces acting on an arterial wall caused by p, p_A, $p_{PL}N_1\kappa_1$, and $N_2\kappa_2$. Now, based on the assumed strain energy function of the arterial wall, discuss the relationship between the vessel diameter, p, $p_A p_{PL}$, $N_1\kappa_1$, and $N_2\kappa_2$.

The determination of the tissue stress σ_T and the membrane tensions N_1, N_2 in the pleura as functions of lung deformation requires information about the stress-strain relationships of the lung parenchyma and the pleura. Relevant literature is discussed in Fung (1990). It is known that the pleural tension is a significant contributor to the pressure-volume relationship of the lung, (may contribute 10–25% of elasticity).

In life, p, p_A, p_{PL}, and p_a (blood pressure in main pulmonary artery), p_v (that in pulmonary vein) are functions of time and space. The problem is greatly simplified in a laboratory condition with an isolated lung at a steady state. Experiments have been done by a number of authors on isolated lobes of lung. Compare your theoretical results with the experimental results given by Lai-Fook, S. J. (1979), "A continuum mechanics analysis of pulmonary vascular interdependence in isolated dog lobes," *J. Appl. Physiol.* 46:419–429.

References

Al-Tinawi, A., Madden, J. A., Dawson, C. A., Linehan, J. H., Harder, D. R., and Rickaby, D. A. (1991) Distensibility of small arteries of the dog lung. *J. Appl. Physiol*, **71** 1714–1722.

Anliker, M. (1972) Toward a nontraumatic study of the circulatory system. In *Biomechanics: Its Foundations and Objectives*, Y. C. Fung, N. Perrone, and M. Anliker (eds.) Prentice-Hall, Englewood Cliffs, NJ, pp. 337–380.

Anliker, M., Histand, M. B., and Ogden, E. (1968) Dispersion and attenuation of small artificial pressure waves in the canine aorta. *Circulation Res.* **23**, 539–551.

Anliker, M., Wells, M. K., and Ogden, E. (1969) The transmission characteristics of large and small pressure waves in the abdominal vena cava. *IEEE Trans. Biomed. Eng.* **BME-16**, 262–273.

Ayorinde, O. A., Kobayashi, A. S., and Merati, J. K. (1975) Finite elasticity analysis of unanesthetized and anesthetized aorta. In 1975 *Biomechanics Symposium*. ASME, New York, p. 79.

Azuma, T., Hasegawa, M., and Matsuda, T. (1970) Rheological properties of large arteries. In *Proceedings of the 5th International Congress on Rheology*, S. Onogi (ed.) University of Tokyo Press, Tokyo and University Part Press, Baltimore, pp. 129–141.

Azuma, T. and Hasegawa, M. (1971) A rheological approach to the architecture of arterial walls. *Jpn. J. Physiol.* **21**, 27–47.

Azuma, T. and Hasegawa, M. (1973) Distensibility of the vein: From the architectural point of view. *Biorheology* **10**, 469–479.

Baez, S., Lamport, H., and Baez, A. (1960) Pressure effects in living microscopic vessels. In *Flow Properties of Blood and Other Biological Systems*, A. L. Copley and G. Stainsby (eds.) Pergamon Press, Oxford, pp. 122–136.

Baldwin, A. L. and Gore, R. W. (1989) Simultaneous measurement of capillary distensibility and hydraulic conductance. *Microvasc. Res.* **38**, 1–22.

Bauer, R. D. and Pasch, T. (1971) The quasistatic and dynamic circumferential elastic modulus of the rat tail artery studied at various wall stresses and tones of the vascular smooth muscle. *Pflügers Arch.* **330**, 335–346.

Bergel, D. H. (1972) The properties of blood vessels. In *Biomechanics: Its Foundations and Objectives*, Y. C. Fung, N. Perrone, and M. Anliker (eds.) Prentice-Hall, Englewood Cliffs, NJ, Chapter 5, pp. 105–140.

Bevan, J. A., Bevan, R. D., Chang, P. C., Pegram, B. L., Purdy, R. E., and Su, C. (1975) Analysis of changes in reactivity of rabbit arteries and veins 2 weeks after induction of hypertension by co-arctation of the abdominal aorta. *Circulation Res.* **37**, 183–190.

Bevan, R. D. (1976) Autoradiographic and pathological study of cellular proliferation in rabbit arteries correlated with an increase in arterial pressure. *Blood Vessels* **13**, 100–124.

Bohr, D. F., Somlyo, A. P., and Sparks, H. V., Jr. (eds.) (1980) *Handbook of Physiology. Sec. 2 The Cardiovascular System. Vol 2: Vascular Smooth Muscle.* American Physiological Society, Bethesda, MD.

Bouskela, E. and Wiederhielm, C. (1979a) Viscoelastic properties of capillaries. (Abstract) *Microvascu. Res.* **17**, No. 3, Part 2, p. S1.

Bouskela, E. and Wiederhielm, C. A. (1979b) Microvascular myogenic reaction in the wing of the intact unanesthetized bat. *Am. J. Physiol.* **237**, H59–H65.

Brankov, G., Rachev, A., and Stoychev, S. (1974) On the mechanical behavior of blood vessels. *Biomechanica* (Bulgarian Academy of Sciences, Sofia) **1**, 27–35.

Brankov, G., Rachev, A., and Stoychev, S. (1976) A structural model of arterial vessel. *Biomechanica* (Bulgarian Academy of Sciences, Sofia) **3**, 3–11.

Burton, A. C. (1944) Relation of structure to function of the tissues of the wall of blood vessels. *Physiol. Rev.* **34**, 619–642.

Burton, A. C. (1947) The physical equilibrium of small blood vessels. *Am. J. Physiol.* **149**, 389–399. See also, ibid, **164**, 319.

Burton, A. C. (1962) Properties of smooth muscle and regulation of circulation. *Physiol. Rev.* **42**, 1–6.

Burton, A. C. (1966) Role of geometry, size, and shape in the microcirculation. *Fed. Proc.* **25**, 1753–1760.

Campbell, J. H. and Campbell, G. R. (1988) *Vascular Smooth Muscle in Culture*, 2 Vols. CRC Press, Boca Raton, FL.

Carew, T. E., Vaishnav, R. N., and Patel, D. J. (1968) Compressibility of the arterial wall. *Circulation Res.* **23**, 61–68.

Caro, C. G., Pedley, T. J., Schroter, R. C., and Seed, W. A. (1978) *The Mechanics of the Circulation.* Oxford University Press, Oxford, 1978.

Cheung, J. B. and Hsiao, C. C. (1972) Nonlinear anisotropic viscoelastic stresses in blood vessels. *J. Biomech.* **5**, 607–619.

Chuong, C. J. and Fung, Y. C. (1983) Three-dimensional stress distribution in arteries. *J. Biomech. Eng.* **105**, 268–274.

Chuong, C. J. and Fung, Y. C. (1984) Compressibility and constitutive equation of arterial wall in radial compression experiments. *J. Biomech.* **17**, 35–40.

Chuong, C. J. and Fung, Y. C. (1986) Residual stress in arteries. In *Frontiers in Biomechanics* G. W. Schmid-Schönbein, S. L.-Y. Woo, and B. W. Zweifach eds. Springer-Verlag, New York, pp. 117–129. Also *J. Biomech. Eng.* **108**, 189–192.

Collins, R. and W. C. L. Hu (1972) Dynamic deformation experiments on aortic tissue. *J. Biomech.* **5**, 333–337.

Cowan, M. J. and Crystal, R. G. (1975) Lung growth after unilateral pneumonectomy: Quantiation of collagen synthesis and content. *Am. Rev. Respir. Disease* **111**, 267–276.

Cox, R. H., Jones, A. W., and Fischer, G. M. (1974) Carotid artery mechanics, connective tissue, and electrolyte changes in puppies. *Am. J. Physiol.* **227**, 563–568.

Cox, R. H. (1974) Three-dimensional mechanics of arterial segments *in vitro*: Method. *J. Appl. Physiol.* **36**, 381–384.

Cox, R. H. (1976a) Mechanics of canine iliac artery smooth muscle *in vitro*. *Am. J. Physiol.* **230**, 462–470.

Cox, R. H. (1976b) Effect of norepinephrine on mechanics of arteries *in vitro*. *Am. J. Physiol.* **231**, 420–425, 1976; **233**, H243–H255, 1977.

Cox, R. H. (1978) Comparison of carotid artery mechanics in the rat, rabbit, and dog. *Am. J. Physiol.* **234**, H280–H288.

Cox, R. H. (1979a) Alterations in active and passive mechanics of rate carotid artery with experimental hypertension. *Am. J. Physiol.* **237**, H597–H605.

Cox, R. H. (1979b) Regional, species, and age related variations in the mechanical properties of arteries. *Biorheology* **16**, 85–94.

Cox, R. H. (1982) Mechanical properties of the coronary vascular wall and the contractile process. In *The Coronary Artery*, S. Kalsner (eds.) Oxford University Press, New York, pp. 59–90.

Crystal, R. G. (1974) Lung collagen: Definition, diversity, and development. *Fed. Proc.* **33**, 2248–2255.

Debes, J. C. (1992) *Mechanics of Pulmonary Parenchyma and Arteries*. Ph.D. Thesis. University of California, San Diego.

Debes, J. C. and Fung, Y. C. (1992) Effect of temperature on the biaxial mechanics of excised lung parenchyma of the dog. *J. Appl. Physiol.* **73**, 1171–1180.

Demiray, H. J. (1972) A note on the elasticity of soft biological tissues. *J. Biomech.* **5**, 309–311.

Doyle, J. M. and Dobrin, P. B. (1971) Finite deformation analysis of the relaxed and contracted dog carotid artery. *Microvasc. Res.* **3**, 400–415.

Doyle, J. M. and Dobrin, P. B. (1973) Stress gradients in the walls of large arteries. *J. Biomech.* **6**, 631–639.

Faulkner, J. A., Weiss, S. W., and McGeachie, J. K. (1983) Revascularization of skeletal muscle transplanted into the hamster cheek pouch: Intravital and light microscopy. *Microvasc. Res.* **26**, 49–64.

Fisher, G. M. and Llaurado, J. G. (1967) Connective tissue composition of canine arteries. Effects of renal hypertension. *Arch. Pathol.* **84**, 95–98.

Folkman, J. (1985) Tumor angiogenesis. *Adv. Cancer Res.* **43**, 175–203.

Frank, O. (1920) Die elastizitat der blutgefasse *J. Biol.* **71**, 255–272.

Frasher, W. G. (1966) What is known about large blood vessels. In *Biomechanics, Proc. Symp.* Y. C. Fung (ed.) ASME, New York, pp. 1–19.

Fronek, K., Schmid-Schönbein, G., and Fung, Y. C. (1976) A noncontact method for three-dimensional analysis of vascular elasticity *in vivo* and *in vitro*. *J. Appl. Physiol.* **40**, 634–637.

Fry, D. L. (1968) Acute vascular endothelial changes associated with increased blood velocity gradients. *Circulation Res.* **22**, 165–197.

Fry, D. L. (1969) Certain histological and chemical responses of the vascular interface of acutely induced mechanical stress in the aorta of the dog. *Circulation Res.* **24**, 93–108.

Fung, Y. C. (1966a) Microscopic blood bessels in the mesentery. In *Biomechanics, Proc. Symp.* Y. C. Fung (ed.) ASME, New York, pp. 151–166.

Fung, Y. C. (1966b) Theoretical considerations of the elasticity of red cells and small blood vessels. *Fed. Proc.* **25**, 1761–1772.

Fung, Y. C. (1967) Elasticity of soft tissues in simple elongation. *Am. J. Physiol.* **28**, 1532–1544.

Fung, Y. C. (1972) Stress–strain-history relations of soft tissues in simple elongation. In *Biomechanics: Its Foundations and Objectives*, Y. C. Fung, N. Perrone and M. Anliker (eds.) Prentice-Hall, Englewood Cliffs, NJ, pp. 181–208.

Fung, Y. C. (1973) Biorheology of soft tissues. *Biorheology* **10**, 139–155.

Fung, Y. C. (1979) Inversion of a class of nonlinear stress–strain relationships of biological soft tissues. *J. Biomech. Eng.* **101**, 23–27.

Fung, Y. C. (1983) What principle governs the stress distribution in living organisms. In *Biomechanics in China, Japan, and USA* Y. C. Fung, E. Fakada, and J. J. Wang (eds.) Science Press, Beijing, China, pp. 1–13.

Fung, Y. C. (1984) *Biodynamics: Circulation.* Springer-Verlag, New York.

Fung, Y. C. (1984) The need for a new hypothesis for residual stress distribution. In *Biodynamics: Circulation.* loc cit, pp. 54–68.

Fung, Y. C. (1990) *Biomechanics: Motion, Flow, Stress, and Growth.* Springer-Verlag, New York.

Fung, Y. C., Zweifach, B. W., and Intaglietta, M. (1966) Elastic environment of the capillary bed. *Circulation Res.* **19**, 441–461.

Fung, Y. C. and Sobin, S. S. (1972) Elasticity of the pulmonary alveolar sheet. *Circulation Res.* **30**, 451–469; 470–490.

Fung, Y. C., Fronek, K., and Patitucci, P. (1979) Pseudoelasticity of arteries and the choice of its mathematical expression. *Am. J. Physiol.* **237**, H620–H631.

Fung, Y. C. and Liu, S. Q. (1989) Change of residual strains in arteries due to hypertrophy caused by aortic constriction. *Circulation Res.* **65**, 1340–1349.

Fung, Y. C. and Liu, S. Q. (1991) Changes of zero stress state of rat pulmonary arteries in hypoxic hypertension. *J. Appl. Physiol.* **70**, 2455–2470.

Fung, Y. C. and Liu, S. Q. (1992) Strain distribution in small blood vessels with zero stress state taken into consideration. *Am. J. Physiol.* **262**, H544–H552.

Gaehtgens, P. and Uekermann, U. (1971) The distensibility of mesenteric venous microvessels. *Pflügers Arch.* **330**, 206–216.

Gore, R. W. (1972) Wall stress: A determinant of regional differences in response of frog microvessels to norepinephrine. *Am. J. Physiol.* **222**, 82–91.

Gore, R. W. (1974) Pressures in cat mesenteric arterioles and capillaries during changes in systemic arterial blood pressure. *Circulation Res.* **34**, 581–591.

Gore, R. W. and Davis, M. J. (1984) Mechanics of smooth muscle in isolated single microvessels. *Ann. Biomed. Eng.* **12**, 511–520.

Gou, P. F. (1970) Strain-energy function for biological tissues. *J. Biomech.* **3**, 547–550.

Gow, B. S. and Taylor, M. G. (1968) Measurement of viscoelastic properties of arteries in the living dog. *Circulation Res.* **23**, 111–122.

Han, H. C. and Fung, Y. C. (1991) Species dependence on the zero stress state of aorta: Pig vs. rat. *J. Biomech. Eng.* **113**, 446–451.

Hardung, V. (1953) Vergleichende messungen der dynamischen elastizitat und viskositat von blutegefassen, Kautschuk und synthetischen elastomeren. *Helv. Physiol. Pharmacol. Acta* **11**, 194–211.

Harkness, M. L. R., Harkness, R. D., and McDonald, D. A. (1957) The collagen and elastin content of the arterial wall in the dog. *Proc. Roy. Soc. London B* **1465**, 541–551.

Hayashi, K., Handa, H., Mori, K., and Moritake, K. (1971) Mechanical behavior of vascular walls. *J. Soc. Mater. Sci. (Jpn.)* **20**, 1001–1011.

Hayashi, K., Sato, M., Handa, H., and Moritake, K. (1974) Biomechanical study of the constitutive laws of vascular walls. *Exper. Mech.* **14**, 440–444.

Hayashi, K., Nagasawa, S., Nario, Y., Okumura, A., Moritake, K., and Handa, H. (1980) Mechanical properties of human cerebral arteries. *Biorheology* **17**, 211–218.

Hayashi, K. and Takamizawa, K. (1989) Stress and strain distributions and residual stresses in arterial walls. In *Progress and New Directions of Biomechanics*, Y. C. Fung, K. Hayashi, and Y. Seguchi (eds.) MITA Press, Tokyo, Japan, pp. 185–192.

Imholz, B. P. M., Wieling, W., Langewouters, G. J., and van Montfrans, G. A. (1991) Continuous finger arterial pressure: Utility in the cardiovascular Laboratory. *Clinical Autonomic Res.* **1**, 43–53.

Intaglietta, M. and Plomb, E. P. (1973) Fluid exchange in tunnel and tube capillaries. *Microvasc. Res.* **7**, 153–168.

Intaglietta, M. and Zweifach, B. W. (1971) Geometrical model of the microvasculature of rabbit omentum from *in vivo* measurements. *Circulation Res.* **28**, 593–600.

Johnson, P. C. (1968) Autoregulatory responses of cat mesenteric arterioles measured *in vivo*. *Circulation Res.* **22**, 199–212.

Johnson, P. C. (1980) The Myogenic Response. In Bohr et al. (1980) *loc. cit.*, pp. 409–442.

Kasyanov, V. (1974) The anisotropic nonlinear model of human large blood vessels. *Mech. Polym. USSR* 874–884.

Kasyanov, V. and Knets, I. (1974) Strain-energy function for large human blood vessels *Mech. Polym. USSR* 122–128.

Kenner, T. (1967) Neue gesichtspunkte und experimente zur beschreibung und messung der arterienelastizitat. *Arch. Kreislaufforschung* **54**, 68–139.

Langewouters, G. J., Wesseling, K. H., and Goedhard, W. J. A. (1984) The static elastic properties of 45 human thoracic and 20 abdominal aortas in vitro and the parameters of a new model. *J. Biomech.* **17**, 425–435. See also, ibid., **8**, 613–620, 1985.

Laszt, L. (1968) Untersuchungen uber die elastischen eigenschaften der blutegasse um ruhe-und im kontraktionszustand. *Angiologica* 5, 14–27.

Lee, J. S., Frasher, W. G., and Fung, Y. C. (1967) Two-dimensional finite deformation experiments on dog's arteries and veins. Rept. No. AFOSR 67-00198, Dept. of Applied Mechanics and Engineering Science—Bioengineering, University of California, San Diego, California.

Lee, J. S., Frasher, W. G., and Fung, Y. C. (1968) Comparison of the elasticity of an artery *in vivo* and in excision. *J. Appl. Physiol.* 25, 799–801.

Lee, Robert M. K. (ed.) (1989) *Blood Vessel Changes in Hypertension: Structure and Function*, Vol. 1 (with articles by R. Cox, H. Ho, M. Kay, M. Mulvany, G. Owens, and C. Triggle). 199 pp., Vol. 2 (T. J. F. Lee, G. Osol, A. L. Loeb and M. J. Peach, R. L. Prewitt, H. Hashimoto and D. L. Stacy, J. H. Lombard, M. J. Mackay, D. W. Cheung, W. J. Stekiel). 177 pp., Vol. 2, CRC Press, Boca Raton, FL.

Litwak, P., Ward, R. S., Robinson, A. J., Yilgor, I., and Spatz, C. A. (1988) Development of a small diameter, compliant vascular prosthesis. In *Tissue Engineering*, R. Skalak and C. F. Fox (eds.) Alan Liss, New York, pp. 25–30.

Liu, S. Q. and Fung, Y. C. (1988) Zero stress states of arteries. *J. Biomech. Eng.* 110, 82–84.

Liu, S. Q. and Fung, Y. C. (1989) Relationship between hypertension, hypertrophy, and opening angle of zero stress state of arteries following aortic constriction. *J. Biomech. Eng.* 111, 325–335.

Liu, S. Q. and Fung, Y. C. (1990) Change of zero stress state of rat pulmonary arteries in hypoxic hypertension. *J. Appl. Physiol.* Submitted.

Liu, S. Q. and Fung, Y. C. (1992) Influence of STZ-induced diabetes on zero stress states of rat pulmonary and systemic arteries. *Diabetes* 41, 136–146.

Maltzahn, W.-W. von, Besdo, D., and Wiemer, W. (1981) Elastic properties of arteries: A nonlinear two-layer cylindrical model. *J. Biomech.* 14, 389–397.

Maltzahn, W. W., Warrinyar, R. G., and Keitzer, W. F. (1984) Experimental measurement soft elastic properties of media and adventitia of bovine carotid arteries. *J. Biomech.* 17, 839–848.

Marquardt, D. W. (1963) An algorithm for least-squares estimation of nonlinear parameters. *J. Soc. Ind. Appl. Math.* 2, 431–441.

Mathieu-Costello, O. and Fronek, K. (1985) Morphometry of the amount of smooth muscle cells in the media of various rabbit arteries. *J. Ultrastruct. Res.* 91, 1–12.

Matsuda, M., Fung, Y. C., and Sobin, S. S. (1987) Collagen and elastin fibers in human pulmonary alveolar mouths and ducts. *J. Appl. Physiol.* 63, 1185–1194.

McDonald, D. A. (1968) Regional pulse-wave velocity in the arterial tree. *J. Appl. Physiol.* 24, 73–78.

Mercer, R. R. and Crapo, J. D. (1990) Spatial distribution of collagen and elastin fibers in the lung. *J. Appl. Physiol.* 69, 756–765.

Meyrick, B. and Reid, L. (1980) Hypoxia-induced structural changes in the media and adventitia of the rat hiller pulmonary artery and their regression. *Am J. Pathol.* 100, 151–178.

Mirsky, I. (1973) Ventricular and arterial wall stresses based on large deformation theories. *Biophys. J.* 13, 1141–1159.

Moritake, K., Handa, H., Hayashi, K., and Sato, M. (1973) Experimental studies of intracranial aneurysms. Some biomechanical considerations of the wall structures of intracranial aneurysms and experimentally produced aneurysms. *Jpn. J. Brain Surg.* 1, 115–123.

Murphy, R. A. (1980) Mechanics of vascular smooth muscle. In *Handbook of Physiology*. American Physiological Society, Bethesda, MD, Sec. 2, Vol. 2, pp. 325–351.

Nagasawa, S., Handa, H., Nario, Y., Okumura, A., Moritake, K., and Hayaski, K. (1982) Biomechanical study on aging changes and vasospasm of human cerebral arteries. *Biorheology* **19**, 481–489.

Nerem, R. M. (1988) Endothelial cell responses to shear stress: Implications in the development of endothelialized synthetic vascular grafts. In *Tissue Engineering*, R. Skalak and C. F. Fox (eds.) Alan Liss, New York, pp. 5–10.

Patel, D. J. and Vaishnav, R. N. (1972) The rheology of large blood vessels. In *Cardiovascular Fluid Dynamics*, D. H. Bergel (ed.) Academic Press, New York, Vol. 2, Chapter 11, pp. 2–65.

Patel, D. J. and Vaishnav, R. N. (eds.) (1980) *Basic Hemodynamics and Its Role in Disease Process*. University Park Press, Baltimore, MD.

Rhodin, J. A. G. (1980) Architecture of the vessel wall. In *Handbook of Physiology, Sec 2 The Cardiovascular Bystem*, Vol. II *Vascular Smooth Muscle*. American Physiological Society Bethesda, MD, pp. 1–31.

Roach, M. R. and Burton, A. C. (1957) The reason for the shape of the distensibility curve of arteries. *Can. J. Biochem. Physiol.* **35**, 681–690.

Roach, M. R. and Buron, A. C. (1959) The effect of age on the elasticity of human iliac arteries. *Can. J. Biochem. Physiol.* **37**, 557–570.

Rushmer, R. F. (1970) *Cardiovascular Dynamics*, 3rd Edition. W. B. Saunders, Philadelphia, p. 196.

Sato, M., Hayashi, K., Niimi, H., Moritake, K., Okumura, A., and Handa, H. (1979) Axial mechanical properties of arterial walls and their anisotropy. *Med. & Biol. Eng. & Comput.*, **17**, 170–176.

Schmid-Schönbein, G. W., Fung, Y. D., and Zweifach, B. W. (1975) Vascular endothelium –leukocyte interaction. *Circulation Res.* **36**, 173–184.

Sharma, M. G. (1974) Viscoelastic behavior of conduit arteries. *Biorheology* **11**, 279–291.

Shepherd, J. T. and Vanhoutte, P. M. (1975) *Veins and Their Control*. W. B. Saunders, Philadelphia.

Smaje, L. H., Fraser, P. A., and Clough, G. (1980) The distensibility of single capillaries and venules in the cat mesentery. *Microvasc. Res.* **20**, 358–370.

Sobin, P. (1977) Mechanical Properties of Human Veins. M. S. Thesis, University of California, San Diego, California.

Sobin, S., Fung, Y. C., Tremer, H. M., and Rosenquist, T. H. (1972) Elasticity of the pulmonary alveolar microvascular sheet in the cat. *Circulation Res.* **30**, 440–450.

Sobin, S. S., Lindal, R. G., Fung, Y. C., and Tremer, H. M. (1978) Elasticity of the smallest noncapillary pulmonary blood vessels in the cat. *Microvasc. Res.* **15**, 57–68.

Sobin, S. S., Fung, Y. C., and Tremer, H. M. (1988) Collagen and elastin fibers in human pulmonary alveolar walls. *J. Appl. Physiol.* **64**, 1659–1675.

Takamizawa, K. and Hayashi, K. (1987) Strain-energy density function and uniform strain hypothesis for arterial mechanics. *J. Biomech.* **20**, 7–17.

Takamizawa, K. and Hayashi, K. (1988) Uniform strain hypothesis and thin-walled theory in arterial mechanics. *Bioreoheology* **25**, 555–565.

Tanaka, T. T. and Fung, Y. C. (1974) Elastic and inelastic properties of the canine aorta and their variation along the aortic tree. *J. Biomech.* **7**, 357–370.

Takamizawa, K. and Matsuda, T. (1990) Kinematics for bodies undergoing residual stress and its applications to the left ventricle. *J. Appl. Mech.* **57**, 321–329.

Vaishnav, R. N., Young, J. T., Janicki, J. S., and Patel, D. J. (1972) Nonlinear anisotropic elastic properties of the canine aorta. *Biophys. J.* **12**, 1008–1027.

Vaishnav, R. N., Young, J. T., and Patel, D. J. (1973) Distribution of stresses and strain-energy density through the wall thickness in a canine aortic segment. *Circulation Res.* **32**, 557–583.

Vaishnav, R. N. and Vossoughi, J. (1983) Estimation of residual strains in aortic segments. In *Biomedical Engineering, II, Recent Developments*, C. W. Hall (ed.) Pergamon Press, New York, pp. 330–333.

Vaishnav, R. N. and Vossoughi, J. (1987) Residual stress ad strain in aortic segments. *J. Biomech.* **20**, 235–239.

Vossoughi, J., Weizsäcker, H. W., and Vaishnav, R. M. (1985) Variation of aortic geometry in various animal species. *Biomedizinische Tech.* **30**, 48–54.

Wesley, R. L. R., Vaishnav, R. N., Fuchs, J. C. A., Patel, D. J., and Greenfield, J. C. Jr. (1975) *Circulation Res.* **37**, 509–520.

Wetterer, E. and Kenner, T. (1968) *Grundlagen der dynamik des arterienpulses*. Springer-Verlag, Berlin.

Wetterer, E., Bauer, R. D., and Busse, R. (1978) New ways of determining the propagation coefficient and the viscoelastic behavior of arteries *in situ*. In *The Arterial System*, E. D. Bauer and R. Busse (eds.) Springer-Verlag, Berlin, pp. 35–47.

Wolinsky, H. and Glagov, S. (1964) Structural basis for the static mechanical properties of the aortic media. *Circulation Res.* **14**, 400–413.

Wylie, E. R. (1966) Flow through tapered tubes with nonlinear wall properties. In *Biomechanics Symposium* (Fung. Y. C. ed.), Am. Soc. Mech. Eng., New York, pp. 82–95.

Xie, J. P., Liu, S. Q., Yang, R. F., and Fung. Y. C. (1991) The zero stress state of rat veins and vena cava. *J. Biochem. Eng.* **113**, 36–41.

Xie, J. P., Zhou, J., and Fung, Y. C. (1993) Bending of blood vessel wall: Stress-strain laws of the intima-media and adventitial layers. *J. Biomech. Eng.* Submitted.

Yen, R. T. and Foppiano, L. (1981) Elasticity of small pulmonary veins in the cat. *J. Biomech. Eng.* **103**, 38–42.

Young, J. T., Vaishnav, R. N., and Patel, D. J. (1977) Nonlinear anisotropic viscoelastic properties of canine arterial segments. *J. Biomech.* **10**, 549–559.

Young, T. (1809) On the functions of the heart and arteries. *Philos. Trans. Roy. Soc. London* **99**, 1–31.

Yu, Q., Zhou, J., and Fung, Y. C. (1993) Neutral axis location in bending and the Young's modulus of different layers of arterial wall. *Am. J. Physiol.*, in press.

Zeng, Y. J., Yager, D., and Fung, Y. C. (1987) Measurement of the mechanical properties of the human lung tissue *J. Biomech. Eng.* **109**, 169–174.

Yager, D., Feldman, H., and Fung, Y. C. (1992) Microscopic vs. macroscopic deformation of the pulmonary alveolar duct. *J. Appl. Phyiol.* **72**, 1348–1354.

CHAPTER 9

Skeletal Muscle

9.1 Introduction

There are three kinds of muscles: skeletal, heart, and smooth. Skeletal muscle makes up a major part of the animal body. It is the prime mover of animal locomotion. It is controlled by voluntary nerves. It has the feature that if it is stimulated at a sufficiently high frequency, it can generate a maximal tension, which remains constant in time. It is then said to be *tetanized*. The activity of the contracting mechanism is then thought to be maximal.

Since a resting skeletal muscle is a viscoelastic material with quite ordinary properties, the really interesting part of the properties of skeletal muscle is the contraction. Hence skeletal muscle is often studied in the tetanized condition. For the frog sartorius muscle at an optimal length, the maximum stress of contraction is about 200 kPa (i.e., about 2 kgf/cm^2), which is much larger than the stress in the same muscle at the same length when unstimulated. Hence the resting stress does not play an important role in skeletal muscle mechanics, except in setting the resting length of the muscle.

Heart muscle is also striated like skeletal muscle, but in its normal function it is never tetanized. Heart muscle functions in single twitches. Each electric stimulation evokes one twitch, which has to run a definite course. Until a certain refraction period is passed, another electric stimulation will not evoke a response. In this way heart muscle is very different from skeletal muscle. Another difference is that the resting heart muscle is stiffer than the resting skeletal muscle. The resting stress in heart muscle is not negligible compared with the stress in the contractile element. Hence the study of heart muscle needs greater attention to the resting state.

Smooth muscles are not striated, and are not controlled by voluntary nerves. There are many kinds of smooth muscles, with widely different mechanical properties.

These three classes of muscles will be discussed in this chapter and Chapters 10 and 11.

9.2 The Functional Arrangement of Muscles

The strategic deployment of muscles to control the mechanisms of animal locomotion is itself a subject of great beauty and importance. Nature has developed many ingenious devices for this purpose. As an introduction to muscle mechanisms, let us consider some patterns of arrangements of muscle fibers when severe geometric constraints are imposed. Consider the claws of a crab. The crab uses muscles in its legs to move its claws. The leg has a rigid shell which is almost completely filled with muscle. It is necessary therefore to arrange the muscle fibers in a pinnate arrangement [see Fig. 9.2:1(a)], so that when contraction occurs, no lateral expansion will occur. If the muscle fibers were arranged parallel as shown in Fig. 9.2:2(a), then a contraction in muscle length would induce an increase of its lateral dimension, because the muscle has to preserve its volume. Since the rigid shell of the crab claw prevents any increase of lateral dimension, a bundle of parallel muscle fibers will not be able to contract at all. Thus the pinnate arrangement of the leg muscle is a necessity for the crab.

Figure 9.2:1(b) shows another interesting arrangement of the muscle in the jaw of a characinoid fish. This fish has a large eyeball. The muscles are arranged in such a manner that when they contract, a component of the tension pulls the muscle away from the eyeball to leave it free from undue stresses. Such beautiful mechanisms are everywhere in nature. Think

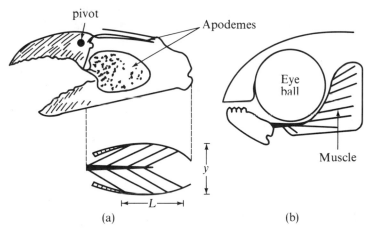

Figure 9.2:1 (a) The skeleton of a crab chela cut open to show the apodemes, and (below) a diagrammatic horizontal section showing the pinnate arrangement of the muscle that closes the chela. (b) A diagram showing the pinnate arrangement of the main jaw muscle of a characinoid fish. From Alexander (1968), by permission.

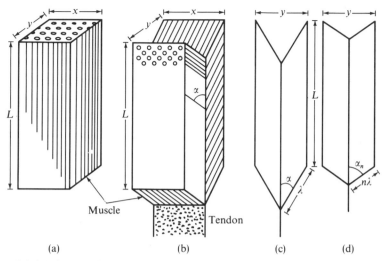

Figure 9.2:2 Diagram illustrating (a) parallel-fibered and (b) pinnate muscles. (c) and (d) illustrate pinnate muscle contraction. From Alexander (1968), by permission.

of our own body. How marvelously our muscles are arranged so that we can do all the things we want to do!

9.3 The Structure of Skeletal Muscle

In the previous section we considered the structure of muscle on a scale visible to the naked eye. At a finer scale we may examine muscle with an optical microscope, with which we can see things down to about 0.2 μm. And we can turn to the electron microscope, which has a theoretical resolution of 0.002 nm, but in practice about 0.2 nm.

What can be seen at various levels of magnification is shown in Fig. 9.3:1. It is seen that the units of skeletal muscle are the muscle fibers, each of which is a single cell provided with many nuclei. These fibers are arranged in bundles or *fasciculi* of various sizes within the muscle. Connective tissue fills the spaces between the muscle fibers within a bundle. Each bundle is surrounded by a stronger connective tissue sheath; and the whole muscle is again surrounded by an even stronger sheath.

A skeletal muscle fiber is elongated, having a diameter of 10–60 μm, and a length usually of several millimeters to several centimeters; but sometimes the length can reach 30 cm in long muscles. The fibers may stretch from one end of muscle to another, but often extend only part of the length of the muscle, ending in tendinous or other connective tissue intersections.

The flattened nuclei of muscle fibers lie immediately beneath the cell membrane. The cytoplasm is divided into longitudinal threads or *myofibrils*,

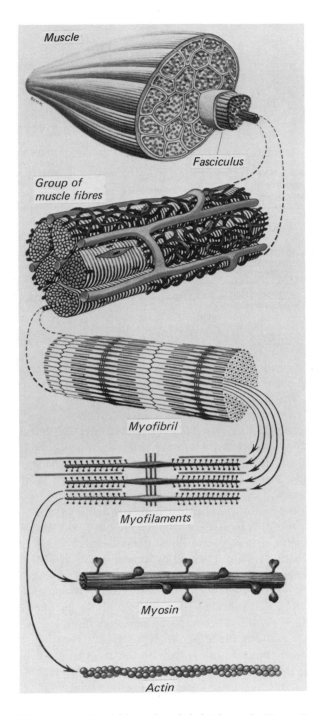

Figure 9.3:1 The organizational hierarchy of skeletal muscle. From *Gray's Anatomy*, 35th British edn. (1973), edited by Warwick and Williams, by permission.

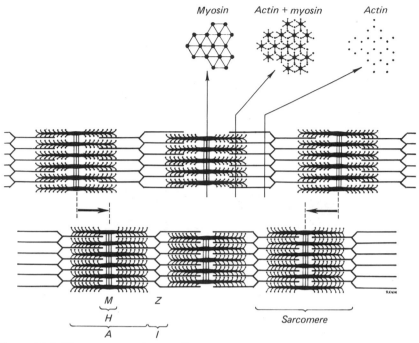

Figure 9.3:2 The structure of a myofilament, showing the spatial arrangement of the actin and myosin molecules. From *Gray's Anatomy*, 35th British edn. (1973), edited by Warwick and Williams, by permission.

each about 1 μm in diameter. These myofibrils are striated when they are stained by dyes and when they are examined optically. Some zones stain lightly with basic dyes such as hematoxylin, rotate the plane of polarization of light weakly, and are called *isotropic* or *I bands*. Others, alternating with the former, stain deeply with hematoxylin and strongly rotate the plane of polarization of light to indicate a highly ordered substructure. They are called *anisotropic* or *A bands*. The *I* bands are bisected transversely by a thin line also stainable with basic dyes: this line is called the *zwischenscheibe* or *Z band*. The *A* bands are also bisected by a paler line called the *H* band. These bands are illustrated in Fig. 9.3:2. If the muscle contracts greatly, the *I* and *H* bands may narrow to extinction, but the *A* bands remain unaltered. The lower part of Fig. 9.3:1 shows what is known about the fine structure of the myofibrils. Each myofibril is composed of arrays of *myofilaments*. These are divided transversely by the *Z* bands into serially repeating regions termed *sarcomeres*, each about 2.5 μm long, with the exact length dependent on the force acting in the muscle and the state of excitation. Two types of myofilament are distinguishable in each sarcomere, fine ones about 5 nm in diameter and thicker ones about 12 nm across. The fine ones are *actin* molecules. The thick ones are *myosin* molecules. The actin filaments are each attached at one end to a *Z* band and are free at the other to interdigitate with the myosin filaments.

The spatial arrangements of these fibers are shown in Fig. 9.3:2. It is seen that the *A* band is the band of myosin filaments, and the *I* band is the band of the parts of the actin filaments that do not overlap with the myosin. The *H* bands are the middle region of the *A* band into which the actin filaments have not penetrated. Another line, the *M band*, lies transversely across the middle of the *H* bands, and close examination shows this to consist of fine strands interconnecting adjacent myosin filaments. The hexagonal pattern of arrangement of these filaments is shown in Fig. 9.3:2. The last two sketches of Fig. 9.3:1 show the structure of actin and myosin filaments in the muscle.

9.4 The Sliding Element Theory of Muscle Action

To explain how the actin and myosin filaments move relative to each other has been the great theme of muscle mechanics research since the 1950s.

Chemical and electron microscopic studies have revealed the fine structure of the myosin filaments. It is shown that each myosin filament consists of about 180 myosin molecules. Each molecule has a molecular weight of about 500 000 and consists of a long tail piece and a "head," which on close examination is seen to be a double structure. On further examination the molecule can be broken into two moieties: *light meromyosin*, consisting of most of the tail, and *heavy meromyosin* representing the head with part of the tail. The myofilament is formed by the tails of the molecules which lie parallel in a bundle, with their free ends directed toward the midpoint of the long axis. The heads project laterally from the filament in pairs, at 180° to each other and at 14.3 nm intervals. Each pair is rotated by 120° with respect to its neighbors to form a spiral pattern along the filament. These heads seem to be able to nod: they lie close to their parent filament in relaxation, but stick out to actin filaments when excited. They are called *cross-bridges*. How do they work is a subject of great interest. In Sec. 9.9 we shall present the hypotheses of the cross-bridge theories. In Sec. 9.10 the evidences in support of the cross-bridge hypothesis are reviewed. In Sec. 9.11 we present the mathematical development of the cross-bridge theory.

9.5 Single Twitch and Wave Summation

Skeletal muscle responds to stimulation by nervous, electric, or chemical impulses. Each adequate stimulation elicits a single twitch lasting for a fraction of a second.

Successive twitches may add up to produce a stronger action. Figure 9.5:1 illustrates the phenomenon. A single isometric twitch is shown in the lower left-hand corner. It is followed by successive twitches at varying frequencies. When the frequency of twitch is 10 per second, the first muscle twitch is not completely over when the second one begins. This results in a stronger

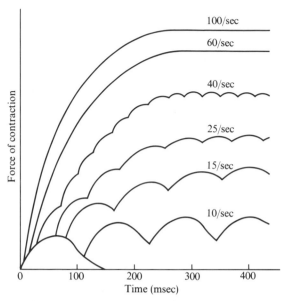

Figure 9.5:1 Wave summation and tetanization.

contraction. The third, fourth, and additional twitches add still more strength. This tendency for summation is stronger if the twitches come at a higher frequency. Finally, a *critical frequency* is reached at which the successive contractions fuse together and cannot be distinguished one from the other. This is then the *tetanized state.* For frequencies higher than the critical frequency further increase in the force of contraction is slight. This kind of reinforcement by successive twitches is called *wave summation.*

9.6 Contraction of Skeletal Muscle Bundles

So far we have considered only the behavior of a single muscle fiber. A muscle has many fibers. They are stimulated by motor neurons. Each motor neuron may innervate many muscle fibers. Not all the muscle fibers are excited at the same time. The total force of contraction of a muscle depends on how many muscle fibers are stimulated. Thus there is an interesting problem of correlating the numbers of excited neurons and muscle fibers with the intensity of contraction of a muscle.

Organizing many muscle fibers into a large muscle is like organizing individual people into a society. The society acquires certain features that are not simply a magnification of individual behavior. One of the features is the precision of response of a muscle to stimuli, and this has to do with the size of the *motor unit,* defined as all the muscle fibers innervated by a single motor nerve fiber. In general, small muscles that react rapidly and whose control is exact have small motor units (as few as two to three fibers in some of the laryngeal muscles), and have a large number of nerve fibers going into each muscle. Large muscles that do not require a fine degree of control, such as the gastrocnemius muscle, may have as many as 1000 muscle fibers in each motor unit. The fibers in adjacent motor units usually overlap, with small bundles of 10 to 15 fibers from one motor unit lying among similar bundles of the second motor unit. This interdigitation allows the separate motor units to acquire a certain harmony in action.

The size of motor units in a given muscle may vary tremendously. One motor unit may be many times stronger than another. The smaller units are more easily excited than the larger ones because they are innervated by smaller nerve fibers whose cell bodies in the spinal cord have a naturally higher level of excitability. This effect causes the gradations of muscle strength. During weak muscle contraction the excitation occurs in very small steps. The steps become progressively greater as the intensity of contraction increases. Further smoothing of muscle action is obtained by firing the different motor units asynchronously. Thus while one is contracting, another is relaxing; then another fires, followed by still another, and so on.

Feedback from muscle cells to the spinal cord is done by the *muscle spindles* located in the muscle itself. Muscle spindles are sensory receptors that exist throughout essentially all skeletal muscles to detect the degree of muscle contraction. They transmit impulses continually into the spinal cord, where they excite the anterior motor neutrons, which in turn provide the necessary nerve stimuli for *muscle tone*, a residual degree of contraction in a skeletal muscle at rest causing it to be taut.

It is easy to see that the total system of nerves and motor units furnishes biomechanics with many interesting problems; but in this chapter we shall concentrate on the mechanics of a single muscle fiber, or often, to a single sarcomere. The mechanics of the sarcomere is the foundation on which mechanics of muscle bundles can be built by the addition of other constituents.

9.7 Hill's Equation for Tetanized Muscle

Hill's equation (due to Archibald Vivian Hill, 1938) is the most famous equation in muscle mechanics. This equation is

$$(v + b)(P + a) = b(P_0 + a), \tag{1}$$

in which P represents tension in a muscle, v represents the velocity of contraction, and a, b, P_0 are constants. It looks very much like a gas law (van der Waal's equation). Roughly, if we ignore the constants a and b on the left-hand side of the equation, then Eq. (1) says that the rate of work done, and hence the rate of energy conversion from chemical reaction, is a constant. This seems reasonable in the tetanized condition, for which the equation is derived.

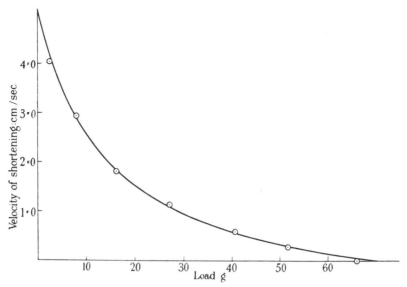

Figure 9.7:1 The experimental data (circles) on force (P, gram weight) and velocity of isotonic shortening (v, cm sec^{-1}) from quick release of frog skeletal muscle in the tetanized condition, as compared with Eq. (1) (solid curve). From Hill (1938), by permission.

Hill's equation refers to the ability of a tetanized skeletal muscle to contract. It is an empirical equation based on experimental data on frog sartorius muscle. A muscle bundle is held fixed in length L_0 (*isometric*: iso, equal; metric, measure) by having its ends clamped. It is stimulated electrically at a sufficiently high voltage and frequency so that the maximum tension P_0 (a function of the muscle length L_0) is developed in the muscle. At this tetanized condition the muscle is released suddenly (by releasing an end clamp) to a tensile force P smaller than P_0. The muscle begins to contract immediately, and the velocity of contraction, v, is measured. The relationship between P and v is plotted on a graph, such as the one shown in Fig. 9.7:1. An empirical equation is then derived to fit the experimental data. The fitting of Eq. (1) with the experimental data is shown in Fig. 9.7:1.

Hill's equation shows a hyperbolic relation between P and v. The higher the load, the slower is the contraction velocity. The higher the velocity, the lower is the tension. This is in direct contrast to the viscoelastic behavior of a passive material, for which higher velocity of deformation calls for higher forces that cause the deformation. Therefore, the active contraction of a muscle has no resemblance to the viscoelasticity of a passive material.

The original derivation of this equation is as follows. Hill writes, first of all, the equation of balance of energy:

$$E = A + H + W, \tag{2}$$

where

E is the rate of energy release,
A is the activation or maintenance heat per unit of time,
W is the rate of work done, equal to Pv,   (3)
H is the shortening heat.

If the muscle is not allowed to shorten, i.e., it is in an isometric condition, then Eq. (2) is reduced to

$$E = A,$$

i.e., the rate of energy release is equal to the activation energy. When the muscle contracts, an additional chemical reaction takes place, and it releases an amount of "extra energy" equal to the sum of the shortening heat and the work done, $H + W$, in Eq. (2). By measuring E and A, Hill identified the term $H + W$, the rate of "extra energy," and showed that it is represented by the empirical equation

$$H + W = b(P_0 - P), \tag{4}$$

where P_0 is the (maximal) isometric tension. He asserts further that empirically,

$$H = av. \tag{5}$$

Combining Eqs. (3), (4), and (5), we obtain

$$H + W = b(P_0 - P) = av + Pv. \tag{6}$$

Rearranging terms, we have

$$(a + P)v = b(P_0 - P) = b(-P - a) + bP_0 + ab.$$

Hence

$$(a + P)(v + b) = b(P_0 + a),$$

which is Eq. (1).

Hill's major contribution lies in his ingenious methods for accurately measuring the heats E, A, W, and H. His method of derivation of Eq. (1) endows a thermomechanical meaning to the constants a and b. Later, by employing a technique of rapid freezing and phosphate analysis, Carlson, Siger (1960), Mommaerts (1954), and Mommaerts et al. (1962) measured E in terms of phosphocreatine split, thus opening the way to biochemical studies of muscle contraction.

Three years earlier than Hill (1938), Fenn and Marsh (1935) presented a relationship between v and P, which is of the form

$$P = A e^{-v/B} + C, \tag{7}$$

where A, B, and C are constants. Although it differs from Eq. (1) considerably in form, it behaves numerically very much the same in the range $0 < P < P_0$, $0 < v < v_0$. On the other hand, Polissar (1952) proposed another form:

$$v = \text{const}(k_L e^{-C_L P} - k_s e^{C_s P}), \tag{8}$$

where k_L, C_L, k_S, and C_S are constants. However, Hill's equation remains the most popular. Many years later, Hill (1970) admitted that it is better to think of Eq. (1) directly as a force-velocity relationship, and not as a thermomechanical expression, because improved experiments do not always support Eqs. (4) and (5). In 1966 Hill changed the constant a in Eq. (1) to α, in order to dissociate it from its former meaning.

Hence, from the point of view of biomechanics, we shall regard Hill's equation as an empirical equation describing the force-velocity relationship of a tetanized skeletal muscle upon immediate release from an isometric condition.

9.7.1 The Dimensionless Form of Hill's Equation

Let us recapitulate: Hill's equation refers to a property of skeletal muscle in the tetanized condition. A muscle bundle is fixed at a certain length, L_0. It is then stimulated electrically at a frequency that is sufficiently high to tetanize the muscle to the maximum tension S_0 (henceforth we shall use S for tension in place of P). In this tentanized condition the muscle is released to a new length L, which is smaller than L_0, or a new tension S which is smaller than S_0. Immediately after release, the velocity of contraction $v = -dL/dt$ and the tension S are measured. Then an empirical relationship between S and v is Hill's equation:

$$(S + a)(v + b) = b(S_0 + a).$$

This can be rewritten as

$$v = b\frac{S_0 - S}{S + a}, \tag{9}$$

or

$$S = \frac{bS_0 - av}{v + b} = a\frac{v_0 - v}{v + b}. \tag{10}$$

When $S = 0$, the velocity v reaches the maximum

$$v_0 = \frac{bS_0}{a}, \tag{11}$$

which is used in Eq. (10). We can write Eqs. (9) and (10) in the nondimensional form

$$\frac{v}{v_0} = \frac{1 - (S/S_0)}{1 + c(S/S_0)} \tag{12}$$

or

$$\frac{S}{S_0} = \frac{1 - (v/v_0)}{1 + c(v/v_0)},\tag{13}$$

where

$$c = \frac{S_0}{a}.\tag{14}$$

It is customary to call v_0 the maximum velocity and designate it by v_{max}. But if a small compressive stress is imposed (as it can be in cardiac muscle; see Sec. 10.5), the velocity of contraction can exceed v_0; so v_0 is not the maximum velocity achievable. Similarly, if a muscle tetanized at length L_0 is stretched a little greater than L_0, the tension can exceed S_0, and S_0 is not the maximum tension achievable.

9.7.2 Behavior of the Constants a, b, S_0, v_0, and c in Hill's Equation

Hill's equation contains three independent constants. Writing it in the form of Eq. (1), the constants are a, b, S_0. Writing it in the form of Eqs. (12) and (13), the constants are S_0, v_0 and $c = S_0/a$. These constants are functions of the initial muscle length L_0, the temperature and composition of the bath, ionic concentration of calcium, drugs, etc.

The maximum isometric tension S_0 depends strongly on L_0. If L_0 is too small or too large, S_0 drops to zero. There is an optimum length L_0 at which S_0 is the maximum. Figure 9.7:2 shows the relationship between S_0 and L_0 (expressed as sarcomere length) for a single fiber of a frog's skeletal muscle. This figure is reproduced from Gordon et al. (1966), who tested single skeletal fibers with two gold leaf markers attached to the surface of the muscle to interrupt the focused light spots from a dual beam oscilloscope, so as to control the amount of light falling on two photocells. By linking the photocell output to the deflection of the spots, the markers can be followed and a signal obtained that is proportional to marker separation. This allows a strain measurement of the middle portion of the fiber, thus avoiding the end effects due to clamping. In this way data of high quality can be obtained.

Figure 9.7:2 shows the result for a frog muscle cell, whose slack sarcomere length is 2.1 μm (the sarcomere length to which the resting muscle returns when unloaded). It is seen that when the sarcomere length is in the range 2.0–2.2 μm, the maximum developed active tension does not depend on the muscle length. When the muscle fiber is maintained at a length outside this range, the maximum developed active tension is smaller. This feature is often explained in terms of the variation of the number of cross-bridges between myosin and actin fibers. If the muscle length is too long, the actin and myosin filaments are pulled too far apart, the cross-bridge number decreases, and the tension drops. If the muscle is too short, the actin filaments

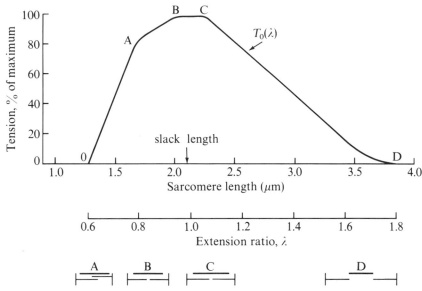

Figure 9.7:2 The steady-state or isometric tension-length curve as found by Gordon et al. (1966) for intact frog skeletal muscle fibers. The slack length (2.1 μm) is the sarcomere length to which the muscle will return when unloaded. It defines the reference point for the extension ratios shown below. The left-hand part of the curve (OA, AB) is commonly referred to as the ascending limb, the central flat portion (BC) is the plateau region, and the right-hand part (CD) is the descending limb. The relative positions of the actin and myosin filaments at different sarcomere lengths from A to D are shown at the bottom. From Gordon et al. (1966), by permission.

interfere with each other; they may shield each other and hinder the function of cross-bridges.

The maximum velocity v_0 does not depend as much on the muscle length L_0. In fact, whether v_0 is independent of L_0 or not was a long controversy in cardiology. For a time it was believed that v_0 is independent of L_0, and hence it was used as an index of "contractility," related to the state of health of the muscle, and its change with drug intervention was used to indicate the effectiveness of drugs. Today it is generally agreed that v_0 does depend on L_0 in some complex but minor way.

The constant c seems to be almost independent of L_0. It follows that the constant a is almost proportional to S_0. The constant b is equal to v_0/c. The value of c for skeletal muscle is in the range of 1.2 to 4.

9.7.3 Other Equivalent Forms

A dimensionless form corresponding to Eq. (7) by Fenn and Marsh is

$$\frac{S}{S_0} = \frac{e^{-\beta(v/v_0)} - e^{-\beta}}{1 - e^{-\beta}}, \tag{15}$$

where β is a constant. An analogous equation is

$$\frac{v}{v_0} = \frac{e^{-\gamma(S/S_0)} - e^{-\gamma}}{1 - e^{-\gamma}}, \tag{16}$$

where γ is a constant.

9.8 Hill's Three-Element Model

Hill's equation was derived from quick-release experiments on a frog sartorius muscle in tetanized condition. It reveals only one aspect of the muscle behavior. It cannot describe a single twitch, nor a wave summation. It cannot even describe the force-velocity relationship when a tetanized muscle is subjected to a slow release, nor to a strain that varies with time. It cannot describe the mechanical behavior of an unstimulated muscle. To remedy these shortcomings a more comprehensive approach is needed. Several methods have been proposed. Of these the best known is the Hill's three-element model.

For 50 years, since 1938, Hill's model of muscle contraction has dominated the field. It is, therefore,useful to know this model and to understand its strength and weakness. In the 50 year period many ideas have been added to the model in order to accommodate newly discovered facts. Originally quite simple, gradually the "series elastic" element became a very complex entity. To explain the single twitch, an "active state" function was introduced, the exact meaning of which turned out to be elusive. The arrangement of the three elements in the model is not unique. These difficulties gradually contributed to the decline of Hill's method.

To make the presentation as brief and definitive as possible, we present a version (Fung, 1970) that uses the language of sliding-element theory, and we base the discussion on a single sarcomere. A muscle is regarded as an accumulation of sarcomeres, although the assumption that all sarcomeres are identical is one of the dubious assumptions.

9.8.1 The Model and the Related Basic Equations

Hill's model (Fig. 9.8:1) represents an active muscle as composed of three elements. Two elements are arranged in series: (a) a contractile element, which at rest is freely extensible (i.e., it has zero tension), but when activated is capable of shortening; and (b) an elastic element arranged in series with the contractile element. To account for the elasticity of the muscle at rest, a "parallel elastic element" is added. This parallel element was ignored in Sec. 9.7. We now include it for the general case.

Most authors would identify the contractile element with the sliding actin-myosin molecules, and the generation of active tension with the number of active cross-bridges between them. Many suggestions have been made

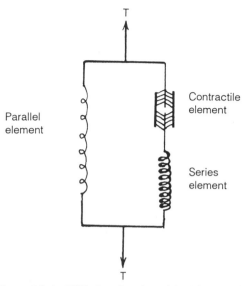

Figure 9.8:1 Hill's functional model of the muscle.

about the structural and ultrastructural basis of the parallel and series elastic elements; but none is definitive. The series elasticity may be due to the intrinsic elasticity of the actin and myosin molecules and cross-bridges, and of the Z band and connective tissues. It may arise also from the nonuniformity of the sarcomeres and nonuniform activation of the myofibril filaments. The parallel elasticity may be due to connective tissues, cell membranes, mitochondria, and collagenous sheaths; but the most unsettling question is whether the actin-myosin complex is truly free in the resting state of the muscle. Some of the complex behavior of the parallel element, its dependence on the contraction history of the muscle, drugs, and temperature, may be due to the residual coupling of actin-myosin complex.

Assuming that the contractile element contributes no tension in a resting muscle, then the stress–strain-history relationship of a resting muscle determines the constitutive equation of the parallel element. The difference between the mechanical properties of the whole muscle and those of the parallel element then characterizes the contractile and series elastic elements in combination. Unfortunately, since these elements are connected in series, the division of the total strain between these two elements is not unique unless some additional assumptions are added.

Let us make the usual assumptions that the contractile element is tension-free when the muscle is in the resting state (a corollary is that the resting state is unique), and that the series elastic element is truly elastic (not viscoelastic). Let us express the geometric change of a muscle sarcomere in terms of the actin and myosin fibers as shown in Fig. 9.8:2, where

M = myosin filament length,
C = actin filament length,

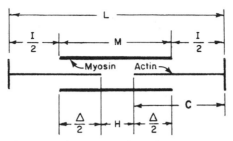

Figure 9.8:2 Geometric nomenclature of various elements of a muscle sarcomere unit.

Δ = insertion of actin filaments, i.e., the overlap between actin and myosin filaments,

H = H bandwidth,

I = I bandwidth,

L = total length of sarcomere,

L_0 = length of sarcomere at zero stress*,

η = extension of the series elastic element in a sarcomere.

We define the *insertion* Δ by the equation

$$\Delta = M - H = 2C - I \tag{1}$$

and the length L by

$$L = M + I = M + 2C - \Delta \qquad \text{(without elastic extension)} \tag{2}$$

and

$$L = M + I + \eta = M + 2C - \Delta + \eta \qquad \text{(with elastic extension).} \tag{3}$$

On differentiating Eq. (3) with respect to time, we obtain the basic kinematic relation

$$\frac{dL}{dt} = -\frac{d\Delta}{dt} + \frac{d\eta}{dt}. \tag{4}$$

Now consider the stress. The strain in the parallel elastic element can be defined as $(L - L_0)/L_0$; but since L_0 is a constant we can simply write[†]

$$T^{(P)} = P(L) \tag{5}$$

for the stress contributed by the parallel elastic element. Similarly, we write the stress contributed by the activated actin-myosin filaments:

$$T^{(s)} = S(\eta, \Delta). \tag{6}$$

$S(\eta, \Delta)$ is identically zero when the muscle is at rest; so that

$$S(\eta, \Delta) \gtrless 0 \qquad \text{when} \quad \eta \gtrless 0$$

and

* Note the specific meaning of L_0 here. It is different from the L_0 in Sec. 9.7.

[†] For simplicity we assume that the parallel element is elastic. But it is very easy to generalize this analysis to the case in which the parallel element is viscoelastic (see Problem 9.3.)

$$S(\eta, \Delta) = 0 \qquad \text{when} \quad \eta = 0. \tag{7}$$

The total tensile stress is the sum of the contributions from the series and parallel elements:

$$T = T^{(p)} + T^{(s)} = P(L) + S(\eta, \Delta). \tag{8}$$

If the stress varies with time, we have

$$\frac{dT}{dt} = \frac{dP}{dL}\frac{dL}{dt} + \frac{\partial S}{\partial \eta}\bigg|_{\Delta}\frac{d\eta}{dt} + \frac{\partial S}{\partial \Delta}\bigg|_{\eta}\frac{d\Delta}{dt}. \tag{9}$$

On substituting Eq. (4) into Eq. (9) we obtain the basic dynamic equation

$$\frac{dT}{dt} = \frac{dP}{dL}\frac{dL}{dt} + \frac{\partial S}{\partial \eta}\bigg|_{\Delta}\left(\frac{dL}{dt} + \frac{d\Delta}{dt}\right) + \frac{\partial S}{\partial \Delta}\bigg|_{\eta}\frac{d\Delta}{dt}$$

$$= \left(\frac{dP}{dL} + \frac{\partial S}{\partial \eta}\bigg|_{\Delta}\right)\frac{dL}{dt} + \left(\frac{\partial S}{\partial \eta}\bigg|_{\Delta} + \frac{\partial S}{\partial \Delta}\bigg|_{\eta}\right)\frac{d\Delta}{\delta t}. \tag{10}$$

Two special cases of great interest are:

Isometric contraction: $L = \text{const.}$ and $dL/dt = 0$,

$$\frac{dT}{dt} = \left(\frac{\partial S}{\partial \eta}\bigg|_{\Delta} + \frac{\partial S}{\partial \Delta}\bigg|_{\eta}\right)\frac{d\Delta}{dt}. \tag{11}$$

Isotonic contraction: $T = \text{const.}$ and $dT/dt = 0$,

$$\left(\frac{dP}{dL} + \frac{\partial S}{\partial \eta}\bigg|_{\Delta}\right)\frac{dL}{dt} + \left(\frac{\partial S}{\partial \eta}\bigg|_{\Delta} + \frac{\partial S}{\partial \Delta}\bigg|_{\eta}\right)\frac{d\Delta}{dt} = 0. \tag{12}$$

Let us illustrate the application of these equations by considering the methods for determining the characteristics of the series elastic element.

9.8.2 The Quick Release Experiment

This experiment consists of three steps:

(1) Preload the muscle in an unstimulated state to a length L_1.
(2) Stimulate isometrically until the tensile stress develops to a specific value T_1.
(3) Suddenly change the tensile stress from T_1 to T_2.

Let us denote the length corresponding to T_2 by L_2, the actin-myosin insertion at T_1 by Δ_1, and the elastic extension of the series element corresponding to T_1 and T_2 by η_1 and η_2, respectively. The hypothesis is made that the process in step (3) takes place so fast that Δ does not change when the stress drops from T_1 to T_2. Then

$$\Delta_2 = \Delta_1, \tag{13a}$$

$$T_1 = P(L_1) + S(\eta_1, \Delta_1), \tag{13b}$$

$$T_2 = P(L_2) + S(\eta_2, \Delta_1). \tag{13c}$$

Substracting, we obtain

$$T_1 - T_2 = P(L_1) - P(L_2) + S(\eta_1, \Delta_1) - S(\eta_2, \Delta_1). \tag{13d}$$

Applying Eqs. (3) and (13a) to the conditions at T_1 and T_2, and noting that M, C, and Δ are unchanged, we have

$$L_1 - L_2 = \eta_1 - \eta_2. \tag{13e}$$

Equations (13d) and (13e) give us $S(\eta_1, \Delta_1) - S(\eta_2, \Delta_1)$ as a function of $\eta_1 - \eta_2$ when T_1, T_2, L_1, L_2, $P(L_1)$, and $P(L_2)$ are measured. To determine $S(\eta_1, \Delta_1)$ as a function of η_1, we need to find a certain muscle length L_2 at which η_2 and $S(\eta_2, \Delta_1)$ are zero. This can be found by trial and error. Suppose we find a special value of $T_2 = T_2^*$ that corresponds to a length $L_2 = L_2^*$, at which

$$T_2^* = P(L_2^*), \tag{13f}$$

then by Eq. (8) we must have

$$S(\eta_2^*, \Delta_2) = 0, \tag{13g}$$

and hence, by Eq. (7),

$$\eta_2^* = 0. \tag{13h}$$

Combining these results with Eqs. (13d) and (13e), we have

$$\eta_1 = L_1 - L_2^*, \tag{13i}$$

$$S(\eta_1, \Delta_1) = T_1 - P(L_1). \tag{13j}$$

Thus the tensile stress S in the series elastic element can be plotted as a function of the extension η_1.

9.8.3 The Isometric-Isotonic Change-Over Method

This method consists of three steps:

(1) Preload at stress T_p, leading to length L_p.
(2) Stimulate isometrically at length L_p until a preassigned afterload T_{aft} is reached.
(3) Isotonic contraction follows with tensile stress remaining at T_{aft}.

According to Eq. (11) and with obvious notation, we obtain at the end of step (2):

$$\frac{dT}{dt}\bigg|_{\text{isometric}, \tau_{aft}} = \left\{ \frac{\partial S}{\partial \eta}\bigg|_{\Delta} (\eta_{aft}, \Delta_{aft}) + \frac{\partial S}{\partial \Delta}\bigg|_{\eta} (\eta_{aft}, \Delta_{aft}) \right\} \times \frac{d\Delta}{dt}\bigg|_{\text{isometric}}, \tag{14a}$$

whereas at the beginning of step (3) according to Eq. (12),

$$\frac{d\Delta}{dt}\bigg|_{\text{isotonic}} = -\left[\frac{dP}{dL} + \frac{\partial S}{\partial \eta}\bigg|_{\Delta}\right]\left[\frac{\partial S}{\partial \eta}\bigg|_{\Delta} + \frac{\partial S}{\partial \Delta}\bigg|_{\eta}\right]^{-1}\frac{dL}{dt}\bigg|_{\text{isotonic}}. \qquad (14b)$$

Under the basic assumption that for an activated muscle $d\Delta/dt$ is a function of the stress S, the insertion Δ, the length L, and the time interval after stimulation t, but nothing else, the rates $d\Delta/dt$ in Eqs. (14a) and (14b) can be equated and we obtain, at the instant of change over, that is,

$$\frac{d\Delta}{dt}\bigg|_{\text{isometric}} = \frac{(dT/dt)|_{\text{isometric}, \tau_{\text{aft}}}}{(\partial S/\partial \eta)|_{\Delta} + (\partial S/\partial \Delta)|_{\eta}} = \frac{d\Delta}{dt}\bigg|_{\text{isotonic}} \qquad (14c)$$

$$= -\left[\frac{dP}{dL} + \frac{\partial S}{\partial \eta}\bigg|_{\Delta}\right]\left[\frac{\partial S}{\partial \eta}\bigg|_{\Delta} + \frac{\partial S}{\partial \Delta}\bigg|_{\eta}\right]^{-1} \cdot \frac{dL}{dt}\bigg|_{\text{isotonic}} \qquad (14d)$$

From the middle term between the $=$ signs in Eq. (14c) and the term in Eq. (14d), we obtain

$$\frac{\partial S}{\partial \eta}\bigg|_{\Delta} + \frac{dP}{dL}\bigg|_{\tau_{\text{aft}}} = -\frac{(dT/dt)|_{\text{isometric}, \tau_{\text{aft}}}}{(dL/dt)|_{\text{isotonic}, \tau_{\text{aft}}}}. \qquad (14e)$$

Since dP/dL is known $\partial S/\partial \eta$ can be computed and plotted as a function of τ_{aft}. The inertial force is assumed neglegible in the above discussion. In practical experiments this is an important factor to be assured of.

9.8.4 Experimental Results on the Series Element

Extensive work by many authors has shown that when $\partial S/\partial \eta$ is plotted against $\tau_{\text{aft}} = S + P$, it turns out to be a straight line. For a given preload, P is a constant; hence a curve of $\partial S/\partial \eta$ vs. S is also a straight line.

To express this result together with the requirement that S must be zero when η vanishes, we write

$$\frac{\partial S}{\partial \eta} = \alpha(S + \beta), \qquad (15)$$

which integrates to

$$S = (S^* + \beta)e^{\alpha(\eta - \eta^*)} - \beta. \qquad (16)$$

Here α, β are constants and S^*, η^* are a pair of corresponding experimental values, $S = S^*$ when $\eta = \eta^*$. The condition $S = 0$ when $\eta = 0$ requires that

$$\beta = \frac{S^* e^{-\alpha\eta^*}}{1 - e^{-\alpha\eta^*}}. \qquad (17)$$

It is sometimes necessary to adjust the experimental values of β, S^*, and η^* so that Eq. (17) is satisfied. Parmley and Sonnenblick (1967) showed that for a cat's papillary muscle,

$$\alpha = 0.4 \text{ per } 1\% \text{ muscle length,}$$
$$\alpha\beta = 0.8 \text{ per } 1\% \text{ muscle length.}$$

9.8.5 Velocity of Contraction

As soon as the length of the muscle is broken down into two components in series, we have to speak of three velocities: the rate of change of the muscle length, that of the contractile element, and that of the series elastic element. They are related by Eq. (4).

In Sec. 9.7 we speak of the velocity-tension relationship. In the three-element model we have two tensions, P and S, and three velocities. Which is related to which? Or more precisely, Hill's equation represents a relationship between which tension and which velocity? In Hill's own experiments the initial length of the muscle was so small that the force in the parallel element, P, was negligible. If P is finite, it is expected not to be related to the velocity as in Hill's equation. Hence it is logical to assume that Hill's equation describes S, the tension in the contractile element. How about velocity? To answer the question, let us consider the redevelopment of tension after a step change in length. If S were related to dL/dt, then since $dL/dt = 0$ after the step change, there can be no change in S, and the phenomenon of tension redevelopment cannot be dealt with. Hence S must be related to $d\Delta/dt$ or $d\eta/dt$, which in this case are equal and opposite. With these considerations, we write Hill's equation [see Eq. (12) of Sec. 9.7] as

$$\frac{1}{v_0}\frac{d\Delta}{dt} = \frac{1 - (S/S_0)}{1 + c(S/S_0)}, \tag{18}$$

where c is a constant, S_0 is the tension in the series element in a tetanized isometric contraction, and v_0 is the velocity of $d\Delta/dt$ when $S = 0$. This equation is expected to be valid for, and only for, a tetanized muscle.

To describe a single twitch or wave summation or other dynamic events in a muscle, Eq. (18) must be modified to include a time factor—a function of time after stimulation. An example of such a modification is given in Chapter 10, Sec. 10.6, for the heart muscle which normally functions in successive single twitches. See Eq. (5) and Eq. (9) of Sec. 10.6. The function $f(t)$ in Eq. (10.6:5) has to be determined by experiments; that given in Eq. (6) of Sec. 10.6 is gleaned from Edman and Nilsson's work. But we shall defer further discussions to Chapter 10.

9.8.6 Tension Redevelopment After a Step Change in Length

To illustrate the application of the preceding equations, consider the redevelopment of tension after a step shortening in length of an isometric tetanized muscle. Let the new length be L and the tension be S_1 at time $t = 0$ immediately after the step change. Let L be sufficiently small so that P is negligible compared with S. Assume that $\partial S/\partial \eta$ is given by Eq. (15) and that $\partial S/\partial \Delta = 0$. Equations (11) and (18) apply. Hence by substitution, Eq. (11) becomes

$$\frac{d\sigma}{dt} = \alpha \gamma v_0 \frac{(1 - \sigma)(\sigma + k)}{\sigma + \gamma}, \tag{19}$$

where

$$\sigma = \frac{S}{S_0}, \qquad k = \frac{\beta}{S_0}, \qquad \sigma_1 = \frac{S_1}{S_0}, \qquad \gamma = \frac{1}{c}. \tag{20}$$

An integration of Eq. (19) yields the time for stress to change from σ_1 to σ:

$$t = \frac{1}{\alpha \gamma v_0} \int_{\sigma_1}^{\sigma} \frac{x + \gamma}{(x + k)(1 - x)} \, dx. \tag{21}$$

The time required for the tension to reach the maximum isometric tension S_0 corresponding to L is

$$t|_{\sigma=1} = \frac{1}{\alpha \gamma v_0} \int_{\sigma_1}^{1} \frac{x + y}{(x + k)(1 - x)} \, dx. \tag{22}$$

Note that the integral is divergent at the upper limit, so that its value is infinite. When $\sigma < 1$ but tends to 1, the time $t(\sigma)$ tends to $-\log(1 - \sigma)$, i.e., it tends to infinity logarithmically.

Since a peak tension is always observed at finite time under isometric conditions, the inaccessibility of maximum isometric tension is an undesirable theoretical difficulty that comes with Hill's model combined with Hill's equation. By modifying the Hill's equation [see, e.g., Eq. (9) of Sec. 10.6], however, this difficulty can be avoided [by using $n < 1$ in Eq. (9) of Sec. 10.6].

9.8.7 Critique of Hill's Model

The basic difficulty with Hill's model is that the division of forces between the parallel and contractile elements and the division of extensions between the contractile and the series elements are arbitrary. These divisions cannot be made without introducing auxiliary hypothesis. Consequently, experimental evaluation of the properties of the elements (contractile, series, and parallel) depends on the auxiliary hypothesis.

For example, Hill assumes that the contractile element is entirely stress free and freely distensible in the resting state, that the series and parallel elements are elastic, and that all sarcomeres are identical. Unfortunately, none of these is true.

Modifications have been proposed. In some, the series and parallel elements are made viscoelastic. In another, these elements are arranged differ-

ently, with one element parallel with the contractile, and another in series
with the combination. But in all models it is necessary to modify Hill's
equation for the contractile element if the muscle is not tetanized. To describe a single twitch, it is necessary to acknowledge the time-dependent
character of the contractile element with respect to stimulation. Furthermore,
the basic nonuniqueness of the division of forces and displacement among the
three elements mentioned above is not resolved.

9.8.8 Further Development

For further progress, one has to explain Hill's equation, to determine whether
the contractile element is entirely stress free and freely distensible in the resting
state or not, and to search for methods to predict the constants involved in
Hill's equation and in the constitutive equations of the parallel and series
elements. For these purposes the basic theory of sliding elements was proposed
by A. F. Huxley and H. E. Huxley in 1954. With the advent of the sliding-
element theory, the main stream of muscle research has been centered on the
theory of cross-bridges. This theory is reviewed in the following three sections.
Disenters include Iwazumi (1970), Noble and Pollack (1977), and Pollack
(1991). A theory by Caplan (1966) derives Hill's equation on the basis of the
principles of optimum design.

The sliding element theory is concerned only with the contractile element.
It does not provide a full constitutive equation of the muscle because it
ignores such details as the cellular structure, mitochondria, membranes,
collagen fibers, capillary blood vessels, etc. They pay no attention to the
details of the transmission of forces from one cell to another in a muscle
bundle, or the effect of deformation of one part of muscle on another. To
account for these is difficult but necessary. This question is discussed in Sec.
9.12. The question of partial activation is discussed in Sec. 9.13. The analysis
of single twitch is discussed in Chapter 10.

9.9 Hypotheses of Cross-Bridge Theory

Since the publication of papers by Hugh Huxley and Hanson (1954) and
Andrew Huxley and Niedergerke (1954), theories and experiments on muscle
contraction have been focused on the cross-bridges. A description of the
cross-bridges is a description of the interaction of actin and myosin filaments.
The myosin filaments is shown to be composed of a light-meromyosin (LMM)
and a heavy-meromyosin (HMM) part, as illustrated in Fig. 9.9:1. The LMM
part of the molecule is bonded into the backbone of the filament, while the
HMM part has a globular head (the S_1 fragment). The interaction between
actin and myosin is visualized as to take place betwen the globular head and

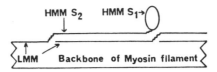

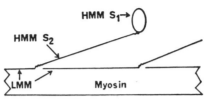

Figure 9.9:1 H. Huxley's concept of the myosin molecules in the thick filament. The light-meromyosin (LMM) part of the molecule is bonded into the backbone of the filament, while the linear portion of the heavy-meromyosin (HMM) component can tilt further out from the filament (by bending at the HMM–LMM junction), allowing the globular part of HMM (i.e., the S_1 fragment) to attach to actin over a range of different side spacings, while maintaining the same orientation. Reproduced from *Science* **164**, 1356–1366, 1969, by permission.

the actin molecule. The interacting part of the globular head of HMM and the actin molecule is called a cross-bridge. The cross-bridge has a length of approximately 19 nm. If the three-dimensional atomic structures of the actin, myosin, and cross-bridge molecules were known, and if statistical mechanics had been so advanced that the dynamics of the cross-bridges can be predicted from the atomic structure of these molecules, then a theory of cross-bridges would require few ad hoc hypothesis. Without definitive structural data and a fundamental atomic approach, the theories have to be stated under various sets of hypotheses. The hypotheses, as Terrell Hill et al. (1975) emphasized, must be self-consistent.

For the kinetic theory at the thermodynamic level, Hill et al. (1975) list the following hypotheses:

(1) The lengths of the actin and myosin filaments are unaltered by stretch or contraction of the muscle.
(2) The cross-bridges are independent force generators.
(3) At any instant of time, with significant probability each cross-bridge is accessible to only one actin binding site.
(4) A cross-bridge can exist in a specified number of biochemical states. Hill et al. (1975) analyzed a six-state cycle model, a five-state cycle model, and a reduced two-state cycle model which was originally proposed by A. F. Huxley (1957). The two-state model permits an attached state and a detached state. In the attached state, the cross-bridge generates a force proportional to its displacement x from a neutral equilibrium position.

(5) The actin-myosin bonding reaction obeys the first-order kinetics. Let $n(x, t)$ be the probability frequency function of the number of attached cross-bridges with displacement x at tim t. The number of attached bridges having x lying in the range x and $x + dx$ is $n(x, t) dx$. Then $n(x, t)$ is assumed to satisfy the following equation (A. Huxley, 1957; Podolsky et al., 1969; A. Huxley and Simmons, 1971; T. Hill et al., 1975):

$$\frac{Dn}{Dt} = \left(\frac{\partial n}{\partial t}\right)_x - v(t)\left(\frac{\partial n}{\partial x}\right)_t = f(x) - [f(x) + g(x)]n. \tag{1}$$

Here v represents the speed of shortening of a half–sarcomere, $f(x)$ represents the forward (bonding) rate parameter, $g(x)$ represents the backward (unbounding) rate parameter, and D/Dt is the material derivative.

(6) The tensile force $F_1(x)$ per cross-bridge as a function of the displacement x from the neutral equilibrium position of the cross-bridge may be represented by a polynomial:

$$F_1(x) = Kx + ax^2 + \cdots. \tag{2}$$

If all terms other than the first on the right-hand side of Eq. (2) vanish, then the force per unit area (stress) is

$$S(t) = \alpha C_1 \int_{-\infty}^{\infty} xn(x, t) dx, \tag{3}$$

where α and C_1 are constants. α represents the level of activation. C_1 is

$$C_1 = \frac{msK}{2L}, \tag{4}$$

where m is the number of cross-bridges per unit volume, s is the saromere length, L is the distance between a successive actin binding site, and K is the spring constant given in Eq. (2).

The question of consistency arises because the functions F_1, $f(x)$, $g(x)$, and the inverses $f'(x)$, $g'(x)$ are not independent of each other. Hill et al. (1975) discussed the constraints exhaustively.

The set of hypotheses listed above are used by Eisenberg and Hill (1978), Zahalak (1981), and Tözeren (1985). Pollack (1990) does not believe them at all. Pollack (1990) builds his theory on the basis of thick elements shortening.

9.10 Evidences in Support of the Cross-Bridge Hypotheses

Early evidences based on electron microscopy and x-ray diffraction patterns of actin and myosin, as well as the biochemical measurements of adenosine triphosphatase activity and its control by calcium ions, have been reviewed by H. E. Huxley (1969). Huxley (1969) provides a tentative solution to the problem of variable filament spacing. The problem arises from the fact that

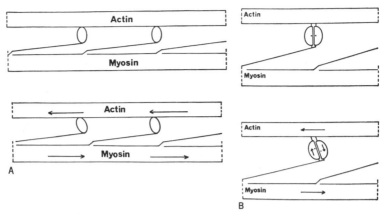

Figure 9.10:1 Huxley's (1969) proposed mechanism. (A) If separation of filaments is maintained by an electrostatic force balance, tilting must give rise to movement of filaments past each other. (B) A small relative movement between two subunits of myosin could give rise to a large change in tilt. Reproduced from *Science* by permission.

muscle is approximately incompressible. When a muscle shortens by a stretch ratio of λ, the lateral dimension will distend by a stretch ratio of $\lambda^{-1/2}$. Thus the spacing between the actin and myosin elements varies during contraction. The question is how can the cross-bridges function with such a large change of spacing. Huxley's answer is that the linear part of the sliding element can tilt out from the thick filament as shown in Fig. 9.9:1, and that the mechanism for producing relative sliding movement by tilting of cross-bridges is that which is shown in Fig. 9.10:1.

These hypotheses were studied at the electron microscopic level by three methods. Huxley (1963) and Moore et al. (1970) used negative staining. Reedy (1968) used thin sections. Heuser and Cooke (1983) used quick-freeze, deep-etch replica techniques. All authors showed photographs of actin and myosin molecules, and cross-bridges. But the light meromyosin and the bending of the linear portion of the heavy-meromyosin were not visible. All authors discussed the limitations of their techniques, and Heuser and Cooke (1983) explained the differences between the results obtained by these three techniques. Although these electron micrographs do not justify all the hypotheses listed in Sec. 9.9, they do not contradict them.

From a theoretical point of view, the hypotheses are justified if they are self-consistent and are capable of predicting the outcome of experiments. Theoretical solutions have been studied by Eisenberg and Hill (1978), Zahalak (1981), Tözeren (1985), and others. The analyses are limited in scope, and have not been evaluated against new experimental results. For example, Sugi and Tsuchiya (1981) presented the change in the ability of frog skeletal muscle fibers to sustain a load during the course of oscillatory length changes or continuous isotonic lengthening following quick increases in load. Whether the details of the muscle mechanics can be predicted by the theory is not known. New mechanical experiments of high precision are few.

Several new *in vitro* expriments are aimed strictly at molecular level. Ishijima et al. (1991) developed a new system for measuring the forces produced by a small number (< 5–150) of myosin molecules interacting with a single actin filament *in vitro*. The technique can resolve forces of less than a piconewton and has a time resolution in the submillisecond range. It can thus detect fluctuations of force caused by individual molecular interactions. From analysis of these force fluctuations, the coupling between the enzymatic ATPase activity of actomyosin and the resulting mechanical impulses can be elucidated. The bending of a microneedle is used as a force transducer. One end of a 10–20 μm long actin filament in solution was caught by N-ethylmaleimide treated myosin on a microneedle. The other end of the filament was brought into contact with a myosin-coated coverslip. In the presence of ATP, actomyosin interactions caused sliding of the filament and bending of the needle. The experimental results imply that the mechanical to chemical coupling ratio is one-to-one in isometric condition, but many-to-one during filament sliding. The authors propose that when work is done at higher velocity, the myosin head directly uses the free energy liberated by the hydrolysis of a single ATP molecule for many working strokes. Alternatively, the energy could be efficiently transferred through actin to multiple heads to generate multiple working strokes.

Taro et al. (1991) developed an *in vitro* motility assay in which single actin filaments move on one or a few heavy-meromyosin (HMM) molecules. This movement is slower than when many HMM molecules are involved. Frequency analysis shows that the sliding speeds distribute around integral multiples of a unitary velocity. This discreteness may be due to differences in the numbers of HMM molecules interacting with each actin filament. The unitary velocity reflects the activity of one HMM molecule. The value of the unitary velocity predicts a step size of 5–20 nm per ATP, which is consistent with the conventional swinging cross-bridge model of myosin function. This conclusion is reached under three additional hypotheses: (1) the actual filaments move at a velocity which is limited by the kinetic properties of the force-generating process because the inertial forces and viscous drag are both negligible. (2) In an ATPase cycle of a myosin head, the duration of the strongly bound state during which the head undergoes a step is a small fraction of the period of one whole cycle. (3) The period of time during which the unitary velocity takes place can be estimated with reasonable accuracy. This time multiplied by the velocity yields the step size.

The length of a single step of sliding associated with the hydrolysis of one ATP molecule is conidered a crucial number for the cross-bridge theory, because muscle contraction is thought to be driven by tilting of the 19 nm long myosin head while attached to actin. Each tilting motion would produce about 12 nm of filament sliding. The results of the experiments mentioned above suggest a varied result. Perhaps the coupling between cross-bridge movement and ATPase activity is not so tight. Many other papers are contributing to this research, some of them are reviewed by H. Huxley (1990) and Higuchi and Goldman (1991). It appears that a multiple power stroke per ATP molecule hydrolyzed is common.

On the biochemical side the role of a calcium ion on the cross-bridge kinetics is known to be important, but very complex. Literature on this subject is huge. Zahalak and Ma (1990) have incorporated a simplified version of the calcium modulated excitation and contraction coupling into a theory of cross-bridge kinetics. The mathematical structure of their final equation is similar to that of the original Huxley theory.

Pollack (1990) proposed a theory of muscle contraction based on the shortening of the thick elements during contraction. This shortening contradicts, of course, the basic hypothesis of Huxley's theory outlined in Sec. 9.9. Pollack (1990) showed a number of electronmicrographs by a number of authors to support his theory. How consistent his hypothesis is with all the known experimental results is unknown.

9.11 Mathematical Development of the Cross-Bridge Theory

Zahalak (1989) shows that the basic equation of the cross-bridge theory given by Eq. (1) of Sec. 9.9 can be integrated with a transformation of independent variables from x, t to ξ, t':

$$\xi = x + \delta(t), \qquad t' = t, \tag{1}$$

where

$$\delta(t) = \int_0^t v(\tau)\, d\tau. \tag{2}$$

With this transformation, the unknown variable $n(x, t)$ becomes $n(\xi, t')$. The rules of partial differentiation yields

$$\frac{\partial n}{\partial t} = \frac{\partial n}{\partial \xi}\frac{\partial \xi}{\partial t} + \frac{\partial n}{\partial t'}\frac{\partial t'}{\partial t} = \frac{\partial n}{\partial \xi}\frac{\partial \delta}{\partial t} + \frac{\partial n}{\partial t'} = v(t)\frac{\partial n}{\partial \xi} + \frac{\partial n}{\partial t'},$$

$$\frac{\partial n}{\partial x} = \frac{\partial n}{\partial \xi}\frac{\partial \xi}{\partial x} + \frac{\partial n}{\partial t'}\frac{\partial t'}{\partial x} = \frac{\partial n}{\partial \xi}. \tag{3}$$

On substituting the above into Eq. (1) of Sec. 9.9, we obtain

$$\frac{\partial n(\xi, t')}{\partial t'} + [f + g]n(\xi, t') = f, \tag{4}$$

where f and g are functions of the variable $\xi - \delta(t')$. Equation (4) is a linear differential equation of $n(\xi, t')$ with a variable coefficient $f + g$. A well-known method of integration factor can be used to solve the equation. The integration factor is

$$Q(\xi, t') = \exp\left[\int_0^{t'} \{f[\xi - \delta(\beta)] + g[\xi - \delta(\beta)]\} \, d\beta\right], \tag{5}$$

where β is a dummy replacement of t' for integration. On multiplying Eq. (4) with $Q(\xi, t')$ and noting that

$$\frac{\partial Q}{\partial t'} = (f + g)Q \tag{6}$$

ones sees that the differential equaton (4) becomes

$$\frac{\partial}{\partial t'}(Qn) = fQ. \tag{7}$$

Integrating with respect to t', dropping the prime on t', using τ as a dummy variable for integration, and dividing the final result by Q, Zahalak obtains

$$n(\xi, t) = Q^{-1}(\xi, t)\int_0^t f[\xi - \delta(\tau)]Q(\xi, \tau) \, d\tau + n_0(\xi)Q^{-1}(\xi, t). \tag{8}$$

Here $n_0(\xi)$ is an integration constant representing the value of $n(\xi, t)$ at $t = 0$. Zahalak interprets $n(x, t)$ as the probability distribution function of finding cross-bridges at the location x and time t, and identifies the Lagrangian stress given by Eq. (3) of Sec. 9.9 as proportional to the first moment of the distribution function $n(x, t)$. He identifies further that the zeroth moment is proportional to the instantaneous stiffness of the muscle (i.e., the force per unit length change in quick stretch); and that the second moment is proportional to the total elastic energy stored in the cross-bridges. He derived equations governing these moments by assuming $n(x, t)$ to be normally distributed with respect to x.

Thus, by specifying the velocity $v(t)$, and the bonding and unbonding rate parameters $f(x)$ and $g(x)$, important features of the contraction process can be analyzed in detail. Zahalak and Ma (1990) introduced calcium excitation into the formulation; Tözeren (1985) introduced partial activation; both without changing the basic mathematical structure. Tözeren (1985) worked out the details when the rate parameters $f(x)$ and $g(x)$ are step functions of x, with a finite number of discontinuities. He obtained a set of equations governing the moments containing information on the discontinuities of f and g, but without the normal distribution hypothesis of Zahalak (1981). This mathematical formalism shifts attention back to biochemistry, by showing that if new theoretical and experimental results can improve quantitative understanding of f, g, and v, then these equations can transform the new understanding into statements on muscle function quantitatively.

Zahalak (1986) has presented a thorough comparison of some of the predictions of the distribution-moment theory with experimental results for the cat soleus muscle subjected to constant stimulation. Many features that are beyond the scope of A. V. Hill's equation can be dealt with the sliding-elements theory. The author says that at this stage the approach described should be viewed more as a philosophy for macroscopic muscle modeling than as a specific final model.

Summarizing the discussions above (Secs. 9.9–9.11), we see that recent work has transformed a brilliant idea proposed in 1957 into a series of fruitful experimental and theoretical research. The ring has not been closed yet. The field remains open for new work.

9.12 Constitutive Equation of the Muscle as a Three-Dimensional Continuum

So far we have considered a single muscle cell, or a single sarcomere. Now we turn to aggregates of cells in a continuum such as a myocardium or a skeletal muscle.

A muscle is a composite material which, in a simplified view, may be regarded as composed of two phases: a phase of contractile material (actin and myosin), and a phase of connective tissues other than the contractile material. As in the general treatment of multiphasic materials (see, e.g., Fung, 1990, *Biomechanics: Motion, Flow, Stress and Growth* p. 300 for a brief outline and references) we view the system as a mixture; and regard each of the two phases as present in the whole space (the hypothesis of equipresence). Hence each point of the tissue is occupied simultaneously by both phases. Let the contractile phase be identified as phase 1, and the connective phase be phase 2. Let the density and volume fractions of the two phases (identified by superscripts 1 and 2) be defined as follows:

$d^{(1)}$, $d^{(2)}$—the true density (mass/material vol);

$\rho^{(1)}$, $\rho^{(2)}$—the phase density (mass/tissue vol) in the mixture;

$\phi^{(1)}$, $\phi^{(2)}$—phase voluem fraction, defined by

$$\phi^{(1)} = \rho^{(1)}/d^{(1)}, \qquad \phi^{(2)} = \rho^{(2)}/d^{(2)}. \tag{1}$$

According to the hypothesis of equipresence, we have

$$\phi^{(1)} + \phi^{(2)} = 1. \tag{2}$$

With respect to a set of rectangular cartesian coordinates, x_1, x_2, x_3, the displacement u_i, velocity v_i, strain e_{ij}, and strain rate $\dot{e}_{ij}$ ($i,j = 1,2,3$) of the two phases can be defined in the usual way in view of the equipresence hypothesis. The law of conservation of mass is expressed by the usual equation of continuity:

$$\frac{\partial \rho^\alpha}{\partial t} + \frac{\partial(\rho^\alpha v_j^\alpha)}{\partial x_j} = 0 \qquad (\alpha = 1, 2). \tag{3}$$

The stress in the muscle is a sum of a part due to the contractile mechanism, $\sigma_{ij}^{(1)}$, a part in the connective tissue, $\sigma_{ij}^{(2)}$, and a pressure p that exists in any incompressible material:

$$\sigma_{ij} = \sigma_{ij}^{(1)} + \sigma_{ij}^{(2)} - p\delta_{ij}. \tag{4}$$

The contractile stress $\sigma_{ij}^{(1)}$ arises from the force generated in the muscle cells. Let S be the magnitude of the fiber stress (newton/m^2) in the muscle cells, and v_i denote the unit-vector field specifying the fiber direction in the muscle. Then $\sigma_{ij}^{(1)}$ may be approximated in the form (Fung 1969, 1984; Chadwick, 1981; Tözeren 1985)

$$\sigma_{ij}^{(1)} = Sv_i v_j \tag{5}$$

Hill's equation and the cross-bridge theory discussed above in Sections 9.7 and 9.9–11 are concerned with the evaluation of S. The basic length of concern is that of the sarcomere. Change of sarcomere length can be computed from the strains of the continuum. If da_i is a vector lying in the direction of a muscle fiber and having the length of a sarcomere in a reference state, and dx_i is the corresponding element after deformation, and if ds_0 and ds represent the lengths of the elements da_i and dx_i, respectively, then by definition (see Eqs. (17) and (18) of Sec. 2.3):

$$ds^2 - ds_0^2 = 2E_{ij}\, da_i\, da_j = 2e_{ij}\, dx_i\, dx_j, \tag{6}$$

where E_{ij} and e_{ij} are Green's and Cauchy's strains, respectively. Hence the stretch ratio of the sarcome length, λ,

$$\lambda = ds/ds_0 \tag{7}$$

is related to the direction cosines, n_i and v_i, in the unstimulated and contracted directions respectively,

$$n_i = da_i/ds_0, \qquad v_i = dx_i/ds \tag{8}$$

by the equations

$$\lambda^2 - 1 = 2E_{ij}n_i n_j, \tag{9}$$

$$1 - \frac{1}{\lambda^2} = 2e_{ij}v_i v_j. \tag{10}$$

These formulas convert the strain measures of the continuum to the stretch ratio of the sarcomeres. From the latter we evaluate the contractile stress, S, according to the cross-bridge theory:

$$S(t) = \text{a functional of } \lambda(t), \text{ and } \lambda(\tau), S(\tau), C_a^{++} \qquad \text{for } 0 \leqslant \tau \leqslant t. \tag{11}$$

From S one obtains the stress in the continuum, $\sigma_{ij}^{(1)}$, according to Eq. (5).

The stress in the connective tissue, $\sigma_{ij}^{(2)}$, is assumed to be a function of the strain and strain history, as in other tissues considered in Chapter 7. The experimental results reviewed in Sec. 10.2 show that the unstimulated myocardium has the same features of a nonlinear elastic response, and a linear viscoelastic memory with a broad relaxation spectrum, so that the bysteresis under cyclic loading is flat over a range of four or more decades of frequencies.

Hence $\sigma_{ij}^{(2)}$, as a function of the strain E_{ij}, is as described in Secs. 7.6–7.11:

$\sigma_{ij}^{(2)}(t)$ = a functional of $E_{ij}(t)$, $E_{ij}(\tau)$, and a relaxation function

$$G(t - \tau) \qquad \text{for } 0 \leqslant \tau \leqslant t. \tag{12}$$

The equation of motion of the tissue as a whole is

$$(\rho \cdot \text{acceleration})_i = \frac{\partial \sigma_{ij}}{\partial x_j}. \tag{13}$$

In a multiphasic medium the motion of the different phases may not be identical (e.g., when one phase is solid and another phase if fluid, see Sec. 8.9 in Fung, 1990, and Sec. 12.10 infra). For a skeletal muscle or a myocardium, however, the muscle cells and the surrounding tissues are so well integrated that their relative motion has probably little significance.

In a mixture theory, fine details of the materials making up each phase are ignored. For example, the interaction between the actin and myosin molecules with other components of the muscle cells, the interaction between cells, and between cells and extracellular matrix, especially the collagen fibrils connecting the cells (Sec. 10.8), are supposed to be taken care of in a global way through the constitutive equations. This is consistent with the continuum approach to real materials. As it is explained in Chapter 2, Sec. 2.1, application of the continuum concept to real materials requires a decision on the range of sizes of the objects to be observed. The mixture theory stated above can be applied to muscles so large that one cannot see the microstructures.

9.13 Partial Activation

Hill's equation refers to a maximally activated (tetanized) muscle shortening under load, and is insufficient for modeling most muscular actions of intact animals whose muscle usually functions at submaximal activation and is as likely to lengthen as to shorten under load. Many authors have studied partially activated human skeletal muscle, and attempted to summarize their experimental results by equations analogous to Hill's. Zahalak et al. (1976) presented their results on the forearm and wrist in the following form:

$$[p_0^+(e) - p] = (k_1 + k_2 p)v \qquad (0 < v < 0.5v_{max}(p)) \tag{1}$$

and discussed other forms in the literature. Here the meaning of the symbols for their experiments on biceps and triceps are as follows:

p = load applied at the wrist/max. of such load,

v = angular velocity of forearm/max of such velocity,

e = smoothed, rectified electromyogram/its maximum,

$$p_0^+(e) = \frac{b}{2m}\left[\left(1 + 4\frac{m}{b^2}\frac{e}{\varDelta}\right)^{1/2} - 1\right], \tag{2}$$

$$e = bp + mp^2 \qquad \text{(for } v < 0) \text{ } (b \geqslant 0), \tag{3}$$

$$e = \Delta(bp + mp^2) \qquad \text{(for } v = 0^+) \text{ } (\Delta \geqslant 1). \tag{4}$$

The $p_0^+(e)$ is an isometric load, and b, m, Δ, k_1, k_2 are empirical constants. For each subject, the fitting was reasonably good, but unfortunately, the values of the empirical constants vary from one individual to another. For the six athetes tested, the range of the constants given in Zahalak (1976) are: for b, 0 to 0.493; for m, 0.494 to 1.12; for Δ, 1.00 to 1.89; for k_1, 0.49 to 1.25, and for k_2, 0 to 6.66.

Problems

9.1 Compare the force transmission in parallel and in pinnate arrangements of muscle fibers, and show the possible mechanical advantage of the pinnate arrangement.

Solution. Refer to Fig. 9.2:2. Let both muscles have the same relaxed dimensions x, y, and L. Let each muscle fiber have a relaxed cross-sectional area A. Then in the *parallel* arrangement [Fig. 9.2:2(a)], the number of muscle fibers in the cross-sectional area xy is xy/A. Let the muscle contract by a *contraction ratio* n (= shortened length/initial length) and generate a force F_n in each fiber. Then

$$\text{Total force} = xyF_n/A.$$

For a *pinnate* arrangement, let α be the angle between the muscle and the tendon, and λ be the relaxed length of each fiber. Then we have [Fig. 9.2:2(c)]

$$\lambda \sin \alpha = y/2.$$

When the fibers contract by a contraction ratio n to a new length $n\lambda$, α is changed to α_n. Since $n\lambda \sin \alpha_n = y/2$, we have $\sin \alpha_n = (\sin \alpha)/n$, and $\cos \alpha_n = (n^2 - \sin^2 \alpha)^{1/2}/n$. If the force in each fiber is F_n, the vertical component transmitted to the tendon is $F_n \cos \alpha_n$. The number of fibers being $xL/(A/\sin \alpha)$, the total vertical force from both sides of the pinnate muscles is $2xL \cdot \sin \alpha \cdot F_n \cdot \cos \alpha_n/A$. Hence the ratio

$$\frac{\text{Total vertical force lifted by the pinnate muscle}}{\text{Total vertical force lifted by the parallel muscle}}$$

$$= \frac{2xL \sin \alpha F_n \cos \alpha_n}{A(xyF_n/A)} = [2L \sin \alpha(n^2 - \sin^2 \alpha)^{1/2}]/ny.$$

This ratio can be either greater than 1 or less than 1 depending on the geometric parameters.

Example. Crab (Carcinus) chela [Fig. 9.2:1(a)]. $\alpha = 30°$, $L/y \doteq 2$. Then at rest $(n = 1)$, the ratio above is $2 \times 2 \times \frac{1}{2}(1 - \frac{1}{4})^{1/2} \doteq \sqrt{3} \doteq 1.73$. At contraction, $n = 0.7$, the ratio becomes 1.39.

9.2 The contractile property of a skeletal muscle is described by Hill's equation. For a muscle that is maximally stimulated, having a tension P and a velocity of contraction v, the power (rate of doing work) is Pv. At what force P can this muscle develop maximum power?

9.3 If the parallel element in Hill's model analyzed in Sec. 9.8 is viscoelastic, we replace Eq. (5) of Sec. 9.8 by

$$T^{(P)} = G * P(L),$$

where $G(t)$ is the normalized relaxation function (so normalized that $G(0) = 1$) and the symbol $*$ represents a convolution integral operation, i.e.,

$$T^{(P)}(t) = P[L(t)] + \int_0^t P[L(t - \tau)] \frac{\partial G(t)}{\partial \tau} d\tau.$$

$P(L)$ then represents the "elastic" response of the resting muscle. See Sec. 7.6, Eqs. (7.6:8) et seq. With this replacement, present a generalization of the analysis given in Sec. 9.8.

References

Alexander, R. M. (1968) *Animal Mechanics.* University of Washington Press, Seattle.

Bergel, D. H. and Hunter, P. J. (1979) The mechanics of the heart, In *Quantitative Cardiovascular Studies*, N. H. C. Hwang, D. R. Gross, and D. J. Patel (eds.) University Park Press, Baltimore, Chapter 4, pp. 151–213.

Caplan, S. R. (1966) A characteristic of self-regulated linear energy converters. The Hill force-velocity relation for muscle. *J. Theor. Biol.* **11**, 63–86.

Carlson, F. D. and Siger, A. (1960) The mechanochemistry of muscular contraction. I. The isometric twitch. *J. Gen. Physiol.* **43**, 33–60.

Eisenberg, E. and Hill, T. L. (1978) A cross-bridge model of muscle contraction. *Progr. Biophys. Mol. Biol.* **33**, 55–82.

Eisenberg, E., Chen, Y., and Hill, T. L. (1980) A cross-bridge model of muscle contraction, quantitative analysis. *Biophys. J.* **29**, 195–227.

Eisenberg, E. and Hill, T. L. (1985) Muscle contraction and free energy transduction in biological systems. *Science* **227**, 999–1006.

Fenn, W. P. and Marsh, B. S. (1935) Muscular force at different speeds of shortening. *J. Physiol.* **85**, 277.

Ferenezi, M. A., Goldman, Y. E., and Simmons, R. M. (1984) The dependence of force and shortening velocity on substrate concentration in skinned muscle fibers from *Rana Temporaria. J. Physiol. (London)* **350**, 519–543.

Ford, L. E., Huxley, A. F., and Simmons, R. M. (1977) Tension responses to sudden length change in stimulated frog muscle fibers near slack length. *J. Physiol.* **269**, 441–515.

Fung, Y. C. (1970) Mathematical representation of the mechanical properties of the heart muscle. *J. Biomech.* **3**, 381–404.

Gordon, A. M., Huxley, A. F., and Julian, F. J. (1966) The variation in isometric tension with sarcomere length in vertebrate muscle fibers. *J. Physiol. (London)* **185**, 170–192.

Heuser, J. E. and Cooke, R. (1983) Actin-myosin interactions visualized by quick-freeze, deep-etch replica technique. *J. Mol. Biol.* **169**, 97–122.

Higuchi, H. and Goldman, Y. E. (1991) Sliding distance between actin and myosin filaments per ATP molecule hydrolyzed in skinned muscle fibers. *Nature* **352**, 352–354.

Hill, A. V. (1938) The heat of shortening and the dynamic constants of muscle. *Proc. Roy. Soc. London B* **126**, 136–195.

Hill, A. V. (1970) *First and Last Experiments in Muscle Mechanics*. Cambridge University Press, Cambridge, U.K.

Hill, T., Eisenberg, E., Chen, Y.-D., and Podolsky, R. J. (1975) Some self-consistent two-state sliding filament models of muscle contraction. *Biophys. J.* **15**, 335–372.

Huxley, A. F. and Niedergerke, R. (1954) Structural changes in muscle during contraction. *Nature* **173**, 971–973.

Huxley, A. F. (1957) Muscle structure and theories of contraction. *Progr. Biophys. Biophys. Chem.* **7**, 255–318.

Huxley, A. F. and Simmons, R. M. (1971) Proposed mechanism of force generation in striated muscle. *Nature (London)* **233**, 533–538.

Huxley, A. F. (1974) Muscular contraction. A review lecture. *J. Physiol.* **243**, 1–43.

Huxley, H. E. and Hanson, J. (1954) Changes in the cross-striations of muscle during contraction and stretch and their structural interpretation. *Nature* **173**, 973–976.

Huxley, H. E. (1957) The double array of filaments in cross–striated muscle. *J. Biophys. Biochem. Cytol.* **3**, 631–643.

Huxley, H. E. (1958) The contraction of muscle. *Sci. Am.* **199**, 67.

Huxley, H. E. (1969) The mechanism of muscular contraction. *Science* **164**, 1356–1366.

Huxley, H. E. (1990) Sliding filaments and molecular motile systems. *J. Biol. Chem.* **265**, 8347– 8350.

Ishijima, A., Doi, T., Sakurada, K., and Yanagida, T. (1991) Sub-piconewton force fluctuations of actomyosin *in vitro*. *Nature* **352**, 301–306.

Iwazumi, T. (1970) A new field theory of muscle contraction. Ph.D. Thesis. University of Pennsylvania, Philadelphia.

Julian, F. J. and Sollins, M. R. (1975) Sarcomere length-tension relations in living rat papillary muscle. *Circulation Res.* **37**, 299–308.

Kishino, A. and Yanagida, T. (1988) Force measurements by micromanipulation of a single actin filament by glass needles. *Nature* **334**, 74–76.

Kreuger, J. E. and Pollack, G. H. (1975) Myocardial sarcomere dynamics during isometric contraction. *J. Physiol.* **251**, 627–643.

Mommaerts, W. F. H. M. (1954) Is adenosine triphosphate broken down during a single muscle twitch? *Nature* **174**, 1083–1084.

Mommaerts, W. F. H. M., Olmsted, M., Seraydarian, K., and Wallner, A. (1962) Contraction with and without demonstratable splitting of energy-rich phosphate in turtle muscle. *Biochim. Biophys. Acta* **63**, 82–92, 75–81.

Moore, P. B., Huxley, H. E., and DeRosier, D. J. (1970) Three-dimensional reconstruction of F-actin, thin filaments, and decorated thin filaments. *J. Mol. Biol.* **50**, 279–295.

Noble, M. I. M. and Pollack, G. H. (1977) Molecular mechanism of contraction. Controversies in research. *Circulation Res.* **40**, 333–342.

Parmley, W. W. and Sonnenblick, E. H. (1967) Series elasticity in heart muscle. *Circulation Res.* **20**, 112–123.

Podolsky, R. J. and Nolan A. C. (1971) In *Contractility of Muscle Cells and Related Processes*. R. J. Podolsky (ed.) Prentice-Hall, Englewood Cliffs, NJ, pp. 247–260.

Podolsky, R. J., Nolan, A. C., and Zavelier, S. A. (1969) Cross-bridge properties derived from muscle isotonic velocity transient. *Proc. Natl. Acad. Sci. U.S.A.* **64**, 504–511.

Polissar, M. J. (1952) Physical chemistry of contractile process in muscle. 1. A physico-chemical model of contractile mechanism. *Am. J. Physiol.* **168**, 766–781.

Reedy, M. K. (1968) Ultrastructure of insect flight muscle: I. Screw sense and structural grouping in the rigor cross-bridge lattice. *J. Mol. Biol.* **31**, 155–176.

Simons, R. M. and Jewell, B. R. (1974) Mechanics and models of muscular contraction. In *Recent Advances in Physiology*, R. J. Linden (ed.) Churchill, London, Vol. 9, pp. 87–147.

Sugi, H. and Tsuchiya, T. (1981) Enhancement of mechanical performance in frog muscle fibers after quick increases in load. *J. Physiol. (London)* **319**, 239–252.

Taro, Q., Uyeda, P., Warrick, H. M., Kron, S. J., and Spudich, J. A. (1991) Quantized velocities at low myosin densities in an *in vitro* motility assay. *Nature* **352**, 307–311.

Tözeren, A. (1983) Static analysis of the left ventricle. *J. Biomech. Eng.* **105**, 39–46.

Tözeren, A. (1985) Constitutive equations of skeletal muscle based on cross-bridge mechanism. *Biophys. J.* **47**, 225–236.

Tözeren, A. (1985) Continuum rheology of muscle contraction and its application to cardiac contractility. *Biophys. J.* **47**, 303–309.

Tözeren, A. (1986) Assessment of fiber strength in a urinary bladder by using experimental pressure volume curves: An analytical method. *J. Biomech. Eng.* **108**, 301–305.

Uyeda, T. Q. P., Warrick, H. W., Kron, S. J., and Spudich, J. A. (1991) Quantized velocities at low myosin densities in an *in vitro* motility assay. *Nature* **352**, 307–311.

Warwick, R. and Williams, P. L. (eds.) (1973) *Gray's Anatomy*, 35th British Edition. W. B. Saunders, Philadelphia.

White, D. C. S. and Thorson, J. (1973) The kinetics of muscle contraction. *Progr. Biophys. Mol. Biol.* **27**, 173–255.

Zahalak, G. I., Duffy, J., Stewart, P. A., Litchman, H. M., Hawley, R. H., and Pasley, P. R. (1976) Partially activated human skeletal muscle: An experimental investigation of force, velocity, and EMG. *J. Appl. Mech.* **98**, 81–86.

Zahalak, G. I. (1981) A distribution-moment approximation for kinetic theories of muscle contraction. *Math. Biosci.* **55**, 89–116.

Zahalak, G. I. and Ma, S.-P. (1990) Muscle activation and contraction: Constitutive relations based directly on cross-bridge kinetics. *J. Biomech. Eng.* **112**, 52–62.

CHAPTER 10

Heart Muscle

10.1 Introduction: The Difference Between Myocardial and Skeletal Muscle Cells

Both myocardial and skeletal muscle cells are striated. Their ultrastructures are similar. Each cell is made up of sarcomeres (from Z line to Z line), containing interdigitating thick myosin filaments and thin actin filaments. The basic mechanism of contraction must be similar in both; but important differences exist.

The most important difference between skeletal and cardiac muscle is the semblance of a *syncytium* in cardiac muscle with branching interconnecting fibers. See Fig. 10.1:1 and compare it with Fig. 9.3:1. The myocardium is not a true anatomical syncytium. Laterally, each myocardial cell is separated from adjacent cells by their respective sarcolemmas (cell membranes). At the ends, each myocardial cell is separated from its neighbor by dense structures, *intercalated disks*, which are continuous with the sarcolemma. Nevertheless, cardiac muscle functions as a syncytium, since a wave of depolarization is followed by contraction of the entire myocardium when a suprathreshold stimulus is applied to any one focus in the atrium. Graded contraction, as seen in skeletal muscle by activation of different numbers of cells, does not occur in heart muscle. All the heart's cells act as a whole (all-or-none response).

The second difference is the abundance of mitochondria (*sarcosomes*) in cardiac muscle as compared with their relative sparsity in skeletal muscle. See Fig. 10.1:2 and compare it with Fig. 9.3:1. Mitochondria extract energy from the nutrients and oxygen and in turn provide most of the energy needed by the muscle cell. On the membranes in the mitochondria are attached the oxidative phosphorylation enzymes. In the cavities between the membranes

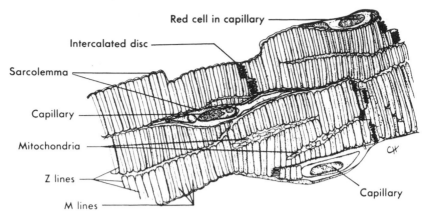

Figure 10.1:1 Diagram of cardiac muscle fibers illustrating the characteristic branching, the cell boundaries (sarcolemmas and intercalated disks), the striations, and the rich capillary supply (approxi. × 3000). From Berne and Levy (1972), by permission.

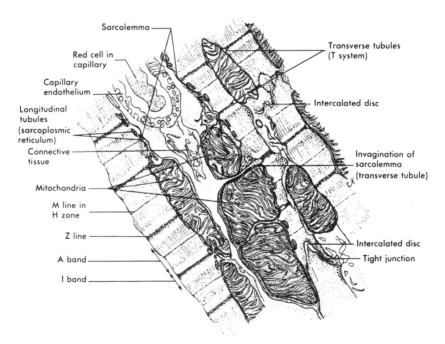

Figure 10.1:2 Diagram of an electron micrograph of cardiac muscle showing large numbers of mitochondria, the intercalated disks with tight junctions (nexi), the transverse tubules, and the longitudinal tubules (approx. × 24 000). From Berne and Levy (1972), by permission.

are dissolved enzymes. These enzymes oxidize the nutrients, thereby forming carbon dioxide and water. The liberated energy is used to synthesize a high energy substance called adenosine triphosphate (ATP). ATP is then transported out of the mitochondrion and diffuses into the myosin-actin matrix to permit the sliding elements to perform contraction and relaxation. The heart muscle relies on the large number of mitochondria to keep pace with its energy needs. The skeletal muscle, which is called on for relatively short periods of repetitive or sustained contraction, relies only partly on the immediate supply of energy by mitochondria. When a skeletal muscle contracts, the remaining energy needed is supplied by anaerobic metabolism, which builds up a substantial oxygen debt. This debt is repaid slowly by the mitochondria after the muscle relaxes. In contrast, the cardiac muscle has to contract repetitively for a lifetime, and is incapable of developing a significant oxygen debt. Hence the skeletal muscle can live with fewer mitochondria than the cardiac muscle must have.

A third difference is the abundance of capillary blood vessels in the myocardium (about one capillary per fiber) as compared with their relatively sparse distribution in the skeletal muscle (see Figs. 10.1:1 and 10.1:2). This is again consistent with the greater need of the myocardium for an immediate supply of oxygen and substrate for its metabolic machinery. With the close spacing of capillaries, the diffusion distances are short, and oxygen, carbon dioxide, substrates, and waste material can move rapidly between myocardial cell and capillary.

The exchange of substances between the capillary blood and the myocardial cells is helped further by a system of longitudinal and transverse tubules; see Figs. 10.1:2 and 10.1:3. These tubules are revealed by electron micrographs. The transverse tubules (T tubules) are deep invaginations of the sarcolemma into the interior of the fiber. The lumina of these T tubules are continuous with the bulk interstitial fluid; hence these tubules facilitate diffusion between interstitial and intracellular compartments. The transverse tubules may also play a part in excitation-contraction coupling. They are thought to provide a pathway for the rapid transmission of the electric signal from the surface sarcolemma to the inside of the fibers, thus enabling nearly simultaneous activation of all myofibrils, including those deep within the interior of the fiber. In mammalian ventricles the transverse tubules are connected with each other by longitudinal branches, forming a rectangular network. In the atria of many mammals the T system is absent or poorly developed.

Mechanically, the most important difference between cardiac and skeletal muscles lies in the importance of the resting tension in the normal function of the heart. The stroke volume of the heart depends on the end-diastolic volume. The end-diastolic volume depends on the stress–strain relationship of the heart muscle in the diastolic condition. Figure 10.1:4 shows the relationship between the end-diastolic muscle fiber length and the end-

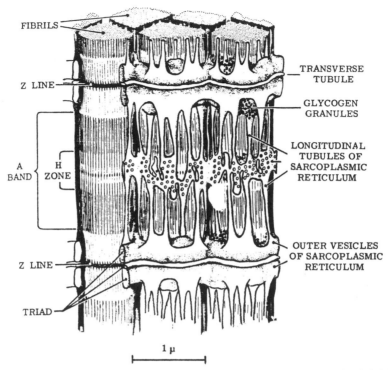

FIBRILS

Z LINE

A
BAND

H
ZONE

Z LINE

TRIAD

TRANSVERSE
TUBULE

GLYCOGEN
GRANULES

LONGITUDINAL
TUBULES OF
SARCOPLASMIC
RETICULUM

OUTER VESICLES
OF SARCOPLASMIC
RETICULUM

1 μ

Figure 10.1:3 The internal membrane system of skeletal muscle. Contraction is initiated by a release of stored calcium ions from the longitudinal tubules. This "trigger" calcium has only a short distance to diffuse in order to reach the myofilaments in the A band. From Peachey (1965).

diastolic and peak systolic pressure in the left ventricle of a dog. If the tension in the muscle is computed from the pressure and is plotted against the length of muscle fibers, a diagram similar to Fig. 10.1:4 will be obtained with the ordinate reading muscle tension and the abscissa reading the muscle length. When such a diagram is compared with Fig. 9.7:2 for the skeletal muscle, two observations can be made. (a) In the normal (physiological) range of muscle length, the resting tension is entirely negligible in the skeletal muscle, but is significant in the heart muscle. (b) Because of the resting tension the operational range of the length of the heart muscle is quite limited, whereas that of the skeletal muscle can be larger. For example, at a filling pressure of 12 mm Hg a normal intact heart will reach its largest developed tension while the sarcomere length is 2.2 μm. If the filling pressure is raised to 50 mm Hg, the sarcomere length of the heart muscle will become 2.6 μm. Further increase in filling pressure will not greatly increase the sarcomere length. On the other hand, for the skeletal muscle the optimum sarcomere length for maximum developed tension is also 2.2 μm; but with stretching a sarcomere

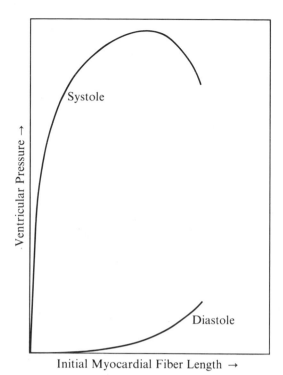

Figure 10.1:4 Relationship of left ventricular end-diastolic myocardial fiber length to end-diastolic and peak systolic ventricular pressure in an intact dog heart. From Berne and Levy (1972), by permission. Figure redrawn from Patterson et al. (1914).

length of 3.65 μm can be obtained quite easily. From these observations we see that while the resting tension can be ignored in normal skeletal muscle mechanics, it cannot be ignored in cardiac mechanics.

10.2 Use of the Papillary or Trabecular Muscles as Testing Specimens

Most of the available information on the mechanical properties of heart muscles has been obtained by testing the papillary muscles or trabecular muscles of the right or left ventricles of the cat, rabbit, ferret, or rat. Since blood perfusion is interrupted, the specimens must be kept alive by diffusion; hence the size of the specimen must be small, usually less than 1 mm in diameter. Use of larger muscle specimens from larger animals must consider blood perfusion. Sometimes smaller specimens from a mouse are used when the method of laser diffraction is chosen as the tool to measure the sarcomere length in the contraction process; then the force and cross-sectional area measurements are very delicate.

The specimen must be soaked in a circulating fluid such as a Ringer Tyrode solution oxygenated by bubbling a gas of 95% O_2, 5% CO_2, and with an ionic content which normally simulates that of the blood plasma as closely as possible. The normal solution contains (in mM) NaCl 140, KCl 5.0, $CaCl_2$ 2.25, Mg SO_4 1.0, NaH_2PO_4 1.0, and acetate 20, buffered to PH 7.4, The solutions used by different authors do differ; and the results do depend on the concentration of Ca^{2+}, etc. So it is important to note the ionic content in comparing data from different experiments.

The papillary and trabecular muscles are chosen as testing specimens because they are somewhat cylindrical in shape. Researchers are hoping that in the testing condition the stress distribution in the specimen is like a uniform uniaxial tensile stress acting in a uniform homogeneous material, so that the analysis can be greatly simplified. Actually, these muscles look like miniaturized thumbs, or at best like the more slender little fingers. One end of the specimen is a tendon which can be clamped into the testing machine. The other end has to be excised from the ventricular wall, and then clamped into the testing mchine. From the point of view of continuum mechanics, the stress distribution in the specimen before and during testing is very complex, and the assumption of uniform uniaxial tensile stress distribution and homogeneous deformation is a poor approximation. Krueger and Pollack (1975) presented evidence to show that the muscles at the clamped ends of the test specimens are severely damaged. Pinto (private communication) showed that deformation in the middle of the specimen is also nonuniform, with a coarse periodicity of roughly 1 mm. Pollack and Krueger (1976) and Ter Keurs et al. (1980) showed that the sarcomeres in the central segments of cardiac muscle preparations shorten between 7% to 20% of their initial lengths during an isometric contraction of a papillary muscle clamped at both ends.

Because of this nonuniform deformation, many experimentors approach papillary experiments with a "localized" approach, in which a central portion of the muscle, away from the ends, is assumed to be "one-dimensional," i.e., where the stress is uniaxial and deformation is uniform, and in that portion the stress and strain are measured or controlled. Several examples of this approach will be presented below.

When papillary muscles are used for mechanical properties determination, preconditioning over a long period of time is recommended. The reason is that the excision of the muscle from the ventricle, and the stopping of blood flow in the muscle are such big disturbances that it takes a long time to reestablish an equilibrium. The active contractile force in periodically stimulated isometric contraction of the muscle specimen usually becomes stronger with time, and a maximum is reached in an hour or more. The maximum contractile force can be maintained steadily in a specimen beating 24 or 36 hrs or longer in the testing machine.

10.3 Use of the Whole Ventricle to Determine Material Properties of the Heart Muscle

The nonsimplicity of the stress and strain distribution in the papillary muscle is the reason why some authors experiment with the whole heart with the assistance of computatonal continuum mechanics in order to determine the mechanical properties of the heart muscle. This approach is, of course, not simple.

Guccione et al. (1991) studied the passive material properties of intact ventricular myocardium with a cylindrical model, and threw considerable light on the effect of the orientation of the muscle fibers in the myocardium.

Taber (1991) analyzed the left ventricle as a thick-walled layered muscle shell.

Finite-element modeling is a major tool in this endeaver. Use of finite-element analysis in experimental studies of the heart is described by McCulloch et al. (1989) and Guccione and McCulloch (1991). The literature is reviewed in the book *Theory of Heart* edited by Glass, Hunter, and McCulloch (1991). Another development of an element for finite deformation is given by Nevo and Lanir (1989).

10.4 Properties of an Unstimulated Heart Muscle

As pointed out in the preceding section, the resting tension in a heart muscle is a significant determinant of the function of the heart, because it determines the end-diastolic volume and hence the stroke volume of the heart. Hence in this section we consider the mechanics of unstimulated heart muscle, with some details of test procedures and results.

A normal heart has a pacemaker, which initiates an electrical potential that causes the muscle to contract. Hence an isolated whole heart can beat by itself. An isolated papillary or trabecula muscle of the ventricles, however, does not have a strong pacemaker and can be tested in the unstimulated state.

From the mechanical point of view, a heart muscle in the resting state is an inhomogeneous, anisotropic, and incompressible material. Its properties change with temperature and other environmental conditions. It exhibits stress relaxation under maintained stretch, and creep under maintained stress. It dissipates energy and has a hysteresis loop in cyclic loading and unloading. Thus, heart muscle in the resting state is viscoelastic.

A papillary or trabecula muscle test specimen is often obtained from the rabbit or cat. The specimen is put in a modified Krebs–Ringer solution bubbled with a 95% O_2 and 5% CO_2 mixture of gases. A specimen with diameter smaller than 1 mm can be kept alive in such a bath for at least 36 hrs without any diminishing of its force of contraction. Larger specimens cannot maintain their viability by diffusion alone without blood circulation, and are not used in *in vitro* experiments.

As in the testing of other biosolids, the specimen must be preconditioned in order to obtain a repeatable stress–strain relationship. After preconditioning, measurement of a reference length (L_{ref}) and reference diameter (d_{ref}) can be made. A reference state is a standard condition, which for convenience is defined as a slightly loaded state of the specimen. In Pinto and Fung (1973), the load is caused by a 12 mg hook hanging from the lower end of the specimen which is suspended freely in a bath containing Kreb–Ringer solution. Thus the reference state referred to here is not a stress-free state, but an arbitrarily defined state convenient for laboratory work. For a very flexible nonlinear material, the specimen is so easily deformed at the zero-stress state that an accurate measure of length and cross section is difficult. The reference state, however, while arbitrary, has the virtue of being definitely measurable. Extrapolation to the zero-stress state can be done afterwards.

10.4.1 Relaxation Test Results

Relaxation test results of Pinto and Fung (1973) are illustrated in Fig. 10.4:1. The specimen was subjected to a step stretch. The force history was recorded and normalized by the force reached immediately after the step. Figure 10.4:1 shows the data obtained at 15°C for five different stretch ratios ($\lambda = 1.05 - 1.30$) performed on a specimen. The ordinate $G(t)$ represents a normalized

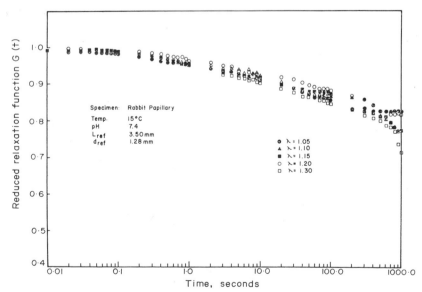

Figure 10.4:1 Reduced relaxation function of rabbit papillary muscle, showing the effect of the stretch ratio λ on $G(t)$. From Pinto and Fung (1973), by permission.

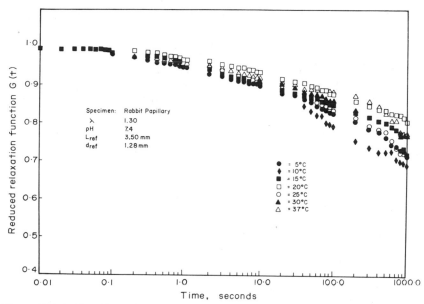

Figure 10.4:2 The effect of temperature on the reduced relaxation function of rabbit papillary muscle. From Pinto and Fung (1973), by permission.

function of time known as the *reduced relaxation function*, which is defined as

$$G(t) = \frac{P(t)}{P_r},\tag{1}$$

where $P(t)$ is the Lagrangian stress (total load divided by the cross-sectional area in the reference state) in the specimen at time t, and P_r is the Lagrangian stress in the specimen reached immediately after the step.

From Fig. 10.4:1 it is seen that $G(t)$ is essentially independent of the stretch λ. This independence is also observed at other temperatures. Figure 10.4:2 shows a plot of $G(t)$ vs. time at $\lambda = 1.30$ for seven temperatures. The trend is not grossly temperature dependent, particularly in periods of time less than 1 sec.

The response of a specimen to a step function in strain is called the relaxation function. For a heart muscle the relaxation function $P(t)$ is in general a function of many variables, such as the time t, the stretch ratio λ, the temperature θ, the pH, and the chemical composition of the fluid environment, signified by the mole number of the ith chemical species, N_i:

$$P(t) = P(\lambda, \theta, \mathrm{pH}, N_i \ldots t).\tag{2}$$

If only λ and θ are varied while pH and the osmolarity of the chemical environment are maintained constant, then $P(t)$ can be written as

$$P(t) = P(\lambda, \theta, t).\tag{3}$$

The experimental results shown in Figs. 10.2:1 and 10.2:2 suggest that T can be written in the form

$$P(\lambda, \theta, t) = G(t)P^{(e)}(\lambda, \theta), \tag{4}$$

where $G(t)$ is a normalized function of time alone and is so defined that $G(0) \equiv 1$. In Eq. (4), $P^{(e)}(\lambda, \theta)$ is called the *elastic response* (Fung, 1972), a stress that is a function of stretch λ and temperature θ. We have seen that Eq. (4) is approximately valid for the artery (Chapter 8) and other tissues (Chapter 7).

If the stretch ratio λ is not a step function, but a continuous function of time, $\lambda(t)$, then Eq. (4) can be generalized into a convolution integral as follows:

$$P(t) = G(t)*P^{(e)}[\lambda(t), \theta], \tag{5}$$

where the * denotes the convolution operator. See Chapter 7, Sec. 7.6.

10.4.2 The Elastic Response $P^{(e)}(\lambda, \theta)$

By definition $P^{(e)}(\lambda, \theta)$ is the tensile stress generated instantly in the tissue when a step stretch λ is imposed on the specimen at temperature θ. Measurement of $P^{(e)}(\lambda, \theta)$ strictly according to the definition is not possible, not only because a true step change in strain is impossible to obtain in the laboratory, but also because in rapid loading, the inertia of the matter will cause stress waves to travel in the specimen so that uniform strain throughout the specimen cannot be obtained. For this reason $P^{(e)}$ is a hypothetical quantity. A rational way to deduce $P^{(e)}$ from a stress–strain experiment is to apply Eq. (5) to a loading–unloading experiment at constant rate, then mathematically invert the operator to compute $P^{(e)}$, i.e., experimentally determine $P(t)$ and compute $P^{(e)}$ and $G(t)$.

To see what can be obtained by attempting a step loading anyway, without regard to its true meaning, a series of "high speed" stretches and releases of varying magnitude were imposed on the papillary muscle. Each quick stretch was followed by a quick release of the same magnitude so that the muscle was unloaded to its reference dimensions (L_{ref} and d_{ref}). The muscle relaxed for about 10 sec following the stretch before it was released. Rise time for these stretches varied from under 1 to 10 ms, depending upon the magnitude of stretch. Results of such a "quick stretch, quick release" test are shown in Fig. 10.4:3. The force-extension curve obtained at the same temperature (20°C) and at a uniform rate of loading and unloading is also shown for comparison.

Figure 10.4:3 shows that the difference between the two sets of curves is considerable. The difference can be considered as the effect of strain rate. Since the relaxation function $G(t)$ is a monotonically decreasing function, stress response to straining at a finite rate is expected to be lower than the quick "elastic" response.

That the stress response in loading and unloading is different even under slow and uniform rates of stretch and release is a feature common to all living

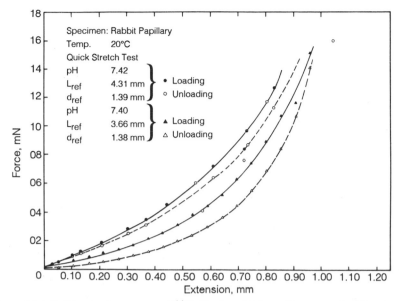

Figure 10.4:3 The elastic response $P^{(e)}(\lambda, \theta)$ of a rabbit papillary muscle. The circular dots refer to results of quick stretch tests. The triangular dots refer to results of loading and unloading at a finite constant rate. From Pinto and Fung (1973).

tissues (see Chapters 7 and 8). Test results for the rabbit papillary muscle stretched at various rates of loading and unloading are shown in Fig. 7.5:2 (Chapter 7), which demonstrates that the stress–strain relationship does not change very much as the strain rate is changed over a factor of 100. Other tests show that the changes remain small in the strain rate range of 10 000. This insensitivity is again in common with other tissues (see Chapters 7 and 8).

To reduce the curves in Fig. 7.5:2 to a mathematical expression, the method of Sec. 7.5 can be used. One observes that if the slope of the stress–strain curve, $dP/d\lambda$, is plotted against the stress P (Fig. 10.4:4), a straight line is obtained for each loading and unloading branch. Denoting the slope of the regression line by α (a dimensionless number), and the intercept on the vertical axis by $\alpha\beta$ (units, 10^3 N/m^2), we can write the stress–strain relationship in uniaxial loading as

$$\frac{dP}{d\lambda} = \alpha(P + \beta), \tag{6}$$

where P is the Lagrangian stress (force/area at zero stress). An integration gives

$$P + \beta = c\, e^{\alpha\lambda}, \tag{7}$$

where c is a constant of integration. To determine c, let a point be chosen on the curve: $P = P^*$ when $\lambda = \lambda^*$. Then

$$P = (P^* + \beta)e^{\alpha(\lambda - \lambda^*)} - \beta. \tag{8}$$

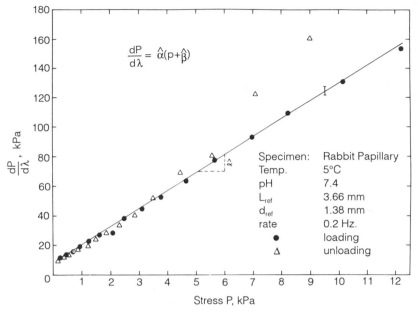

Figure 10.4:4 The exponential law of the stress–strain relationship for the rabbit papillary muscle. From Pinto and Fung (1973).

This procedure is valid only in the range in which Eq. (6) can be trusted. Usually Eq. (6), and hence Eq. (8), do not apply in the region $P \to 0$, $\lambda \to 1$. For this reason the point (P^*, λ^*) is usually chosen at the upper end of the curve (a large stress in the physiological range).

The need to exclude the origin is an unsatisfactory feature of Eq. (8). Use of a uniaxial strain-energy function, which is an exponential function of the square of the strain (see Sec. 8.9) will remedy this situation. Table 10.4:1 lists some typical values of α and β. An extensive listing given in Pinto and Fung (1973) shows that for the papillary muscle α and β are rather insensitive to strain rates and temperature within the range of $1 < \lambda < 1.3$ and 5–37°C.

10.4.3 Oscillations of Small Amplitude

Figure 10.4:5 shows the frequency response of a papillary muscle subjected to a sinusoidal strain of an amplitude equal to 4.06% of the reference length superposed on a steady stretch with a stretch ratio of 1.22. The ordinate shows the dynamic stiffness, which has been normalized with respect to the stiffness at 0.01 Hz. It is seen that in a frequency range that varied 10 000-fold, the dynamic stiffness increased by less than a factor of 2. This modest frequency dependence is again in accord with the general behavior of other biosolids (see Secs. 7.6 and 7.7).

TABLE 10.4:1 Values of α and β from Eq. (6) for Rabbit Papillary Muscle. From Pinto and Fung (1973)[a]

Rate (Hz)	Loading			Unloading		
	$\alpha \pm SE$	β (kPa)	P^* (kPa)	$\alpha \pm SE$	β (kPa)	P^* (kPa)
0.01	11.46 ± 0.11	5.7	7.3	17.08 ± 0.45	0.1	6.5
0.02	12.33 ± 0.22	4.9	7.6	17.47 ± 0.62	0.6	6.8
0.05	12.56 ± 0.14	5.3	8.1	16.95 ± 0.44	1.3	7.4
0.08	12.70 ± 0.20	5.9	8.1	17.27 ± 0.49	1.8	7.6
0.10	12.73 ± 0.13	5.1	8.2	17.08 ± 0.39	2.5	7.7
0.20	13.37 ± 0.14	4.3	8.5	17.84 ± 0.38	2.4	8.2
0.50	12.18 ± 0.37	5.4	6.5	16.00 ± 0.57	3.3	5.3
0.80	12.94 ± 0.14	4.8	6.5	17.47 ± 0.39	3.0	5.5
1.00	13.12 ± 0.16	2.4	6.5	16.99 ± 0.36	2.9	5.4
2.00	13.72 ± 0.17	4.2	6.9	18.43 ± 0.53	1.2	6.0
5.00	14.00 ± 0.30	2.0	7.4	17.34 ± 0.45	0.2	6.2
10.00	13.52 ± 0.14	1.7		16.98 ± 0.33	0.5	

[a] L_{ref}, 3.66 mm; d_{ref}, 1.38 mm; max. stretch, 30%; $\lambda^* = 1.25$; pH, 7.4; temp., 37°C.

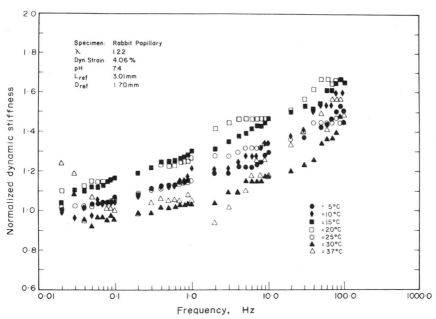

Figure 10.4:5 The frequency response of a rabbit papillary muscle to sinusoidal oscillations of small amplitude. From Pinto and Fung (1973).

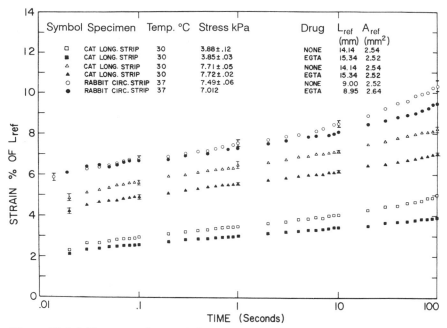

Figure 10.4:6 The creep characteristics of the papillary muscle of the cat and rabbit at different levels of stress and different temperatures, with and without EGTA. The application of 2 μM EGTA to the bath (0.76 g/liter) abolishes spontaneous contractile activity of the muscle. Chealating calcium by EGTA reduces creep strain over a given time. From Pinto and Patitucci (1977), by permission.

10.4.4 Creep Tests

Creep (deformation with time under constant load) of the cardiac muscle over short durations of time has relevance to intact heart behavior and is of particular importance to hypertrophied heart. Figure 10.4:6 shows the creep characteristics of four rabbit papillary muscles. The ordinate is the strain expressed as a percentage of the reference length. It is seen that for a given stress and temperature, creep is essentially a linear function of log time for the the first 10 min. Beyond 10 min it increases more rapidly with log t.

10.4.5 Summary

In summary, it is seen that in the resting state, the mechanical behavior of heart muscle is quite similar to that of other living tissues. For relaxation under constant strain, the reduced relaxation function $G(t)$ is independent of the stretch ratio for strains up to 30% of muscle length. For time periods comparable with the heartbeat (1 sec), the function $G(t)$ is independent of temperature in the range 5–37°C. This permits the application of the theory of quasi-linear viscoelasticity (Sec. 7.6) to the heart in the end-diastolic condition.

When a heart muscle is subjected to a cyclic loading and unloading at a constant strain rate, the stress–strain curve stabilizes into a unique hysteresis loop, which is independent of temperature (in the range 5–37°C) and is affected by the strain rate only in a minor way. Thus the pseudoelasticity concept (Sec. 7.7) is applicable. For the loading curve, the stress varies approximately as an exponential function of the strain. Unloading follows a similar curve with different characteristic constants.

In harmonic oscillations of small amplitude, the dynamic stiffness increases slowly with increasing frequency, by a factor of 2 when the frequency is increased from 0.01 to 100 Hz. In creep tests the heart muscle shows considerable creep strain under a constant load.

The relative insensitivity of stresses to strain rate in the cyclic process of stretching and release, and the small effect of frequency on dynamic stiffness and damping, can be explained by a continuous relaxation spectrum as presented in Sec. 7.6.

10.5 Force, Length, Velocity of Shortening, and Calcium Concentration Relationship for the Cardiac Muscle

The rest of this chapter is concerned with active contraction of the heart muscle. Since the heart works in single twitches, Hill's equation does not apply. But many authors believe that Hill's equation can be made to work for single twitches by the introduction of a *characteristic function of time* to describe the *active state* of the muscle. Available data on this approach are reviewed in Sec. 10.6. The advantage of this approach is to make the formalism of Hill's three element model available to general heart analysis—a step very useful to cardiology. However, the data of Sec. 10.6 were collected in papillary muscle experiments in which the nonuniformity of strain in the test specimen discussed in Sec. 10.2 was not taken into account. Hence the accuracy of the constants quoted in Sec. 10.6 may be questioned; although the general approach is valuable.

In the present section, some data based on newer experiments which did pay attention to the caution discussed in Sec. 10.2 are reviewed. The clamped ends are excluded from the length measurement. The strains in the middle partion of the muscle are assumed to be uniform. The term "segmental length" is used to indicate the length of this middle portion.

Huntsman et al. (1983) measured the force-length relationship of the central segments of ferret papillary muscles at 27°C. Martyn et al. (1983) published the experimental relationship between the shortening velocity at a very light load and the muscle length and the extracellular calcium at various times of releasing the muscle from an isometric twitch to a 1 mN load ($\leqslant 3\%$ of the

maximum force) on the same muscle. They used a small loop of a fine coiled
wire and a uniform oscillating magnetic field to measure the cross-sectional
area of a chosen segment of the papillary muscle, and used the area to assess
the length of the segment. Figure 10.5:1 shows their results on the maximum
total forces developed in isometric contractions from given segmental lengths
at various levels of calcium concentration in the bathing solution. Normally,
the Ca^{2+} concentration was 2.25 mM. The data indicate that increasing Ca^{2+},
at least up to 4.5 mM, causes an increase in force, a shift of the zero-force
intercept to the left, and a change of the shape of the total force-length curve.
For the experiments presented in Fig. 10.5:1, the zero-force intercepts in 4.5,
2.25, and 1.125 mM Ca^{2+} solutions were found to be 67%, 68%, and 74% of
the maximum segmental length. The resting tension at various lengths is

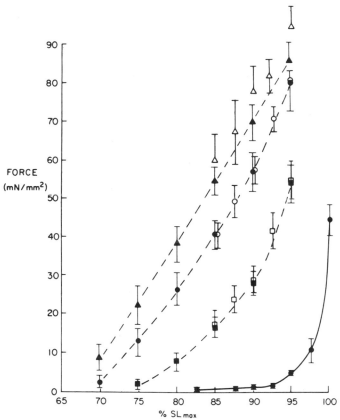

Figure 10.5:1 Total force-segmental length relations obtained at 1.125 (□), 2.25 (○),
and 4.50 (△) mM Ca^{2+} concentrations. At all concentrations, auxotonic (contracting
against feedback control at specified force) data are shown with *filled symbols*, whereas
isometric data (segmental length feedback controlled) are shown with *open symbols*.
Mean values $\pm$ SE are shown ($n = 9$). Reproduced from Huntsman et al. (1983), with
permission of the authors and the American Physiological Society.

plotted by the solid curve at the bottom, and it is seen that the resting tension is a small fraction of the total force for segmental length less than 95% of the maximum, SL_{max}. Since the parallel element of the cardiac muscle is viscoelastic, the calculation of the force of active contraction (the difference between the total force and the force in the parallel element) should follow the equations given in Sec. 9.8, with P viscoelastic.

The results shown in Fig. 10.5:1 for the ferret papillary muscle may be compared qualitatively with the results shown in Fig. 9.7:2 for the frog skeletal muscle. The segmental length cannot be directly translated to sarcomere length, but it is known that the sarcomere length is approximately equal to 2.4 μm at the maximum segmental length achievable for the experiment. Thus the range of sarcomere length in which contractile force can be generated is considerably shorter in the cardiac muscle than in the skeletal muscle.

Martyn et al.'s (1983) results on the shortening of a ferret papillary muscle when it was released from an isometric twitch to a light load of 1 mN are shown in Fig. 10.5:2. The segmental length was held constant until the time the load clamp was initiated. Load clamps were initiated from 95% of the

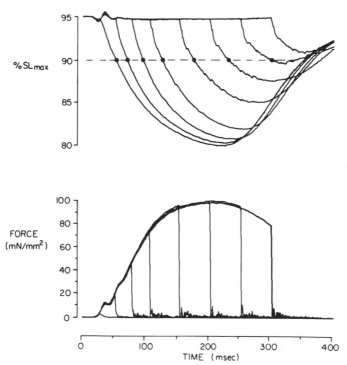

Figure 10.5:2 Representative traces of segment length (*top*) and force (*bottom*) for a series of load releases initiated at various times during a segment isometric twitch at a length of 95% SL_{max}. Dashed line (*top*) intersects SL traces at 90% SL_{max} (*closed circles*). Extracellular calcium was 2.25 mM. Reproduced from Martyn et al. (1983), with permission of the authors and the American Physiological Society.

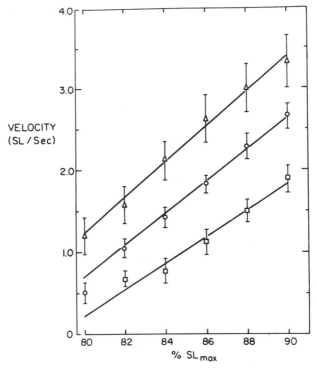

Figure 10.5:3 Variation of the segment velocity with the length of the segment and the calcium concentration in the Tyrode solution. The symbols for Ca are 1.125 ($\square$), 2.25 ($\bigcirc$), and 4.5 ($\triangle$) mM. Each symbol represents mean value ($\pm$ SE) of measurements taken from 5 ($\square$), 8 ($\bigcirc$), and 6 ($\triangle$) muscles. Reproduced from Martyn et al. (1983), by permission.

maximum segment length. Such clamps took 5 ms to complete. Then the velocity of shortening was computed electronically at various percentage of SL_{max} (e.g., at 90% SL_{max}), or at various time after stimulation (e.g., at 100 ms). The time at 100 ms line intersects the shortening after release curves of Fig. 10.5:2 at a number of points marked by x. At each x, there is a velocity of shortening and a length of segment as % SL_{max}. A plot of this pair of numbers is shown in Fig. 10.5:3 for a calcium concentration of 2.25 mM. Repetition of experiments at other calcium concentrations yielded data shown by the other two curves in Fig. 10.5:3. The least squares regression lines are plotted. This figure reveals the influence of calcium concentration and length of cardiac muscle on the contraction velocity at 100 ms after stimulation.

Comparison of Figures 10.5:2 and 10.5:3 with Figures 9.7:1 and 9.7:2 show that the force and velocity levels are lower and twitches are slower in the cardiac muscle than in the skeletal muscle. The dependence of the unloaded velocity on calcium concentration is much stronger in the cardiac muscle than in the skeletal muscle.

Similar data of high quality have been published by Ter Keurs et al. (1980), whose results are stated in terms of sarcomere length.

10.6 The Behavior of Active Myocardium According to Hill's Equation and Its Modification

For the purpose of analyzing the dynamics of the heart, we need to know the constitutive equation of the myocardium in systole and diastole and the states in between. The constitutive equation describes the force-velocity-length-and-time relationship for the muscle. It contains certain material constants. The change of these constants in health and disease, and with the muscle's environmental conditions (e.g., ionic composition, presence of inotropic agents, and temperature) and neural and humoral controls, will supply the information desired for clinical applications.

It was thought that Hill's three-element model (Sec. 9.8), together with a description of the contractile element by Hill's equation, properly modified for the active state, would be sufficient to provide a constitutive equation for the heart muscle. It was further hoped that the active state function and the constants in Hill's equation can be predicted or experimentally determined according to the sliding-element theory with a comprehensive knowledge of the cross-bridges. Roughly speaking, this idea was confirmed when the muscle is tested at a length so small that the resting tension in the parallel element (see Sec. 9.8) is negligible compared with the active tension (setting $P = 0$ in the equations of Sec. 9.8). This condition is reviewed in the present section.

Trouble appeared, however, when experiments were done on muscles at such a length that the resting tension is not negligible. To account for the rest tension, there are two different but equivalent ways. One is to use the model discussed in Sec. 9.8, and shown in Fig. 10.6:1(B). The other is shown in Fig. 10.6:1(C). These two models are called the "parallel resting tension" and the "series resting tension" models, respectively. By identifying the experimental results with these models, one can evaluate the properties of the contractile element. It was then found that the contractile element property is model-dependent, the Hill's equation does not apply, the force-velocity relationship is complex, and the active state is hard to define. Thus the handling of the parallel element has not been successful. This leads one to question the Hill's three-element model, and to search for an alternative method of approach, which will be discussed in the next two sections.

10.6.1 Active Muscle at Shorter Length with Negligible Resting Tension

The relative magnitudes of resting tension and active tension in a heart muscle are shown in Fig. 10.1:5. If the muscle length is sufficiently small, the resting

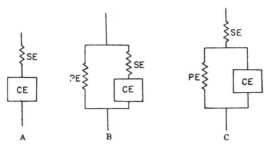

Figure 10.6:1 Mechanical analogs of mammalian papillary muscle. (A) Simple model with only a contractile element (CE) and a series elastic element (SE). (B) Parallel resting tension model in which a parallel elastic element (PE) bears resting tension. (C) Series resting tension model in which a series elastic element (SE) also bears resting tension.

tension is negligible. Then Hill's model is reduced to the two-element model of Fig. 10.6:1(A). In this case, if the series element is assumed elastic, then its elastic characteristics can be determined by the methods described in Sec. 9.8, namely, either by the quick-release or by the isometric-isotonic-change-over method. With the series elastic element's property determined, the contractile element behavior can be identified with the experimental data in a simple and unambiguous manner.

In the following discussion, we shall summarize the experimental results using the mathematical analysis presented in Sec. 9.8. The symbol S denotes tension in the series element. P denotes tension in the parallel element, which is assumed to be zero in the present section. The total tension $T = S + P$ is therefore equal to S. The length of the muscle is denoted by L. The elastic extension of the series element is η. The "insertion" or summation of the overlap between actin and myosin fibers is Δ.

10.6.2 The Series Element

Extensive work on the series element was reported by Sonnenblick (1964), Parmley and Sonnenblick (1967), Edman and Nilsson (1968), and others. Their results may be summarized by the equation

$$\frac{dS}{d\eta} = \alpha(S + \beta), \tag{1}$$

where α and β are constants. Thus the stiffness $dS/d\eta$ is linearly related to the tension S. See Fig. 10.6:2. For the cat's papillary muscle with a cross-sectional area $A_0 = 0.98$ mm^2, subjected to preload of 5 mN and an elastic extension of 4%–5% of the initial muscle length at a developed tension of 98 mN, corresponding to a stress of approximately 100 kPa, the value of α is 0.4 for η measured in % muscle length, and $\beta = 20$ mN.

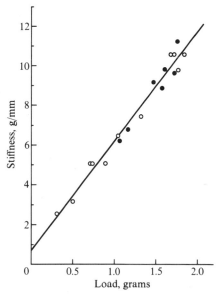

Figure 10.6:2 The straight-line relationship between the elastic modulus and tension in the series element. Rabbit papillary muscle. Open symbols: Measurements carried out during the rising phase of the isometric twitch. Filled symbols: those measured during the decay phase. Contraction frequency, 36 per min. Temp., 31°C. From Edman and Nilsson (1972), by permission.

Equation (1) integrates to

$$S = (S^* + \beta e^{\alpha(\eta - \eta^*)}) - \beta, \tag{2}$$

where $S = S^*$ when $\eta = \eta^*$. The condition $S = 0$ when $\eta = 0$ implies

$$\beta = \frac{S^* e^{-\alpha\eta^*}}{1 - e^{-\alpha\eta^*}}, \tag{3}$$

which requires that the point (S^*, η^*) be related to β in a specific way. Equations (1) and (2) also apply to many connective tissues (see Secs. 7.5 and 8.3), as well as to the unstimulated papillary muscle (Sec. 10.2).

10.6.3 The Contractile Element

Edman and Nilsson (1968, 1972) showed that if the force-velocity relationship of a papillary muscle is measured at any specific length of the contractile element, reached at a given time after stimulus, then it can be fitted by Hill's equation. Their experimental procedure is illustrated in Fig. 10.6:3. They used a small preload of 1 mN on rabbit papillary muscles with diameters in the range 0.5–1.0 mm and length in the range 3.8–7.0 mm. Each specimen was stimulated at a frequency of 30–48 per min and allowed to contract isometri-

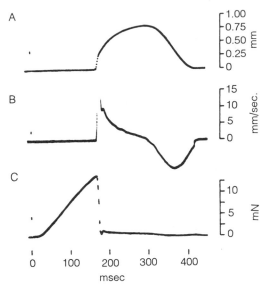

Figure 10.6:3 Oscilloscope records of release experiment. (A) Shortening. (B) Velocity of shortening (electricical differentiation). (C) Tension. Stimulation signal at zero time. Temp., 29.5°C. Contraction frequency, 30 per min. From Edman and Nilsson (1968), by permission.

cally. At a selected instant of time after stimulation, the load is suddenly released to a lower value [Fig. 10.6:3(C)], and the isotonic contraction of the muscle is recorded [Fig. 10.6:3(A)]. The velocity of contraction is obtained by differentiating the displacement signal electronically [Fig. 10.6:3(B)]. The velocity of shortening at a fixed sarcomere length (occurring at a fixed time interval after stimulation) is plotted against the load in Fig. 10.6:4, to which Hill's equation is fitted by the least squares method. The correlation is better than 0.99. Figure 10.6:4 shows the force-velocity relation at several instants after stimulus. It is seen that the hyperbolic relations at different instants of time are parallel to each other, indicating that the muscle's ability to shorten and ability to generate tension vary with time in the same way. The constants a and b in Hill's equation (Eq. (1)) of Sec. 9.7 are listed in the figure.

In a later paper, Edman and Nilsson (1972) demonstrated the importance of "critical damping" in the recording instrument on the shape of the force-velocity curve. They installed a dash pot containing silicone oil to damp out the oscillation of the lever. A 1 mm wide by 14 mm long duralumin rod extended from the lever to the damping fluid. A thin disk of aluminum (diam. 3.7 mm, thickness 0.4 mm) was attached perpendicular to the end of the rod. The degree of damping was varied by using silicone oils of different viscosity. With a 60-stoke silicone fluid, the damping is considered "critical." Figure 10.6:5 shows the effect of damping on the force-velocity curve. The critically damped curve can be fitted with Hill's equation; the substantially under-damped ones cannot. The reason for this is not entirely clear.

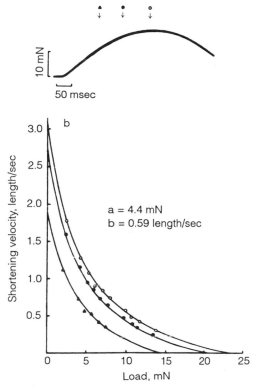

Figure 10.6:4 Force-velocity curves (lower diagram) defined at different times indicated by arrows in the isometric myogram (upper diagram). Curves drawn according to Hill's equation for values of a and b listed. The velocity values at a given load refer to the same length of the contractile unit (with variations $<0.5\%$ of resting muscle length) in all curves of the same experiment. Contraction frequency, 45 per min. Temp., 27.5°C. From Edman and Nilsson (1968), by permission.

We can condense Edman and Nilsson's results by modifying Hill's equation:

$$v = \frac{b(S_0 - S)}{a + S} \tag{4}$$

to the form (Fung, 1970)

$$v = \frac{b[S_0 f(t) - S]}{a + S}, \tag{5}$$

where $f(t)$ is a function of time after stimulation. We shall call $f(t)$ the *characteristic function of time*, or the *active state* function. Edman and Nilsson's results suggest that $f(t)$ can be represented by a half-since wave which is normalized to a unit amplitude:

$$f(t) = \sin\left[\frac{\pi}{2}\left(\frac{t + t_0}{t_{ip} + t_0}\right)\right]. \tag{6}$$

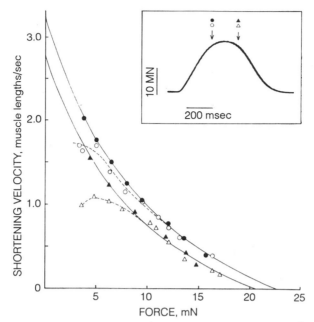

Figure 10.6:5 Force-velocity curves derived from quick release recordings carried out during the rising and falling phases of the isometric twitch. Arrows on top of inserted isometric myogram indicate times for collection of experimental data. Filled symbols: lever movements were critically damped. Open symbols: lever movements were underdamped. Solid lines (referring to filled symbols) have been fitted according to Hill's equation, using the following values of the constants a and b: $\bullet$, $a = 11.7$ mN, $b = 1.68$ lengths/sec; $\blacktriangle$, $a = 11.7$ mN, $b = 1.57$ lengths/sec. Correlation coefficient between experimental points and data predicted from the hyperbolic curve; $\bullet$, 0.999 $\blacktriangle$, 0.998. Dashed lines have been fitted by eye to data obtained from underdamped releases. From Edman and Nilsson (1968), by permission.

In Eqs. (5) and (6) the constants a and b are functions of muscle length L at the time of stimulation, S_0 is the peak tensile stress arrived at in an isometric contraction at length L, t is the time after stimulus, t_0 is a phase shift related to the initiation of the active state at stimulation, and t_{ip} is the time to reach the peak isometric tension after the instant of stimulation. The velocity v is the velocity of the contractile element, $d\Delta/dt$. The constant $a(L)$ is known empirically to be proportional to S_0, and can be written as

$$a(L) = \gamma S_0(L). \tag{7}$$

The numerical parameter γ is of the order of 0.45.

10.6.4 Brady's Test of the Validity of Hill's Equation

Brady (1965) plotted the ratio $(S_0 - S)/v$ against S. For Hill's equation we have

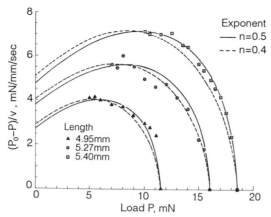

Figure 10.6:6 Comparison of Brady's data (1965) on rabbit papillary with the theoretical curve given by Eq. (9). Ordinate, mN/mm per sec. Abscissa, mN. $\gamma = 0.45$. From Fung (1970). In this figure S is written as P.

$$\frac{S_0 - S}{v} = \frac{a + S}{b},\qquad(8)$$

and Brady's plot should be a straight line. But the experimental data plotted in Fig. 10.6.6 do not fall on such a straight line.

The deviation is principally caused by the deviation of the force-velocity relationship from being a hyperbola in the high tension, low velocity region. Such deviations are seen in many other publications. For example, Fig. 10.6:7 shows the data of Ross et al. (1966) on the dog's heart. The data near the toe ($S/S_0 \sim 1$) show a sigmoid relationship.

Fung (1970) showed that Edman and Nilsson's curves (Figs. 10.6:3 and 10.6:4) as well as those of Brady and Ross et al. can be represented by a modified Hill's equation:

$$v = \frac{B[S_0 f(t) - S]^n}{a + S},\qquad(9)$$

where v is the contractile element velocity, $f(t)$ is the characteristic time function given in Eq. (6), and n is a numerical factor. B is, of course, different from b both in numerical value and in units. Figures 10.6:6 and 10.6:7 show the fitting of Eq. (8) to the data of Brady and Ross et al. In Fig. 10.4:6, Eq. (9) is presented in a dimensionless form:

$$v = \gamma v_{max} \frac{(1 - \sigma)^n}{\gamma + \sigma},\qquad(10)$$

where

$$v_{max} = \frac{B S_0^{n-1}}{\gamma},\qquad \sigma = \frac{S}{S_0},\qquad(11)$$

$$a = \gamma S_0, \quad \text{and} \quad v \to v_{max} \quad \text{when} \quad \sigma \to 0.\qquad(12)$$

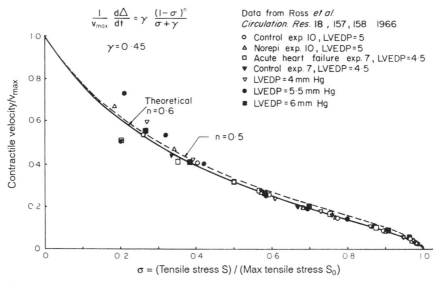

Figure 10.6:7 Comparison of the force-velocity data of a dog's left ventricle obtained by Ross et al. (1966) with theoretical curves given by Eq. (9). $\gamma = 0.45$, $n = 0.6, 0.5$. From Fung (1970). LVEDP = left ventricular end diastolic pressure in mm Hg. (1 mm Hg = 133.32 N/m^2).

Here v represents $d\Delta/dt$. In Eq. (10), as well as in Figs. 10.4:5 and 10.4:6, we have set the characteristic time function $f(t)$ of Eq. (9) to unity, $f(t) = 1$, because the experiments were done at a time near the peak isometric tension, i.e., $t \rightarrow t_{ip}$ in Eq. (6). Further, with Eqs. (9) and (10) the right-hand side becomes a complex number if $S > S_0 f(t)$ or $S > S_0$ and $n \neq 1$. In that case, we define the right-hand side of Eqs. (9) and (10) as the real part of the complex number. With these equations, a good fitting of the curves in Figs. 10.6:3 and 10.6:4 is demonstrated in Fung (1970). Incidentally, the introduction of the exponent n with a value less than 1 removes the difficulty mentioned in Sec. 9.8, namely, that the time required for the redevelopment of tension after a step shortening in length of an isometric tetanized muscle to reach a peak is infinity if Hill's equation holds strictly. See Eq. (22) of Sec. 9.8. The exponent n was originally introduced for the purpose to remove this difficulty.

All the formulas of this section apply only to papillary muscles of shorter lengths in which resting tension is negligible. Since they are good empirical formulas summarizing experimental data, they should be derivable from theories such as sliding elements, cross bridges or their alternatives. Such a theoretical derivation has not yet been done.

10.6.5 Stimulated Papillary Muscle at Lengths at Which Resting Tension Is Significant

For cardiac muscle at such a length that the resting tension must be taken into account, the analysis of the experimental data according to the three-element model becomes more complex (and nonunique). At such lengths, Brady (1965) found no hyperbolic relation between force and the velocity of contraction. Hefner and Bowen (1967) and Noble et al. (1969) found bell-shaped force-velocity curves and concluded that the maximum velocity of shortening occurs for loads appreciably larger than zero. They found near-hyperbolic behavior, however, for large resting tensions. Pollack et al. (1972) observed that the stiffness of the series elastic element depends on the time after stimulation.

When the same experimental data were analyzed by two different but equivalent three-element models as shown in Figs. 10.4:1(B) and (C), the derived contractile element force-velocity relations were found to be different for the two models. Thus the results are model-dependent and are not unique. Therefore, it does not represent an intrinsic property of the material.

There are other complications. When quick stretch experiments were done (Brady, 1965), it was found that the modulus of elasticity of the series elastic element in quick stretch is different from that in quick release. Thus the series spring must be interpreted as viscoelastic or viscoplastic.

These complications are associated with the inherent inadequacy of Hill's model to handle the viscoelastic behavior of the parallel element, and the nonuniqueness in the separation of the force into parallel and contractile components, and the displacement into contractile and series elements. Are the contractile elements stress free when the muscle is unstimulated? In smooth muscles a resting state cannot be defined uniquely (see Secs. 11.4 and 11.6); hence the contractile mechanism is not "free" between twitches. Furthermore, the separation of displacements between the series and contractile elements depends on the assumption of perfect elasticity of the series element. This amounts to a definition, and cannot be tested uniquely by experiment.

Hence the success of Hill's model is incomplete. The following sections are intended for the development of alternatives.

10.7 Pinto's Method

Pinto (1987) proposed to retain Hill's three-element model but modify the way by which the mechanical properties of the parallel element is measured. He proposed the following expression to describe the active contraction of the heart muscle:

$$S(\lambda, t) = A(\lambda)t^{\nu}e^{-\delta t}, \tag{1}$$

where $S(\lambda, t)$ represents the Lagrangian stress developed by the muscle through *active contraction only*, λ represents the stretch ratio measured relative to a reference state at which S is zero (in the passive state), and t is time. The parameter ν was introduced to account for the contraction delay. The parameter δ was introduced to account for the inotropic state of the muscle at a given physiological condition. $A(\lambda)$ is related to the amplitude of contraction force.

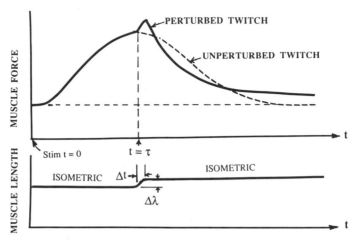

Figure 10.7:1 Schematic of an isometric quick-stretch experiment proposed by Pinto and Boe (1991). Reproduced by permission.

The author shows that this expression can fit the experimental data on a length-tension relationship, time course in single twitches, a quick-stretch response, and a quick-release response. Extension to isovolumic contraction was accomplished by a different set of constants of A, γ, and δ.

The total stress is a sum of the stress in the parallel element, $P(\lambda)$, and that in the contractile element:

$$T(\lambda, t) = P(\lambda, t) + A(\lambda)t^{\nu}e^{-\delta t}. \tag{2}$$

For the passive elasticity, Pinto and Boe (1991) proposed a method which consists of imposing isometric transients (such as quick-stretch or quick-release) on a muscle bundle during the contraction pahse and observing the difference between the developed force level of a normal twitch and that of a perturbed twitch. This is shown schematically in Fig. 10.7:1. The idea is based on Eq. (2). The time course of the active tension development is given by the second term, which is $S(\lambda, t)$ of Eq. (1). The total tension is measured in the experiment. The passive viscoelasticity, $P(\lambda, t)$, is what one wishes to deduce. By a sudden small step change of λ at time τ, the active tension changes by an amount which can be computed from Eq. (1):

$$\frac{dA}{d\lambda}\Delta\lambda\tau^{\nu}e^{-\delta\tau} \tag{3a}$$

at the instant of time $t = \tau$, and then after τ, by the amount

$$\left(\frac{dA}{d\lambda}\Delta\lambda\right)t^{\nu}e^{-\delta t}. \tag{3b}$$

The disturbance, Eq. (3b), has a course that is exactly the same as the undisturbed twitch. The parallel element, however, is subjected to a step increase of stretch $\Delta\lambda$ in a short time interval of Δt, during which the tension is increased by an amount ΔP. After words, because of the viscoelasticity, the tension in the parallel element becomes

$$\Delta PG(t - \tau), \tag{4}$$

where $G(t - \tau)$ is the normalized relaxation function of the tissue. The courses of time given by Eqs. (3) and (4) are different. Their difference is the difference of the two curves shown in Fig. 10.7:1. $\Delta P/\Delta\lambda$ gives the Young's modulus. $G(t - \tau)$ gives a normalized relaxation function. Pinto and Boe (1991) tested pig papillaries of the right ventricle and showed that the Young's modulus and the relaxation function so determined is *independent* of the instant of time τ (during a twitch) when the perturbation is imposed, suggesting the correctness of the scheme. The characters of the length-tension curve and the relaxation function are similar to those described in Sec. 10.4. This method is based on Hill's model, but does not require testing of an isolated papillary muscle in a passive state. Hence it is a better representation of the model.

10.8 Micromechanical Derivation of the Constitutive Law for the Passive Myocardium

Lanir (1983) has formulated a three-dimensional material law which is based on the microstructure and mechanical properties of the actual elements of the tissue. In Horowitz et al. (1988), Lanir and his colleagues extended the analysis to the myocardium with special attention to the collagen fibers. Robinson et al. (1983) and Caulfield and Borg (1979) have shown that fine collagen fibers tether myocardial cells. Lanir et al. worked out a detailed model as shown in Fig. 10.8:1. The collagen fibers are assumed to have a linear stress–strain relationship when they are straight, and to carry no load when they are wavy. On assuming a statistical distribution of the waviness (curvature) of the fibers, the probability of the number of taut collagen fibers can be computed as a function of the strain. On such a basis the authors computed the strain-energy function of the myocardium. They showed how to fix the constants so that the experimental stress–strain curves can be fitted. There are, however, no data available with regard to the waviness of the collagen fibers, nor on the mechanical properties of these fibers. The theory provides a framework, but more validation is needed.

Humphrey et al. (1990, 1991) proposed a form of a constitutive equation for the passive myocardium not quite on the basis of the microstructure, but on the observation that the tissue is transversely isotropic. The restrictions imposed by the condition of transverse isotropy on the constitutive equation has been determined by Green and Adkins (1960). The conclusion is that the strain-energy function may be expressed as a polynomial of the strain components e_{ij} in the form

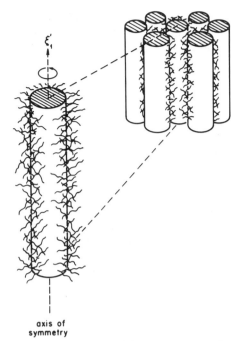

axis of
symmetry

Figure 10.8:1 Schematic representation of the arrangement of muscle fibers and their surrounding matrix of collagen fibers assumed by Horowitz et al. (1988). Reproduced by permission.

$$W(I_1, I_2, I_3, K_1, K_2) \tag{1}$$

in which I_1, I_2, I_3 are the first, second, and third invariants of the strain tensor, and K_1, K_2 are

$$K_1 = e_{33}, \qquad K_2 = e_{31}^3 + e_{32}^2, \tag{2}$$

if the material is transversely isotropic with respect to the x_3 axis, and x_1, x_2 are any pair of rectangular axes in a plane perpendicular to x_3. Humphrey et al. take x_3 in the direction of the muscle fiber, and assume that a simplified version will suffice for the myocardium:

$$W(I_1, \lambda_3). \tag{3}$$

They say that on the basis of their preliminary experimental data, W can be expressed in the form

$$\begin{aligned} W = {} & C_1(\lambda_3 - 1)^2 + C_2(\lambda_3 - 1)^3 + C_3(I_1 - 3) \\ & + C_4(I_1 - 3)(\lambda_3 - 1) + C_5(I_1 - 3)^2, \end{aligned} \tag{4}$$

where $C_1, \ldots, C_5$ are material constants. Humphrey et al. showed how to identify these constants on the basis of experimental data on resting heart muscle specimens.

If one uses Lanir et al. or Humphrey et al's results to characterize the mechanical properties of the parallel element of Hill's three-element model (Sec. 10.6), then one has to answer the criticism named in Sec. 10.6.5. The evaluation remains to be done. On theoretical ground, one may anticipate the need for some further generalization to take the viscoelastic characteristics of the myocardium into account. Furthermore, throughout the thickness of the ventricle the fiber directions change significantly. Hence these fibers are not necessarily transversely isotropic. If x_3 is the direction of the muscle fiber, the shear stresses e_{31}, e_{32} do not vanish in general in the myocardium, as the experimental results of Waldman et al. (1985, 1988, 1991) show. These have to be accounted for in the theory.

10.9 Other Topics

We study the cross bridge theory in order to gain some understanding of the actin-myosin interaction. Success in cross bridge theory will enable us to know the events in a sarcomere. Then the mechanics of myocardial cells must be studied. The sarcomeres have to work with other components of the muscle cells. The stress and movement of the sarcomeres must be transmitted to the cell membrane, which alone connects the interior of the cell with the exterior. The function of the cell depends on its shape, stress, and strain. Is there a sufficiently high internal pressure in the cell to keep the cell membrane taut (in tension)? Is there other structural members (stress fibers) that help maintain the cell shape? Cell shape depends on cell volume. What are the mechanisms that regulate the cell volume? How is the cell volume influenced by the stress and strain of the tissue?

Then we must study the interstitial space between the cells. The extra-cellular matrix has a unique structure. It contains molecules such as integrins, fibronectin, etc., which are also factors that determine the abhesive properties of the cells and the shapes of the cells. There are also collagen and elastin fibers embedded in and supported by ground substances. These fibers are believed to provide mechanical strength to the structure. And there are fibroblasts. Furthermore, the interstitial space has an interstitial pressure which is very important with regard to the movement of fluid in the interstitial space and the transport of matter. What determines the interstitial pressure? What role does it play in maintaining the integrity of the cells?

A tissue is a union of cells and extracellular matrix. An organ is made of tissues. The mechanics of the heart must deal with all the issues named above. Of these, I think the most important topic that is getting little attention is the cell pressure. The cell pressure and interstitial pressures are connected with swelling. Swelling is a classical subject. It is one of the classical signs of disease. Borelli (1608–1679), in his book *De Motu Animalium*, which was published in 1680, insisted that muscles work by swelling. Soft tissues are normally swollen. Mow, Lai, Lanir, Bogan, and others have analyzed swelling from the point of view of multiphasic mixture theory. For the heart, however, I believe that a cellular theory needs to be developed.

In heart muscle, perhaps every phase may be considered as incompressible. The static pressure in an incompressible material is an arbitrary number in the constitutive equation: a number which is not defined by strain. To determine the swelling pressure in the cells, and in the tissue, one has to consider the chemical potential of every phase, and the appropriate boundary conditions. In other words, the static pressure, which enters the free energy as a Lagrange multiplier, can be determined only through the equations of motion, continuity, and boundary conditions, together with the constitutive equations of every phase. If a cell membrane is involved, then the movement of fluid through the membrane constitutes an essential boundary condition. The biology of mass transport through the membrane holds the key to the pressure story.

Bogen (1987) introduced an analysis of tissue swelling to cardiac mechanics. He has considered both a one-fluid-compartment model and a multiple-fluid-compartment model. The simplest model illustrates his approach. Consider a model with incompressible solid elastic elements floating in an incompressible fluid. Let the strain energy function per unit volume of the mixture at its zero stress state be given by Eqs (9) and (10) of Sec. 7.5, pp. 275, 276:

$$\rho_0 W = \frac{\mu}{k}(\lambda_1^k + \lambda_2^k + \lambda_3^k) + p(\lambda_1 \lambda_2 \lambda_3 - 1), \tag{1}$$

where $\lambda_1, \lambda_2, \lambda_3$ are the principal stretch ratios, μ and k are elastic constants, and p is a Lagrange multiplier interpreted as pressure. The condition of incompressibility is $\lambda_1 \lambda_2 \lambda_3 = 1$. If a unit cube of the material swells isotropically to a volume λ_s^3 due to an addition to fluid into the elastic matrix, then the strain energy for the swollen material, $\rho_0 W^{(s)}$ per unit swollen volume becomes

$$\rho_0 W^{(s)} = \frac{\mu \lambda_s^{-3}}{k}[\lambda_1^k + \lambda_2^k + \lambda_3^k] + p\lambda_s^{-3}(\lambda_1 \lambda_2 \lambda_3 - 1). \tag{2}$$

Now, introduce a new symbol

$$\Lambda_i = \frac{\lambda_i}{\lambda_s} \tag{3}$$

to represent the principal extension with respect to the swollen reference configuration. The Eq. (2) may be written as

$$\rho_0 W^{(s)} = \frac{\mu \lambda_s^{k-3}}{k}[\Lambda_1^k + \Lambda_2^k + \Lambda_3^k] + p\lambda_s^{k-3}(\Lambda_1 \Lambda_2 \Lambda_3 - 1). \tag{4}$$

Bogen then proceeds to consider various boundary-volume problems such as triaxial isotropic swelling, uniaxial stretching, biaxial stretching, etc., to determine the fluid pressure, and gradually adding greater complexity to the model by considering the solid phase as springs, membranes, and tubes.

The pressure p is calculated from the principle of virtual work. Given a volume V of tissue, the work required to inject an extra volume of fluid, ΔV, is $p\Delta V$ and is equal to the sum of the increment of strain energy and the work of the external loads:

$$p\Delta V = \Delta(\rho_0 W^{(s)})V + \Delta W_e \tag{5}$$

in which $\Delta(\rho_0 W^{(s)})$ is the increment in strain-energy density of the swollen material (per unit swollen volume), ΔW_e is the increment of external work due to forces acting on this boundary of V. When the boundary forces are zero, ΔW_e is zero. Let the expansion be isotropic relative to the swollen volume of $V = 1$, so that

$$\Lambda_1 = \Lambda_2 = \Lambda_3 = \Lambda,$$
$$\Delta V = \Lambda_1 \Lambda_2 \Lambda_3 = \Lambda^3, \tag{6}$$

then Eqs. (5) and (4) yields

$$p = \frac{d(\rho_0 W^{(s)})}{d(\Lambda^3)} = \frac{d(\rho_0 W^{(s)})}{d\Lambda}\frac{d\Lambda}{d(\Lambda^3)}. \tag{7}$$

$$p = \mu\lambda_s^{k-3}. \tag{8}$$

Bogen (1987) shows how the swelling pressure is affected by the assumption of different types of structure of the solid matrix in the tissue.

This example shows the role of boundary conditions in determining the pressure p. It is clear that the complexities of the real biology need to be taken into account in order to reach a better understanding. The example given in Chapter 12, Sec. 12.10, shows how a triphasic theory can be developed. The examples presented in Chapter 9 of *Biomechanics: Motion, Flow, Stress, and Growth*, (Fung 1990) show how important the interstitial pressure is.

We conclude this section with an entirely different matter: on the contribution of epicardium to cardiac mechanics—to remind us that to see a whole picture requires attention to all the corners.

Humphrey and Yin (1988) called attention to the possible significance of the epicardium to cardiac mechanics. Pericardium is a thin serous membrane composed of mesothelial cells, a ground substance matrix, and a fibrous network of elastin and collagen. The pericardium forms a double-walled sac enveloping the heart: a visceral layer, or epicardium, adheres intimately to the surface of the heart, a parietal layer encloses the heart, a narrow space separates the parietal layer and the visceral layer. The enclosed space is clled the pericardial space, which contains a small amount of lubricating fluid.

The mechanical properties of the parietal pericardium have been measured by Lee et al. (1985) and others (see Lee et al. for references), and those of the visceral pericardium were measured by Humphrey and Yin (1988). Generally speaking, the characteristics are highly nonlinear and anisotropic, in a way similar to the skin (Chapter 7). Clearly, the epicardium is capable of carrying large in-plane loads. Moreover, atrial epicardium behaves more like parietal pericardium than myocardium. Both parietal and atrial epicardium are compliant and seemingly isotropic over large ranges of equibiaxial stretch, but stiffen rapidly and anisotropically near the limit of their distensibility. In contrast, excised myocardium stiffens, and is anisotropic at all values of equibiaxial stretch (Yin et al., 1987).

The significance of epicardium on cardiac mechanics is obvious when the heart dimension increases beyond a certain limit. In Sec. 10.3 it is pointed out that there are many reasons to determine heart muscle behavior by testing the whole ventricle. It is important not to forget this very thin layer of tissue wrapping the heart muscle on the outside.

Problems

10.1 Let the left ventricle be approximated by a thick-walled hemispherical shell of uniform thickness. The valves and valve ring will be assumed to be infinitely compliant. In diastolic condition the left ventricle is subjected to an internal pressure p_i and an external pressure p_0. p_i is a function of time. Ignoring inertia effects, what is the average stress in the wall?

10.2 Let the inner radius of the hemisphere be a_0 and the wall thickness be h_0 at the no-load condition, $p_i = p_0 = 0$. Assume that the average strain is related to the average stress in the same way as stress is related to strain in the uniaxial case [see Eq. (8) in Sec. 10.4]. Under the assumptions of Problem 10.1, what will the radius $a(t)$ and the wall thickness $h(t)$ be as functions of time t in response to $p_i(t)$ and p_0? Assume the wall mateiral to be incompressible. See Fig. P10.2.

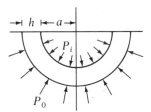

Figure P10.2 A ventricle subjected to internal and external pressures.

10.3 At time $t = 0$, the valves are closed and every muscle fiber in the shell begins to contract. Assume that the plane of the valves remain plane and the volume enclosed by the hemispherical shell remains constant. Compute the course of pressure change $p_i(t)$ under the assumption that the equations of Sec. 10.6 are applicable.

10.4 Under the same assumptions of Problems 10.1–10.3, but considering an electric event different from that of Problem 10.3, let the wave of depolarization begin at the axis of symmetry $\theta = 0$ (see Fig. P10.4) and spread with a constant angular velocity in the circumferential direction, $d\theta/dt = c$, a constant. As the muscle fibers shorten, the shell will be distorted. It is necessary to consider bending of the shell wall in this case. Develop an approximate theory for the change of shape of the shell as a function of time, assuming that the equation of Secs. 10.4 and 10.6 are applicable. Make a reasonable assumption about the compliance of the valves.

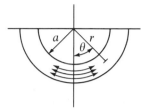

Figure P10.4 A wave of depolarization propagating circumferentially.

10.5 Consider another electric event: the wave of depolarization beginning at the inner wall and propagating radially at a constant velocity V. As in Problem 10.4, determine the history of the shape change of the shell. In the real left ventricle the wave of depolarization is determined by conduction pathways [see Scher, A. M. and Spach, M. S. (1979), Cardiac depolarization and repolarization and the electrocardiogram. In *Handbook of Physiology*, Sec. 2., *The Cardiovascular System*, Vol. 1, *The Heart*, R. M. Berne, N. Sperelakis, and S. R. Geiger, (eds.) American Physiol. Soc., Bethesda, MD, pp. 357–392]. See Fig. P10.5.

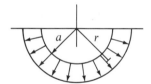

Figure P10.5 A wave of depolarization propagating radially.

10.6 Discuss the improvements that are necessary in order to bring the idealized analyses of preceding problems closer to reality.

10.7 (a) In what ways do the structure and function of a heart muscle differ from those of a skeletal muscle?
(b) Why must Hill's equation be modified for heart muscle?
(c) In which way has Hill's equation been successfully modified for the heart muscle? Under what circumstance is the modified Hill's equation valid?

10.8 Design a single purpose instrument to measure the force-velocity-time relationship of a muscle, or its electric activity, or its work output. The purpose can be one of the following.
(a) An accurate force-velocity-time relationship in single twitch after stimulation for the purpose of fundamental research.
(b) To measure the work output of a muscle for the purpose of industrial safety and health assessment.
(c) To clarify the relative function of a group of muscles such as those on the abdomen or on the back in one particular posture.

(d) To determine the fraction of muscle fibers in a skeletal muscle that are stimulated in certain motion.

State your chosen objective. Sketch your design. Explain how it works. Discuss its pros and cons.

References

Abbott, B. C. and Mommaerts, W. F. H. M. (1959) Study of inotropic mechanisms in the papillary muscle preparation. *J. Gen. Physiol.* **42**, 533–551.

Allen, D. G. (1985) The cellular basis of the length-tension relaton in cardiac muscle. *J. Mol. Cell Cardiol.* **17**, 821–840.

Berne, R. M. and Levy, M. N. (1972) *Cardiovascular Physiology*, 2nd edition. C. V. Mosby, St. Louis.

Bogen, D. K. (1987) Strain-energy descriptions of biological swelling. I: Single Fluid Compartment Models; II: Multiple Fluid Compartment Models. *J. Biomech. Eng.* **109**, 252–262.

Borelli, Giovanni Alfonso (1680) *De Motu Animalium*, first half published posthumously in 1680, second half published in 1681. Translated by Paul Maquet under the title of *On The Movement of Animals*. Springer-Verlag, Berlin (1989).

Bornhorst, W. J. and Mirandi, J. E. (1969) Comparison of Caplan's irreversible thermodynamics theory of muscle contraction with chemical data. *Biophys. J.* **9**, 654–665.

Brady, A. J. (1965) Time and displacement dependence of cardiac contractility: Problems in defining the active state and force-velocity relations. *Fed. Proc.* **24**, 1410–1420.

Brady, A. (1979) Mechanical properties of cardiac fibers. In *Handbook of Physiology*, Sec. 2, *The Circulation System*, Vol. 1: *The Heart*. American Physiological Society, Bethesda, MD, Chap. 12, pp. 461–474.

Brutsaert, D. I. and Sonnenblick, E. H. (1969) Force-velocity-length-time relations of the contractile elements in heart muscle of the cat. *Circulation Res.* **24**, 137–149.

Brutsaert, D. L., Victor, A. C., and Ponders, J. H. (1972) Effect of controlling the velocity of shortening on force-velocity-length and time relations in cat papillary muscle velocity clamping. *Circulation Res.* **30**, 310–315.

Caulfield, J. B. and Borg, T. K. (1979) The collagen network of the heart. *Lab. Invest.* **40**, 364–372.

Daniels, M., Noble, M., ter Keurs, H., and Wohlfart, B. (1984) Velocity of sarcomere shortening in rat cardiac muscle: Relationship to force, saromere length, calcium, and time. *J. Physiol.* **355**, 367–381.

Edman, K. A. P. and M. Johannsson (1976) The contractile state of rabbit papillary muscle in relation to stimulation frequency. *J. Physiol.* **245**, 565–581.

Edman, K. A. P. and Nilsson, E. (1968) The mechanical properties of myocardial contraction studied at a constant length of the contractile element. *Acta Physiol. Scand.* **72**, 205–219.

Edman, K. A. P. and Nilsson, E. (1972) Relationships between force and velocity of shortening in rabbit papillary muscle. *Acta Physiol. Scand.* **85**, 488–500.

Ford, L. E., Huxley, A. F., and Simmons, R. M. (1981) The relation between stiffness and filament overlap in stimulated frog muscle fibers. *J. Physiol.* **311**, 219–249.

Frank, J. S. and Langer, G. A. (1974) The myocardial interstitium: Its structure and its role in ionic exchange. *J. Cell Biol.* **60**, 596–601.

Fung, Y. C. (1970) Mathematical representation of the mechanical properties of the heart muscle. *J. Biomech.* **3**, 381–404.

Fung, Y. C. (1971a) Comparison of different models of the heart muscle. *J. Biomech.* **4**, 289–295.

Fung, Y. C. (1971b) Muscle controlled flow. In *Proc. 12th Midwest Mechanics Conf.* University of Notre Dame Press, Notre Dame, IN, pp. 33–62.

Fung, Y. C. (1972) Stress–strain-history relations of soft tissues in simple elongation. In *Biomechanics, Its Foundations and Objectives*, Y. C. Fung, N. Perrone, and M. Anliker (eds.) Prentice-Hall, Englewood Cliffs, NJ, pp. 191–208.

Gay, W. A. and Johnson, E. A. (1967) Anatomical evaluation of the myocardial length-tension diagram. *Circulation Res.* **21**, 33–43.

Glass, L., Hunter, P., and McCulloch, A. (eds.) (1991) *Theory of Heart.* Springer-Verlag, New York.

Green, A. E. and Adkins, J. E. (1960) *Large Elastic Deformations.* Oxford University Press, London.

Guccione, J. M. and McCulloch, A. (1991) Finite element modeling of ventricular mechanics. In *Theory of Heart*, Glass et al. (eds.) pp. 121–144.

Guccione, J. M., McCulloch, A. D., and Waldman, L. K. (1991) Passive material properties of intact ventricular myocardium determined from a cylindrical model. *J. Biomech. Eng.* **113**, 42–55.

Hefner, L. L. and Bowen, T. E., Jr. (1967) Elastic components of cat papillary muscle. *Am. J. Physiol.* **212**, 1221–1227.

Hill, A. V. (1949) The abrupt transition from rest to activity in muscle. *Proc. Roy. Soc. London B* **136**, 399–420.

Horowitz, A., Lanir, Y., Yin, F. C. P., Perl, M., Sheinman, I., and Strumpf, R. K. (1988) Structural three-dimensional constitutive law for the passive myocardium. *J. Biomech. Eng.* **110**, 200–207.

Hort, W. (1960) Makroskopische und mikrometrische Untersuchungen am Myodard verschieden stark gefüllter linker Kammern. *Virchows Arch [Pathol Anat.]* **333**, 523–564.

Huisman, R. M., Sipkema, P., Westerhof, N., and Elzinga, G. (1980) Comparison of model used to calculate left ventricle wall force. *Med. Biol. Eng. Comput.* **18**, 122–144.

Humphrey, J. D. and Yin, F. C. P. (1988) Biaxial mechanical behavior of excised epicardium. *J. Biomech. Eng.* **110**, 349–351.

Humphrey, J. D., Strumpf, R. H., and Yin, F. C. P. (1990) Determination of a constitutive relation for passive myocardium. I. A nero-functional form, II. Parameter identification. *J. Biomech. Eng.* **112**, 333–339, 340–346.

Humphrey, J., Strumpf, R., Halperin, H., and Yin, F. (1991) Toward a stress analysis in the heart. In *Theory of Heart*, Glass et al. (eds.) Springer-Verlag, New York, pp. 59–75.

Huntsman, L. L., Rondinone, J. F., and Martyn, D. A. (1983) Force-length relations in cardiac muscle segments. *Am. J. Physiol.* **244**, H701–H707.

Huxley, H. E. (1957) The double array of filaments in cross-striated muscle. *J. Biophys. Biochem. Cytol.* **3**, 631–648.

Huxley, H. E. (1963) Electron microscope studies on the structure of natural and synthetic protein filaments from striated muscle. *J. Mol. Biol.* **7**, 281–308.

Huxley, H. E. (1969) The mechanism of muscular contraction. *Science* **164**, 1356–1366.

Jewell, B. R. (1977) A reexamination of the influence of muscle length on myocardial performance. *Circulation Res.* **40**, 221–230.

Korecky, B. and Rakusan, K. (1983) Effects of hemodynamic load on myocardial fiber orientation. In *Cardiac Adaptation to Hemodynamic Overload, Training, and Stress, International Erwin Riesch Symp., Tübingen, September* 19–22, 1982, Dr. S. Steinkopff Verlag.

Kreuger, J. W. and Pollack, G. H. (1975) Myocardiac sarcomere dynamics during isometric contraction. *J. Physiol. (London)* **251**, 627–643.

Lanir, Y. (1983) Constitutive equatons for fibrous connective tissue. *J. Biomech.* **16**, 1–12.

Lee, M.-C., LeWinter, M. M., Freeman, G., Shabetai, R., and Fung, Y. C. (1985) Biaxial mechanical properties of the pericardium in normal and volume overload dogs. *Am. J. Physiol.* **249**, H222–H230.

Martyn, D. A., Rondinone, J. F., and Huntsman, L. L. (1983) Myocardial segment velocity at a low load: Time, length, and calcium dependence. *Am. J. Physiol.* **244**, H708–H714.

McCulloch, A. D., Smail, B. H., and Hunter, P. J. (1989) Regional left ventricular epicardial deformation in the passive dog heart. *Circulation Res.* **64**, 721–733.

Nevo, E. and Lanir, Y. (1989) Structural finite deformation model of the left ventricle during diastole and systole. *J. Biomech. Eng.* **111**, 343–349.

Noble, M. I. M., Bowen, T. E., and Hefner, L. L. (1969) Force-velocity relationship of cat cardiac muscle, studied by isotonic and quick-release techniques. *Circulation Res.* **24**, 821–834.

Parmely, W. W. and Sonnenblick, E. H. (1967) Series elasticity in heart muscle; its relation to contractile element velocity and proposed muscle models. *Ciruclation Res.* **20**, 112–123.

Parmley, W. W., Brutsaert, D. L., and Sonnenblick, E. H. (1969) Effects of altered loading on contractile events in isolated cat papillary muscle. *Circulation Res.* **24**, 521–532.

Patterson, S. W., Piper, H., and Starling, E. H. (1914) The regulation of the heart beat. *J. Physiol.* **48**, 465–513.

Peachey, L. D. (1965) The sarcoplasmic reticulum and transverse tubles of the frog sartorius. *J. Cell Biol.* **25**, 209–231.

Pietrabissa, R., Montevecchi, F. M., and Fumero, R. (1991) Mechanical characterization of a model of a multicomponent cardiac fibre. *J. Biomed. Eng.* **13**, 407–414.

Pinto, J. G. and Fung, Y. C. (1973) Mechanical properties of the heart muscle in the passive state. *J. Biomech.* **6**, 596–616.

Pinto, J. G. and Fung, Y. C. (1973) Mechanical properties of stimulated papillary muscle in quick-release experiments. *J. Biomech.* **6**, 617–630.

Pinto, J. G. and Patitucci, P. (1977) Creep in cardiac muscle. *Am. J. Physiol.* **232**, H553–H563.

Pinto, J. G. (1987) A constitutive description of contracting papillary muscle and its implications to the dynamics of the intact heart. *J. Biomech. Eng.* **109**, 181–191.

Pinto, J. G. and Boe, A. (1991) A method to characterize the passive elasticity incontracting muscle bundles. *J. Biomech. Eng.* **113**, 72–78.

Pollack, G. H., Huntsman, L. L., and Verdugo, P. (1972) *Circulation Res.* **31**, 569–579.

Robinson, T. F. (1983) The physiological relationship between connective tissue and contractile elements in heart muscle. *The Einstein Q.* **1**, 121–127.

Robinson, T. F., Cohen-Gould, L., and Factor, S. M. (1983) Skeletal framework of mammalian heart muscle. *Lab. Invest.* **49**, 482–498.

Ross, Jr., J., Covell, J. W., Sonnenblick, E. H., and Braunwald, E. (1966) Contractile state of the heart. *Circulation Res.* **18**, 149–163.

Schmid-Schönbein, G. W., Skalak, R. C., Engelson, E. T., and Zweifach, B. W. (1986) Microvascular network anatomy in rat skeletal muscle. In *Microvascular Network: Experimental and Theoretical Studies*, A. S. Popel and P. C. Johnson (eds.) Karger, Basel, pp. 38–51.

Schmid-Schönbein, G. W., Skalak, T. C., and Sutton, D. W. (1989) Bioengineering analysis of blood flow in resting skeletal muscle. In *Microvascular Mechanics*, J.-S. Lee and T. C. Skalak (eds.) Springer-Verlag, New York, pp. 65–99.

Sommer, J. R. and Johnson, E. A. (1979) Ultrastructure of cardiac muscle. In *Handbook of Physiology*, Sec. 2, *The Cardiovascular System*, Vol. 1: *The Heart*. American Physiological Society, Bethesda, MD, Chap. 5, pp. 113–186.

Sonnenblick, E. H. (1964) Series elastic and contractile elements in heart muscle: Changes in muscle length. *Am. J. Physiol.* **207**, 1330–1338.

Sonnenblick, E. H., Ross, J. Jr., Covell, J. W., Spotnitz, H. M., and Spiro, D. (1967) Ultrastructure of the heart in systole and diastole: Changes in sarcomere length. *Circulation Res.* **21**, 423–431.

Taber, L. A. (1991) On a nonlinear theory for muscle shells: Part I: Theoretical Development. Part II: Applicaton to the Beating Left Ventricle. *J. Biomech. Eng.* **113**, 56–62.

Ter Keurs, H. E. D. J., Rijnsburger, W. H., van Heuningen, R., and Nagelsmit, M. (1980) Tension development and sarcomere length in rat cardiac trabecular. *Circulation Res.* **46**, 703–714.

Ter Keurs, H. E. D. J., and Tyberg, J. V. (eds.) (1987) *Mechanics of the Circulation*, Martininus Nijhoff, Pub.

Waldman, L. K. (1991) Multidimensional measurement of regional strains in the intact heart. In *Theory of Heart*, Glass et al. (eds.) Springer-Verlag, New York, pp. 145–174.

Waldman, L. K., Fung, Y. C., and Covell, J. W. (1985) Transmural myocardial deformation in the canine left ventricle: Normal *in vivo* three-dimensional finite strains. *Circulation Res.* **57**, 152–163.

Waldman, L. K., Nosan, D., Villarreal, F. J., and Covell, J. W. (1988) Relation between transmural deformaton and local myofiber direction in canine left ventricle. *Circulation Res.* **63**, 550–652.

Warwick, R. and Williams, P. L. (eds.) *Gray's Anatomy*. 35th British Edition. W. B. Saunders, Philadelphia.

Whalen, W. J., Nair, P., and Ganfield, R. A. (1973) Measurements of oxygen tension in tissues with a micro oxygen electrode. *Microvasc. Res.* **5**, 254–262.

Yin, F. C. P. (1981) Ventricular wall stress. *Circulation Res.* **49**, 829–842.

Yin, F. C. P., Strumpf, R. K., Chew, P. H., and Zeger, S. L. (1987) Quantification of the mechanical properties of noncontracting canine myocardium under simultaneous biaxial loading. *J. Biomech.* **20**, 577–589.

Zahalak, G. I. (1986) A comparison of the mechanical behavior of the cat soleus muscle with a distribution-moment model. *J. Biomech. Eng.* **108**, 131–140.

CHAPTER 11

Smooth Muscles

11.1 Types of Smooth Muscles

Muscles in which striations cannot be seen are called smooth muscles. Smooth muscles of the blood vessels are called vascular smooth muscles. That of the intestine is intestinal smooth muscle. Different organs have different smooth muscles: there are sufficient differences among these muscles anatomically, functionally, mechanically, and in their responses to drugs to justify studying them one by one. But there are also common features. All muscles contain actin and myosin. All rely on ATP for energy. Changes in the cell membrane induce Na^+ and K^+ ion fluxes and action potentials. The Ca^{++} flux furnishes the excitation-contraction coupling. These properties are similar in all muscles.

In analyzing the function of internal organs, we need the constitutive equations of the smooth muscles. The constitutive equations are unknown at this time. We outline in this chapter some of the basic features that are known, and on which constitutive equations will be built.

11.1.1 Cell Dimensions

Smooth muscle cells are generally much smaller than skeletal and heart muscle cells. Table 11.1:1 gives some typical dimensions obtained by electron microscopy (see Burnstock, 1970, for original references).

TABLE 11.1:1 Size of Smooth Muscle Cells

Tissue	Animal	Length (μm)	Diameter in nuclear region (μm)
Intestine	Mouse	400	
Taenia coli	Guinea pig	200 (relaxed)	2–4
Nictitating membrane	Cat	350–400	
Vas deferens	Guinea pig	450	
Vascular smooth muscle			
Small arteries	Mouse	60	1.5–2.5
Arterioles	Rabbit	30–40	5

11.1.2 Arrangement of Muscle Cells Within Bundles

There are a variety of patterns of smooth muscle packing. In guinea pig vas deferens, the thick middle portion of the cell that contains the nucleus lies adjacent to the long tapering ends of surrounding cells. In many blood vessels, the muscle cells meet in end-to-end fashion. In the uterus of the pregnant female, interdigitation between cells is common. In taenia coli each muscle cell is surrounded by 6 others, but there is an irregular longitudinal splicing of neighboring cells, so that each muscle cell is surrounded by about 12 others over its length. The cells are not straight, but are bent and interwoven with each other. The separation of neighboring muscle cells is generally between 500 and 800 Å in most organs. Basement membrane material and sometimes scattered collagen filaments fill the narrow spaces between muscle cells within bundles.

In the extracellular space, a variety of materials, including collagen, blood vessels, nerves and Schwann cells, macrophages, fibroblasts, mucopolysaccharides, and elastic tissue. Abundant "micropinocytotic" or "plasmalemmal" vesicles are seen in the extracellular space, some of which appear to be connected with an intracellular endoplasmic tubular system. These fine structures are probably important in facilitating ion exchange across the cell membrane during depolarization, and in creating excitation-contraction coupling.

The size of extracellular space from various measurements is shown in Table 11.1:2. Again see Burnstock (1970) for references to original papers. Note the difference in extracellular space between the visceral and the vascular smooth muscles of large arteries. The latter are often called *multi-unit smooth muscles*, each fiber of which operates independently of the others and is often innervated by a single nerve ending, as in the case of skeletal muscle fibers. See the sketch in Fig. 11.1:1. They do not exhibit spontaneous contractions. Smooth muscle fibers of the ciliary muscle of the eye, the iris of the eye, and the piloerector muscles that cause erection of the hairs when stimulated by the sympathetic nervous sytem, are also multi-unit smooth muscles.

TABLE 11.1:2 Extracellular Space in Smooth Muscles

Tissue	Extracellular space (% of total) measured by	
	Electron microscope	Uptake of solutes (inulin)
Visceral smooth muscle		
Guinea pig taenia coli	12	30–39
Cat intestine	9	
Mouse vas deferens	12 (adult)	
	50 (neonatal)	
Cat uterus		31–40
Vascular smooth muscle		
Pig carotid artery	39	
Mouse femaral artery	30	
Rat aorta		35
Rabbit aorta		62
Dog carotid artery	25	

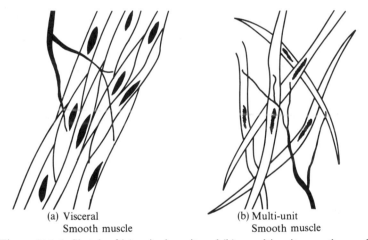

(a) Visceral
Smooth muscle

(b) Multi-unit
Smooth muscle

Figure 11.1:1 Sketch of (a) a single-unit and (b) a multi-unit smooth vessel.

In contrast, the visceral smooth muscle cells are crowded together, and behave somewhat like those of the heart muscle. They are usually stimulated and act as a unit. Conduction is from muscle fiber to muscle fiber. They are usually spontaneously active. A sketch is given in Fig. 11.1:1.

11.2 The Contractile Machinery

Smooth muscles have actin and myosin. Figures 11.2:1 and 11.2:2 show the electron micrographs of a vascular smooth muscle cell of a rabbit vein.

Figure 11.2:1 An electron micrograph of a vascular smooth muscle cell from the rabbit portal-anterior mesenteric vein. Longitudinal section. Two dense bodies (DB) are included in the section. The arrow near the center of the figure points to thin filaments attaching to one of the dense bodies. Many cross-bridges are evident on the thick filament (TF), traversing the center of the section. Intermediate size filaments (IF) may be seen arranged obliquely to the left of the upper dense body. From Ashton, Somlyo, and Somlyo (1975), reproduced by permission.

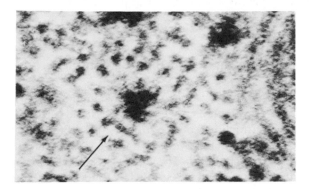

Figure 11.2:2 Transverse section of vascular smooth muscle cell from rabbit portal-anterior mesenteric vein. The thick filament in the center of the figure is surrounded by an array of thin filaments. The arrow points to an area where cross-bridges may be seen linking the thick to thin filaments. From Ashton, Somlyo, and Somlyo (1975), reproduced by permission.

The thick filaments shown in Fig. 11.2:1, comprised of myosin molecules, have a transverse dimension of 14.5 nm. Cross-bridges extending from the thick filament are evident in these figures. X-ray diffraction suggests that the repeat distance for the cross-bridge is 14.4 nm, identical to that of striated muscle. The length of the thick filament in vascular smooth muscle is 2.2 μm, which is larger than the 1.6 μm length of striated muscle by a factor of $2.2/1.6 = 1.4$. The significance of the longer myosin filaments is that, under the assumption that the cross-bridge of smooth muscle is the same as that in striated muscle, the tension generated per filament in smooth muscle is larger than that in striated muscle by about 40%. This is because the cross-bridges are arranged in parallel and at equal spacing; hence the sum of forces is proportional to the myosin fiber length.

The thick filaments of vascular smooth muscle appear to be arranged in groups of three to five adjacent filaments that terminate in the same transverse serial section. The thin filaments are attached to *dense bodies* throughout the sarcoplasm and are also attached to the plasma membrane in numerous areas (dark areas). These are analogs of a sarcomere arrangement. The lack of a periodic sarcomere structure is probably responsible for the slow action of the smooth muscle.

In vascular smooth muscle, thin filaments, composed of actin and tropomyosin, are far more numerous than thick filaments, giving a thin-to-thick ratio of 15:1. The thin filaments in vascular smooth muscle have an average diameter of 6.4 mm.

The contractile proteins are thus quite similar to those of striated muscle. For the electrophysiology and biochemistry of muscle contraction, the reader is referred to the books of Guyton (1976), Bohr et al. (1980), and Burnstock (1970). These works, however, do not treat the mechanics of contraction in sufficient depth. Hence in the following sections we appeal directly to laboratory experiments to learn something about the quantitative aspects of muscle contraction.

11.3 Rhythmic Contraction of Smooth Muscle

Spontaneous contraction is a phenomenon common to many muscular organs. Long ago, Engelmann (1869) and Bayliss and Starling (1899) determined that the contractions of the ureter and intestine are myogenic. Burnstock and Prosser (1960) have shown that step stretch can alter the cell membrane potential (excitability). The length-tension relationship for spontaneous contraction of taenia coli was studied by Bülbring and Kuriyama (1963) and Mashima and Yoshida (1965). Their results do not agree with respect to the role of passive tension during a spontaneous contraction. Golenhofen (1964) studied sinusoidal length oscillations and found two frequencies for maximum tension amplitude. He concluded that a mechanical coupling between muscle units may play a role in synchronization. Golenhofen (1970) also studied the influence of various environmental factors such as temperature, pH, glucose concentration, and CO_2 tension on spontaneous contraction. Since the time duration of a single isometric spontaneous contraction is on the order of 1 min, Golenhofen has used the term *minute rhythm* to describe these contractions.

11.3.1 Wave Form of Contraction in Taenia Coli

Details of the contraction process of the taenia coli muscle from the cecum of guinea pigs are presented by Price, Patitucci, and Fung (1977). To avoid the effect of drugs, the animals were sacrificed by a sharp blow to the head which fractured the spinal cord. The *in vivo* length, L_{ph} (read the subscript "ph" as "physiological"), of 10 mm or more was marked on the specimen before dissection. Immediately after dissection the specimen was mounted in a test chamber at a length of approximately L_{ph} and 37°C. In a few minutes the

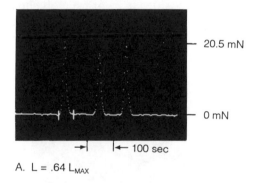

A. $L = .64\ L_{MAX}$

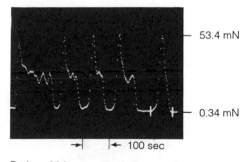

B. $L = .88\ L_{MAX}$

Figure 11.3:1 Computer display of spontaneous contractions in real time for $L < L_{max}$. Time increases from left to right. Each point represents a reading of force on muscle. Time between points is 2 sec. (A) Contractions for $L = 0.64\ L_{max}$. (B) Contractions for $L = 0.88\ L_{max}$. Reference dimensions for specimen are: $L_{max} = 13$ mm, $A_{max} = 0.36$ mm^2, $L_{ph} = 10$ mm. For taenia coli of guinea pig. From Price, Patitucci, and Fung (1977), by permission.

muscle began to contract spontaneously and rhythmically. Testing could begin after letting the muscle contract for an hour or more. The solution in the bath, environment control, test equipment, and test procedure are similar to those used for heart muscle (Sec. 10.2).

When taenia coli is tested isometrically at various lengths, the tension history as shown in Figs. 11.3:1 and 11.3:2 is obtained. Each dot on the photographs represents the level of force that is sampled every 2 sec. Figure 11.3:1 shows the muscle response when the length is shorter than the length for maximum activity. Figure 11.3:2 shows the muscle response at the length for maximum activity (upper) and the response at extremely stretched lengths (lower). The maximum tension (107 mN) for a long time after stretch is also marked for Fig. 11.3:2. The transient nature of these records arose because the muscle needed time to readjust to a change of length. To obtain the traces of these figures the muscle length change was accomplished in 100 sec. Then the new length was held constant for 1000 sec; and the force history was recorded, part of which is shown in these figures. The lower trace of Fig. 11.3:2 shows a record that began immediately after the length change.

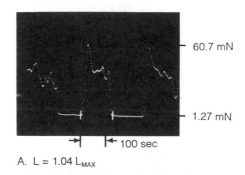

A. $L = 1.04 L_{MAX}$

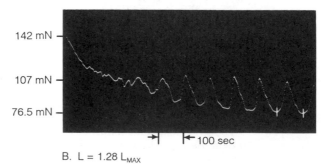

B. $L = 1.28 L_{MAX}$

Figure 11.3:2 Computer display of spontaneous contractions in real time for $L \geq L_{max}$. Same specimen as Fig. 11.3:1. (A) Contractions for $L = 1.04 L_{max}$. (B) Contractions for $L = 1.28 L_{max}$. From Price, Patitucci, and Fung (1977), by permission.

These figures tell us that the wave form of the tension in taenia coli in the isometric condition varies with the muscle length. The values of the tension (especially the maximum and minimum) depend on the muscle length, as can be seen from the numbers marked on these curves.

11.3.2 Response to Step Change in Length or Tension

The stress response of taenia coli to a step stretch in length is shown in Fig. 11.3:3. The ordinate, $G(t)$, is the ratio of the tension in the muscle at time t divided by the tension σ_{ss} immediately following the stretch, which ends at time t_{ss} (read the subscript "ss" as "step stretch"). The abscissa is time on a log scale. The response can be divided into two phases, the latent period for $t < 1$ sec and the minute rhythm for $t > 1$ sec. During the latent phase, the stress decreases monotonically to less than 40% of the peak stress (σ_{ss}) that occured at t_{ss}. At 1 sec after stretch, the minute rhythm begins. The minute rhythm is capable of decreasing the tension to a negligibly small value

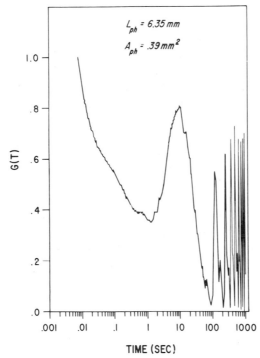

Figure 11.3:3 Stress response to a 10% of L_{ph} step change in length. The response $G(t)$ is the ratio $\sigma(t)/\sigma_{ss}$, where $\sigma(t)$ is the stress at time t following the stretch, and σ_{ss} is the stress at the end of ramp (10 ms after beginning of stretch). Time is displayed on a logarithmic scale. $\sigma_{ss} = 133$ kPa. For taenia coli of guinea pig. From Price, Patitucci, and Fung (1977), by permission.

for $t > 100$ sec. Gradually the minute rhythm tends to a steady state. In Fig. 11.3:3 there are eight cycles of contraction from 100 to 1000 sec, with approximately 112 sec per cycle.

On the other hand, if a taenia coli muscle is tested isotonically at a constant tension, it does not generate a large amplitude length oscillation. Instead, the length change appears as irregular oscillations of small amplitude with periods much less than 1 min. Thus, in the isotonic state, the minute rhythm is lost and the contractions are weaker and faster.

11.3.3 Active Tension

For a muscle at length L and cross-sectional area A, let the maximum stress in a cycle of spontaneous contraction be denoted by $\sigma(L)_{max}$ and the minimum stress by $\sigma(L)_{min}$. The stress, σ, is the sum of two parts: the active tension produced by the contractile elements of the smooth muscle cells and the passive stress due to stretching of the connective tissue in the muscle. Although it is difficult to evaluate these two parts exactly, an indication can be obtained. Following Fung (1970), we define the "active" stress $S(L, t)$ as the difference between $\sigma(L, t)$ and $\sigma(L)_{min}$; $\sigma(L, t)$ being the stress at length L and time t. Then the maximum active stress is given by $S(L)_{max} = \sigma(L)_{max} - \sigma(L)_{min}$.

If the values of $S(L)_{max}$ are plotted as a function of length, then an upper bound, S_{ub}, is reached at a certain length, L_{max}. This length is used as the reference length because it is characteristic of the contractile element, and it is easily measured. The corresponding reference cross-sectional area, A_{max}, is used to compute the stress on the specimen in the Lagrangian sense. The results of a typical experiment are shown in Fig. 11.3:4. This figure shows the functional dependence of $S(L)_{max}$, $\sigma(L)_{max}$, and $\sigma(L)_{min}$ on muscle length. In addition to the existence of an optimum length, L_{max}, for tension development, it shows an almost symmetrical decrease in active tension as the length varies in either direction from L_{max}. As the length increased from L_{max}, the values of $\sigma(L)_{min}$ increase from a relatively small value in a nonlinear fashion. For muscle lengths above approximately 110% of L_{max}, $\sigma(L)_{min}$ and $\sigma(L)_{max}$ increase very rapidly. If the length is decreased to about 20% below L_{max}, the value of $\sigma(L)_{min}$ falls to zero. The length range for active tension production is on the order of $\pm 50\%$ from L_{max}. Some statistical data for the five specimens of taenia coli of guinea pig tested over the entire range of muscle length are as follows. The mean value of $S(L)_{max}$ at L_{max} is 160.1 ± 8.38 (SE) kPa. The mean value of $\sigma(L)_{min}$ at L_{max} is 9.3 ± 3.36 (SE) kPa.

11.3.4 Summary

Spontaneous contraction occurs in taenia coli in the isometric condition when the environment is favorable. A step stretch in length produces a

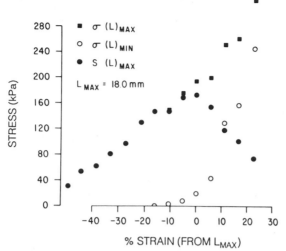

Figure 11.3:4 Dependence of $\sigma(L)_{max}$, $\sigma(L)_{min}$, and $S(L)_{max}$ on muscle length. L_{max} is defined as the length at which $S(L)_{max}$ is maximum. Strain from L_{max} is $(L - L_{max})/L_{max}$. Specimen dimensions are $L_{max} = 18$ mm, $A_{max} = 0.32$ mm^2, and $L_{ph} = 10$ mm. For taenia coli of guinea pig. From Price, Patitucci, and Fung (1977), by permission.

period of relaxation in which the muscle behaves as a resting tissue before the spontaneous contractions begin. The minute rhythm is abolished if the load becomes isotonic, which induces small amplitude oscillations with frequencies in the order of 10–20 per min. Other experiments have shown that a quick succession of stretches will reduce the tension in minute rhythm, and vibrations of appropriate frequency and amplitude will inhibit contraction tension.

The spontaneous maximum active tension in each cycle, $S(L)_{max}$, depends on the muscle length, L. There exists an optimal length, L_{max}, at which the tension $S(L)_{max}$ has an absolute maximum (upper bound). This feature is reminiscent of Figs 9.7:2 and 10.1:4 for the striated muscles.

11.4 The Property of a Resting Smooth Muscle: Ureter

In the striated muscle, it is usually assumed that the contractile element offers no resistance to elongation or shortening when the muscle is in the resting state. This is the basic reason for separating the muscle force into "passive" and "active" components, or, in the terminology of Hill's three-element model discussed in Chapters 9 and 10, the "parallel" and "series" elements and the "contractile" element. Usually implied in this definition is the uniqueness of the passive elements, that they have constitutive equations independent of the active state of the contractile element. This assumption might be acceptable for skeletal muscle, but is sometimes questioned for the heart (see Sec. 10.5).

For smooth muscles it seems to be entirely doubtful. For example, in the taenia coli smooth muscle, spontaneous contractions can be arrested by several methods, but different methods lead to somewhat different mechanical behavior of the muscle. Therefore the contractile element cannot be assumed to be freed up completely when the spontaneous contraction is arrested.

The lack of a unique resting state of a smooth muscle means that the constitutive equation may not be considered as the sum of two components: passive and active. One would have to experiment on the active muscle and deduce its constitutive equation directly. Nevertheless, it is still interesting to study the properties of smooth muscle in the resting states, not only because they are physiological, but also because they furnish a base upon which the active state can be better understood.

11.4.1 The Relative Magnitude of Active and Resting Stresses in Ureter

The relative importance of the resting and active forces in a smooth muscle may be illustrated by Fig. 11.4:1, from data obtained on the dog's ureter. The test specimen was stimulated electrically in an isometric condition at

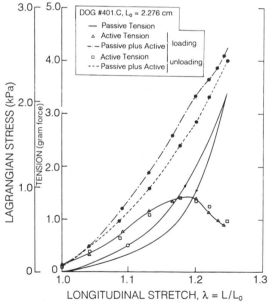

Figure 11.4:1 Graph showing typical relationships between total tension, resting tension, and active tension in the dog's ureter for both loading and unloading. Optimal stimulus: square-wave dc pulse, 9 V, 500 ms, 1 pulse/twitch. Temp.: 37°C. Stretching rate, 2% of L_0 min^{-1}. Releasing rate, same. For each twitch, time from stimulation to peak tension: 1.195 ± 0.097 sec, time of 90% relaxation from peak tension to 10% of peak: about 1.35 sec. Delay time between stimulation and onset of mechanical contraction 297 ± 105 ms. From Yin and Fung (1971).

various levels of stretch. When unstimulated and quiescent the ureter does not contract spontaneously, and hence is resting. A resting ureteral muscle is viscoelastic, hence it is necessary to distinguish loading (increasing stretch) and unloading (decreasing stretch) processes. It was found that each twitch requires one stimulus for most animals, except for the rabbit, for which two consecutive pulses are required to elicit the maximal response. The maximum tension or Lagrangian stress developed in a twitch at specific lengths of the ureter is plotted as the ordinate in Fig. 11.4:1. Solid lines refer to the resting state and broken lines refer to the stimulated state, whereas the triangular and square symbols represent the active tension (total tension minus resting tension). The open circle represents the *in situ* value of stretch. It is seen that the ureter developes the maximum active tension at such a muscle length that there is a considerable amount of resting tension. The magnitude of the maximum active tension is about equal to the resting tension at such a muscle length. This is in sharp contrast with skeletal muscle (see Sec. 9.1) and heart muscle (see Fig. 10.1:4), in which the resting stress is either insignificant or relatively minor compared with the maximum active tension in a normal physiological condition.

11.4.2 Cyclically Stretched Ureteral Smooth Muscle

The ureter is a thick-walled cylindrical tube. In the normal relaxed condition it contracts to such an extent that its lumen becomes practically zero. In the animal the lumen is increased with the passage of urine, and the stretching of the ureteral wall induces active response in the form of peristalsis. Electric events accompany the ureteral contraction, which in turn can be elicited by electric stimulation. In the pelvis of the kidney and in the ureter, there are pacemakers; but specimens that are segments taken from the ureter apparently do not have pacemakers strong enough to make them spontaneously contracting. Ureteral segment specimens remain resting until stimulated.

The mechanical properties of ureter in the longitudinal direction can be determined by using isolated segments of intact ureters subjected to simple elongation. Properties in the circumferential direction can be determined by using slit ring segments subjected to simple elongation. The test equipment, specimen preparation, preconditioning, experimental procedure, and bath composition are similar to those described in Sec. 10.2.

For simple elongation tests the specimen is first loaded at a given strain rate to some final stress level and then unloaded at the same rate. Figure 11.4:2 shows some typical results. The stress–strain curves are quite similar to those discussed in Chapters 7–10 for other soft tissues. If the rate of change of stress with respect to stretch is plotted against the stress, we obtain the results shown in Fig. 11.4:3. Over a range of stretch ratios this appears to be representable as a straight line, so that as in Sec. 7.5, we can write the relationship between the Lagrangian stress T (tension divided by the initial

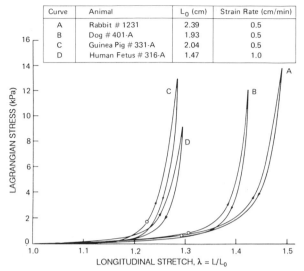

Curve	Animal	L_0 (cm)	Strain Rate (cm/min)
A	Rabbit # 1231	2.39	0.5
B	Dog # 401-A	1.93	0.5
C	Guinea Pig # 331-A	2.04	0.5
D	Human Fetus # 316-A	1.47	1.0

Figure 11.4:2 Typical ureteral stress–strain curves in loading and unloading for species studied. Open circles denote the *in situ* stretch value. Temp.: 22°C. From Yin and Fung (1971), by permission.

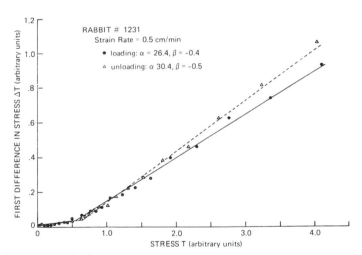

Figure 11.4:3 Sketch of typical results obtained from curve A of Fig. 11.4:2. Rabbit ureter. Solid and dotted lines are least-squares fit to the data on the loading and unloading processes, respectively. Two different straight line segments are used to fit the data in the lower (1.0–1.37) and higher (1.37–1.5) ranges of λ in each case. From Yin and Fung (1971), by permission.

cross-sectional area), and the stretch ratio, λ, by the equation

$$T = (T^* + \beta)e^{\alpha(\lambda - \lambda^*)} - \beta \qquad (\lambda_1 \le \lambda \le \lambda_2), \tag{1}$$

where α, β, T^*, and λ^* are experimentally determined constants. T^* and λ^* are a pair of specified stress and strain constants, and α and β are the elasticity parameters in the range of λ tested. α represents the slope of the curve of $dT/d\lambda$ vs. T; β represents (approximately) the intercept on the T axis, α is dimensionless, and β has the units of stress. In Fig. 11.4:3 the finite difference ΔT, instead of $dT/d\lambda$, is plotted against T. According to Eq. (1),

$$\Delta T_i = T_{i+1} - T_i = (e^{\alpha\Delta\lambda} - 1)(T_i + \beta), \tag{2}$$

where T_i represents the ith tabular value of T listed at equal intervals of λ, and $\Delta\lambda$ is the size of the λ interval chosen. It can be easily proved that the slope of the curve, ΔT_i vs. T_i, or $\tan\theta$, is given by

$$\tan\theta = e^{\alpha\Delta\lambda} - 1, \tag{3}$$

from which α can be easily calculated.

Tabulated results for α and β for the ureters of the dog, rabbit, guinea pig, and man are given in Yin and Fung (1971). It is shown for the rabbit ureter that the *in situ* value of λ falls into the range in which the data can be fitted by an exponential curve of the form of Eq. (1). The hysteresis, as revealed by the difference in the values of α for loading and unloading, is small. Paired statistical analysis of the results in the higher ranges of λ obtained by using the student t test ($P = 0.05$) showed the following: (a) the distal segments have a significantly higher value of α than the proximal segments; (b) varying the strain rate from 0.03 to 3 muscle lengths per min produces no significant differences in the elasticity parameters during either loading or unloading in dog's ureter; and (c) an obstructed-dilated ureteral segment has a significantly lower value of α than a normal segment.

The strain rate effect on α, however, depends on the animal. In the range of λ and $d\lambda/dt$ tested, α of the guinea pig ureter depends on the strain rate, but α of the dog ureter does not.

11.4.3 Stress Relaxation in Ureter

The relaxation test is done by stretching a specimen to a desired strain level, then holding the strain constant and measuring the change of force with time. Typical relaxation curves for circumferential segments of the dog's ureter are shown in Fig. 11.4:4, where the abscissa represents the time elapsed from the end of stretch, and the ordinate represents the ratio of the stress at time t to that at 0.1 sec after the end of stretch. Note how fast and thoroughly the tissue relaxed! At 10 sec, 50% to 65% of initial stress is lost; at 1000 sec (not shown in this figure) 70% to 80% of the stress is gone. Such thorough relaxation is not seen in other connective tissues such as the skin and mesentery.

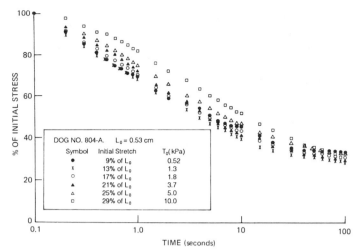

Figure 11.4:4 Stress relaxation of circumferential segments of a dog's ureter, showing dependence on amount of initial stretch. Temp.: 37°C. Initial strain rate 20 L_0/min. From Yin and Fung (1971), by permission.

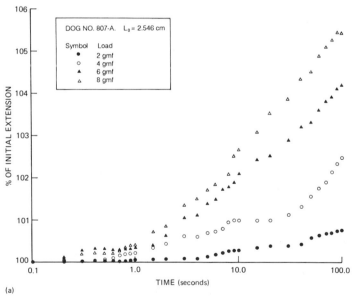

(a)

Figure 11.4:5 (a) Short-term creep results from a dog ureter specimen showing load dependence. Temp.: 37°C. From Yin and Fung (1971). (b) Long-term creep data from rabbit specimens. Temp.: 37°C. From Yin and Fung (1971), by permission.

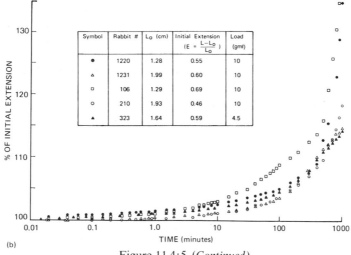

Symbol	Rabbit #	L_0 (cm)	Initial Extension $(E = \frac{L-L_0}{L_0})$	Load (gmf)
●	1220	1.28	0.55	10
△	1231	1.99	0.60	10
□	106	1.29	0.69	10
○	210	1.93	0.46	10
▲	323	1.64	0.59	4.5

(b)

Figure 11.4:5 (*Continued*)

11.4.4 Creep of Ureter

The counterpart of relaxation is creep, the continued elongation under a fixed load. Typical creep characteristics of the ureter are illustrated in Fig. 11.4:5. The creep rate depends clearly on the stress level.

Additional data on the resting and active properties of the pig ureter are given by Mastrigt (1985).

11.5 Active Contraction of Ureteral Segments

A ureter will respond to a suitable electric stimulation by a single twitch. During the twitch the length or tension of the ureter can be disturbed in various controlled manners to reveal the intrinsic behavior of the muscle. The methods discussed in Secs. 9.8 and 10.6 can be applied to the ureter. The results obtained from these experiments, the force-velocity relationship of a ureteral segment released from an isometric twitch, is discussed below. It will be seen that A. V. Hill's equation (Eq. (9.7:13))

$$\frac{S}{S_0} = \frac{1 - (v/v_0)}{1 + c(v/v_0)} \tag{1}$$

applies quite well. Here v is the velocity of contraction, S is the active tension in the muscle after release, S_0 is the active tension in the muscle immediately prior to release, v_0 is the velocity of the contraction if S were zero, and c is a dimensionless constant. The experimental results show that v_0 is the largest if the release took place early in the rise portion of the contraction cycle. Further, if tension is released from an isometric contraction at a fixed time in the rise portion of the twitch, the largest v_0 is obtained when the muscle length is in the range of $0.85–0.90\ L_{max}$, where L_{max} is the muscle length which yields the largest active tension in isometric contraction. Interestingly, the *in vivo* length of the ureter also lies in this range: $0.85–0.90\ L_{max}$.

Some details of the experiments reported by Zupkas and Fung (1985) are outlined below. Dog ureter from an anesthetized animal was excised and tested in a modified Krebs' solution using the "Biodyne" machine referred to by Pinto and Fung (1973), see p. 464. The tissue was preconditioned, and periodically stimulated electrically at an interval of 40 sec between pulses, a voltage of 30 V, and a duration of 0.50 sec. The difference between the maximum total tension in the twitch and the minimum (the "resting") total tension is defined as the "active" tension.

11.5.1 Single Twitch Characteristics

The "tension curve" of Fig. 11.5:1 shows the course of the tension development after stimulation when the length of the ureter was held constant. The "displacement curve" of Fig. 11.5:1 shows the course of shortening of the ureter after stimulation when the tension was held constant. It is seen that in the isometric case the tension development has a delay of about half a sec, then rises to a peak in about 3 sec, and then relaxes gradually to its original level of resting tension. In the isotonic case, the shortening has no latent

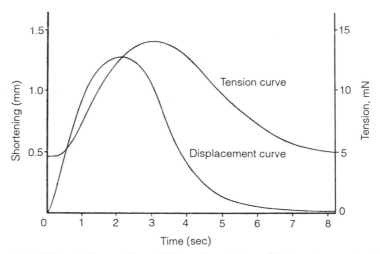

Figure 11.5:1 Typical isometric and isotomic twitches of dog ureter segments. The tension curve shows total tension, the minimum of which is the resting tension, or tension in the parallel element of the ureter, at a length chosen for the experiment. The difference between the total tension and the resting tension is the active tension, S. The displacement curve shows the shortening of the ureteral segment. From Zupkas and Fung (1985), reproduced by permission.

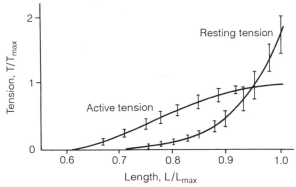

Figure 11.5:2 The length-tension relationship of the dog ureter. The active tension reaches the maximum, S_{max}, when the length of the ureter segment is L_{max}. Tension is normalized by dividing with S_{max}. Length is normalized by dividing with L_{max}. The magnitude of the resting tension exceeds S_{max} beyond the place where the two curves cross. From Zupkas and Fung (1985), by permission.

period, reaches the peak within 2 sec, then relaxes and completes the cycle in about 7 sec.

11.5.2 Length-Tension Relationship in Isometric Twitch

Figure 11.5:2 shows the active tension developed in isometric twitch when the length was varied. The active tension is the difference between the total tension (not shown) and the resting tension; and it reaches its maximum when the length is L_{max}. It is seen that the active tension in an isometric twitch increases almost linearly with increasing length until L/L_{max} is about 0.85, and that in this region the active tension is considerably larger than the resting tension. However, as the stretch in the ureter segment approaches L_{max}, the rate of increase of the maximum active tension slows, and reaches a peak value of S_{max}. No appreciable decrease in the level of active tension exists for stretches past L_{max}. The resting tension exceeds the active tension when $L/L_{max} > 0.95$.

11.5.3 Force-Velocity Relationships of Quick Release at Different Times During an Isometric Twitch

The quick release results shown in Fig. 11.5:3 are very similar to those shown in Secs. 9.8 and 10.6. Hill's equation (Eq. (1)) fits the data quite well. Typical values of the constants S_0, v_0, and c are listed in Table 11.5:1. The maximum velocity of shortening occurred at 75–100 ms after release. The greatest values of v were obtained for releases made in the rising portion of the contraction

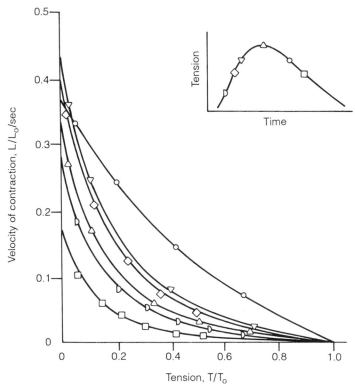

Figure 11.5:3 Force-velocity relationships obtained by quick release of tension during a twitch of a dog ureter specimen no. 11780. Each release was made at a certain time after stimulation as indicated by the various symbols. The active tension in the specimen was S_0 immediately before the release, and was a constant S immediately afterwards. The maximum velocity of contraction was obtained immediately after release; and was plotted vs. S/S_0. Curves are those of Hill's equation with constants listed in Table 11.5:1 for specimen no. 11780. From Zupkas and Fung (1985), by permission.

TABLE 11.5:1 Force-Velocity Data from Quick Release of Tension in Dog Ureter No. 11780, Preloaded 0.29 g to a Testing Length of 15.50 mm, Clamped, Stimulated, then Released at Various Times Shown in the Table

Experiment no.	Release time (sec)	T_0 (mN)	V_0 (L/L_0/sec)	c
011780	0.80	5.3	0.370	1.07
	1.19	12.3	0.440	5.78
	1.42	12.9	0.450	6.16
	2.62	13.3	0.360	8.19
	4.00	11.3	0.280	8.02
	4.80	10.8	0.170	9.47

cycle, at times varying from 37% to 60% of the time to peak isometric tension at different ureter segment lengths. Releases made before or after this period attain reduced levels of v_0. In particular, v_0 was lower for release made exactly at the instant of time of the peak tension development than for releases made in the rise portion of the twitch.

11.5.4 Force-Velocity Relationship as a Function of Muscle Length or Preload

The ability of a ureter to contract can be measured by two quantities: the maximum achievable active tension and the maximum achievable velocity of contraction. Figures 11.5:1 and 11.5:2 tell us that the active tension is a function of two variables: the length of the muscle and the time after stimulation; the maximum is obtained when the length is L_{max} and the time is that of the peak twitch. Additional data in Zupkas and Fung (1985) show that L_{max} is, on the average, 1.147 ± 0.102 times longer than the *in situ* length of the ureter. The velocity of contraction achieved by quick release of tension in an isometric twitch should also be a function of two variables: the length of the muscle being stimulated isometrically and the time after stimulation. The results reported in Fig. 11.5:3 tell us that at fixed length the time of release for higher velocity of contraction lies in the rise portion of the twitch. Hence by fixing a time of release after stimulation in the rise portion of the twitch we can examine the effect of muscle length on the velocity of contraction. Our results are shown in Fig. 11.5:4. The length, or the stretch level of the ureter before release, is shown in the inset of Fig. 11.5:4 by its relative position on a typical length-tension relationship. The maximum velocity, v_0, is listed in Table 11.5:2 as a function of the ureteral segment length, L_0, at the instant of release. The values of v_0 reached a peak for L/L_{max} in the range of 0.85 to 0.90. The correlation coefficient between v_0 and L_0/L_{max} was 0.75 for $L_0 < 0.875L_{max}$, and -0.85 for $L_0 > 0.875L_{max}$. The *in vivo* length of the ureter segments was also in this range: $L = 0.85L_{max}$ to $0.90L_{max}$.

Summarizing, we see that the active contraction of the ureter has features in common with those of the heart muscle. Both respond to a pacemaker. Both can be represented by A. V. Hill's three-element model under suitable interpretation. The total tension is the sum of the tension in the parallel element, P, and that in the series element, S. Analysis of the active tension in the contractile element and the velocity of contraction upon release to a lower tension on the basis of Hill's model shows that the Hill's equation, modified and interpreted as in Sec. 10.6, fits the data pretty well, provided that the parallel element is taken to be elastic. However, at the muscle length for maximum active tension generation, the tension in the parallel element of ureter far exceeds the active tension, whereas the parallel element tension of the heart muscle is far less than the muscle active tension. The skeletal muscle has negligible parallel element tensions at lengths yielding the maximum active tension.

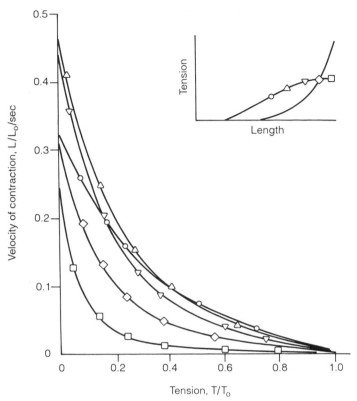

Figure 11.5:4 Force-velocity relations of a dog ureter specimen no. 121379 released at a given time after stimulation. The specimen was first stretched to various lengths marked by the symbols indicated on the curve shown in the inset at the upper right corner, then stimulated and released. The measured velocity of shortening was plotted vs. the tension ratio S/S_0, i.e., released tension/tension before release. Solid curves are Hill's equations with constants listed in Table 11.5:1 for specimen 121379. From Zupkas and Fung (1985), by permission.

TABLE 11.5:2 Force-Velocity Data from Quick Release of Tension in Dog Ureter No. 121379, Preloaded (i.e., stretched) to Various Length, Clamped, Stimulated, then Released (at 1.58 sec after stimulation), from S_0 to S

Testing length (mm)	L_0/L_{max}	Preload (mN)	T_0 (mN)	V_0 (L/L_0/sec)	c
14.3	0.773	2.4	7.2	0.310	2.20
15.2	0.822	4.9	12.2	0.460	4.57
16.5	0.892	6.2	13.7	0.440	5.78
17.6	0.951	10.9	20.0	0.300	7.57
18.3	0.990	16.5	28.1	0.220	23.57

These data are in agreement with those of Yin and Fung (1971) on a dog ureter and Weiss et al. (1972) on a cat ureter.

11.6 Resting Smooth Muscle: Taenia Coli

Taenia coli muscle of the intestine differs from the ureter in that normally it contracts spontaneously. To test taenia coli in a resting state it is necessary to suppress its spontaneous activity. The suppressed state depends on the method of suppression, and is not unique.

Three convenient methods can be used to suppress the spontaneous activity in taenia coli: (a) use calcium-free EGTA solution; (b) use epinephrine; and (c) lowering the temperature to less than 20°C. The first two methods consist of a change in the bathing fluid for the muscle. In normal experiments, the specimen is bathed in a physiological solution that has the following composition (mM): NaCl, 122; KCl, 4.7; $CaCl_2$, 2.5; $MgCl_2$, 1.2; KH_2PO_4, 1.2; $NaHCO_3$, 15.5; and glucose, 11.5; bubbled with O_2 (95%) and CO_2 (5%). The calcium-free EGTA solution is obtained by replacing $CaCl_2$ with 2 mM EGTA (disodium ethylene glycol-bis-(β-aminoethylether)-N, N'-tetraacetic acid). The epinephrine treated solution is obtained by injecting adrenalin chloride solution directly into the bath to obtain an initial concentration of 0.1 mg/ml.

These methods have different ways of acting on the actin-myosin coupling. Removal of calcium ion can depolarize the cell membrane, and this action is temprature dependent. Bülbring and Kuriyama (1963) have shown that 20°C is a critical temperature for spontaneous electric activity in taenia coli. They also showed that inactivation of taenia coli by epinephrine is associated with an increased membrane potential and a block of action potential; epinephrine increases the amount of creatine phosphate and ATP, and the utilization of these substances in hyperpolarization of the cell membrane.

11.6.1 Relaxation After a Step Stretch

We have already described the step-stretch test in Sec. 11.3 to study the spontaneous contraction of taenia coli. We now use the same method to study the resting state of taenia coli. The tissue was stretched at a constant strain rate from time $t = 0$ to time $t_{ss} = 10$ ms (the subscript "ss" stands for the time to "step stretch"), and the length was maintained constant thereafter. The stress at t_{ss} is denoted as σ_{ss}. The stress $\sigma(t)$ at time $t > t_{ss}$ divided by σ_{ss} is defined as the *normalized relaxation function*, $G(t)$.

On a spontaneously contracting specimen, the step-stretch test begins at the end of a contraction cycle when the tension was zero. $G(t)$ for various degrees of stretch of a spontaneously contracting specimen is shown in Fig. 11.6:1. A latent period occurs immediately following stretch, in which the

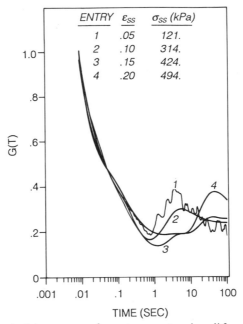

Figure 11.6:1 Step-stretch response of spontaneous taenia coli for various amounts of stretch. Reference dimensions: $L_0 = 6.17$ mm and $A_0 = 0.198$ mm². From Price, Patitucci, and Fung (1979), by permission.

response is a monotonically decreasing function of time. During the latent period the membrane action potential is absent. The latent period ends approximately 1 sec after the initiation of stretch due to the resumption of membrane electric activity and the onset of contraction. Although the contractile response varies with the amount of stretch, $G(t)$ in the latent period appears to be independent of stretch. The delay time, or the time to the onset of contraction, increases with increasing stretch. The strain ε_{ss} in Fig. 11.6:1 is defined as the change of length divided by L_0, the longest length of the specimen under a preload of 1 mN, while the muscle contracted spontaneously.

When the bath was replaced by the calcium-free EGTA solution, the normalized relaxation functions became those shown in Fig. 11.6:2. The step size for all the curves in this figure was 10% of the initial length L_0 (arbitrarily defined as the length of the tissue in the bath under a load of 1 mN). The relaxation function $G(t)$ is now seen to be monotonically decreasing, with no resumption of spontaneous contraction. The effect of decreasing temperature after calcium removal is shown by the curves 3 and 4 in the figure. It is seen that removal of calcium ions suppresses spontaneous activity, increases the stiffness of the taenia coli (as reflected in the higher values of σ_{ss}), and decreases the rate of relaxation.

When the temperature is lowered from 37°C, the stiffness and the relaxation rate both decrease. If temperature is lowered while the specimen is

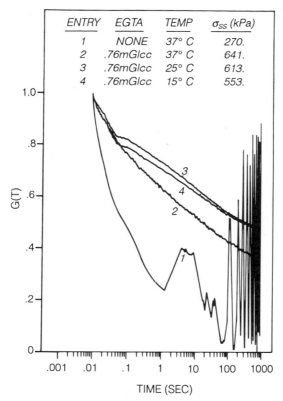

ENTRY	EGTA	TEMP	σ_{ss} (kPa)
1	NONE	37° C	270.
2	.76mGlcc	37° C	641.
3	.76mGlcc	25° C	613.
4	.76mGlcc	15° C	553.

TIME (SEC)

Figure 11.6:2 Step-stretch response of guinea pig taenia coli before (curve No. 1) and after (curves No. 2, 3, 4) calcium removal. Each response is for a stretch of 10% L_0. Entries 3 and 4 show effect of lower temperature after calcium removal. Reference dimensions for entry 1, $L_0 = 7.35$ mm and $A_0 = 0.34$ mm^2; for entry 2, $L_0 = 8.10$ mm and $A_0 = 0.31$ mm^2; for entry 3, $L_0 = 9.19$ mm and $A_0 = 0.27$ mm^2; and for entry 4, $L_0 = 9.25$ mm and $A_0 = 0.27$ mm^2. From Price, Patitucci, and Fung (1979), by permission.

bathed in normal physiological saline, then σ_{ss} decreases. For $\varepsilon_{ss} - 0.10$ (a 10% stretch), σ_{ss} at 37, 25, and 15°C is 133, 136, and 59 kPa, respectively. Spontaneous contractions remain at 25°C, but are abolished at 15°C.

The response before and after injection of epinephrine into the bath is shown in Fig. 11.6:3. Curve no. 5 refers to a spontaneously contracting specimen subjected to a step stretch of 10% of L_0 in a bath without epinephrine injection. Entries 1 through 4 refer to specimens in a bath after the injection of epinephrine. At the same strain step ε_{ss}, the values of σ_{ss} are lower after epinephrine injection, whereas the relaxation function $G(t)$ continues to be monotonically decreasing. Note that in this case $G(t)$ becomes less than 5% at 100 sec. It seems that at large time the stress will be relaxed to almost zero. Such a behavior is seen in ureteral smooth muscle (see Fig. 11.4:4), but is not seen in other tissues such as the arteries, veins, skin, mesentery, and striated muscles.

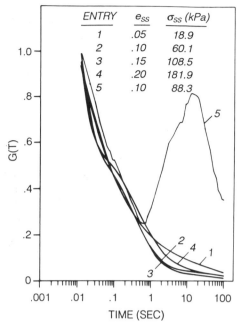

Figure 11.6:3 Step-stretch respone before and after treatment with epinephrine. Entry 5 is the response to a stretch of 10% L_0 with no epinephrine. Entries 1–4 are responses to step stretch in 0.1 mg ml epinephrine. Reference dimensions for entry 5, $L_0 = 4.75$ mm and $A_0 = 0.338$ mm²; for entries 1–4, $L_0 = 5.65$ mm and $A_0 = 0.283$ mm². Taenia coli of guinea pig. From Price, Patitucci, and Fung (1979), by permission.

Response of a tissue to a step change in length may be represented by the simplified formula

$$K(\varepsilon, t) = G(t)T^{(e)}(\varepsilon),\qquad(1)$$

as discussed in Secs. 7.6, 8.3, and 10.4. Here $G(t)$ is the normalized relaxation function, which is assumed to be a function of time, and is independent of the strain. $T^{(e)}(\varepsilon)$ is a function of strain alone. This assumption is supported by the experimental data shown in Fig. 11.6:3. Resting taenia coli seems to obey such a relation. The mathematical form for $G(t)$ has been discussed in Sec. 7.6. Although curves as shown in Fig. 11.6:3 can be represented by a sum of a few exponential functions, the simultaneous requirement of the rate-insensitivity feature shown in Fig. 11.6:4 for cyclic loading suggests the use of a continuous spectrum. The form proposed by the author (Fung, 1972) is

$$G(t) = \frac{1}{A}\left[1 + c\int_{\tau_1}^{\tau_2}\frac{1}{\tau}e^{-(t/\tau)}d\tau\right],\qquad(2)$$

where c, τ_1, and τ_2 are constants and A is the value of the quantity in the brackets at $t = 0$:

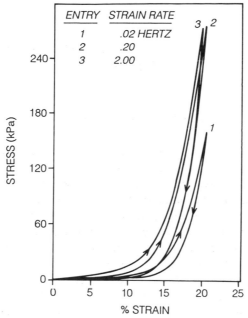

Figure 11.6:4 Stress–strain relationship of taenia coli in epinephrine solution at various strain rates. Peak strain is 0.205 for each test. Reference dimensions are $L_{max} = 12.15$ mm and $A_{max} = 0.493$ mm^2. From Price, Patitucci, and Fung (1979), by permission.

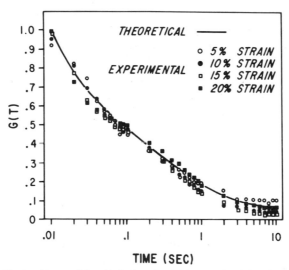

Figure 11.6:5 Comparison of theoretical and experimental values of $G(t)$ of taenia coli smooth muscle in epinephrine treated solution. Specimen is the same as in Fig. 11.5:3. Parameters defining theoretical $G(t)$ are $c = 2.25$, $\tau_1 = 0.003$ sec, and $\tau_2 = 3.38$ sec. From Price, Patitucci, and Fung (1979), by permission.

$$A = 1 + c \int_{\tau_1}^{\tau_2} \frac{1}{\tau} d\tau = 1 + c \log_e \left(\frac{\tau_2}{\tau_1} \right). \tag{2a}$$

A comparison of the values of $G(t)$ predicted by Eq. (2) with those by experiments is shown in Fig. 11.6:5, and is seen to be satisfactory.

11.6.2 Cyclic Loading and Unloading Tests

The specimen is stretched at a constant rate (dL/dt) to a specified length and immediately reversed at the same strain rate, resulting in a cyclic triangular strain history. The reslting stress–strain relationship of taenia coli in epinephrine treated solution is illustrated in Fig. 11.6:4. Here the stress is Lagrangian (force divided by initial cross-sectional area), and the strain is $\Delta L/L_0$, L_0 being the length of muscle in epinephrine at 1 mN preload. The features shown in Fig. 11.6:4 are similar to those of other soft tissues discussed in Chapter 7. The stress is a nonlinear function of strain and depends on the strain rate. The relationship is different in loading from that in the unloading process. At higher strain rates (above 0.2 Hz), the loops are almost independent of the frequency. This last mentioned property is decisive in selecting the relaxation spectrum presented in Eq. (2).

Information on the "elastic response" $T^{(e)}(\varepsilon)$ can be extracted from experimental results as shown in Fig. 11.6:4. Assuming that at sufficiently high rates of stretching the stress–strain relationship is essentially independent of the strain rate, then the stress corresponding to the strain on the loading curve (increasing strain) can be considered to approximate $T^{(e)}(\varepsilon)$. Under such an assumption our problem is to find a mathematical expression for $T^{(e)}(\varepsilon)$. A replot of the loading curves shown in Fig. 11.6:4 on a log–log scale yields the results shown in Fig. 11.6:6. Each curve in Fig. 11.6:6 can be represented by two straight line segments. Because a straight line on a log–log plot represents a power law, we see that the data in Fig. 11.6:6 can be expressed by the equations

$$T = \beta \varepsilon^\alpha \qquad \text{for} \quad 0 < \varepsilon < \varepsilon^*, \tag{3}$$

and

$$T = \beta' \varepsilon^{\alpha'} \qquad \text{for} \quad \varepsilon > \varepsilon^*, \tag{4}$$

where α and α' are the respective slopes of the log–log plot, and β and β' are constants determined by a known point on the curve. Let the intersection of the two straight lines be the point $T = T^*$, $\varepsilon = \varepsilon^*$, then

$$\beta = T^*/(\varepsilon^*)^\alpha, \qquad \beta' = T^*/(\varepsilon^*)^{\alpha'}. \tag{5}$$

Experimental values of α for guinea pig taenia coli range from 1.77 to 5.21, which may be compared with the values of α found for the aorta, i.e., 1.23–3.05 (Tanaka and Fung, 1974).

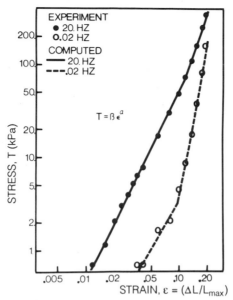

Figure 11.6:6 Log–log graph of stress vs. increasing strain for epinephrine treated taenia coli specimens from guinea pigs. Experimental points are shown for the same specimen as in Fig. 11.5:5. The computed curves are based on Eqs. (3) and (4) with the following constants and standard errors: At 0.02 Hz, $\alpha = 1.77 \pm 0.213$, $\beta = 0.193 \pm 0.043$, $\alpha' = 5.21 \pm 0.29$, $\beta' = 534.0 \pm 28.0$, $\varepsilon^* = 0.1006$, $T^* = 0.0033$. At 20 Hz, $\alpha = 2.04 \pm 0.041$, $\beta = 5.31 \pm 0.389$, $\alpha' = 2.80 \pm 0.108$, $\beta' = 27.50 \pm 0.53$, $\varepsilon^* = 0.117$, $T^* = 0.065$. The units of β, β', and T^* are MPa. From Price, Patitucci, and Fung (1979), by permission.

11.6.3 Relative Magnitude of Active and Passive Stresses in Taenia Coli

If the tension in an epinephrine treated resting taenia coli at various lengths is compared with the minimum tension of spontaneously contracting muscle at the same lengths, it is found that they are approximately equal. This is shown in Fig. 11.6:7. If the resting tension is compared with the active tension shown in Fig. 11.3:4, it is seen that the resting tension is not large at L_{max}, the length at which the maximum active tension is generated. Unlike the ureter, taenia coli operates normally in a range of length in which the resting tension is negligible compared with the active tension.

11.6.4 Summary

The main lesson we have learned from the study of resting smooth muscles is that the passive state may depend on the method by which the spontaneous

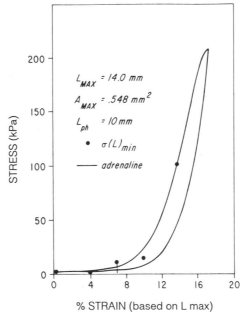

% STRAIN (based on L max)

Figure 11.6:7 Comparison of stress in epinephrine relaxed muscle and $\sigma(L)_{min}$. Length changes are shown as percent strain from L_{max}. A length-tension test was first performed on a spontaneous specimen; then a stress–strain test (0.01 Hz) was done after the addition of epinephrine (0.1 mg/ml). Taenia coli of guinea pig. From Price, Patitucci, and Fung (1979), by permission.

activity is suppressed. This implies that the contractile mechanism is not entirely freed up when the electric activity is arrested. This is the case with taenia coli, but not with ureter. The point is, however, that the conventional concept of separating the mechanical action of a muscle to "parallel", "series", and "contractile" elements (see Sec. 9.8) may fail with smooth muscles, because the parallel element is inseparable from the contractile element in the resting condition.

Of the properties of resting ureter and taenia coli, the most remarkable is the thorough stress relaxation under a constant strain. After a long time following a step change in length, the stress may relax to almost zero. This implies that the geometry of these organs whose structure is dominated by smooth muscles could be quite plastic in its behavior, and moldable by environmental forces and constraints.

Although the constitutive equation that mathematically describes the contraction process of smooth muscles is still unknown, a vast amount of information has been accumulated with respect to the fine structure of the muscles, as well as their electrical activity, metabolic characteristics, pharmacological responses, innervation, growth, and proliferation. The reader may be referred to the books by Huddart (1975), Huddart and Hunt (1975), Wolf and Werthessen (1973), and Aidley (1971), as well as those mentioned in Secs. 11.1 and 11.2. The recent volume of *Handbook of Physiology*, Sec. 2, Vol. 2, *Vascular Smooth Muscle* (ed. by Bohr et al., 1980) contains a wealth of material and should be consulted.

11.7 Other Smooth Muscle Organs

In quick release experiments on the portal vein segments, Hellstrand and Johannson (1975) found a peak plateau of velocity of contraction late in the rise portion of the contrction cycle, similar to the ureteral behavior.

The mechanical properties of the urinary bladder have been studied by van Mastrigt (1979), Uvelius (1979), and Ekstrom and Uvelius (1981).

The intestinal smooth muscles have been studied exhaustively from the point of view of nervous control and phamacology. For the mechanical aspects, see Aberg and Axelsson (1965).

The vascular smooth muscle has a huge literature (see Chapter 8), but a constitutive equation describing its mechanical property does not exist. Mulvaney (1979), Murphy et al. (1974), and Peiper et al. (1975) have presented useful data.

Problems

11.1 Describe the similarities and differences between the skeletal muscle, heart muscle, and smooth muscles.

11.2 Write down and explain Hill's equation for muscle contraction. Explain the meaning of the symbols, under what conditions is the equation derived, under what conditions does it fail? How could it be applicable to smooth muscles? With what kind of modification? To which smooth muscle has it been shown to apply?

11.3 Consider an approximate theory of ureteral peristalsis. The urine is a Newtonian fluid and the Reynolds number of flow is much less than one. The ureteral wall is incompressible; its mechanical properties in the resting state have been described in Sec. 11.4. By a wave of contraction of the ureteral muscle the urine is sent through the ureter from the kidney to bladder. Under normal conditions urine is moved one bolus at a time; the muscle contraction is strong enough to completely close the lumen of the ureter at the ends of the bolus. In the diseased condition of hydroureter, the ureter is dilated, and the force of contraction is not enough to close the lumen.

Interaction of fluid pressure and muscle tension must be considered. The course of contraction is quite similar to a single twitch of the heart papillary muscle, hence, as an approximation we may use the equations presented in Sec. 10.4 to describe the active ureteral muscle contraction. Assume that the fluid bolus is axisymmetric and is so slender that its diameter is much smaller than its length. Let the geometry of the fluid bolus and the ureter be as shown in Fig. P11.3. Use polar coordinates (r, θ, x) with the x axis coinciding with the axis of the ureter, write down the equation of equilibrium of the tube wall (the Laplace equation given

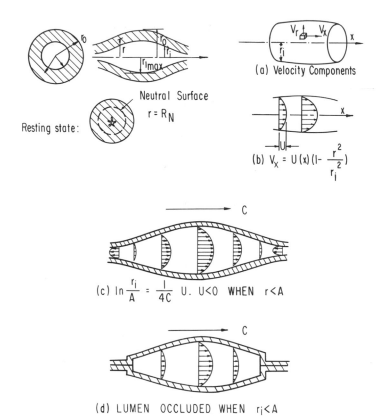

(a) Velocity Components

Resting state:

Neutral Surface
$r = R_N$

(b) $V_x = U(x)(1- \dfrac{r^2}{r_i^2})$

c

(c) $\ln \dfrac{r_i}{A} = \dfrac{1}{4C} U$. $U < 0$ WHEN $r < A$

c

(d) LUMEN OCCLUDED WHEN $r_i < A$

Figure P11.3(a) and (b): Notations and coordinates for the analysis of ureteral peristalsis. (c) and (d): Fluid velocity distribution as demanded by the equation of continuity.

in Sec. 1.9 of Chapter 1, p. 14), the equation of motion of the fluid, the equation of continuity for the conservation of mass, and the appropriate boundary conditions. Solve the equations to obtain the velocity profile, and the radial velocity at the wall. Let $U(x, t)$ be the axial velocity on the centerline and r_i the radius of the inner wall. The velocity components are shown in Figs. P11.3(a) and (b). Show that

$$\frac{\partial r_i(x, t)}{\partial t} = -\frac{r_i}{4} \frac{\partial U(x, t)}{\partial x}. \tag{1}$$

In a steady peristaltic motion for which the whole pattern moves in the x direction at a constant velocity, c, r_i, and U are functions of the single variable $x - ct = \zeta$, and Eq. (1) can be integrated to

$$\log \frac{r_i(\zeta)}{A} = \frac{1}{4c} U(\zeta) \tag{2}$$

with an integration constant A. Thus the velocity U is positive (agreeing with the direction of propagation of the peristaltic wave) when $r_i > A$; it is negative when $r_i < A$. Backward flow (U negative) occurs if the tube is open with a radius less than A; it does not occur if the ureter is closed both at the front and rear as shown in Figs. P11.3(c) and (d).

To determine the shape of the bolus we must consider the action of the muscle. Formulate the mathematical problem and work out the details (cf. Fung, Y. C., 1971a and 1971b, and Jaffrin and Shapiro, 1971).

References

Aaberg, A. K. G. and Axelsson, J. (1965) Some mechanical aspects of intestinal smooth muscle. *Acta Physiol. Scand.* **64**, 15–27.

Aidley, C. J. (1971) *The Physiology of Excitable Cells.* Cambridge University Press, Cambridge, U. K.

Ashton, F. T., Somlyon, A. V., and Somlyo, A. P. (1975) The contractile apparatus of vascular smooth muscle: Intermediate high-voltage sterco electron microscopy. *J. Mol. Biol.* **98**, 17–29.

Bayliss, W. M. and Starling, E. H. (1899) The movements and innervation of the small intestine. *J. Physiol.* (*London*) **24**, 99–143.

Bohr, D. F., Somlyo, A. P., and Sparks, H. V., Jr. (1980) *Handbook of Physiology*, Sec. 2, *Cardiovascular System*, Vol. 2., *Vascular Smooth Muscle.* American Physiological Society, Bethesda, MD.

Bülbring, E. and Kuriyama, H. (1963) Effects of changes in the external sodium and calcium concentration on spontaneous electrical activity in smooth muscle of guinea pig taenia coli. Also, adrenaline in relation to the degree of stretch. *J. Physiol.* (London) **166**, 29–58; **169**, 198–212.

Bülbring, E., Brading, A. F., Jones, A. W., and Tomita, T. (eds.) (1970) *Smooth Muscle.* Arnold, London.

Burnstock, G. and Prosser, C. L. (1960) Responses of smooth muscles to quick stretch; relation of stretch to conduction. *Am. J. Physiol.* **198**, 921–925.

Burnstock, G. (1970) Structure of smooth muscle and its innervation. In *Smooth Muscle*, Bülbring, E. et al. (eds.) Arnold, London, Chap. 1, pp. 1–69.

Cox, R. H. (1975–1978) Arterial wall mechanics and composition and the effects of smooth muscle activation. *Am. J. Physiol.* **229**, 807–812 (1975); **230**, 462–470 (1976); **231**, 420–425 (1976); **233**, H248–H255 (1977); **234**, H280–H288 (1978).

Dobrin, P. B. (1973) Influence of initial length on length-tension relationship of vascular smooth muscle. *Am. J. Physiol.* **225**, 664–670.

Ekstrom, J. and Uvelius, B. (1981) Length-tension relations of smooth muscle from normal and denervated rat urinary bladders. *Acta Physiol. Scand.* **112**, 443–447.

Engelmann, T. W. (1869) *Pflügers Arch.* **2**, 664–670.

Fung, Y. C. (1971a) Muscle controlled flow. In *Developments in Mechanics, Proc. 12th Midwest Mechanics Conference*, pp. 33–62. University of Notre Dame Press, South Bend, IN.

Fung, Y. C. (1971b) Peristaltic pumping: A bioengineering model. In *Urodynamics: Hydrodynamics of the Ureter and Renal Pelvis*, S. Boyarsky, C. W. Gottschalk, E. A. Tanago, and P. D. Zimsking (eds.), pp. 177–198. Academic Press, New York.

Golenhofen, K. (1964) "Resonance" in the tension response of smooth muscle of guinea-pig's taenia coli to rhythmic stretch. *J. Physiol.* (London) **173**, 13–15.

Golenhofen, K. (1970) Slow rhythms in smooth muscle (minute-rhythm). In *Smooth Muscle* (ed. by Bülbring, E. et al.). Arnold, London. pp. 316–342.

Gordon, A. R. and Siegman, M. H. (1971) Mechanical properties of smooth muscle. I. Length-tension and force-velocity relations. *Am. J. Physiol.* **221**, 1243–1254.

Guyton, A. C. (1976) *Textbook of Medical Physiology*. W. B. Saunders, Philadelphia.

Hellstrand, P. and Johansson, B. (1975) The force-velocity relation in phasic contractions of venous smooth muscle. *Acta Physiol. Scand.* **93**, 157–166.

Hill, A. V. (1938). Proc. Roy. Soc. London, Ser. B **126**, 136–195.

Huddart, H. (1975) *The Comparative Structure and Function of Muscle*. Pergamon, New York.

Huddart, H. and Hunt, S. (1975). *Visceral Muscle. Its Structure and Function*. Blackie, Glasgow.

Jaffrin, M. Y. and Shapiro, A. H. (1971) Peristaltic pumping. *Annual Rev. Fluid Mech.* **3**, 13–36.

Johnson, P. C. (ed.) (1978) *Peripheral Circulation*. Wiley, New York.

Kurihara, S., Huriyama, H., and Magaribuchi, T. (1974) Effect of rapid cooling on the electric properties of the smooth muscle of the guinea-pig urinary bladder. *J. Physiol.* (London) **238**, 413–426.

Lowy, J. and Mulvaney, M. J. (1973) Mechanical properties of guinea pig taenia coli muscles. *Acta Physiol. Scand.* **88**, 123–136.

Mastrigt, R. van (1979) Contractility of the urinary bladder. *Urol. Int.* **34**, 410–420.

Mastrigt, R. van (1985) Passive properties of the smooth muscle of the pig ureter. In *Urodynamics*, W. Lutzeyer and J. Hannappel (eds.), pp. 1–12. Springer-Verlag, Berlin.

Mastrigt, R. van. (1985) The propagation velocity of contractions of the pig ureter in vitro. In *Urodynamics*, W. Lutzeyer and J. Hannappel (eds.), pp. 126–128. Springer-Verlag, Berlin.

Merrillees, N. C. R., Burnstock, G., and Holman, M. E. (1963) Correlation of fine structure and physiology of the innervation of smooth muscle in the guinea pig vas deferens. *J. Cell Biol.* **19**, 529–550.

Mulvaney, M. J. (1979) The active length-tension curve of vascular smooth muscle related to its cellular components. *J. Gen. Physiol.* **74**, 85–104.

Murphy, R. A., Herlihy, J. T., and Megerman, J. (1974) Force generating capacity of arterial smooth muscle. *J. Gen. Physiol.* **64**, 691–705.

Peiper, U., Laven, R., and Ehl, M. (1975) Force-velocity relationships in vascular smooth muscle. The influence of temperature. *Pflügers Arch. Eur.* **356**, 33–45.

Price, J. M., Patilucci, P., and Fung, Y. C. (1977) Mechanical properties of taenia coli smooth muscle in spontaneous contraction. *Am. J. Physiol.* **233**, C47–C55.

Price, J. M., Patitucci, P., and Fung, Y. C. (1979) Mechanical properties of resting taenia coli smooth muscle. *Am. J. Physiol.* **236**, C211–C220.

Price, J. M. and Davis, D. L. (1981) Contractility and the length-tension relation of the dog anterior tibial artery. *Blood Vessels* **18**, 75–88.

Siegman, M. J., Butler, T. M., Moores, S. U., and Davies, R. E. (1976) *Am. J. Physiol.* **231**, 1501–1508.

Tanaka, T. T. and Fung, Y. C. (1974) Elastic and inelastic properties of the canine aorta and their variation along the aortic tree. *J. Biomech.* **7**, 357–370.

Uvelius, B. (1979) Shortening velocity, active force, and homogeneity of contraction during electrically evoked twitches in smooth muscles from rabbit urinary bladders. *Acta Physiol. Scand.* **106**, 481–486.

Weiss, R., Basset, A., and Hoffman, B. F. (1972) Dynamic length-tension curves of cat ureter. *Am. J. Physiol.* **222**, 388–393.

Wolf, S. and Werthessen, N. T. (1973) *The Smooth Muscle of the Artery.* Plenum, New York.

Yin, F. C. P. and Fung, Y. C. (1971) Mechanical properties of isolated mammalian ureteral segments. *Am. J. Physiol.* **221**, 1484–1493.

Zupkas, P. F. and Fung, Y. C. (1985) Active contractions of ureteral segments. *J. Biomech. Eng.* **107**, 62–67.

CHAPTER 12

Bone and Cartilage

12.1 Introduction

Bone works in the small strain range; yet its biology is very sensitive to the strain level. Its constitutive equation is linear with respect to the strain, and the strain-displacement relationship is also linear; but the relationship is anisotropic. In this chapter the mechanical properties of bone are described with an emphasis on biology.

Cartilage is related to bone. Bone is calcified cartilage. The articular cartilage has a unique quality of having a very small coefficient of friction for relative motion between two pieces of cartilage. In arthroidal joints, cartilage has a unique superior quality of lubrication and shock absorption. These qualities are due largely to the multiphasic structure of the cartilage. The structure is a composite of fluids, ions, and solids. Biological tissues are all multiphasic: and articular cartilage has been studied more thoroughly. This gives us an opportunity to learn about Mow, Lai, and Hou's triphasic theory of such tissues.

The presentation below is aimed at the basic features of the mechanics of bone and cartilage. References are selected from a very extensive literature. As an introduction, a sketch of bone anatomy and material composition is given.

12.1.1 The Anatomy of a Long Bone

Figure 12.1:1 shows a sketch of a long bone. It consists of a shaft (diaphysis) with an expansion (metaphysis) at each end. In an immature animal, each metaphysis is surmounted by an epiphysis, which is united to its metaphysis

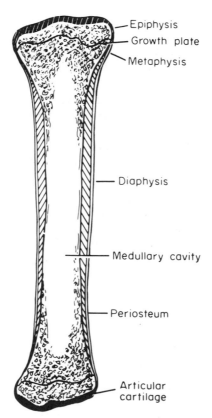

Epiphysis
Growth plate
Metaphysis

Diaphysis

Medullary cavity

Periosteum

Articular cartilage

Figure 12.1:1 The parts of a long bone. From Rhinelander (1972). Reproduced by permission.

by a cartilaginous growth plate (epiphyseal plate). At the extremity of each epiphysis, a specialized covering of articular cartilage forms the gliding surface of the joint (articulation). The coefficient of dry friction between the articulate cartilages of a joint is very low (can be as low as 0.0026, see Sec. 12.9, probably the lowest of any known solid material); hence the cartilage covering makes an efficient joint.

The growth plate, as its name indicates, is the place where calcification of cartilage takes place. At the cessation of growth, the epiphyses, composed of cancellous bone, become fused with the adjacent metaphyses. The outer shell of the metaphyses and epiphyses is a thin layer of cortical bone continuous with the compactum of the diaphysis.

The diaphysis is a hollow tube. Its walls are composed of dense cortex (compactum), which is thick throughout the extent of the diaphysis but tapers off to become the thin shell of each metaphysis. The central space (medulla or medullary cavity) within the diaphysis contains the bone marrow.

Covering the entire external surface of a mature long bone, except for the articulation, is the periosteum. The inner layer of the periosteum contains

the highly active cells that produce circumferential enlargement and remodeling of the growing long bone; hence it is called the osteogenic layer.
After maturity, this layer consists chiefly of a capillary blood vessel network.
The outer layer of the periosteum is fibrous and comprises almost the entire
periosteum of a mature bone. In the event of injury to a mature bone, some
of the resting cells of the inner periosteal layer become osteogenic.

Over most of the diaphysis, the periosteum is tenuous and loosely attached,
and the blood vessels therein are capillary vessels. At the expanded ends of
long bones, however, ligaments are attached firmly and can convey blood
vessels of larger size. The same is true at the ridges along the diaphyses,
where heavy fascial septa are attached.

Thus when we speak about the mechanical properties of bone, we must
specify which part of the bone we are talking about. The figures quoted in
Table 12.1:1 refer to the cortical region of the diaphysis, measured from
specimens machined from the compactum, with a cross-sectional area of at
least several mm². Thus they represent the average mechanical properties
of that part of the bone.

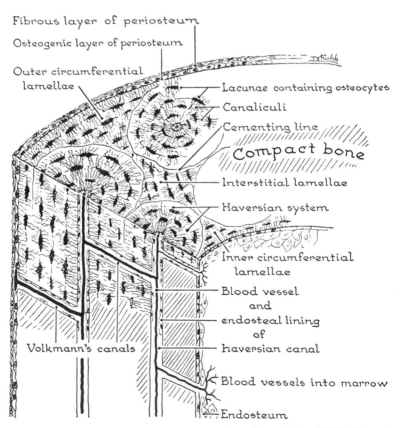

Figure 12.1:2 The basic structure of compact bone. From Ham (1969). Reproduced
by permission.

When examined microscopically, the bone material is seen to be a composite. Figure 12.1:2 shows Ham's (1969) sketch of the basic structure of compact bone. The basic unit is the Haversian system or osteon. In the center of an osteon is an artery or vein. These blood vessels are connected by transverse channels called Volkmann's canals. The rectangular pattern shown in Fig. 12.1:2 is an idealization; in actual bone they are more or less oblique.

About two-thirds of the weight of bone, or half of its volume, is inorganic material with a composition that corresponds fairly closely to the formula of hydroxyapatite, $3Ca_3(PO_4)_2 \cdot Ca(OH)_2$, with small quantities of other ions. It is present as tiny crystals, often about 200 Å long, and with an average cross section of 2500 Å^2 (about 50 × 50 Å) (Bourne, 1972). The rest of the bone is organic material, mainly collagen. The hydroxyapatite crystals are arranged along the length of the collagen fibrils.

Groups of collagen fibrils run parallel to each other to form fibers in the usual way. The arrangements of fibers differ in different types of bone. In woven-fibered bone the fibers are tangled. In other types of bone the fibers are laid down neatly in lamellae. The fibers in any one lamella are parallel to each other, but the fibers in successive lamellae are almost perpendicular to each other. As is seen in Fig. 12.1:2, the lamallae in osteones are arranged in concentric (nearly circular) cylindrical layers, whereas those near the surface of the bone are parallel to the surface (e.g., the outer circumferential lamellae).

12.1.2 Bone as a Composite Material

Bone material is a composite of collagen and hydroxyapatite. Apatite crystals are very stiff and strong. The Young's modulus of fluorapatite along the axis is about 165 GPa. This may be compared with the Young's modulus of steel, 200 GPa, Aluminum, 6061 alloy, 70 GPa. Collagen does not obey Hooke's law exactly, but its tangent modulus is about 1.24 GPa. The Young's modulus of bone (18 GPa in tension in human femur) is intermediate between that of apatite and collagen. But as a good composite material, the bone's strength is higher than that of either apatite or collagen, because the softer component prevents the stiff one from brittle cracking, while the stiff component prevents the soft one from yielding.

The mechanical properties of a composite material (Young's modulus, shear modulus, viscoelastic properties, and especially the ultimate stress and strain at failure) depend not only on the composition, but also on the structure of the bone (geometric shape of the components, bond between fibers and matrix, and bonds at points of contact of the fibers). To explain its mechanical properties, a detailed mathematical model of bone would be very interesting and useful for practical purposes (see Problems 12.1–12.3). That such a model cannot be very simple can be seen from the fact that the strength of bone does correlate with the mass density of the bone; but only loosely. Amtmann (1968, 1971), using Schmitt's (1968) extensive data on the distribution of strength of bone in the human femur, and Amtmann and Schmitt's (1968) data on the distribution of mass density in the same bone (determined by radiography), found that the correlation coefficient of strength and density is only 0.40–0.42. Thus one would have to consider the structural factors to obtain a full understanding of the strength of bone. Incidentally, Amtmann and Schmitt's data show that both density and strength are nonuniformly distributed in the human femur: the average density varies from 2.20 to 2.94 from the lightest to the heaviest spot, while the strength varies over a factor of 1.35 from the weakest place to the strongest place.

12.2 Bone as a Living Organ

The most remarkable fact about living bone is that it is living, and this is made most evident by the blood circulation. Blood transports materials to and from bone, and bone can change, grow, or be removed by resorption; and these processes are stress dependent.

That mechanical stresses modulate the change, growth, and resorption of bone has been known for a long time. An understressed bone can become weaker. An overstressed bone can also become weakened. There is a proper range of stresses that is optimal for the bone. Evidence for these biological effects of stresses are prevalent in orthopedic surgery and rehabilitation. Local stress concentration imposed by improper tightening of screws, nuts, and bolts in bone surgery, for example, may cause resorption, and result in loosening of these fasteners in the course of time.

Many authors, looking toward nature, feel that the evolution process has resulted in an optimum design of bone: optimum in the sense familiar to engineers designing light weight structures such as airplanes and space craft. This includes (a) the general shaping of the structure to minimize stresses while transmitting prescribed forces acting at specified points, (b) distribution of material to achieve a minimum weight (or volume, or some other pertinent criteria). Some well-known theories of optimum design include (a) the theory of uniform strength, that every part of the material be subjected to the same maximum stress (maximum normal stress if the material is brittle, maximum shear stress if the material is ductile) under a specific set of loading conditions, and (b) the theory of trajectorial architecture, which would put material only in the paths of transmission of forces, and leave voids elsewhere.

The idea of optimum design may be illustrated by a few examples. Let us consider the design of a thin-walled submarine container to enclose a volume V and to resist an external pressure p while maintaining the internal pressure at atmospheric. Any one of the shells sketched in Fig. 12.2:1(a) can be designed to meet this objective. If the same material is used for the construction

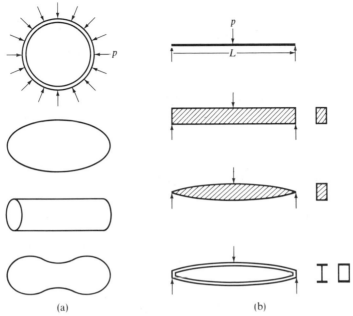

Figure 12.2:1 Examples of optimal design. (a) Shells resisting an external pressure. (b) Equally strong beams supporting a load P over a span L.

of these shells, the spherical shell will be the lightest and most economical in the use of the material. The cylindrical shell will be at least twice as heavy. The egg-shaped shell lies in between, while the biconcave shell will be the heaviest. Hence if the weight of the structure is the criterion, then the spherical shell is the optimal.

Next consider the design of a beam to support a load P over a span L; Fig. 12.2:1(b). To design such a beam, we first compute the bending moment in the beam. Let the moment at station x be $M(x)$. Then the maximum bending stress at any station x can be computed from the formula $\sigma = Mc/I$, where c is the distance between the neutral axis and the outer fiber of the beam, and I is the area moment of inertia of the cross section. For a specific material of construction, σ must be smaller than the permissible design stress (yield stress or ultimate failure stress). If the beam cross section is uniform, the critical condition is reached only at one point: under the load P. Greater economy of material can be obtained by designing a beam of variable cross section, with I/c varying with x in the same manner as $M(x)$. Then the stress will reach the critical condition in every cross section simultaneously. The latter design is optimal in the sense of economy of material. Next, we can compare different shapes of beam cross section. We see that for the same value of I/c, it is most economical to concentrate the material at the outer flanges, as sketched in Fig. 12.2:1(b). In this sense, then, the last entry is the minimum weight design.

The design illustrated in the last drawing of part (b) of Fig. 12.2:1 is an example of "load trajectory" design. As illustrated in this figure, the upper flange is subjected to compression while the lower flange is subjected to tension. The compressive and tensile forces are transmitted along these "trajectories" (flanges). The significance of this remark will become evident when we consider bone structure below.

Note that it is imperative to state clearly the conditions under which an optimal design is sought. An optimal design for one set of design conditions may not be optimal under another set of design conditions.

For bone, Roux (1895, p. 157) formulated the *principle of functional adaptation*, which means the "adaptation of an organ to its function by practicing the latter," and the *principle of maximum–minimum design*, which means that a maximum strength is to be achieved with a minimum of constructional material. He assumes that through the mechanism of hypertrophy and atrophy, bone has functionally adapted to the living conditions of animals and has achieved the maximum–minimum design. Considerable studies have been done since Roux's formulation to show that this is true

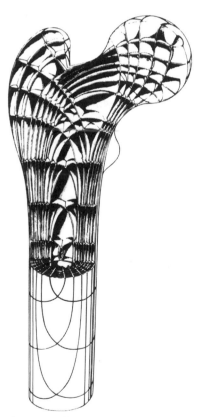

Figure 12.2:2 Theoretical construction of the three-dimensional trajectorial system in a femur model. From Kummer (1972), by permission.

(see, for example, Pauwels, 1965, Kummer, 1972). This has resulted in many beautiful illustrations that tell us how wonderful our bones are.

Let us quote one example. Roux (1895) suggests that spongy bone represents a trajectorial structure. Pauwels (1948) demonstrated that the architecture of substantia spongiosa is indeed trajectorial. Kummer's (1972) theoretical construction of a three-dimensional trajectorial system in a femur model is shown in Fig. 12.2:2, and it resembles the real bone structure quite closely.

12.3 Blood Circulation in Bone

The discussions of the preceding sections suggest the importance of blood circulation on the stress-dependent changes that take place in bone. Let us review the vascular system in bone in this section. Because of its hardness and opaqueness, it is not easy to investigate blood flow in bone. But by methods of injection (with ink, polymer, dye, radio-opaque or radioactive material), thin sectioning, calcium dissolution, microradiography, and electron microscopy, much has been learned about the vasculature in bone. A rich collection of interesting photographs can be found in Brookes (1971) and Rhinelander (1972). Figure 12.3:1 shows the vascular patterns in bone cortex sketched by Brookes (1971). Starting from the bottom of the figure, it is seen that the principal artery enters the bone through a distinct foramen. Within the medulla the artery branches into ascending and descending medullary arteries. These subdivide into arterioles that penetrate the endosteal surface to supply the diaphyseal cortex.

At the top of Fig. 12.3:1 can be seen the articular cartilage. Beneath it are the epiphyseal arteries. Then there is the growth cartilage, below which are three types of bone: the endochondral bone, the endosteal bone, and the periosteal bone. The orientations of vessels in these three types of bone are different (according to Brookes, 1971). In the endochondral bone the vessels point upward and outward; at mid-diaphyseal levels the vessels are transverse, while inferiorly they point downward and outward; that is, the cortical vessels have a radiate, fan-shaped disposition when viewed as a whole. This vascular pattern is evident in bone formed by periosteal apposition. The center of radiation of the vessels in periosteal bone corresponds with the site of primary ossification of the shaft. The center of radiation of the vessels in endosteal bone lies outside of the shaft. It is emphasized that communicating canals are frequent, but these are by no means wholly transverse, nor so numerous as to obliterate the three primary patterns observed in the cortex of long bones.

A fuller view of the blood supply to a long bone is shown in Fig. 12.3:2. The top half is similar to Fig. 12.3:1, except that the venous sinusoids are added. The principal nutrient vein and its branches are shown more fully in Fig. 12.3:2. The numerous metaphyseal arteries are shown here to arise

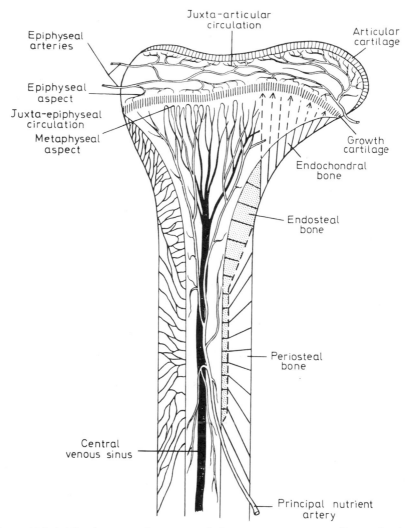

Figure 12.3:1 The three vascular patterns in bone cortex, corresponding to the three types of bone present. The left-hand side shows vessels. The right-hand side sketch shows an idealized pattern. From Brookes (1971), by permission.

from periarticular plexuses, and to anastomose with terminal branches of the ascending and descending medullary arteries. In the middle part of the figure are shown the periosteal capillaries, which are present on all smooth diaphyseal surfaces where muscles are not firmly attached.

Anatomical variations are great between mammalian species and long bones in the same species. These sketches illustrate only the general concepts. The vascular patterns shown in Figs. 12.3:1 and 12.3:2 do not conform to the

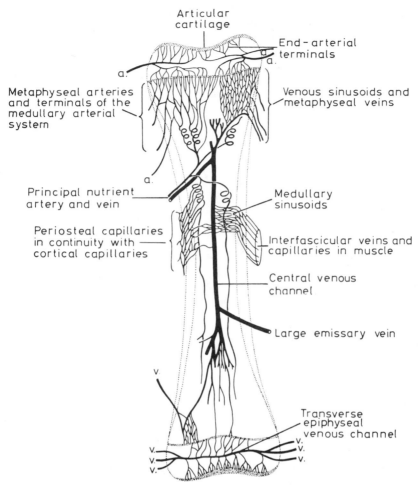

Figure 12.3:2 Vascular organization of a long bone in longitudinal section. From Brookes (1971), by permission.

sketch shown in Fig. 12.1:2, which is a stylized traditional view that has to be modified in its details as far as blood vessel pattern is concerned.

Hemodynamics of bone is difficult to study because of the smallness of the blood vessels and their inaccessibility for direct observation. Measurements have been made on the temperature distribution and temperature changes in bone; and indirect estimates of blood flow rate have been made under the assumption that heat transfer is proportional to blood flow. Perfusion with radioactive material and measurement of retained radioactivity in bone is another indirect approach. Some data are given in Brookes (1971). Direct measurements of flow and pressure are obviously needed.

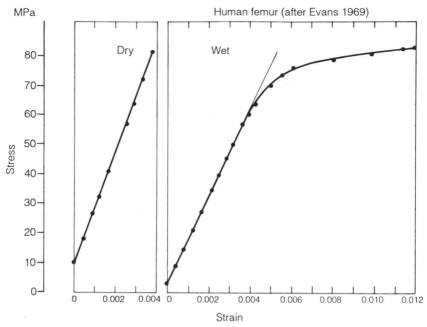

Figure 12.4:1 Stress–strain curves of human femoral bone. Adapted from Evans (1969).

12.4 Elasticity and Strength of Bone

Bone is hard and has a stress–strain relationship similar to many engineering materials. Hence stress analysis in bone can be made in a way similar to the usual engineering structural analysis. Figure 12.4.1 shows the stress–strain relationship of a human femur subjected to uniaxial tension. It is seen that dry bone is brittle and fails at a strain of 0.4%; but wet bone is less so, and fails at a strain of 1.2%. Since the range of strain is so small, it suffices to use the infinitesimal strain measure (the strain defined in the sense of Cauchy):

$$\varepsilon_{ij} = \frac{1}{2}\left(\frac{\partial u_i}{\partial x_j} + \frac{\partial u_j}{\partial x_i}\right), \tag{1}$$

where x_1, x_2, x_3 are rectangular cartesian coordinates, u_1, u_2, u_3 are displacements referred to x_1, x_2, x_3, and ε_{ij} represents the strain components. Figure 12.4:1 suggests that Hooke's law is applicable for a limited range of strains. For uniaxial loading below the proportional limit, the stress σ is related to the strain ε by

$$\sigma = E\varepsilon, \tag{2}$$

where E is the Young's modulus.

Table 12.4:1 gives the mechanical properties of wet compact bones of some animals and man. It is seen that the ultimate strength and ultimate

TABLE 12.4:1 Mechanical Properties of Wet Compact Bone in Tension, Compression, and Torsion Parallel to Axis. Data Adopted from Yamada (1970) 1970

Bone	Horses	Cattle	Pigs	Human (20–39 yrs.)
		Ultimate Tensile Strength (MPa)		
Femur	121 ± 1.8	113 ± 2.1	88 ± 1.5	124 ± 1.1
Tibia	113	132 ± 2.8	108 ± 3.9	174 ± 1.2
Humerus	102 ± 1.3	101 ± 0.7	88 ± 7.3	125 ± 0.8
Radius	120	135 ± 1.6	100 ± 3.4	152 ± 1.4
		Ultimate Percentage Elongation		
Femur	0.75 ± 0.008	0.88 ± 0.020	0.68 ± 0.010	1.41
Tibia	0.70	0.78 ± 0.008	0.76 ± 0.028	1.50
Humerus	0.65 ± 0.005	0.76 ± 0.006	0.70 ± 0.033	1.43
Radius	0.71	0.79 ± 0.009	0.73 ± 0.032	1.50
		Modulus of Elasticity in Tension (GPa)		
Femur	25.5	25.0	14.9	17.6
Tibia	23.8	24.5	17.2	18.4
Humerus	17.8	18.3	14.6	17.5
Radius	22.8	25.9	15.8	18.9
		Ultimate Compressive Strength (MPa)		
Femur	145 ± 1.6	147 ± 1.1	100 ± 0.7	170 ± 4.3
Tibia	163	159 ± 1.4	106 ± 1.1	
Humerus	154	144 ± 1.3	102 ± 1.6	
Radius	156	152 ± 1.5	107 ± 1.6	
		Ultimate Percentage Contraction		
Femur	2.4	$1.7 + 0.02$	1.9 ± 0.02	1.85 ± 0.04
Tibia	2.2	1.8 ± 0.02	1.9 ± 0.02	
Humerus	2.0 ± 0.03	1.8 ± 0.02	1.9 ± 0.02	
Radius	2.3	1.8 ± 0.02	1.9 ± 0.02	
		Modulus of Elasticity in Compression (GPa)		
Femur	9.4 ± 0.47	8.7	4.9	
Tibia	8.5		5.1	
Humerus	9.0		5.0	
Radius	8.4		5.3	
		Ultimate Shear Strength (MPa)		
Femur	99 ± 1.5	91 ± 1.6	65 ± 1.9	54 ± 0.6
Tibia	89 ± 2.7	95 ± 2.0	71 ± 2.8	
Humerus	90 ± 1.7	$86 + 1.1$	59 ± 2.0	
Radius	94 ± 3.3	93 ± 1.8	64 ± 3.2	
		Torsional Modulus of Elasticity (GPa)		
Femur	16.3	16.8	13.5	3.2
Tibia	19.1	17.1	15.7	
Humerus	23.5	14.9	15.0	
Radius	15.8	14.3	8.4	

strain in compression are larger than the corresponding values in tension for all the bones, whereas the modulus of elasticity in tension is larger than that in compression. The difference in the mechanical properties in tension and compression is caused by the nonhomogeneous anisotropic composite structure of bone, which also causes different ultimate strength values when a bone is tested in other loading conditions. Thus, for adult human femoral compact bone, the ultimate bending strength is 160 MPa, and the ultimate shear strength in torsion is 54.1 ± 0.6 MPa, whereas the modulus of elasticity in torsion is 3.2 GPa.

It is also well known that the strength of bone varies with the age and sex of the animal, the location of the bone, the orientation of the load, the strain rate, and the test condition (whether it is dry or wet). The strain rate effect may be especially significant, with higher ultimate strength being obtained at higher strain rate (see Schmitt, 1968). Yamada (1970), Evans (1973), Crowningshield and Pope (1974), and Reilly and Burstein (1974) have presented extensive collections of data.

The strength and modulus of elasticity of spongy bone are much smaller than those of compact bone. Again, see Yamada (1970) for extensive data on human vertebrae.

12.4.1 Anisotropy of Bone

Lotz et al. (1991) analyzed the anisotropic mechanical properties of the metaphyseal bone. Cowin (1988), Lipson and Katz (1984), and Reilly and Burstein (1975) analyzed these properties of the diaphyseal bone. Both used the transverse isotropic model. Major differences exist:

	Diaphyseal cortical shell	Metaphyseal cortical shell
$E_{longitudinal}$(MPa)	17 000	9650
$E_{transverse}$(MPa)	11 500	5470
ρ density (g cm^{-3})	1.95	1.62
Reference	Reilly et al.	Lotz et al.

Here E is the Young's modulus, ρ is the mass density, and the subscripts refer to directions.

12.4.2 Failure Criteria of Bone

Lotz et al. (1991) used von Mises' yield criterion for cortical bone, and von Mises and Hoffman's yield criterion for trabecular bone. The Hoffman (1967) failure theory assumes linear terms to account for different tensile and compressive strengths, and has been demonstrated to fit experimental trabecular bone data well (Stone et al., 1983). Assuming isotropy, the criterion is given by

$$C_1(\sigma_2 - \sigma_3)^2 + C_2(\sigma_3 - \sigma_1)^2 + C_3(\sigma_1 - \sigma_2)^2 + C_4\sigma_1 + C_5\sigma_2 + C_6\sigma_3 = 1, \tag{3}$$

where

$$C_1 = C_2 = C_3 = \frac{1}{2S_t S_c}, \tag{4}$$

$$C_4 = C_5 = C_6 = \frac{1}{S_t} - \frac{1}{S_c}. \tag{5}$$

Here σ_1, σ_2, σ_3 are the principal stresses, and S_t and S_c are the ultimate strengths in tension and compression, respectively. If S_t and S_c are equal, then Eq. (1) reduces to the von Mises yield criterion. These criteria will overestimate the strength under hydrostatic compression. Lotz et al. (1991) found that the strains at failure predicted by the von Mises criterion do not correspond well with measured values, but yield and fracture were accurately predicted for two femora tested.

12.5 Viscoelastic Properties of Bone

Probably beginning with Wertheim (1847), many authors have written about the viscoelastic properties of the bone. The list includes at least the names of A. A. Rauber, R. Smith, R. Walmsley, W. T. Dempster, R. T. Liddicoat, S. B. Lang, H. S. Yoon, J. L. Katz, F. Bird, H. Becker, J. Healer, M. Messer, A. A. Lugassy, E. Korostoff, D. Keiper, J. D. Curry, J. H. McElhaney, J. Black, W. Bonfield, C. H. Li, A. Ascenzi, E. Bonucci, P. Frasca, F. A. Meyers, S. S. Sternstein, and F. C. Cama. See Lakes and Katz (1979), Evans (1973), Katz and Mow (1973), Cowin et al. (1987), and Johnson and Katz (1987) for reviews, summaries, references, and data. In the following, let us focus on the constitutive equations.

Lakes, Katz, and Sternstein (1979) presented the details of a testing equipment and examples of results. Lakes and Katz (1979, part II) discussed various physical processes contributing to viscoelasticity of bone, including thermo-elastic coupling, piezoelectric coupling, motion of fluid in canals in bone, inhomogeneous deformation in osteons, cement lines, lamellae, interstitium, and fibers, and molecular modes in collagen. They then proposed a constitutive equation for wet human compact bone measured in torison at body temperature.

Let $G(t)$ be the relaxation function as defined in Sec. 7.6, and $S(\tau)$ be the relaxation spectrum defined by Eq. (7.6:28), Lakes and Katz (1979, part III), after synthesizing all the mechanisms participating in bone viscoelasticity, proposed the following "triangle spectrum" for the wet cortical bone:

$$S(\tau) = \frac{H(\tau)}{\tau},$$

$$H(\tau) = \log \tau \quad \text{for} \quad \tau_1 \leqslant \tau \leqslant \tau_2 \tag{1}$$

$$= 0 \quad \text{for} \quad \tau < \tau_1, \tau > \tau_2.$$

$H(\tau)$ is the notation used by Lakes and Katz (1979) who did not normalize the relaxation function (i.e., the condition $G(0) = 1$ is not imposed). As a practical generalization, Lakes and Katz considered $H(\tau)$ to be a sum of several terms proportional to $\log \tau$ with a different range of τ_1, τ_2 for each term.

Certain nonlinearity exists in the viscoelasticity of bone. Lakes and Katz (1979) approached the nonlinearity three ways: (1) The quasi-linear viscoelasticity as described in Sec. 7.6 is used, in which the "elastic" stress is a nonlinear function of the strain, but the memory (the relaxation function) is linear. (2) The multi-integral method of Green and Rivlin (1956) is used, in which the stress at time t is expressed as a function of strain history $\varepsilon(\tau)$ in the form

$$\sigma(t) = \int_{-\infty}^{t} G_0(t - s) \frac{d\varepsilon}{ds} ds$$

$$+ \int_{-\infty}^{t} \int_{-\infty}^{t} G_1(t - s_1, t - s_2) \frac{d\varepsilon}{ds_1} \frac{d\varepsilon}{ds_2} ds_1 ds_2$$

$$+ \cdots. \tag{2}$$

(3) The multi-integral method of Pipkin and Rogers (1968) is used, in which the first term represents the stress response to a single step increase in strain, the second term represents the stress response to two steps of strain increase at different times, etc. The first approach was illustrated in Chapter 7, Sec. 7.6. A limited application of the third approach to tendons and ligaments was done by Wineman, Ragajapol, Dai, Johnson, and Woo, see Sec. 7.6. When a single term is used, the result is equivalent to the quasi-linear approach. Some experiments were done by Young, Vaishnav, and Patel (1977) on a dog aorta to identify the second term of the Green–Rivlin approach. No sufficiently detailed experiment is known to have identified the higher order integrals in the Green–Rivlin and Pipkin–Rogers approach.

12.6 Functional Adaptation of Bone

The most frequently used method to verify whether hypertrophy or atrophy has occurred in a bone due to its use or disuse is by means of x ray, which measures the opacity of bone, which in turn is proportional to the mineral content of the bone. Another way is to measure wave transmission velocity and vibration modes of the bone as a means to determine the density of the bone. Results obtained by these methods have generally supported the idea of functional adaptation.

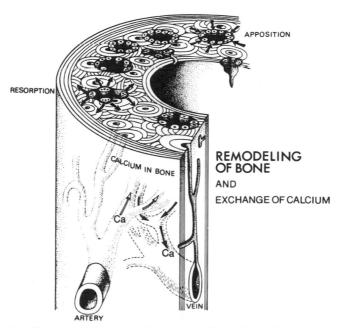

APPOSITION

RESORPTION

CALCIUM IN BONE

REMODELING
OF BONE

AND

EXCHANGE OF CALCIUM

Ca

Ca

VEIN

ARTERY

Figure 12.6:1 Conceptual drawing of the remodeling of bone by apposition, resorption, and calcium exchange. From Kummer (1972), by permission.

Changes in bone may take place slowly (in months or years) due to the action of bone cells (osteoclasts for resorption; osteoblasts for apposition), or rapidly (in days) due to the uptake or output of mineral salts. These processes are illustrated in a sketch made by Kummer (1972); see Fig. 12.6:1.

Julius Wolff (1884) first advanced the idea that living bones change according to the stress and strain acting in them. Change in the external shape of bone is called *external* or *surface remodeling*. Changes in porosity, mineral content, x-ray opacity, and mass density of bone are called *internal remodeling*. Both types occur during normal growth. But they also occur in mature bone.

After Wolff, the phenomenon of stress-controlled bone development was described by Glücksmann (1938, 1939, 1942) and Frost (1964). Evans (1957) reviewed the literature and concluded that clinical and experimental evidence indicated that compressive stress stimulates the formation of new bone and is an important factor in fracture healing. Dietrick et al. (1948) conducted an experiment on remodeling in humans by immobilizing some volunteers from the waist down in plaster casts for periods of from 6 to 8 weeks. During this study their urine, feces, and blood were analyzed for organics such as creatine and inorganics such as calcium and phosphorous. Four days after the plaster casts were removed the subjects resumed normal activity. The chemical analysis indicated that during the immobilization, their bodies suffered a net loss of bone calcium and phosphorous. After normal activity had been resumed, the mineral loss phenomenon was reversed and the body regained calcium and phosphorous.

A similar net loss of calcium was reported by Mack et al. (1967) for astronauts subjected to weightlessness. Kazarian and von Gierke (1969) reported on immobilization studies with rhesus monkeys. Wonder et al. (1960) studied the mouse and the chicken in hypergravity. Hert et al. (1971) studied intermittent loading on rabbits. The results are the same: subnormal stresses cause loss of bone strength, radiographic opacity, and size. Hert et al. concluded further that intermittent stress is a morphogenetic stimulus to functional adaptation of bone, and that the effect of compressive stress is the same as that of tensile stress.

Pauwels (1948) and Amtmann (1968) observed the bone structure of humans and showed that the distribution of material and strength is related to the severity of stresses in normal activity. Woo et al. (1976) and Torino et al. (1976) showed that remodeling due to rigid plate fixation in dogs occurs by thinning of the femoral diaphysis cortex rather than by induced osteoporosis in the cortex. In other words, it is primarily surface remodeling.

How can we put these concepts into a mathematical form so that bone modeling can be studied and predicted? Cowin and Hegedus (1976) expressed the ideas as follows. Bone is considered to be constituted of three basic materials: the bone cells, the extracellular fluid, and the solid extracellular material called bone matrix. The bone matrix is porous. The extracellular fluid is in contact with blood plasma, which supplies the material necessary for the synthesis of bone matrix. Let ξ be the matrix volume fraction and γ be the local mass density of the bone matrix. Let ε_{ij} be the strain tensor. Then at a constant temperature, one assumes the existence of constitutive equations of the form

$$\dot{\xi} = \frac{1}{\gamma} c(\xi, \varepsilon_{ij}) \tag{1}$$

and

$$\sigma_{ij} = \xi C_{ijkl}(\xi)\varepsilon_{kl}, \tag{2}$$

where the superposed dot indicates the material time rate, σ_{ij} is the stress tensor, $c(\xi, \varepsilon_{ij})$ represents the rate at which bone matrix is generated by chemical reaction, and $\xi C_{ijkl}(\xi)$ is a tensor of rank four representing the elastic constants of the bone matrix. The elastic constants are assumed to depend on the volume fraction of the bone matrix. These equations may represent internal remodeling of the bone when the functions $c(\xi, \varepsilon_{ij})$ and $C_{ijkl}(\xi)$ are determined.

For external remodeling, one needs to describe the rate at which bone material is added or taken away from the bone surface. Cowin and Hegedus (1976) suggested the following form. Let x_1, x_2, x_3 be a set of local coordinate axes with an origin located on the bone surface, with x_3 normal to the surface of the bone and x_1, x_2 tangent to it. Let the strains in the $x_1 x_2$ plane be ε_{11}, $\varepsilon_{22}, \varepsilon_{12}$. If external modeling is linearly proportional to the strain variation, then one might express the rate of increase of the surface in the x_3 direction in the form

$$U = k_{11}(\varepsilon_{11} - \varepsilon_{11}^0) + k_{22}(\varepsilon_{22} - \varepsilon_{22}^0) + k_{12}(\varepsilon_{12} - \varepsilon_{12}^0), \qquad (3)$$

where k_{11}, k_{22}, k_{12} and ε_{11}^0, ε_{22}^0, ε_{12}^0 are constants. If the right-hand side is positive, the surface grows by deposition of material. If the right-hand side is negative, the surface resorbs. But this equation is untenable, because it does not incorporate the idea that tensile stress and compressive stress have the same effect with regard to bone remodeling (Hert et al., 1971). Nor does it include remodeling due to surface traction (normal and shear stress acting on the surface) as is often found under orthopedic prosthesis (Woo et al., 1976). To include these effects we may assume

$$U = k_{ij}[\varepsilon_{ij}^2 - (\varepsilon_{ij}^0)^2], \qquad (4)$$

where the indexes range over 1, 2, 3. The summation convention is used.

With these constitutive equations, the stress and strain distribution in a bone can be determined by the methods of continuum mechanics, and the remodeling can be predicted. However, a more complete theory is given below.

12.6.1 Tensorial Wolff's Law

Cowin et al. (1992) reasoned that the equations governing the temporal changes in the architecture of bone must be nonlinear tensor equations in order to take into account the feedback nature between stress and growth. Their analysis puts the tensorial character of remodeling in the clearest perspective. Without explaining the background, derivation, and meaning, we present Cowin et al.'s mathematical constitutive equations below:

$$\mathbf{T} = \beta_1 \mathbf{I} + \beta_2 \mathbf{E} + \beta_3 \mathbf{K} + \beta_4 \mathbf{K}^2 + \beta_5(\mathbf{KE} + \mathbf{EK}) + \beta_6(\mathbf{K}^2\mathbf{E} + \mathbf{EK}^2), \quad (5)$$

$$\dot{\mathbf{K}} = \alpha_1 \mathbf{I} + \alpha_2 \mathbf{K} + \alpha_3 \mathbf{K}^2 + \alpha_4 \mathbf{E} + \alpha_5(\mathbf{KE} + \mathbf{EK}) + \alpha_6 \mathbf{K}^2\mathbf{E} + \mathbf{EK}^2), \quad (6)$$

$$\mathrm{tr}\,\dot{\mathbf{K}} = 0, \quad \text{where} \quad \dot{\mathbf{K}} = 0 \quad \text{when} \quad \mathbf{E} = \mathbf{E}_0, \qquad (7)$$

$$\dot{v} - \dot{v}_0 = \text{function of } (\mathrm{tr}\,\mathbf{E}, \mathrm{tr}\,\mathbf{K}^2, \mathrm{tr}\,\mathbf{K}^3, \mathrm{tr}\,\mathbf{EK}, \mathrm{tr}\,\mathbf{EK}^2, v - v_0),$$

$$\text{where} \quad \dot{v} - \dot{v}_0 = 0 \quad \text{when} \quad \mathbf{E} = \mathbf{E}_0. \qquad (8)$$

In these equations, $\mathbf{T}$ is the stress tensor, $\mathbf{K}$ is the deviatoric part of the normalized *facric tensor* (see *Biomechanics*, Fung, 1990, p. 510, for definition and references) which describes the geometric pattern of the trabecular bone, $\mathbf{E}$ is the strain tensor, $\mathbf{E}_0$ is a specific strain characterizing the homeostatic state, v is the solid volume fraction of the trabecular bone, v_0 is a reference value, and all the α's and β's are functions of $v - v_0$, $\mathrm{tr}\,\mathbf{K}^2$, and $\mathrm{tr}\,\mathbf{K}^3$, but α_1, α_2, α_3, and β are also functions of $\mathrm{tr}\,\mathbf{E}$, $\mathrm{tr}\,\mathbf{EK}$, and $\mathrm{tr}\,\mathbf{EK}^2$. These equations look formidable, but they actually state the evolution of remodeling in the simplest way.

12.6.2 The Mechanism for the Control of Remodeling

Piezoelectricity has been proposed as a mechanism for bone to sense stress and cause remodeling. Fukada (1957, 1968) discovered piezoelectricity in bone and later identified it as due to collagen. Becker and Murray (1970) reported that an electric field is capable of activating the protein-synthesizing organelles in osteogenic cells of frogs. It is also known that the presence of an electric field near polymerizing tropocollagen will cause the fibers to orient themselves perpendicular to the line of force. Bassett and Pawlick (1964) reported that if a metal plate is implanted adjacent to a living bone, a negative charge on the plate will induce the deposition of new bone material on the electrode. Thus it is possible that piezoelectricity lies behind the remodeling activities.

Biochemical activity of calcium is another possible mechanism. Justus and Luft (1970) have shown that straining the bone increases calcium concentration in the interstitial fluid. They showed that this is due to a change in the solubility of the hydroxyapatite crystals in response to stresses.

This discussion would be woefully amiss if we did not recall also that growth in general is modulated by endocrine. The endocrine STH (somatotropic or growth hormone) affects all growing tissues; it increases all cell division and all subsidiary processes, such as total protein synthesis, net protein synthesis, total turnover (likewise for lipid and carbohydrate metabolism), and tissue growth. The endocrine ACH (adrenal-cortical hormone) has the opposite effect; it affects all tissues and it decreases all cell division and subsidiary processes. The endocrine T_4 (thyroxine) also affects all tissues. Estrogens selectively decrease cell division and subsidiary processes; they affect cartilage and lamellar bone. Furthemore, the normality of bone is significantly influenced by vitamins A, C, and D, and calcitonin. Thus, bone is a complex biochemical entity. Yet the importance of the mechanical aspects is obvious: not only to intellectual speculation, but also to clinical applications in orthopedics. If we know the mechanical aspects well, then we can control bone remodeling through mechanical stresses that can be applied through exercises, either voluntarily or with the help of mechanical devices. Exercise is man's most basic approach to health.

12.6.3 Osseointegration in Skeletal Reconstruction

Brånemark et al. (1977) first used titanium fixtures to support fixed dental prostheses and over the years have achieved a long term success rate of stable, useful prostheses well over 95% and a demonstrated useful life of 25 years (Rydevik et al., 1990). In recent years, titanium screw fixtures have been used in head and neck, eye and ear, hearing aids, hand, knee, etc. Brånemark discovered bone integration with titanium in 1959 in his research on bone marrow microcirculation using a titanium device which is a modification of the so-called rabbit ear chamber. He exploited this discovery in many surgical procedures. In practice, the cleanliness of the surface of the metal before contact with tissue is extremely important.

12.7 Cartilage

Cartilage and bone have the same three tissue elements: cells, intercellular matrix, and a system of fibers. In man during early fetal life, the greater part of the skeleton is cartilaginous. In adult life, cartilage persists on the articulating surfaces of synovial joints, in the walls of the thorax, larynx, trachea, bronchi, nose, and ears, and as isolated small masses in the skull base.

The matrix in which the cartilage cells (chondrocytes) are embedded varies in appearance and nature. Acordingly, there are *hyaline cartilage* (from *hyalos*, meaning glass), *white fibrocartilage* (containing much collagen), and *yellow elastic fibrocartilage* (containing a rich elastin network). A *cellular cartilage*, with only fine partitions of matrix separating the cells, is normal in the fetus.

Costal, nasal, tracheo-bronchial and all temporary cartilages, as well as most articular cartilages, are of the hyaline variety. It has long been known that if the surface of articular cartilage is pierced by a pin, then after withdrawal, a longitudinal "split-line" remains, and that on any particular joint surface, the pattern of the split-lines follows the predominant direction of the collagen bundles in the cartilage. The proportion of collagen in the matrix increases with age.

White fibrocartilage is found in intervertebral disks, articular disks, and the lining of bony grooves that lodge tendons. Yellow elastic fibrocartilage is found in the external ears, larynx, epiglottis, and the apices of the arytenoids.

The location of these cartilages suggests their function. The intervertebral disk bears the load imposed by the spine; it is elastic and makes the spine flexible. The cartilage at the ends of the ribs gives the ribs the desired mobility. The articular cartilage at the ends of long bones provides lubrication for the surfaces of the joints, and serves as a shock absorber to impact loads and as a load bearing surface in normal function. Sometimes it is its elasticity and rigidity that seems to be needed. In the bronchioles of man there is very little cartilage; but in diving animals such as the seal there is a considerable amount of cartilage in the bronchioles: its function seems to keep the bronchioles from collapsing too soon when the animals dive into deep water. By this mechanism these animals avoid nitrogen sickness. If man dives too deeply into the water, the bronchioles collapse before the alveoli do, so gas is trapped in the lung, and nitrogen has to be absorbed into the bloodstream, causing bends. For a seal the alveoli can empty first, and nirogen trapping is avoided.

Thus, cartilage is biologically active, and rheologically unique.

12.8 Viscoelastic Properties of Articular Cartilage

Cartilage is rather porous, and the interstitium is filled with fluid. Under stress, fluid moves in and out of the tissue, and the mechanical properties of cartilage change with the fluid movement.

A simple method of demonstrating the viscoelasticity of cartilage is the indentation test, which can be done *in situ* to mimic physiological conditions. Upon unloading, an instantaneous recovery is followed by a time-dependent one. The recovery will not be complete if the test is done with the tissue exposed to air; but if the specimen is completely immersed in a bath, complete recovery can be achieved by fluid resorption during unloading.

For a rigorous theoretical analysis of the indentation test, it is better to begin with a uniaxial loading on a flat, plate-like specimen. (However, the preparation of test specimens is then quite difficult.) In the following we shall quote a report by Woo et al. (1979), which also contains references to vibration tests and theoretical analyses by other authors.

Articular cartilage specimens were taken from the humeral heads of bovine animals. This joint surface is large (about 9–10 cm in diameter) and is relatively flat, so that cartilage-bone plugs, 1.25 cm in diameter, can be cored. A die cutter with sharp razor blades formed into a standardized dumbbell shape was used to stamp each plug perpendicularly through the articular cartilage surface to the subchondral bone along the split line, or the zero-degree ($0°$) direction. The cartilage-bone plug was then placed on a sledge microtome to obtain slices 250–325 μm thick. The specimen's test section has dimensions $1 \times 4.25 \times 0.25 - 0.325$ mm.

An Instron testing machine coupled with a video dimensional analyzer (VDA) system was used for strain rate dependent and cyclic tests. A special solenoid device provided rapid stretch (up to 30 mm/sec) for relaxation tests. This test equipment was not entirely satisfactory because it was found that stress relaxation in articular cartilage is very rapid. For example, comparing the load values recorded using a storage oscilloscope and that of a strip chart recorder, the differences in peak forces could be as much as 25%, although the time lag for the strip chart recorder was less than 250 ms. However, the mechanical vibration of the test system prevented the authors from using the oscilloscope data.

Test results are stated in terms of the stretch ratio, λ,

$$\lambda = \frac{\text{deformed gage length}}{\text{initial gage length}},$$

the Green's strain, E,

$$E = \tfrac{1}{2}(\lambda^2 - 1),$$

the Lagrangian stress, T,

$$T = \frac{\text{force}}{\text{initial undeformed cross-sectional area}},$$

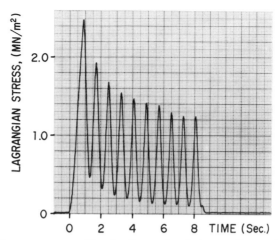

Figure 12.8:1 Stress response of bovine femural articular cartilage subjected to cyclic stretching between $\lambda = 1.07$–1.10. Temp.: 37°C. Specimen immersed in saline. From Woo et al. (1979), by permission.

and the Kirchhoff stress, S,

$$S = \frac{T}{\lambda}.$$

Test specimens were immersed in normal saline solution at 37°C. Four specimens were subjected to cyclic tests at different strain rates varying from 0.04%/sec to 4%/sec, and it was found that the hysteresis loops and the peak stresses increased slightly with increased strain rates. Thus articular cartilage is moderately strain rate sensitive. Figure 12.8:1 shows the preconditioning behavior of a cartilage subjected to a cyclic stretching between $\lambda = 1.07$ and 1.10. It is seen that after 10 cycles of stretching the stress-time curve tends asymptotically to a steady state cyclic response.

The results of uniaxial tensile stress relaxation studies are summarized in Fig. 12.8:2. The *experimental reduced relaxation function* is defined as $G(t) = T(t)/T$ ($t \sim 250$ ms). The true *reduced relaxation function* $T(t)/T(t = 0)$ cannot be obtained experimentally. At a small extension ($\lambda = 1.05$), the stress reached its relaxed state within 15 min. However, for higher extensions (λ between 1.16 and 1.29), the stress relaxation did not level after 100 min. These characteristics are similar to that which are shown in Fig. 7.5:4 in Chapter 7 for the mesentery. Different reduced relaxation functions are needed for the small stretch and moderate stretch regimes. However, as shown in Fig. 12.6:2, in each regime the reduced relaxation function depends on λ only to a small degree. By ignoring this dependence, Woo et al. (1979) used the quasi-linear model of viscoelasticity (Chapter 7, Sec. 7.6) to characterize the articular cartilage. In brief, this model assumes that the relaxation function, K, is dependent on both strain and time and can be written as

$$K(t) = G(t) * S^e[E(t)], \tag{1}$$

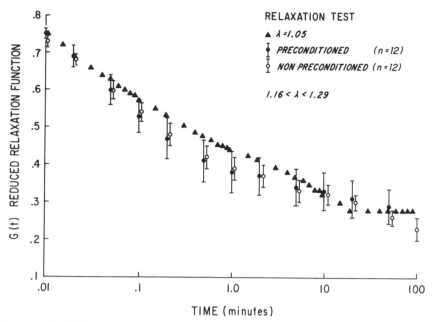

Figure 12.8:2 The reduced relaxation function of articular cartilage at a lower stretch ($\lambda = 1.05$) and at higher stretches, $\lambda = 1.16$–1.29. From Woo et al. (1979), by permission.

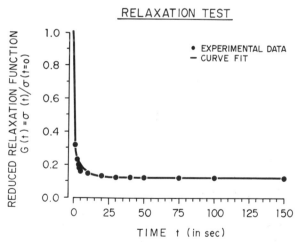

Figure 12.8:3 The experimental reduced relaxation function of articular cartilage compared with a theoretical expression given in Eq. (3) with the constants c, τ_1, τ_2 determined empirically. From Woo et al. (1979), by permission.

where S^e is the "elastic response" and $G(t)$ is the "reduced relaxation function" with $G(0) = 1$. The stress–strain-history integral takes the forms

$$S(t) = \int_{-\infty}^{t} G(t - \tau)\dot{S}^e(\tau)\,d\tau$$

$$= S^e[E(t)] - \int_0^t \frac{\partial G(t - \tau)}{\partial \tau} S^e(\tau)\,d\tau. \tag{2}$$

Once the mechanical property functions in this model, $G(t)$ and $S^e[E(t)]$, are determined, the time dependent stress, $S(t)$, with known history of strain, $E(t)$, can be determined by Eq. (2).

Relaxation test data for a $0°$ mid-zone cartilage specimen are shown in Fig. 12.8:3, in which the experimental values for $G(t)$ were calculated using stress data normalized by an extrapolated value for the stress at $t = 0$. The form for $G(t)$ used in the model was taken to be Eq. (36) of Sec. 7.6:

$$G(t) = \left[1 + C\left\{E_1\left(\frac{t}{\tau_2}\right) - E_1\left(\frac{t}{\tau_1}\right)\right\}\right] \Big/ \left[1 + C\log\left(\frac{\tau_2}{\tau_1}\right)\right], \tag{3}$$

where E_1 is the *exponential integral* function, and C, τ_1, and τ_2 are material constants. C, τ_1, and τ_2 were determined in the least square sense using a computer program based on the method of Powell (1965); the values found were $\tau_1 = 0.006$ sec, $\tau_2 = 8.38$ sec, and $C = 2.02$.

The cyclic stress data (Fig. 12.8:1) corresponding to a saw-tooth series of ramp extensions for the same specimen was used to determine the elastic response, $S^e\{E\}$. Here S^e was assumed to be a power series in E of the form

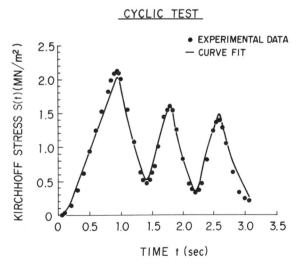

Figure 12.8:4 Comparison of experimental data on stress response to the first three cycles of loading and unloading as given in Fig. 12.8:1 with theoretical expressions given in Eqs. (1)–(4) with the constants c, τ_1, τ_2, a_1, a_2 determined empirically. From Woo et al. (1979), by permission.

$$S^e = S^e\{E\} = \sum_{i=1}^{n} a_i E^i. \tag{4}$$

Substitution of Eq. (4) into Eq. (2) with $G(t)$ now known in the form of Eq. (3) yields a linear set of equations that can be solved for the constants, a_i. By numerical integration and least squares procedure, the constants found for $n = 2$ are $a_1 = 30$ MN/m² and $a_2 = 56$ MN/m². Figure 12.8:4 compares Kirchhoff stress-time curves measured experimentally and computed from the quasi-linear viscoelastic model. Similar agreement was obtained for a higher order of n.

The S^e so computed is considerably higher than the measured stress level at 250 ms when the specimen is subjected to a ramp stretching, indi-

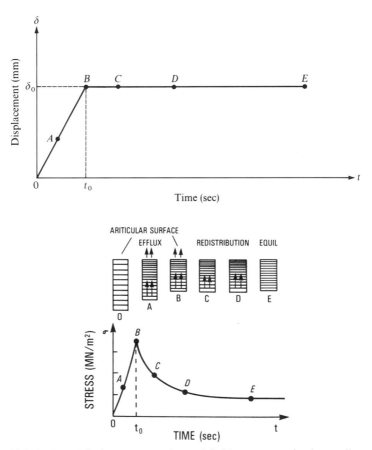

Figure 12.8:5 Top: Displacement, strain, and fluid movement in the cartilage in the free-draining confined-compression stress-relaxation experment. O, A, B defines the compressive phase and B, C, D, E defines the stress-relaxation phase. Bottom: The corresponding stress history. Middle diagrams: Horizontal lines depict the compressive strain field, and arrows indicate the fluid velocity. Stress relaxation times of the order of 1–5 sec are usually observed. From Mow et al. (1980), by permission.

cating that a significant amount of stress relaxation has taken place in the first 250 ms. This method of determining the elastic response from cyclic stretching data is quite effective. The rapid relaxation of articular cartilage at very small time is believed to be caused by the fluid movement from the tissue when it is first subjected to stress.

A similar rapid stress relaxation was found by Mow et al. (1977), who tested articular cartilage plugs in compression. They attributed the rapid relaxation to the ease of extrusion of fluid in the tissue. Figure 12.8:5 shows the result of Mow et al. (1980) on the stress response to a ramp-step strain history in compression and their concept of dynamic strain distribution and fluid movement. A detailed theoretical analysis is given in their 1980 paper.

12.9 The Lubrication Quality of Articular Cartilage Surfaces

Articular cartilage, which is the bearing material lining the bone in synovial joints, exhibits unique and extraordinary lubricating properties. It possesses a "coefficient of friction" (the resistance to sliding between two surfaces divided by the normal force between the surfaces) that is many times less than man's best artificial material. It exhibits almost two orders of magnitude less friction than most oil lubricated metal on metal. Mature cartilage in man, with little or no regenerative ability, maintains a wear life of many decades.

Earlier experiments on the synovial joint tribology have usually taken either of two forms. One approach has sought to experimentally simulate intact synovial joint conditions *in vitro* by preserving the natural cartilage geometry, kinematics, and chemical environment. This approach is exemplified by Linn (1967) and is shown schematically in Fig. 12.9:1, where an excised

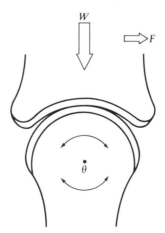

Figure 12.9:1 Schematic diagram of a joint friction experiment using an isolated animal synovial joint.

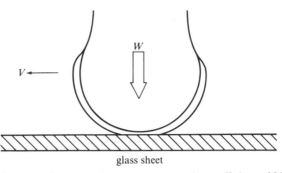

Figure 12.9:2 Concept of an experiment to measure the coefficient of friction between articular cartilage and glass.

intact joint, devoid of surrounding connective tissue, capsule, or restraining ligaments, is tested *in vitro*. Figure 12.9:2 shows a diagrammatic sketch of another approach, in this case representative of the work of McCutchen (1959), where excised segments of articular cartilage are tested upon planar surfaces of another (man-made) material (usually glass). These experimental approaches tend to complement one another, the intact joint tests being

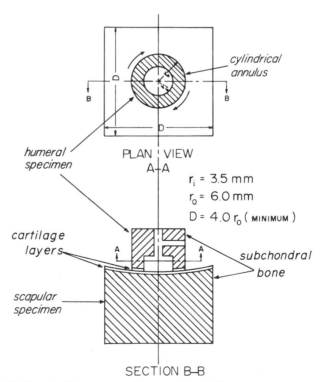

Figure 12.9:3 Sketch of the test specimen used by Malcom to measure the coefficient of friction between articular cartilage surfaces. From Malcom (1976), by permission.

more physiologically representative of the *in vivo* situation, while the planar sliding studies are more tribologically definitive. They each have their limitations, however: the geometric and kinematic complexity of the intact tests inhibit parameter isolation, making analysis difficult, while the geometrically more tractable planar sliding studies lack the cartilage against cartilage interfacial contact condition.

Malcom (1976) presented an alternative approach that is much easier to analyze. Figure 12.9:3 shows a schematic diagram of the experimental specimen configuration used by Malcom. It consists essentially of an axisymmetric cylindrical cartilage annulus tested in contact with another conformationally matching layer of cartilage. The articular cartilage specimens are obtained from opposing, geometrically matched regions of bovine scapular and humeral joint surfaces. A specially designed electromechanical instrument generated specified mechanical and kinematic conditions by loading and rotating the cylindrical specimen and measuring the resultant cartilage frictional and deformational responses. Experimental monitoring and data recording were done on-line via a PDP 8/E computer. The specimen environment was regulated throughout the experiment by maintaining the cartilage in a thermally controlled bath of whole bovine synovial fluid or buffered normal saline.

Figure 12.9:4 shows the experimentally obtained shear stress history on nine cartilage specimen pairs after they were suddenly loaded to a static

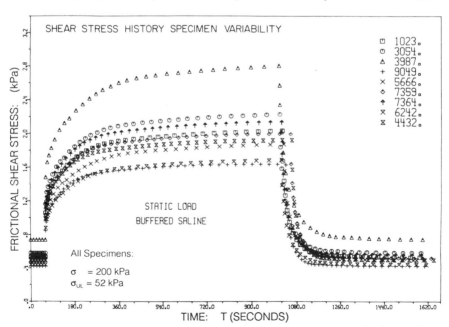

Figure 12.9:4 History of variation of the shear stress between two articular cartilage surfaces when they are subjected to a step loading and rotation, and at a later instant of time a step unloading to a lower level of normal stress and continued rotation. From Malcom (1976), by permission.

normal stress of 200 kPa and were continuously rotated at 0.7 rad/sec. The shear stress approached asymptotically a value lying in the range 1.5–2.8 kPa after 1000 sec of loading. Upon a sudden unloading to a normal stress of 52 kPa, the shear stress decreased and reached asymptotically a value in the range 0.4–0.8 kPa. The curves show the typical specimen response diversity when a group of different biological specimens are tested under identical conditions.

The gradual increase in friction after a step loading and rotation is accompanied by compression of the specimen and extrusion of interstitial fluid from the cartilage, as one could have predicted according to the results shown in Fig. 12.8:5 and discussed in the preceding section. Figure 12.9:5 shows the cartilage normal compression histories for the nine specimens, which were obtained simultaneously with the data in Fig. 12.9:4. The compressive strain is seen to be directly related to the increase in frictional shear between the cartilage surfaces.

When tested in different rates of rotation, it was found that the asymptotic values of the shear stress (and therefore the coefficients of friction) are relatively insensitive to the sliding velocity within the range 1–400 mm/sec. When the sliding velocity further decreases, the friction coefficient increases. At the static condition, a considerably larger coefficient of friction was measured. A summary of results for four of the loading conditions and lubricating

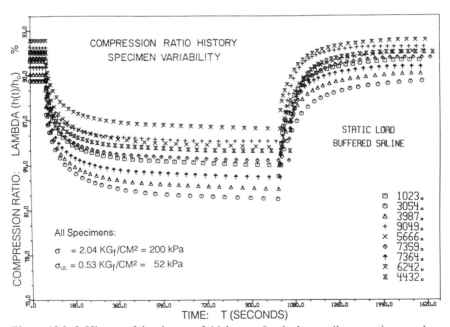

Figure 12.9:5 History of the change of thickness of articular cartilage specimens taken simultaneously with the shear stress record shown in Fig. 12.7:4. From Malcom (1976), by permission.

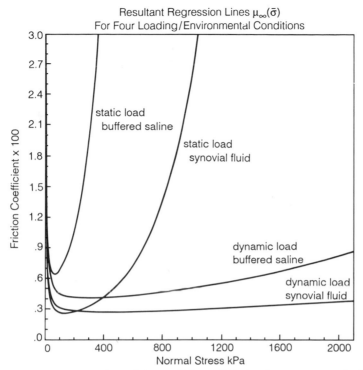

Figure 12.9:6 The mean values of the coefficient of friction between two articular cartilage surfaces as a function of the normal stress between the surfaces. The coefficient of friction depends on whether the measurements were made in a static or dynamic condition, and on the bathing fluid, whether it is buffered saline or synovial fluid. From Malcom (1976), by permission.

bath combinations is shown in Fig. 12.9:6. The y axis represents the steady state friction coefficient, $\mu_\infty = \tau_\infty/\sigma$, where σ is the value of the static or dynamic (1 cycle/sec square wave) normal stress, and τ_∞ is the asymptotic frictional shear stress. The actual curves shown are nonlinear regression line fits to the resultant experimental data (57 to 112 observations per curve). Note particularly the low magnitude of the friction coefficients that were obtained. For the least efficient lubrication mode of statically loaded cartilage in buffered isotonic saline lubricant, the mean friction coefficient ranged from 0.0065 to 0.030 at normal stresses of 70–370 kPa, respectively. The most efficient lubrication combination was cyclic dynamic loading in a whole bovine synovial fluid lubricating both. Under these conditions, the data show a mean minimum friction coefficient of 0.0026 at a normal stress of 500 kPa and a maximum coefficient of 0.0038 at a maximum normal stress of 2 MPa. These values may be compared with other typical engineering values of μ of 0.5–2.0 or higher for clean metal surfaces, 0.3–0.5 for oil lubricated metal, and 0.05–0.1 for teflon coated surfaces. The superiority of the synovial joint is truly striking.

A change of environmental factor can significantly change the coefficient of friction between articular cartilage surfaces. Especially noteworthy is the effect of hydrogen ion concentration in the bath. The cartilage's compressibility (for specimens with a porous boundary without impediment to fluid extrusion) suddenly increases when pH values become less than 5, and correspondingly the coefficient of friction between cartilage surfaces increases at pH less than 5. This finding is in conflict with the results of Linn and Radin (1968), which showed that deformation increased for pH $\sim$ 5.5 and below, but the μ is maximum at pH 5.0. Malcom explained this difference in terms of a "ploughing" effect in Linn and Radin's experiments and its absence in his.

Malcom's experimental results are essentially in agreement with those of McCutchen (1959) and the synovial joint pendulum studies by Unsworth et al. (1975); but his control of experimental condition was unsurpassed.

Why is articular joint so efficient in lubrication? There have been several theories which we now list:

Fluid Transport Effects. There is no doubt that cartilage supports load imposed on it by hydrostatic pressure of the fluid in the composite material, rather than by elastic recoil of the solid matrix material. Friction is low when the cartilage is filled with a normal amount of fluid. Friction increases when fluid is squeezed out. McCutchen (1959) advocates that the synovial joint is so effective because the time required for the fluid to be squeezed out is long, and as soon as the load is removed the compressed cartilage rebounds quickly to absorb synovial fluid back into it. Analyses of fluid movement in squeezed film have been given by Fein (1967), Mow (1969), and Kwan (1984).

Fluid Mechanical (Lubrication Layer) Effects. The synovial fluid is drawn between sliding surfaces by virtue of its viscosity, and the high pressure generated within it would support the load (MacConaill, 1932). By a reason which we have elaborated earlier in Chapter 5, Sec. 5.5, in connection with the lubrication layer theory for the red blood cells, the fluid layer in the synovial joint will have a lubrication effect. This effect is caused by the change in velocity profile in the lubrication layer. In addition, the rheological properties due to the high concentration of hyaluronic acid may also contribute to this mechanism by providing a high viscosity when the shear rate drops to zero and by shear thinning at higher shear rates (see Chapter 6, Sec. 6.7). Finally, the normal stress effects generated by the long chain polymer molecules will also help support the load (Ogston and Stanier, 1953).

Boosted Lubrication Effect. Walker et al. (1968) and Maroudas (1967) hypothesize that as the articulating surfaces approach each other, water passes into the cartilage, leaving a concentrated pool of proteins to support the load and lubricate the surface. A theoretical analysis of this was done by Lai and Mow (1978).

Articular Cartilage Effects. There are two kinds. First, the cartilage deforms under loading so that it can redistribute the load as mechanical contact stresses and fluid pressure dictate (Dowson, 1967). Second, there could be a lubricating molecular species that interacts with the cartilage and acts as a boundary lubricant insulating and coating the surfaces (Radin et al., 1970).

Some of these theories overlap, or are, at times, contradictory. The final resolution awaits further research.

In diseased states, the lubrication quality of cartilage deteriorates. An experiment on sliding friction between a normal bovine cartilage annulus and a pathological human tibial plateau specimen showed a significant increase in the measured interfacial friction force. For the osteoarthritic case an average friction coefficient μ of $0.01-0.09$ was measured across the normal force range for static and dynamic loading in a buffered saline environment.

12.10 Constitutive Equations of Cartilage According to a Triphasic Theory

Lai, Hou, and Mow (1991) derived the constitutive equations for the swelling and deformation of articular cartilage, meniscus, and intervertebral disk on the basis of a triphasic theory. A brief presentation of the salient features of the theory is given below. As an introduction to their theory, the reader may find it helpful to read about the general equations of mass transport in a structure consisting of a solid immobile matrix and a fluid mixture phase presented in terms of chemical potential in *Biomechanics: Motion, Flow, Stress, and Growth* (Fung 1990), Sec. 8.9. The chemical potential and activity of gases, liquids, solutions, and mixtures are discussed in that reference, Sec. 8.4, and can be extended to solids, ions, and fixed charges.

The three phases in cartilage identified by Lai et al. (1991) are: (1) a solid phase of collagen and proteoglycan extracellular matrix; (2) the interstitial water; and (3) the Na^+ and Cl^- ions of NaCl. All phases are assumed incompressible. Let ϕ^α denote the volume fraction of the phase α, with α identified with $s, w, +, -$ named above; i.e., solid, water, Na^+ ions, and Cl^- ions, respectively. The volume fraction is defined by the equations

$$\rho^\alpha = \phi^\alpha d^\alpha, \tag{1}$$

where d^α is the true mass density of the phase no. α, and ρ^α is the phase density in the tissue (equal to the mass of phase α divided by the tissue volume). Then, since there is no void,

$$\phi^s + \phi^w + \phi^+ + \phi^- = 1. \tag{2}$$

The equation of continuity of the phase α (α stands for s, or w, or $+$, or $-$) is, in a rectangular cartesian frame of reference with coordinates x_1, x_2, x_3,

$$\frac{\partial \rho^\alpha}{\partial t} + \frac{\partial(\rho^\alpha v_j^\alpha)}{\partial x_j} = 0, \tag{3}$$

where v_j^α is the jth velocity component of the phase α. The repetition of index j means summation over $j = 1, 2, 3$. Since every phase is incompressible and d^α are constant, one can substitute Eqs. (1) into (3), adding, and using Eq. (2) to obtain an equation which is, in vector notation,

$$\text{div}(\phi^s v^s + \phi^w v^w + \phi^+ v^+ + \phi^- v^-) = 0. \tag{4}$$

12.10.1 The Electric Charges

The existence of fixed negative charges on the glycosaminoglycan chains of the proteoglycan molecules is recognized. Let $\tilde{c}^+$, $\tilde{c}^-$, and $\tilde{c}^F$ represent the charge density (charges expressed in Avogadro numbers of electrons per unit tissue volume) for the cations, anions, and fixed charges, respectively. Then the *electroneutrality condition* is expressed by the equation

$$\tilde{c}^+ = \tilde{c}^- + \tilde{c}^F. \tag{5}$$

With M^+ and M^- denoting the atomic weights of Na^+ and Cl^-, respectively, the charges per unit tissue volume and the mass densities are related by

$$\rho^+ = M^+ \tilde{c}^+ \quad \text{and} \quad \rho^- = M^- \tilde{c}^-. \tag{6}$$

Thus, from the equations of continuity for ρ^+ and ρ^-, one obtains

$$\frac{\partial \tilde{c}^+}{\partial t} + \text{div}(\tilde{c}^+ v^+) = 0 \tag{7a}$$

and

$$\frac{\partial \tilde{c}^-}{\partial t} + \text{div}(\tilde{c}^- v^-) = 0. \tag{7b}$$

In Lai et al.'s model, the fixed charges are assumed to be unchanged during deformation of the tissue. Hence

$$\frac{\partial \tilde{c}^F}{\partial t} + \text{div}(\tilde{c}^F v^s) = 0. \tag{8}$$

Combination of Eqs. (5)–(8) yields

$$\text{div}[\tilde{c}^+(v^+ - v^s) - c^-(v^- - v^s)] = 0. \tag{9}$$

The quantity in [ ] multiplied by the Faraday constant is the electric current density.

12.10.2 The Equations of Motion

The vector equation of motion of any phase is of the form

(mass density)(acceleration) = divergence of stress tensor

+ external body force

+ momentum per unit volume supplied to
the phase by other phases.

Hence Lai et al. write, in tensor notation for the ith component of the vectors, the following equation of motion for the phase α:

$$\rho^\alpha \left(\frac{\partial v_i^\alpha}{\partial t} + v_j^\alpha \frac{\partial v_i^\alpha}{\partial x_j} \right) = \frac{\partial \sigma_{ij}^\alpha}{\partial x_j} + p^\alpha b_i^\alpha + \pi_i^\alpha, \tag{10}$$

in which the stress of phase α is defined for that phase with respect to the tissue as a whole, b^α is the external body force per unit mass of phase α, and π^α is the momentum supplied to phase α by other phases. The external body forces for the solid and water phases are zero, that for the ions may be finite if an external electric field acts. By the conservation of linear momentum for the mixture, they obtain

$$\pi^s + \pi^w + \pi^+ + \pi^- = 0. \tag{11}$$

12.10.3 Derivation of Constitutive Equations

The rest of the theoretical development is similar to the general theory of mass transport according to irreversible thermodynamics, of which a succinct description and a list of references are given in Chapter 8 of *Biomechanics: Motion, Flow, Stress, and Growth*, by Fung (1990). One begins with the first law of thermodynamics which relates the internal energy of the system with the mechanism of doing work on and transporting heat to the system. Then the entropy change in a given piece of material is examined, and is separated into two parts, one part is due to the flux of entropy through the boundary, another part is due to internal entropy production. The second law of thermodynamics asserts that the entropy production is non-negative. Next, the entropy production is identified as a sum of the products of *fluxes* and the conjugate *generalized forces*. Then one assumes that the fluxes and forces are related by phenomenological laws which are restricted by the non-negative requirement of the entropy production. These phenomenological laws are the constitutive equations of the material. Linearity is often assumed for simplicity. In Chapter 8 of Fung (1990), it is shown how Fick's law of diffusion, Starling's law of membrane filtration, and Darcy's law of porous media are derived this way. In every case, it was found that the derivatives of chemical potential are the driving forces of the conjugate fluxes. Lai, Hou, and Mow (1991) have shown that in applying the triphasic theory to the articular cartilage, meniscus, and intervertebral disk, they can derive the constitutive equations, and clarify the change of dimension in equilibrium free-swelling, the change of stress in equilibrium confined-swelling, the Donnon osmotic pressure, the cartilage stiffness, the chemical-expansion stress, and the isometric transient swelling.

Because of the limitation of space, the long and numerous equations of Lai et al. will not be reproduced here. Only a few key equations are given below. The energy equation for the phase α reads, in the notations used in this book:

$$\rho^\alpha \frac{D^\alpha U^\alpha}{Dt} = \sigma_{ij}^\alpha V_{ij}^\alpha - \operatorname{div} J_q + \varepsilon^\alpha \tag{12}$$

in which U is the internal energy, V_{ij}^α is the strain rate tensor, J_q is the heat flux vector, and ε^α is the part of energy supplied to the α component from the other components which contributes to the increase of internal energy of phase α. The "material derivative" D^α/Dt and the strain rate tensor are defined by

$$\frac{D^\alpha}{Dt} = \frac{\partial}{\partial t} + v_j^\alpha \frac{\partial}{\partial x_j}, \tag{13}$$

$$V_{ij}^\alpha = \frac{1}{2}\left(\frac{\partial v_i^\alpha}{\partial x_j} + \frac{\partial v_j^\alpha}{\partial x_i}\right). \tag{14}$$

The entropy production rate is

$$\frac{D_i S_v}{Dt} = \sum_{\alpha=s,\,w,\,+,\,-}\left[\rho^\alpha \frac{D^\alpha S_v}{Dt} + \operatorname{div}\left(\frac{J_q^\alpha}{T}\right)\right] \geqslant 0 \tag{15}$$

in which S_v is the entropy per unit volume. The internal energy and entropy are related to the Helmholtz free energy F by the following equation for each phase α:

$$F^\alpha = U^\alpha - TS_v^\alpha. \tag{16}$$

The free energy is a function of the temperature T, strain e_{ij}, the phase densities of water ρ^w, and ions ρ^+, ρ^-; and the density of the fixed charge $\tilde{c}^F$. With these relations, and with a great deal of work, the entropy production rate $D_i S_v/Dt$ given by Eq. (15) can be reduced to the following form

$$\frac{D_i S_v}{Dt} = \sum_k J_k X_k. \tag{17}$$

in which J_k, $(k = 1, 2 \ldots K)$ are the generalized fluxes, and X_k the generalized forces. If J_k is unrestricted then the non-negative requirement of the entropy production can be satisfied by requiring

$$X_k = 0 \tag{18}$$

If the generalized fluxes are related to the generalized forces, then the linear phenomenological laws may take the form

$$J_k = \sum_{m=1}^K L_{km} X_m. \tag{19}$$

where L_{km} are constants. The constants L_{km} are subjected to Onsager's principle that $L_{km} = L_{mk}$, and the non-negative entropy production restriction. Equations (18) and (19) are the desired constitutive equations.

12.10.4 Other Theories

An extensive new mathematical development of the multiphasic tissue subjected to finite deformation is given by Holmes (1986) and Kwan et al. (1990). Many practical and theoretical problems in physiology and medicine have been studied by the biphasic and triphasic theory, see papers by Holmes, Lai, Mow, and their associates in the list of references. A thorough analysis of the viscoelastic behavior of articular cartilage is given by Mak (1986) on the basis of bi-phasic theory, in which the viscoelasticity of the solid matrix is taken into account, whereas the interstitial fluid is considered as incompressible and inviscid.

12.11 Tendons and Ligaments

In the musculoskeletal system, the function of tendons is to transmit forces from muscles to bones in order to move the bones, whereas that of the ligaments it to join bones togther to form joints. A functional unit of a skeletal muscle is bone-tendon-muscle-tendon-another bone. A functional unit of a ligament is bone-ligament-another bone. Tendons and ligaments serve in this capacity by their tensile strength. This strength is derived from collagen. The collagen molecules, fibrils, and fibers have been discussed in Chapter 7. Collagen in blood vessels is discussed in Chapter 8. Theories that aim at deriving the constitutive equations of tendons and ligaments from their microstructures are mentioned in Sec. 7.12. Theories that originate in rational mechanics are discussed in Sect. 7.6. Hence, as far as the representation of mechanical properties by constitutive equations is concerned, tendons and ligaments fall in line with other soft tissues as discussed in Chapter 7.

Research on ligaments, tendons, and menisci is advancing vigorously because of its clinical importance.

An excellent general reference is the *Handbook of Bioengineering*, edited by Skalak and Chien (1987), which contains an article by Viidik on the properties of tendons and ligaments. Another general reference is the book *Biomechanics of Diarthrodial Joints*, edited by Mow, Ratcliffe, and Woo (1990). This book contains an article by Viidik on the structure and function of normal and healing tendons and ligaments, a paper by Frank and Hart on the biology of tendons and ligaments, and a paper by Woo, Weiss, and MacKenna on biomechanics and morphology of the medial collateral and anterior cruciate ligaments. The anterior cruciate ligament and its replacements are discussed further by Butler and Guan, and Markolf, Gorek, Kabo, Shapiro, and Finerman. Meniscus are discussed by Arnoczky, and Kelly, Fithian, Chern, and Mow. You will see that the biology and clinical problems are made clear by biomechanics. See also *Basic Orthopaedic Biomechanics*, edited by Mow and Hayes (1991).

The book *Injury and Repair of Musculoskeletal Soft Tissues*, edited by Woo and Buckwalter (1988), is a remarkable first exposition of the biology and clinical problems of ligaments and tendons. It is not yet bolstered by biomechanics. One could anticipate a deepened and enlarged role of biomechanics in future treatment of injury and repair.

Concerning the engineering of living tissues, there is a book *Tissue Engineering*, edited by Skalak and Fox (1988), and another book by the same name edited by Woo and Seguchi (1989). But, newer results are coming out fast. Tissue engineering has a lot to do with the reaction of cells to stresses. Some references to cell mechanics are given in Chapter 4, Sec. 4.11.

Tissue remodeling in bone, blood vessels, the heart, and other organs is discussed in Fung (1990). The effect of immobilization and exercise on tissue remodeling of tendons and ligaments is discussed in Woo et al. (1982).

These references are mostly reviews with comprehensive lists of references.

Problems

12.1 Assume that in a piece of bone the collagen fibrils and apatite crystals are perfectly bonded parallel to each other. Derive the Young's modulus of the bone in uniaxial tension or compression in a direction parallel to the fibrils. Derive also the Young's modulus in the direction perpendicular to the fibrils, as well as the shear modulus parallel and perpendicular to the fibrils. Express the results in terms of the elastic moduli of the individual components and the ratio of the volumes of the collagen and apatite (see Currey, 1964).

12.2 Relax the perfect bond hypothesis of the previous problem. Introduce some model of bonding between the collagen and apatite to explore the effect of bonding on the mechanical properties of the composite material.

12.3 Use the results of Problems 12.1 and 12.2 to establish a model of an osteon. Then build up a model of a bone by a proper organization of the osteons. What additional experiments should be done to verify whether the model is satisfactory or not? A successful model will then be able to correlate the mechanical properties of the bone with the mass density and structure and composition of the bone, and will be useful in applications to surgery, medicine, nutrition, sports, etc. For some early attempts, which are not quite successful, see Currey (1964), Young (1957), and Knese (1958).

12.4 Consider a long bone of hollow circular cylindrical cross section with inner radius R_i and outer radius R_o. Let the bone be subjected to an axial compressive load P. Assume that Eq. (3) of Sec. 12.6 applies. Show that the inner and outer radii will change according to

$$R_o(t) = R_o(0) + k_o a(1 - e^{-bt}),$$
$$R_i(t) = R_i(0) + k_i a(1 - e^{-bt}),$$

where a and b are constants;

$$b = \frac{2P}{\pi E}\left\{\frac{k_o R_o(0) + k_i R_i(0)}{[R_o^2(0) - R_i^2(0)]^2}\right\},$$

$$a = \frac{R_o^2(0) - R_i^2(0)}{2[k_o R_o(0) - k_i R_i(0)]}\left\{1 - \frac{\varepsilon_{11}^0 \pi E}{P}[R_o^2(0) - R_i^2(0)]\right\}.$$

The constants k_i and k_o are the values of k_{11} on the inner and outer radii, respectively. The limits are

$$R_o(\infty) = R_o(0) + k_o a, \qquad R_i(\infty) = R_o(0) - k_i a.$$

12.5 Discuss the preceding problem if the constitutive Eq. (4) of Sec. 12.6 applies.

12.6 A steel plate is attached to a bone by means of screws, thus inducing a compressive stress between the plate and the bone. Discuss the remodeling of the bone under the plate if (a) the screws are tightened uniformly so that compressive stress under the plate is uniform, or (b) one screw is given an extra turn or two so that a nonuniform compressive stress results.

12.7 Consider the interaction of the screw threads with the bone in the situations named in Problem 12.6. Discuss the remodeling of the bone around the screws.

12.8 A long bone is subject to a bending moment M. Discuss the remodeling of the bone under bending in the context of Eqs. (1), (2), and (4) of Sec. 12.6.

12.9 Describe the structure of bone, including the blood circulation system therein. Discuss the relevance of blood microcirculation to the healing of a fractured bone.

12.10 Many people have suggested that some tissues are porous media, i.e., they are solids containing some fluids and that the movement of the fluids obeys Darcy's law. Darcy invented the law to describe the movement of water in soil, see *Biomechanics: Motion, Flow, Stress, and Growth* (Fung, 1990), Chapter 8, Sec. 8.6. In order to obtain some idea as to how such a material may function under an external load, consider the following idealized problem. Let a flat layer of the porous medium be of uniform thickness h_0 initially, and infinite in extent in the x, z plane, free on one surface ($y = h_0$), and attached to an infinite rigid base on the other surface ($y = 0$). Let this porous layer be loaded suddenly by a large rigid cylinder, parallel to y, of radius a ($a \gg h_0$), and with a total of force of W acting in the y direction. Under the load the porous layer will deform and fluid in the pores will move. Assume that the solid matrix material of the porous layer is linearly elastic and obeys Hooke's law. Assume that the fluid movement in the porous medium obeys Darcy's law:

$$\langle v_f \rangle = -\frac{K}{\mu}(\nabla\langle p \rangle - \rho_f \mathbf{g}),$$

where $\langle v_f \rangle$ is the macroscopic velocity of the fluid, ρ_f is the density of the fluid, μ is the shear viscosity of the fluid, $\langle p \rangle$ is the macroscopic fluid pressure, $\mathbf{g}$ is the gravitational acceleration, ∇ is the gradient operator, and K is a constant called the *permeability* of the porous medium. The cylinder and the base are assumed impermeable. The normal stress acting on the cylinder at the interface with the porous layer is equal to the sum of the stress in the matrix material and the fluid pressure. On the free surface of the layer there is nothing to prevent the fluid from moving in or out of the surface.

Write down the equations of conservation of mass and momentum. Assuming $a \gg h_0$, and derive an approximate equation that governs the thickness distribution h as a function of location x and time t. If the load W is applied as a step function, determine $h(x,t)$.

12.11 Generalize the investigation proposed in the preceding problem along the following lines. (a) The force imposed by the cylinder consists of a normal force W parallel to y and a shear force S parallel to x. (b) The matrix solid behaves like most biological solids with an elastic modulus proportional to the normal stress. See Sec. 7.5, Eq. (3), p. 274. (c) The cylinder is replaced by a sphere.

12.12 Tissues of brain, liver, kidney are rich in capillary blood vessels. If the stress of the tissue is nonuniform, the blood flow in the capillaries is going to be affected. Formulate a theory of the mechanical properties of the tissue affected by the fluid movement. Analyze and compare the following alternatives: (1) A biphasic mixture theory. (2) A Darcy porous medium. (3) A system composed of a viscous fluid in tubular pores in an anisotropic deformable body advocated by H. S. Lew and Y. C. Fung (1970).

References

Amtmann, E. (1968) The distribution of breaking strength in the human femur shaft. *J. Biomech.* **1**, 271–277.

Amtmann, E. (1971) Mechanical stress, functional adaptation, and the variation of structure of the human femur diaphysis. *Ergebnisse Anat. Entwicklungsgeschichte* **44**, 7–89.

Amtmann, E and Schmitt, H. P. (1968) Über die Verteilung der Corticalisdichte im menschlichen Femurschaft und ihre Bedeutung für die Bestimung der Knochenfestigkeit. *Z. Anat. u. Entwickl.-ges.* **127**, 25–41.

Basset, C. A. L. and Pawlick, R. J. (1964) Effect of electrical currents on bone *in vivo*. *Nature* **204**, 652–653.

Becker, R. O. and Murray D. G. (1970) The electrical control system regulating fracture healing in amphibians. *Clin. Orthopedics* **73**, 169–198.

Bourne, G. H. (ed.) (1972) *The Biochemistry and Physiology of Bone*, 2nd edition, Vol. 1: *Structure*. Vol. 2: *Physiology and Pathology*. Vol. 3: *Development and Growth*. Academic Press, New York.

Brånemark, P.-I., Hansson, B. O., Breine, U., Lindström, J., Hallén, O., and Öhman, A. (1977) *Osseointegrated Implants in the Treatment of the Edentulous Jaw*. Almquist and Wiksell, Stockholm, 132 pp.

Brannan, E. W., Rockwood, C. A., and Potts, P. (1963) The influence of specific exercises in the prevention of debilitating musculoskeletal disorders. *Aerospace Med.* **34**, 900–906.

Brookes, M. (1971) *The Blood Supply of Bone. An Approach to Bone Biology*. Butterworths, London.

Carter, D. R. and Hayes, W. C. (1977) The compressive behavior of bone as a two-phase porous material. *J. Bone Joint Surg.* **49A**, 954–962.

Carter, D. R., Harris, W. H., Vasu, R., and Caler, W. E. (1981) The mechanical and biological response of cortical bone to *in vivo* strain histories. In *Mechanical Properties of Bone*, S. Cowin ed. AMD Vol. 45, American Society of Mechanical Engineering, New York, pp. 81–92.

Carter, D. R., Fyhrie, D. P., and Whalen, R. T. (1987) Trabecular bone density and loading history: Regulation of connective tissue biology by mechanical energy. *J. Biomech.* **20**, 785–794.

Carter, D. R. (1987) Mechanical loading history and skeletal biology. *J. Biomech.* **20**, 1095–1109.

Carter, D. R., Orr, T. E., Fyhrie, D. P., and Schurman, D. J. (1987) Influences of mechanical stress on prenatal and postnatal skeletal development. *Clin. Orthopaedics* **219**, 237–250.

Carter, D. R. and Wong, M. (1988) Mechanical stresses and endochondralossification in the chondroepiphysis. *J. Orthopaedic Res.* **6**, 148–154.

Cassidy, J. J. and Davy, D. T. (1985) Mechanical and architectural properties in bovine cancellous bone. *Trans. Orthopaedic Res. Soc.* **31**, 354.

Churches, A. E. and Howlett, C. R. (1981) The response of mature cortical bone to controlled time-varying loading. In *Mechanical Propoerties of Bone* S. Cowin (ed.) AMD Vol. 45. American Society of Mechanical Engineering, New York, pp. 69–80.

Cowin, S. C. and Hegedus, D. M. (1976) Bone remodeling. *J. Elasticity* **6**, 313–325, 337–352.

Cowin, S. C and Nachlinger, R. R. (1978) Bone remodeling III. *J. Elasticity* **8**, 285–295.

Cowin, S. C. and Van Buskirk, W. C. (1978) Internal bone remodeling induced by a medullary pin. *J. Biomech.* **11**, 269–275.

Cowin, S. C. and Van Buskirk, W. C. (1979) Surface remodeling induced by a medullary pin. *J. Biomech.* **12**, 269–276.

Cowin, S. C. (ed.) (1981) *Mechanical Properties of Bone*, ASME Publication No. AMD Vol. 45.

Cowin, S. C. (1983) The mechanical and stress adaptive properties of bone. *Ann. Biomed. Eng.* **2**, 263–295.

Cowin, S. C. (1984) Modeling of the stress adaptation process in bone. *Cal. Tissue Int.* **36** (Suppl.), S99–S104.

Cowin, S. C., Hart, R. T., Balser, J. R., and Kohn, D. H. (1985) Functional adaptation in long bones: Establishing *in vivo* values for surface remodeling rate coefficients. *J. Biomech.* **18**, 665 684.

Cowin, S. C. (1986) Wolff's law of trabecular architecture at remodeling equilibrium. *J. Biomech. Eng.* **108**, 83–88.

Cowin, S. C and Van Buskirk, W. C. (1986) Thermodynamic restrictions on the elastic constants of bone. *J. Biomech. Eng.* **108**, 83–88.

Cowin, S. C., Van Buskirk, W. C., and Ashman, R. B. (1987) Properties of bone. In *Handbook of Bioengineering*, R. Skalak and S. Chien (eds.) McGraw-Hill, New York, pp. 2.1–2.27.

Cowin, S. C. (1988) Strain assessment by bone cells. In *Tissue Engineering*, R. Skalak and C. F. Fox (eds.) Alan Liss, New York, pp. 181–188.

Cowin, S. C., Sadegh, A. M., and Luo, G. M. (1992) An evolutionary Wolff's law for trabecular architecture. *J. Biomech. Eng.* **114**, 129–136.

Crowningshield, R. D. and Pope, M. H. (1974) The response of compact bone in tension at various strain rates. *Ann. Biomed. Eng.* **2**, 217–225.

Culmann, C. (1866 and 1875) *Die Graphische Statik.* 1st edition, Meyer und Zeller, Zurich.

Currey, J. D. (1964) Three analogies to explain the mechanical properties of bone. *Biorheology* **2**, 1–10.

Dietrick, J. E., Whedon, G., and Shorr, E. (1948) Effects of immobilization upon various metabolic and physiological functions of normal man. *Am. J. Med.* **4**, 3–36.

Dintenfass, L. (1963) Rheology of synovial fluid and its role in joint lubrication. *Proc. Int. Congress Rheolog.* **4**, 489.

Dowson, D., Longfield, M., Walker, P., and Wright, V. (1968) An investigation of the friction and lubrication in human joints. *Proc. Institution Mech. Eng. (London)* **181**, Part 3J, 45–54.

Dowson, D. and Whoms, T. L. (1968) Effect of surface quality upon the traction characteristics of lubricated cylindrical contacts. *Proc. Institution Mech. Eng. (London)* **182**, Part 1, 292–299.

Evans, F. G. (1957) *Stress and Strain in Bones. Their Relation to Fractures and Osteogenesis.* C. C. Thomas, Springerfield, II.

Evans, F. G. (1969) The mechanical properties of bone. *Artificial Limbs* **13**, 37–48.

Evans, F. G. (1973) *Mechanical Properties of Bone.* Charles C. Thomas, Springfield, IL.

Fein, R. S. (1967) Are synovial joints squeeze film lubricated? *Proc. Inst. Mech. Eng.* **181**, 125–128.

Firoozbakhsh, K. and Cowin, S. C. (1980) Devolution of inhomogeneities in bone structure— Predictions of adpative elasticity theory. *J. Biomech. Eng.* **102**, 287–293.

Frost, H. M. (1964) *The Laws of Bonre Structure.* Charles C. Thomas, Springfield, IL.

Fukada, E. and Yasuda, I. (1957) Piezoelectric effect of bone. *J. Physical. Soc. Jpn.* **12**, 1158–1162.

Fukada, E. (1968) Mechanical deformation and electrical polarization in biological substances. *Biorheology* **5**, 199–208.

Fung, Y. C. (1972) Stress–strain history relations of soft tissues in simple elongation. In *Biomechanics: Its Foundations and Objectives.* Prentice-Hall, Englewood Cliffs, NJ, pp. 181–208.

Fung, Y. C. (1990) *Biomechanics: Motion, Flow, Stress, and Growth.* Springer-Verlag, New York.

Fyhrie, D. P. and Carter, D. R. (1985) A unifying principle relating stress state to trabecular bone morphology. *Trans. Orthopaedic Res. Soc.* **31**, 337.

Gjelsvik, A. (1973) Bone remodeling and piezoelectricity. I & II. *J. Biomech.* **6**, 69–77, 187–193.

Glücksmann, A. (1938) Studies on bone mechanics *in vitro.* I. Influence of pressure on orientation of structure. *Anat. Record* **72**, 97–115.

Glücksmann, A. (1939) II. Role of tension and pressure in chodrogenesis. *Anat. Record* **73**, 39–55.

Glücksmann, A. (1942) The role of mechanical stress in bone formation *in vitro.* J. *Anat.* **76**, 231–239.

Ham, A. W. (1969) *Histology,* 6th edition. Lippincott, Philadelphia.

Harrigan, T. and Mann, R. W. (1984) Characterization of microstructural anisotropy in orthotropic materials using a second rank tensor. *J. Mater. Sci.* **19**, 761–767.

Hayes, W. C. and Snyder, B. (1981) Toward a gnemtitative formulation of Wolff's law in trabecular bone. In *Mechanical Properties of Bone*, S. C. Cowin ed. AMD Vol. 45. American Society of Mechanical Engineers, New York.

Hegedus, D. H. and Cowin, S. C. (1976) Bone remodeling II: Small strain adaptive elasticity. *J. Elasticity* **6**, 337–352.

Hert, J. A., Liskova, M., and Landa, J. (1971) Reaction of bone to mechanical stimuli. Part 1. Continuous and intermittent loading of tibia in rabbit. *Folia Morphol.* **19**, 290–317.

Hert, J., Sklenska, A., and Liskova, M. (1971) Reaction of bone to mechanical stimuli. Part 5. Effect of intermittent stress on the rabbit tibia after resection of the pripheral nerves. *Folia Morphol.* **19**, 378–387.

Hoffman, O. (1967) The brittle strength of orthotropic materials. *J. Composite Mater.* **1**, 200–207.

Holmes, M. H. (1986) Finite deformation of soft tissue: Analysis of a mixture model in uniaxial compression. *J. Biomech. Eng.* **108**, 372–381.

Hong, S. Z., Wu, Z. K., and Zu, C. M. (1987) Experiments on human vertebrae cervical. *Chin. J. Biomed. Eng.* **6**, 75–83.

Johnson, M. W. and Katz, J. L. (1987) Electromechanical effects in bone. In *Handbook of Bioengineering*, R. Skalak and S. Chien (eds.) McGraw-Hill, New York, pp. 3.1–3.11.

Jones, H. H., Priest, J. D., Hayes, W. C., Tichemor, C. C., and Nagel, D. A. (1977) Humeral hypertrophy in response to exercise. *J. Bone Joint Surg. A* **59**, 204–208.

Justus, R. and Luft, J. H. (1970) A mechanochemical hypothesis for bone remodeling induced by mechanical stress. *Calcified Tissue Res.* **5**, 222–235.

Katz, J. L. and Mow, V. C. (1973) Mechanical and structural criteria for orthopaedic implants. *Biomat. Med. Dev. Art. Organs* **1**, 575–638.

Kazarian, L. E. and van Gierke, H. E. (1969) Bone loss as a result of immobilization and chelation. *Clin. Orthopedics* **65**, 67–75.

Knese, K.-H. (1972) *Knochenstruktur als Verbundbau*. G. Thieme, Stuttgart.

Kummer, B. K. F. (1972) Biomechanics of bone: Mechanical properties, functional structure, and functional adaptation. In *Biomechanics: Its Foundations and Objectives*, Y. C. Fung, N. Perrone, and M. Anliker (eds.) Prentice-Hall, Englewood Cliffs, NJ, pp. 237–271.

Kwan, M. K., Lai, W. M., and Mow, V. C. (1984) Fundamentals of fluid transport through cartilage in compression. *Ann. Biomedical Eng.* **12**, 537–558.

Kwan, M. K., Lai, W. M. and Mow, V. C. (1990) A finite deformation theory for cartilage and other soft hydrated connective tissues. I. Equilibrium results. *J. Biomechanics.* **23**, 145–155.

Lai, W. M., Hou, J. S., and Mow, V. C. (1991) A triphasic theory for the swelling and deformation behaviors of articular cartilage. *J. Biomech. Eng.* **113**, 245–258.

Lakes, R. S., Katz, J. L., and Sternstein, S. (1979) Viscoelastic properties of wet cortical bone—I. Torsional and biaxial Studies. *J. Biomech.* **12**, 657–678.

Lakes, R. S. and Katz, J. L. (1979) Viscoelastic properties of wet cortical bone. II. Relaxation mechanisms. III. A nonlinear constitutive equation. *J. Biomech.* **12**, 679–687, 689–698.

Lanyon, L. B. and Baggott, D. G. (1976) Mechanical function as an influence on the structure and form of bone. *J. Bone Joint Surg. B* **58**, 436–443.

Lew, H. S. and Fung, Y. C. (1970) Formulation of a statistical equation of motion of a viscous fluid in an anisotropic nonrigid porous solid. *Int. J. Solids Struct.* **6**, 1323–1340.

Linn, F. C. (1967) Lubrication of animal joints: I. The arthrotripsometer. *J. Bone Joint Surg. A* **49**, 1079–1098.

Linn, F. C. and Radin, E. L. (1968) Lubrication of animal joints: III. The effect of certain chemical alterations of the cartilage and lubricant. *Arth. Rheum.* **11**, 674–682.

Lotz, J. C., Gerhart, T. N., and Hayes, W. C. (1991) Mechanical properties of metaphysical bone in the proximal femur. *J. Biomech.* **24**, 317–329.

Lotz, J. C., Cheal, E. J., and Hayes, W. D. (1991) Fracture prediction for the proximal femur using finite element models. Part I: Linear analysis. Part II: Nonlinear analysis. *J. Biomech. Eng.* **113**, 353–365.

MacConaill, M. A. (1932) The function of intra-articular fibrocartilages, with special reference to the knee and inferior radio-ulnar joints. *J. Anat.* **66**, 210–227.

Mack, P. B., La Change, P. A., Vost, G. P., and Vogt, F. B. (1967) Bone demineralization of the foot and hand of Gemini IV, V, and VII astronauts during orbital flight. *Am. J. Roentgenol.* **100**, 503–511.

Mak, A. F. (1986) The apparent viscoelastic behavior of articular cartilage—The contributions from the intrinsic matrix viscoelasticity and interstitial fluid flows. *J. Biomech. Eng.* **108**, 123–130.

Malcom, L. L. (1976) Frictional and deformational responses of articular cartilage interfaces to static and dynamic loading. Ph.D. thesis, University of California, San Diego, La Jolla, California.

Maroudas, A. (1967) Hyaluronic acid films. *Proc. Inst. Mech. Eng.* **181**, 122–124.

Martin, B. (1972) The effects of geometric feedback in the development of osteoporosis. *J. Biomech.* **5**, 447–455.

Martin, R. B. (1984) Porosity and specific surface of bone. *CRC Crit. Rev. Biomed. Eng.* **10**, 179–222.

McCutchen, C. W. (1959) Mechanism of animal joints. *Nature* **184**, 1284–1285.

McCutchen, C. W. (1962) The frictional properties of animal joints. *Wear* **5**, 1.

McCutchen, C. W. (1967) Lubrication and wear in living and artificial joints. *Proc. Inst. Mech. Eng.* **181**, 55, Part 3J.

Merz, W. A. and Schenk, R. K. (1970) Quantitative structural analysis of human cancellous bone. *Acta Anat.* **75**, 54–66.

Mow, V. C. (1969) The role of lubrication in biomechanical joints. *J. Lubr. Technol. Trans. ASME* **91**, 320–329.

Mow, V. C., Lipschitz, H., and Glimcher, M. J. (1977) Mechanisms of stress relaxation in articular cartilage *Trans. Ortho. Res. Soc.* **2**, 75.

Mow, V. C., Kuei, S. C., Lai, W. M., and Armstrong, C. G. (1980) Biphasic creep and stress relaxation of articular cartilage in compression: Theory and experiments. *J. Biomech. Eng. Trans. ASME* **102**, 73–84.

Mow, V. C. and Lai, W. M. (1980) Recent developments in synovial joint biomechanics. *SIAM Rev.* **22**, 275–317.

Mow, V. C., Ratcliffe, A., and Woo, S. L.-Y. (eds.) (1991) *Biomechanics of Diarthroidal Joints*, Vols. 1 and 2. Springer-Verlag, New York.

Mow, V. C. and Hayes, W. C. (1991) *Basic Orthopaedic Biomechanics*. Raven Press, New York.

Oda, M. (1976) Fabrics and their effects on the deformation behaviors of sand. Department of Foundation Engineering, Saitama University.

Oda, M., Konishi, J., and Nemat-Nasser, S. (1980) Some experimentally based fundamental results on the mechanical behavior of granular materials. *Geotechnique* **30**, 479–495.

Ogston, A. G. and Stanier, J. E. (1953) The physiological function of hyaluronic acid in synovial fluid: Viscous, elastic, and lubrication properties. *J. Physiol.* **119**, 244–252, 253–258.

Patwardham, A. G., Bunch, W. H., Meade, K. P., Vanderby, R., and Knight, G. W. (1986) A biomechanical analog of curve progression and orthotic stabilization in idiopathic scoliosis. *J. Biomech.* **19**, 103–117.

Pauwels, F. (1948) Die Bedeutung der Bauprinzipien der Stütz- und Bewegungsapparatus für die Beanspruchung der Röhrenknochen. *Z. Anat.* **114**, 129–166.

Pauwels, F. (1950) Die Bedeutung der Muskelkräfte für der Regelung der Beanspruchung des Röhrenknochens während der Bewegung der Glieder. *Z. Anat.* **115**, 327–351.

Pauwels, F. (1968) *Gesammelte Abhandlungen zur funktionellen Anatomie des Bewegungsapparates.* Springer-Verlag, New York.

Powell, M. J. D. (1965) A method for minimizing a sum of species of nonlinear functions without calculating derivatives. *Computer J.* **7**, 303–307.

Radin, E. L., Swann, D. A., and Weisser, P. A. Separation of a hyaluronate–free lubricating factor from synovial fluid. (1970) *Nature* **228**, 377.

Reilly, D. T. and Burstcin, A. II. (1974) The mechanical properties of cortical bone. *J. Bone Joint Surg. A* **56**, 1001–1022.

Rhinelander, F. W. (1972) Circulation of bone. In *The Biochemistry and Physiology of Bone*, 2nd edition, G. H. Bourne (ed.) Academic, New York, pp. 2–78.

Roux, W. (1895) *Gasammelte Abhandlungen über Entwicklungsmechanik der Organismen* Vols. I and II. Engelmann, Leipzing.

Rydevik, B., Bränemark, P.-I., and Skalak, R. (eds.) (1990) *International Workshop on Osseointegration in Skeletal Reconstruction and Joint Replacement.* Institute for Applied Biotechnology, Göteborg, Sweden.

Schmitt, H. P. (1968) Über die Beiehungen zwischen Dichte und Festigkeit des Knochens am Beispiel des menschlichen Femur. *Z. Anat.* **127**, 1–24.

Sedlin, E. (1985) A rheological model for cortical bone. Suppl. 83, *Acta Scand. Ortho.* **36**.

Skalak, R. and Chien, S. (eds.) (1987) *Handbook of Bioengineering.* McGraw-Hill, New York.

Skalak, R. and Fox, C. F. (eds.) (1988) *Tissue Engineering.* Alan Liss, New York.

Spilker, R. L., Jakobs, D. M., and Schultz, A. B. (1986) Material constants for a finite element model of the intervertebral disk with a fiber composite amulus. *J. Biomech. Eng.* **108**, 1–11.

Stone, J. L., Beaupre, G. S., and Hayes, W. O. (1983) Multiaxial strength characteristics of trabecular bone. *J. Biomech.* **16**, 743–752.

Stone, J. L., Snyder, B. D., Hayes, W. C., and Strang, G. L. (1984) Three-dimensional stress morphology analysis of trabecular bone. *Trans. Orthopedic Res. Soc.* **30**, 199.

Tencer, A. F., Ahmed, A. M., and Burke, D. C. (1982) Some static mechanical properties of the lumbar intervertebral joint, intact and injured. *J. Biomech. Eng.* **104**.

Torino, A. J., Davidson, C. L., Klopper, P. J., and Linclau, L. A. (1976) Protection from stress in bone and its effects: Experiments with stainless steel and plastic plates in dogs. *J. Bone Joint Surg. B* **58**, 107–113.

Torzilli, P. A. and Mow, V. C. (1972) On the fundamental fluid transport mechanisms through normal and pathological articular cartilage during function, parts I and II. *J. Biomech.* **9**, 541–522 (this is in error), 587–606.

Turner, C. H. (1989) Yield behavior of bovine cancellous bone. *J. Biomech. Eng.* **11**, 257–260.

Unsworth, A., Dowson, D., and Wright, V. (1975) The frictional behavior of human synovial joints—Part I: Natural joints. *J. Lub. Tech. Trans. ASME* **97**, 369–376.

Walker, P. S., Dowson, D., Longfield, M. D., and Wright, V. (1968) "Boosted lubrication" in synovial joints by fluid entrapment and enrichment. *Ann. Rheum. Dis.* **27**, 512–520.

Wertheim, G. (1847) Memoire sur l'elasticité et la cohésion des principaux tissus du corps humain. *Ann. Chim. Phys.* **21**, 385–414.

Whitehouse, W. J. (1974) The quantitative morphology of anisotropic trabecular bone. *J. Microsc.* **101**, 153–168.

Whitehouse, W. J. and Dyson, E. D. (1974) Scanning electron microscope studies of trabecular bone in the proximal end of the human femur. *J. Anat.* **118**, 417–444.

Wolff, J. (1869) Über die bedeutung der Architektur der spongiösen Substanz. *Zentralblatt für die medizinische Wissenschaft. VI. Jahrgang.* pp. 223–234.

Wolff, J. (1870) Über die innere Architektur der Knochen und ihre Bedeutung für die Frage vom Knochenwachstum. *Arch. pathol. Anat. Physiol. klinische Medizin* (*Virchovs Arch.*) **50**, 389–453.

Wolff, J. (1884) Das Gesetz der Transformation der inneren Architektur der Knochen bei pathologischen Veränderungen der äusseren Knochenform. *Sitz, Ber. Preuss. Akad. Wiss. 22. Sitzg., phys.-math. Kl.*

Wolff, J. (1892) *Das Gesetz der Transformation der Knochen.* Hirschwald, Berlin.

Wonder, C. C., Briney, S. R., Kral, M., and Skavgstad, C. (1960) Growth of mouse femurs during continual centrifugation. *Nature* **188**, 151–152.

Woo, S. L.-Y., Akeson, W. H., Coutts, R. D., Rutherford, L., Doty, D., Jemmott, G. F., and Amiel, D. (1976) A comparison of cortical bone atrophy secondary to fixation with plates with large differences in bending stiffness. *J. Bone Joint Surg. A* **58**, 190–195.

Woo, S. L.-Y., Simon, B. R., Kuei, S. C., and Akeson, W. H. (1979) Quasi-linear viscoelastic properties of normal articular cartilage. *J. Biomech. Eng.* **102**, 85–90.

Woo, S. L.-Y., Gomez, M. A., Woo, Y.-K., and Akeson, W. H. (1982) Mechanical properties of tendons and ligaments. II. The relationships of immobilization and exercise on tissue remodeling. *Biorheology* **19**, 397–408.

Woo, S. L.-Y. and Buckwalter, J. A. (eds.) (1988) *The Injury and Repair of Musculoskeletal Soft Tissues.* American Academy of Orthopedic Surgeons, Park Ridge, IL.

Woo, S. L.-Y. and Seguchi, Y. (eds.) (1989) *Tissue Engineering.* ASME BIO Vol. 14. American Society of Mechanical Engineers, New York.

Woo, S. L.-Y., and Wayne, J. S. (1990). Mechanics of the anterior cruciate ligament and its contribution to knee kinematics. *Appl. Mech. Rev.* **43**, S142–S149.

Yamada, H. (1970) *Strength of Biological Materials*, translated by F. G. Evans. Williams and Wilkins, Baltimore.

Young, J. Z. (1957) *The Life of Mammals.* Oxford University Press, London.

Young, J. T., Vaishnav, R. N., and Patel, D. J. (1977). Nonlinear anisotropic viscoelastic properties of canine arterial segments. *J. Biomech.* **10**, 549–559.

Author Index

Subject Index